PARAMECIUM – A Current Survey

PARAMECIUM

A Current Survey

edited by

W. J. VAN WAGTENDONK

Professor of Biochemistry, Division of Natural Sciences and Mathematics, Talladega College, Talladega, Alabama, U.S.A.

ELSEVIER SCIENTIFIC PUBLISHING COMPANY

Amsterdam / London / New York

1974

ELSEVIER SCIENTIFIC PUBLISHING COMPANY
335 Jan van Galenstraat
P.O. Box 211, Amsterdam, The Netherlands

AMERICAN ELSEVIER PUBLISHING COMPANY, INC.
52 Vanderbilt Avenue
New York, N.Y. 10017

Library of Congress Card Number: 73–83398

ISBN 0–444–41147–X

With 304 illustrations and 53 tables

Printed in The Netherlands

List of Contributors

H. M. BUTZEL, JR., Department of Biological Sciences, Union College, Schenectady, N.Y. 12308 (U.S.A.)

S. DRYL, Department of Biology, M. Nencki Institute of Experimental Biology, Polish Academy of Science, ul. Pasteura 3, Warszawa (Poland)

C. F. EHRET, Division of Biological and Medical Research, Argonne National Laboratory, Argonne, Ill. 60439 (U.S.A.)

I. FINGER, Biology Department, Haverford College, Haverford, Penn. 19041 (U.S.A.)

E. D. HANSON, Shanklin Laboratory, Wesleyan University, Middletown, Conn. 06457 (U.S.A.)

M. KANEDA, Shanklin Laboratory, Wesleyan University, Middletown, Conn. 06457 (U.S.A.)

E. W. MCARDLE, Division of Biological and Medical Research, Argonne National Laboratory, Argonne, Ill. 60439 (U.S.A.)

A. T. SOLDO, Research Laboratories of the Veterans Administration Hospital and the Department of Biochemistry, University of Miami School of Medicine, Miami, Fla. 33125 (U.S.A.)

W. J. VAN WAGTENDONK, P.O. Box 206, Cedar Mountain, N.C. 28718 (U.S.A.)

E. VIVIER, Laboratory of Protistology and Electron Microscopy, Department of Biology, University of Lille, Lille (France)

Foreword

Few organisms acquire the cachet of being important beyond themselves by becoming popular research tools. *Paramecium* is a member of this hallowed company, joining man (as seen by clinical investigators); the mouse, rat, and guinea pig; *Drosophila*, the silkworm, and a few other insects; and, among protozoa, *Tetrahymena, Ochromonas, Crithidia;* the phytoflagellates *Euglena* and *Ochromonas*, several bacteria, notably what may be summarized as the E–E effect (what holds for *E. coli* holds for elephants). Dr. Van Wagtendonk has amply earned the right to be impresario for *Paramecium:* his demonstration that *Paramecium* and the guinea pig show an as yet mysterious dependence on plant sterols is at last being followed up by the bio-chemical community. This monograph (or rather polygraph with a small "p") is welcome in presenting *Paramecium* as a metabolizing as well as gene-swapping animal.

The officers of the Society of Protozoologists, deeming that a comprehensive work on *Paramecium* would be of great service to the field, enthusiastically supported the planning of the present volume. Its success will be gauged by how effectively it stimulates its own obsolescence.

Haskins Laboratories at Pace University,
New York, N.Y. (U.S.A.)

SEYMOUR H. HUTNER
Chairman, Editorial Board for
Special Publications,
Society of Protozoologists

Preface

Antonie van Leeuwenhoek observed *Paramecium* for the first time around 1678. Since then, this animal with its world-wide distribution has been a favorite subject for research in many areas. The early studies were concerned with structure, comparative morphology, and physiological problems such as the coordinated beating of the cilia, their role in locomotion and feeding, and the avoiding reaction evoked when *Paramecium* encounters an obstacle or unfavorable environment. The vast early literature is cited in the treatises by Kalmus, and Wichterman.

Sonneborn's discovery of mating types, antigens, and endosymbiotes (first thought to be cytoplasmic genes or plasmagenes) provided a great stimulus for research with this organism. *Paramecium* has many advantages as a subject for laboratory investigations. Large populations of all the strains of *Paramecium* can be maintained indefinitely under laboratory conditions in bacterized media, while a certain number of strains can be grown under axenic conditions.

As an eukaryotic unicellular organism *Paramecium* occupies an important niche between the prokaryotic and the multicellular organisms. *Paramecium* is highly differentiated, far more complex than viruses and bacteria, but less complex than multicellular organisms. Mechanisms of differentiation, cytoplasmic inheritance, and control over the expression of different genes, radiation genetics and mutations can be more easily studied with *Paramecium*. Clones of *Paramecium* pass through phases of sexual immaturity, maturity, and senescence, ending in clonal death. This makes the organism an excellent subject for the study of aging at the cellular level.

Paramecium, in common with other free-living ciliates possesses a well developed capacity for reception of external stimuli, and a highly differentiated system of motor organelles which offer extraordinary possibilities of obtaining responses to various changes in the external medium. For many behavioral studies *Paramecium* is exceptionally well suited because its size not only allows an easy observation of its behavior, but microsurgical procedures can be easily performed. Similarly, *Paramecium* is a research organism *par excellence* for studies on the molecular basis of contractile systems, and the excitability of the protozoan cell and its various cellular components.

Many strains of *Paramecium* harbor specific endosymbiotes. Biochemical, nutritional, and electron-microscopical studies might reveal the mechanism of infection and subsequent adaptation of the invader to the environment of the host, and above all, might clarify the enigmatic relationship between the genic composition of the host and its ability to maintain a specific endosymbiote. Since several strains of endosymbiote-bearing strains of *Paramecium* can be grown axenically, the data thus obtained, can be unambiguously interpreted.

Nutritional studies have revealed that *Paramecium* has a lipid requirement that can be satisfied only with C-24 alkyl-substituted sterols, possessing a Δ^5 or Δ^7 unsaturara-ration, and a fatty acid. Further studies might provide an insight into the significance of such lipids in the metabolism of the cell.

Several review articles and books have discussed the advances made in elucidating certain aspects of the life cycle, genetics, development, evolution, mating type-substances and mating-type determination, antigens, endosymbiotes, and the ultrastructure of *Paramecium*. *Paramecium*, in sum, is a basic resource in experimental biology.

Topics not discussed *in extenso* in these reviews were selected for inclusion in this monograph. The contributors were chosen for their competence in the respective fields and were left free to write their contributions in their own style. Only minor editoral changes were made. I am grateful for the cheerful cooperation of the contributors.

The assembly of the Bibliography on *Paramecium* since 1953 was a major undertaking, in which the editor was ably and untiringly assisted by Mrs. Betty Stauderman and Mrs. Helen Davis. Mrs. Betty Manson, Chief Librarian, and Mrs. Ruth Scott, Medical Librarian, both at the Veterans Administration Hospital, Miami, Fla., were of great assistance in verifying and retrieving many of the entries. To all my sincere gratitude.

Very special thanks are due to Professor L. N. Seravin of the Biological Institute of Leningrad State University, Leningrad, U.S.S.R., who so graciously accepted the task of assembling the bibliography of the papers published in Russian since 1953.

Undoubtedly, considering the large number of research papers published since 1953, omissions have occurred. These are the sole responsibility of the Editor.

The final phase of proof reading the entire monograph was performed when I was at the M. Nencki Institute of Experimental Biology, Warszawa, Poland, under the auspices of an exchange program between The National Academy of Sciences, U.S., and the Polish Academy of Sciences. I thank Professor Dr. Stanislaw Dryl, Head of the Department of Cell Biology for his hospitality, and Mr. Henry Adler, Librarian of the Institute, for his assistance in checking the correctness of the references.

Finally, I would like to express my thanks to the Special Publications Committee of the Society of Protozoologists for their confidence in my editorial capabilities, and to Elsevier Scientific Publishing Company for their encouragement and assistance.

W. J. Van Wagtendonk

Contents

Morphology, Taxonomy and General Biology of the Genus
Paramecium

E. VIVIER (Lille, France)

Mating Type Determination and Development in *Paramecium aurelia*

H. M. BUTZEL, JR. (Schenectady, N.Y., U.S.A.)

Surface Antigens of *Paramecium aurelia*

I. FINGER (Haverford, Pa. U.S.A.)

Behavior and Motor Response of *Paramecium*

S. DRYL (Warszawa, Poland)

Growth Patterns and Morphogenetic Events in the Cell Cycle of *Paramecium aurelia*

M. KANEDA AND E. D. HANSON (Middletown, Conn., U.S.A.)

The Structure of *Paramecium* as Viewed from its Constituent Levels of Organization

C. F. EHRET AND E. W. MCARDLE (Argonne, Ill., U.S.A.)

Nutrition of *Paramecium*

W. J. VAN WAGTENDONK (Miami, Fla., U.S.A.)

Intracellular Particles in *Paramecium*

A. T. SOLDO (Miami, Fla., U.S.A.)

Bibliography 1953-1973

W. J. VAN WAGTENDONK (Miami, Fla., U.S.A.)

pp. 443–492

Subject Index

pp. 493–499

Morphology, Taxonomy and General Biology of the Genus *Paramecium*

E. VIVIER

Professor of Animal Biology, Laboratory of Protistology and Electron Microscopy, Department of Biology, University of Lille (France)

I. INTRODUCTION

Species of *Paramecium* are the ciliate Protozoa that are most widely known, particularly by students, and they are, as well as the amoebae, often mentioned as examples of unicellular animals. This is due to simple reasons:

It is very easy to find them in nature.

Their size is comparatively large which makes it easy to observe them.

It is possible to raise them without much difficulty and much precaution in a laboratory.

These factors have brought about many studies that have made paramecium one of the most popular protozoa.

However, in spite of the very favorable conditions, many points in the research conducted with paramecia, if one wants to go thoroughly into them, remain obscure or are known but superficially. *Paramecium* is still a material for research that is far from being exhausted. It can and it must give us much information not only about the biology of the genus and that of the ciliate Protozoa but also about the biology of unicellular organisms and about basic biology itself.

Paramecia seem to have been studied and described, for the first time, by Antony van Leeuwenhoek (195, 196). In 1718, Joblot (173) gave a description together with a sketch, of an animalcule, in the shape of a slipper, which he called "chausson"; it seems to be the first reference to paramecium having the shape of the sole of a shoe. The scientific name *Paramecium* comes from Hill (125) who tried to give a definition of it and differentiated four types. Yet, those descriptions have but little interest, and it was not until O. F. Müller (219, 220), who applied to the Infusoria the rules of nomenclature established by Karl von Linné, that a precise scientific basis for the genus *Paramecium*, erroneously spelled "Paramaecium" by this author, could be obtained; it is in this study of Müller that the first species was given a name: *Paramecium aurelia*. Therefore, the full name of the genus, with authorship and date, is correctly written *Paramecium* O. F. Müller, 1773.

Of course, a few errors and inaccuracies are found in the first investigations, and numerous successive observations and restatements were necessary to attain a clearer and sounder view of the species belonging to the genus *Paramecium*. The first important survey of these infusoria was made by Ehrenberg (76) who tried to sort out, among the numerous previous descriptions (56 species of *Paramecium* had been created since Müller), those which were right, rejecting the homonymies or the species that did not really belong to the genus; before that, he (75) had identified the second correct species: *Paramecium caudatum*. In the same period, the third common species, *Paramecium bursaria*, was given a name by Focke (93).

Among the other great surveys of these organisms, we must mention those of Stein (290), of Claparède and Lachmann (49), of Maupas (204–206). But it was Wenrich (312) who defined precisely the eight species of the genus *Paramecium* known and accepted by the end of the first quarter of the present century.

Aside from those studies of systematics, studies were being developed in cytology and general biology which were to reach their full extent during the 20th century. Undoubtedly, Maupas initiated these studies with his precise morphological research of the ciliates and particularly *Paramecium* (204). A few years later, the same author (206), thanks to his remarkably precise and extensive investigations of conjugation in ciliates which were to stand out as a landmark in the morphological knowledge of the phenomenon by laying the foundations of our present knowledge, opened up new biological perspectives which were to initiate numerous exciting investigations. From then on, many scientists were to advance our knowledge of the genus *Paramecium* in various directions.

After Kalmus' monograph (181), the survey of Wichterman (317), *The Biology of Paramecium*, gave an excellent and complete account of all the works on that genus and their results; there the reader will find a complete documentation and a detailed bibliography. Much has been achieved since then; new techniques of investigations have been used: particularly electron microscopy and biochemistry have contributed to the domains of cytology, physiology and genetics.

In this chapter, we are conducting a survey of the present knowledge of the morphology, taxonomy and general biology of the genus *Paramecium*. Of course, we shall not be able to go into the details of every point, the more so as several of them are dealt with by specialists in the following chapters. We shall merely give the general outlines of the present data and problems, insisting, here and there, on certain points that might not be taken into consideration in the other chapters, because of their specialization or their isolation in the whole of the knowledge, but which, nevertheless, are of real originality and interest, and can, in the future, be at the origin of new research discoveries.

II. PARAMECIUM IN NATURE AND IN THE LABORATORY

Paramecium species are certainly the most widely spread ciliates in the world and the easiest to be raised in a laboratory.

A. Paramecium in Nature: Ecology and Adaptation

Paramecium is generally found in ponds, pools, lakes, slow flowing rivers and brooks, that is, in any fresh water, stagnant or quiet, that constitutes a favorable biotope. We know no species in sea water, though some of them can be found in brackish water, or in interior ponds containing some salt.

The main factor conditioning the presence of *Paramecium* is its food, but different other factors also have an influence: pH, oxygenation, etc. Feeding essentially on bacteria, *Paramecium* is very numerous among the leaves and other vegetal debris putrifying along the banks, in shallow water and near the bottom. They are very seldom found in deep water.

As aerobic organisms, *Paramecium* is usually near the surface, and they are seldom to be found at a depth of more than 20 cm.

Paramecia prefer the pH very near neutrality (6.5–7.5) and they are rarely to be found in acid water (peat-bogs, ponds with many duckweeds, etc.). However, it is always possible to find some species in media that are more or less different from the best conditions, and *Paramecium* reveals extremely important possibilities of adaptation. Those abilities have been tested experimentally and the possibility of adapting *Paramecium* (or other ciliates) to salt solutions has been revealed by different authors, among others Jollos (174), Frisch (94), Orlova (235), Beale (5), Génermont (106). Génermont has studied the adaptation of *Paramecium aurelia* to sodium chloride and calcium chloride. He has shown that adaptation can be achieved in several stages, each comprising at least a 48 hours' stay in solutions of increasing concentration: $0.04N$-$0.08N$-$0.1N$ ... after each transfer the death-rate is low, and after a three days' stay in a medium at $0.08N$, for instance, *Paramecium* is highly resistant to sodium chloride. The determination of the death rate, according to the concentration, shows the great possibilities of survival of the adapted lines which are far superior to those which can be calculated from *Paramecium* of the same pure lines that are not adapted. Those paramecia adapted to sodium chloride are hardly different from normal paramecia in their morphological aspect; the only perceptible variations are a period of pulsation of the contractile vacuoles longer than that of the normal paramecium (no doubt, precisely, in relation to the osmotic pressure of the medium), and a rate of multiplication slightly inferior to that of the non-adapted paramecia (4 divisions a day at 27 °C instead of 4.5 or 5 at the same temperature).

Similar observations have been made regarding temperature tolerance. Poljansky and Posnanskaya (247) have shown that *P. caudatum* could live and multiply from 0 °C to 30 °C and that it was possible to cultivate *P. caudatum* at 0 °C after the organism had been adapted progressively to this temperature.

The great ability of adaptation explains the extremely wide distribution of the different species on all the continents.

B. Paramecium in the Laboratory

A considerable number of techniques of raising *Paramecium* have been reported in the literature. We do not intend to discuss these. The reader will find precise, valuable and extensive information in the publications of Hyman (143, 144), Sonneborn (285) and Wichterman (317). We shall mention only the principal methods of obtaining cultures of paramecia and the problems involved.

The first operation consists in obtaining a stock. Since encystment and consequently excystment of *Paramecium* have never been proven, it is therefore necessary to collect living samples from natural sources. Those samples necessarily contain not only different species of ciliate protozoa, but also, most often, rhizopods and flagellates, and also varied Invertebrata among which generally rotifer and crustaceans prevail. If the latter are numerous, it is essential to transfer the paramecia rapidly into a proper medium.

Even if the populations of *Paramecium* can be kept easily in a laboratory by periodical transfer into hay infusion, many experiments require cultures of pure lines on more defined media and under controlled conditions.

1. Culture techniques
Several types of cultures can be distinguished:

a. Mass cultures, pure for the species. This is the easiest method, since it requires only a precise control of the species at the beginning. The culture can include different varieties and individuals with various sexual tendencies; consequently, aside from a normal vegetative multiplication, sexual phenomena (conjugation, autogamy) may take place. Such cultures are generally well suited for research on the morphology and the general physiology of paramecia.

b. Pure line cultures. They are started from one individual, taken either directly from nature, or from a mass-culture.

Several categories of pure line cultures can be distinguished:

i. Pure line "stock". In the population obtained by vegetative multiplication, sexual phenomena and varied nuclear alterations can take place (conjugation, autogamy) according to the species and varieties; they can cause the appearance of physiological or genetic differentiations; nevertheless, the various genes present in the original individuals conform to those existing in the first individual.

ii. Clone. The clone can be considered as a pure line when only vegetative multiplication takes place to the exclusion of any sexual phenomena. However, in some species (as will be discussed in the section on sexual phenomena), intraclonal conjugations can apparently take place, without any perceptible variations of the initial characters of the clone. The problem of the purity of the clone can therefore be considered in relation with the stability of its evolution.

The term "synclone" is generally applied to the group of the two clones constituted by the descendants of a couple of conjugants (Figs. 1, 2).

iii. Caryonide. This is a pure line whose individuals have a macronucleus derived from the same original macronucleus, whose starting point is an ascendant having its macronucleus formed directly by one of the products of the division of the syncaryon,

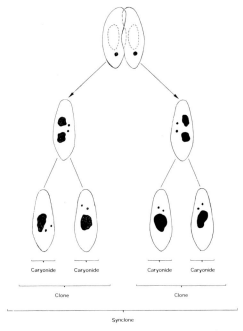

Fig. 1. Diagram showing the origin of the units of culture: caryonides, clones and synclones of *P. aurelia*, originating from one pair of animals.

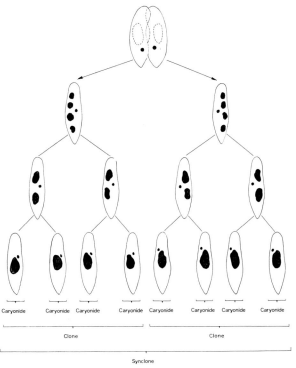

Fig. 2. Diagram showing the origin of the units of culture: caryonides, clones and synclones of *P. caudatum*, originating from one pair of animals.

References p. 74–86

as the consequence of a sexual phenomenon (conjugation or autogamy). Therefore, the number of caryonides to be established from an exconjugant depends on the species considered (see the morphological phenomena of conjugation, p. 51): whereas only 2 caryonides can be formed from an exconjugant of *P. aurelia* or *P. bursaria* (Fig. 1), four can be established by *P. caudatum* (Fig. 2).

The definition of a caryonide is therefore extremely precise; there can exist no problem about it. In some very critical genetic studies, it can be useful to consider units of culture still more reduced and strict: for instance, as is the case of sub-caryonides (Génermont, 106) of decreasing degrees that correspond to the lines established from individuals obtained from the first divisions after the isolation of the caryonide; the precise determination of the different degrees at the level of each unit of culture is then essential.

iv. Homozygous lines. It is possible to consider a population consisting of individuals of the same genetic constitution and homozygotic in all their genes as a pure line. Such a unit of culture is possible only in a few particular cases, e.g., stocks capable of autogamy. The line is consequently established from one individual having just undergone such a nuclear reorganization. However, the purity of such a line with regard to its homozygotic character is quite theoretical; indeed, other researchers have shown the existence of a non-nuclear heredity in the ciliates and the existence of multiple variations of the DNA in the nuclei or chromosomes. According to our present knowledge, these variations do not seem to follow precise rules.

In research investigations the most widely used method of culture is undeniably the clone culture. Yet, it is the most difficult one to delimit, since one of the essential criteria, the exclusively vegetative multiplication, can be perturbed by the appearance, unknown to the observer, of sexual and parasexual phenomena such as intraclonal conjugation or autogamy, without modifying some basic characters of the line. The problem is now to know to what extent we must consider that the clone is still genuine, at what moment the notion of clone must be replaced by that of the synclone (without any crossed fecundation); for, obviously, physiological and probably genetic variations have appeared before sexual differences are evident.

If the establishment of a type of culture presents some problems, its development presents some also. Observations can then reveal some interesting points. My purpose is not to relate all those investigations which have been made recently; I shall merely recall two of their aspects: on the one hand, the complexity of the factors which can influence the division and therefore the growth of the population; on the other hand, some special forms of behavior of the individuals in the culture vessels (aggregation into a ring).

2. *The factors controlling division*

The different factors that may influence the division can, a priori, be divided into several categories:

i. factors that are naturally fixed: light, humidity, O_2 and CO_2 pressure, genetic factors;

ii. variable factors that can be controlled: volume of the medium, the number of generations, bacterial density (food), pH, temperature, group effect;

iii. unknown factors that therefore cannot be controlled.

Those different factors can influence the duration of generation and the fission rate. Some research in progress (unpublished, Laboratoire de Biologie animale, Faculté des Sciences, Amiens) dealing with the different factors that can be controlled has yielded some interesting results. The experiments, carried out under rigorously controlled conditions, and statistically analyzed, established:

(a) The interfission period is significantly different from the first through the fifth generation (no observations have been made beyond the fifth generation). In particular the duration of the first generation is always far longer than that of the others.

(b) The length of the interfission period varies significantly as a function of the volume of the medium. For a certain generation the interfission period increases proportionately to the volume of the medium. When the volume is very large, the interfission period becomes so long that the ciliate dies before dividing. However, a second factor, the density of the food supply in the medium, can intervene at this time. Experiments in progress show that:

(a) In the same volume the interfission period is longer when the number of bacteria present is small.

(b) The interfission period is shorter in 0.1 ml than in 0.5 ml when the same number of bacteria is present. The smallest volumes seem to be most favorable for rapid fission.

Experiments on the influence of an eventual group effect also show that, when the conditions of culture are strictly controlled, there does not seem to be any influence of that factor on the duration of the generation of the individuals. Consequently, there does not seem to be, contrary to some opinions advanced before, any reciprocal influence of the individuals on the duration of their division, but only influences of the conditions of culture: particularly, the volume of the medium and the density of bacteria.

3. Principal culture methods

The application of those different data concerning the type of culture medium and the conditions of growing paramecia make three essential methods of experimental cultures possible:

a. A method with periodical reisolation. This method is carried out in depression slides or in bowls, with regular transfers into fresh medium, with a determined frequency (every day, every 3 days or more) according to the rate of multiplication of the paramecia and of the contemplated experiments.

b. A method of culture in a limited medium. This method consists of culturing *Paramecium* in a known bacterized medium (for instance, lettuce infusion inoculated with *Klebsiella aerogenes*) of a determined volume. The population advances then according to Fig. 3. After a period of adaptation, the logarithmic growth phase ensues; the maximum population is reached after a few days; after a period of

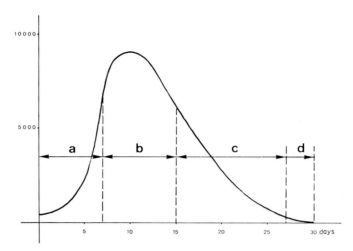

Fig. 3. Curve representing the evolution of a population of *P. caudatum* in a limited medium (10 ml), not replenished (in a test-tube). On the Y-axis: total number of individuals. a = period of logarithmic increase; b = period of stabilization and of balancing; c = period of decline; d = period of disappearance of the last individuals (it can be longer or shorter).

stability the population declines caused by the impoverishment of the medium; this decline unavoidably ends with the death of the individuals. It is therefore necessary, in such a case, to transfer the culture every 2 or 3 weeks.

c. A method of culture in renewed medium. This method consists in making a culture in a bacterized medium that is partly and regularly renewed, keeping a constant total volume: adding some fresh medium every day and removing the equivalent quantity of old medium. The quantity of fresh medium added to the remaining quantity of old

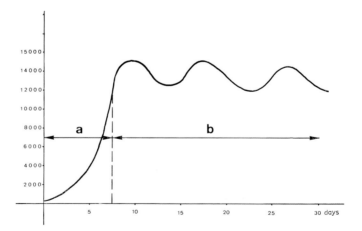

Fig. 4. Curve representing the evolution of a population of *P. caudatum* in a limited medium (10–15 ml) partly and daily replenished. On the Y-axis: total number of individuals. a = period of logarithmic increase; b = period of balancing with rhythmical variations.

medium conditions the rate of division and makes it possible to obtain an important population density. This remains relatively stable during very long periods (Fig. 4); some precise reports on the relative volumes of medium to be renewed in the case of the culture of *P. aurelia* have been made by Sonneborn (285), and different investigations using this technique have been made with *P. caudatum* (Vivier, 302; Vivier and Mallinger, 309).

Nevertheless, however stable the condition of the culture may be, experience has shown that the rate of division varies in time; it reaches maxima and minima, with periods of crisis at variable intervals, more or less regular and important. Those variations, or rhythms of division, have been well-known since Woodruff's investigations (322–324) on *P. aurelia*, and have been found again in the other studied species; they seem to be connected with genetic or physiological factors independent of the conditions of culture.

4. The behavior of the individuals in culture

The behavior of paramecia in culture poses many complex problems. One example is the gathering of the individuals in a ring near the surface of the culture liquid (Fig. 5).

This phenomenon has been investigated for many years. Moore (218) and Dembowsky (59) postulated that the amassment was caused by geotaxis, although they

Fig. 5. Culture of *P. caudatum* in a test tube: agglomeration of animals in a ring (arrow) near of the surface of the medium.

admitted that other factors might interfere. The formation of such rings, independent of the shape and size of the culture vessel, can be used as a natural means of concentration of the organisms (Sonneborn, 285). This aggregation proved to be even more interesting since Vivier (302) demonstrated that the formation of the aggregates depended upon a particular state of sexual reactivity of the organisms. Some simple experiments by Estève (81) have shown that a negative geotaxis can hardly explain the formation of these aggregates. However, the influence of the atmosphere above the culture medium, as well as that of the nutritional state, must be taken into consideration. In fact, various questions can be asked, some concerning the very shape of the aggregation (why, for example, a ring and not a disk), others with regard to the position of the ring as a function of the medium and the culture conditions.

Two hypotheses which have been proposed to explain the formation of a ring come to mind: thigmotaxis might govern the ring formation, or the agglomeration might be a more general reaction due to the behavior of any animal confined to limited space and moving around the periphery of this space. This question has not been solved.

The second question concerns the position of the ring. This ring forms a few millimeters below the surface of the liquid, and forms within a span of 20–40 minutes after the introduction of bacterial food in a partially exhausted medium. Estève (81, 83) has been able to establish that the partial pressure of oxygen influences the establishment of the ring. This depends on the physiological state of the paramecia, as determined particularly by their nutritional state. The organisms congregate at the level where the optimal partial pressure of oxygen exists.

The above example demonstrates how an apparently simple fact can be extremely complex with regard to its cause and the interactions that come into play.

III. GENERAL MORPHOLOGY, STRUCTURE AND ULTRASTRUCTURE

Observations with the light microscope had established the general morphology of *Paramecium*. During the past twenty years great progress has been made in the studies on morphology, thanks to the emergence of new techniques. The use of these techniques has revolutionized and completely renewed our knowledge, especially in the field of cytology. It must be noted, however, that *Paramecium* which perhaps was considered to be well-known and too common has been seldom used for observations with these new techniques; instead, strangely enough, some species of extremely rare protozoa, or protozoa presenting exceptional characteristics have preferentially profited by these new techniques, and, as a result, might in certain aspects be better known now than *Paramecium*.

A. Techniques of Observation

It is out of the question to present a detailed survey of all the means of observation that an investigator of *Paramecium* can have at his disposal. In fact, he will find all the necessary information in specialized books. But it is necessary to give a summary of the different techniques that can be used and of the results that can be obtained.

1. Methods for the study of living specimens

The observation *in vivo* has been and remains an essential technique of great interest. Only this technique makes it possible to verify the exact shape and size of the examined individuals, as well as the existing natural deformations; it is thus the basic method for the determination of the species and the definition of the general morphological characters.

During the last few years, two instruments, very precious for the protozoologist, have been added to the stereoscopic microscope and to the usual light microscope: the phase contrast microscope and the interference contrast microscope. These microscopes make it possible to see, within the limits of resolving power of an instrument using visible light, all the cytological details, internal as well as on the surface of the individuals, without having to resort to staining processes.

The essential difficulty of the microscopic examination *in vivo* is due to the mobility of the animals that move rapidly in the observation field. Thanks to two procedures, we can remove this difficulty: The first one consists in using chemicals which increase the viscosity of the medium, in particular some compounds derived from cellulose (methyl-cellulose or Methocel, sodium-carboxy-methyl-cellulose). The second one consists in using a mechanical process of compression of *Paramecium* sufficient only to restrain them in their movements or to immobilize them. For that purpose, micro-compression chambers exist, which are very convenient and adequate (Wichterman, 317).

Although these methods are valuable, they are not perfect; in fact, the chemicals increasing the viscosity are slightly toxic, and compression can upset some biological phenomena such as division and conjugation. Nevertheless, it is sometimes indispensable to use them in order to make some observations or to facilitate microphotography or microcinematography.

2. Methods for the examination of fixed specimens

These methods are by necessity adapted to the mode of observation, either light microscopy or electron microscopy. The latter is indispensable for modern cytological research. The study of a protozoan, like any other study on the cellular level, cannot be envisioned without electron microscopy.

a. Light microscopy. This classical technique will always enable the investigator to make numerous observations on intact animals, as well as on sectioned animals. During the past years, more refined, precise and specific methods of preparation of the specimens have been added to those that have been known for a long time. We will only discuss the newer techniques, because the classical ones can be found in the cytological and histological textbooks and manuals.

The examination of intact individuals can be made on a glass slide; the animals are generally embedded in a medium composed of gelose, gelatin, or albumin, after adequate fixation. Those preparations permit the use of numerous stains: silver impregnation for the examination of superficial structures, or histological coloration for the examination of internal structures (Hematoxylin, Feulgen staining). One of the most interesting techniques is that of impregnation with Protargol (Bodian, 15, 16), modified by Kirby (189) and successively perfected by Dragesco (69) and by Tuffrau (299, 300). This method, in the opinion of some workers, has advantageously replaced the technique of impregnation of Klein (190) and that of Chatton and Lwoff (41) to reveal the superficial structures (kineties, silverline system) as well as the numerous, deep, subpellicular fibrillar structures.

The examination of sections is valuable for cytological and especially cytochemical studies. The precision of the observation can be improved considerably when the fixed organisms are embedded in agar before the paraffin coating. Only then is it possible to position the organisms and the sections in a perfect way. This method is particularly recommended for those cytochemical investigations where it is necessary to exactly localize a certain component. Enzymatic research techniques have progressed in the last years, however, very little research has been carried out on *Paramecium*. The applications of these enzymological techniques has lagged far behind, mainly due to the fact that these techniques can also be used with a much greater sophistication on the electron microscope level.

b. Electron microscopy. The use of electron microscopy has considerably expanded these last few years and the techniques of preparing the specimens have made considerable progress in quality, reliability and versatility, especially when we compare the present results with those of the earlier experiments.

At present, two processes are available: (i) transmission electron microscopy, which used very thin sections or preparations transparent to electrons, and (ii) scanning electron microscopy which studies complete specimens.

i. Transmission electron microscopy. With the present techniques of preparation, we can obtain very satisfactory results. The most common technique, which can be considered as a standard method, is the one that uses a double fixation: glutaraldehyde followed by osmium tetraoxyde in phosphate buffer at pH 7.2; the specimens are then embedded in epoxy resin (Epon or Araldite) and thin sections are stained either with uranyl acetate or lead citrate, or successively with both chemicals. This technique, elaborated in 1962–1963, makes possible an excellent preservation of ultrastructures; it also gives remarkably sharp images and can be perfectly reproduced. In the case of cytochemical investigations, it is sometimes useful to use a water soluble compound as an embedding medium; glycol methacrylate is now mostly used; thanks to it, enzyme-digestions on very thin sections are possible.

Single organisms can be embedded in the resin, but it can be convenient to embed several of them together. Several processes of concentration can be used; such as microinclusion in agar, similar to the one used in light microscopy. Other techniques have been proposed: the use of nucleohistones (138) or the preinclusion in clots of fibrin (31), the latter being particularly convenient and efficient.

As mentioned before, most of the cytochemical techniques have been adapted to electron microscopy; for example, the detection of some organic components, especially polysaccharides with the technique of Seligman, modified by Thiery (298) (Fig. 28). Detection of enzymes, especially phosphatases (Fig. 27), can be carried out with the technique of Gomori, adapted by Novikoff (229) and Essner and Novikoff (80). Staining techniques for determining the presence and the localization of the nucleic acids RNA and DNA are not perfectly specific. On the other hand, enzymatic digestion by ribonuclease or deoxyribonuclease, and especially high resolution autoradiography after incorporation of tritiated uridine or tritiated thymidine, give excellent results (Fig. 30).

The technique of negative staining with phosphotungstic acid can yield excellent information about fibrillar and other surface structures after the cells have been adequately treated (lysis, or crushing the cell). This technique has thus far been little used, but it has yielded interesting results in the hands of Pitelka (246) and Hufnagel (140).

ii. Scanning electron microscopy. Scanning electron microscopy, recently perfected, makes it possible to get spectacular images of the superficial morphology with a resolving power that can now reach about 150 Å and with a considerable depth of field.

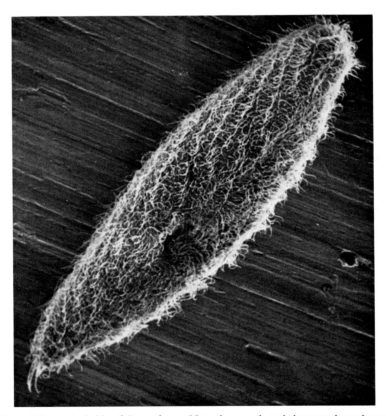

Fig. 6. View of the ventral side of *P. caudatum*. Note the mouth and the metachronal waves of the cilia. Scanning microscopy, × 500.

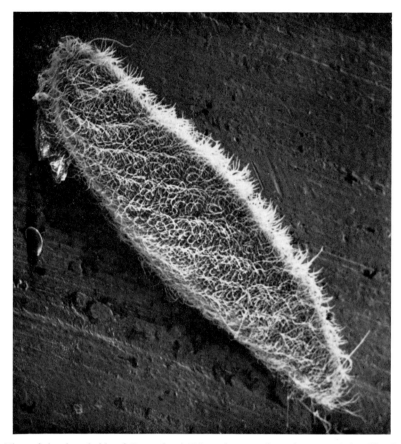

Fig. 7. View of the dorsal side of *P. caudatum.* Note the metachronal waves of the cilia. Scanning microscopy, × 500.

The preparation of specimens can be made with the technique perfected by Small and Marszalek (276) which consists of a fixation in the Párducz mixture, followed by freeze drying before shadowing the specimen.

B. Morphology and Structure

Since structure and ultrastructure are covered in detail in other chapters, I shall consider only those points which are necessary to the understanding of the systematics or the general biology of the whole organism.

1. General organization (Figs. 6–8)

Paramecium is a long-shaped, unicellular organism whose length is variable, according to the species, from 80 to 350 microns, and whose width is variable from 40 to 80 microns. The cross section is more or less circular, often flattened in its anterior part. The general aspect is that of the sole of a shoe, more or less wide and more or less tapering in its extremities.

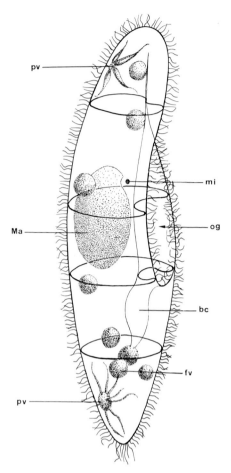

Fig. 8. Diagram showing the general organization of a *Paramecium* as seen from the right side. bc = buccal cavity; fv = food vacuole; Ma = macronucleus; mi = micronucleus; pv = pulsatile (contractile) vacuole; og = oral groove.

The orientation is easy, thanks to the presence of the mouth, which conventionally determines the ventral side; the anterior part is, of course, the part moving ahead in normal movement.

On the ventral side, there is a large flattening: the oral groove, which starts at the anterior part and spreads down to half or two-thirds of the animal, curving slightly to the left.

The mouth is situated in the far posterior part (vestibulum) of this groove. Behind the mouth is the buccal cavity, which is directed towards the posterior part, at the end of which the alimentary vacuoles (cytostome) are formed in a cytoplasmic area (cytopharynx) limited by special fibrillar systems.

On the ventral side, behind the mouth, halfway between it and the posterior part, the cytopyge or cytoproct is situated. Here the wastes resulting from digestion in the alimentary vacuoles are excreted.

References p. 74–86

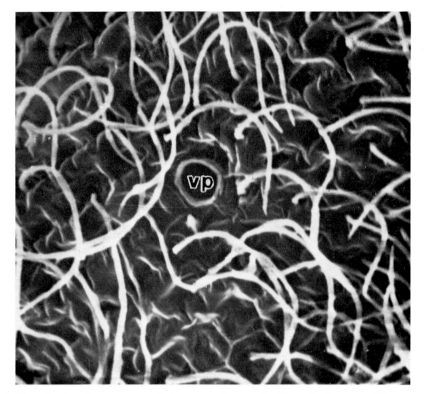

Fig. 9. View showing the open pore of the contractile vacuole. Note the "argyrome" hexagons and cilia. Scanning microscopy, × 5,700.

On the dorsal side, an even surface, we can see only the orifices of the contractile vacuoles, usually two, an anterior one and a posterior one (Fig. 9).

The whole surface of the body is covered with cilia, in longitudinal rows or kineties. The kineties are regularly extended from the fore-part to the hind-part of the dorsal side but are slightly curved on the ventral side; those of the right side are opposed to those of the left side of the mouth on a suture line that goes from the anterior end, along the mouth and the cytopyge down to the posterior end. A particular ciliature exists inside the buccal cavity (see below) and a small tuft of longer and motionless cilia is present at the posterior end (Fig. 10). Inside, a certain number of organelles can easily be seen (Fig. 8).

a. The nuclear apparatus, which includes a large macronucleus and one or several micronuclei: the latter are small and difficult to observe "*in vivo*" but are visible after using adequate cytochemical techniques.

b. The contractile vacuoles, generally two, an anterior and a posterior one. They have alternate movements of contraction (systoles) and dilatation (diastoles) and are connected with systems of collecting canals.

c. The food vacuoles, which start from the bottom of the buccal cavity and corre-

Fig. 10. View of the posterior end of *P. caudatum* showing "argyrome" hexagons and elongated cilia. Scanning microscopy, × 2,200.

spond to a phenomenon of phagocytosis. Once out of the buccal system, the vacuoles make a circuit in the cytoplasm; that circuit is more or less regular and often takes them to the anterior part before bringing them back to the posterior part where the cytopyge is situated and where they open themselves to expel their wastes.

Beside these fundamental and characteristic organelles of *Paramecium*, there exist in the cytoplasm all the usual cellular inclusions: mitochondria, Golgi bodies, reserve material, etc. as well as several other special structures such as trichocysts, fibrillar systems, etc.

2. Cortical morphology

Since Maupas' studies (204) which described the pellicle of *Paramecium* as consisting of an assemblage of rectangular or rhomboedric fields, numerous observations have clarified the superficial morphology, both pellicular and subpellicular. The main perfections are due to Klein (191), von Gelei (99, 101), Lund (200), Taylor (297), Wichterman (317), Yusa (329), Ehret and Powers (78), Párducz (240, 241) and Roque (255). All the essential data concerning those structures have been obtained by using the technique of silver impregnation. The investigations were refined using electron microscopy. The main contributions are those of Metz *et al.* (210), Sedar and Porter

(270), Ehret and Powers (79), Schneider (261), Vivier and André (306), Ehret and De Haller (77), Stewart and Muir (291), Schneider and Wohlfarth-Bottermann (266), Pitelka (246), Hufnagel (140), and Jurand and Selman (177).

Here, we shall merely give a short synopsis of all these publications.

   a. Pellicular structures. Silver impregnation reveals, on the surface of *Paramecium*, as well as of all the ciliates, several structural types (Fig. 11):

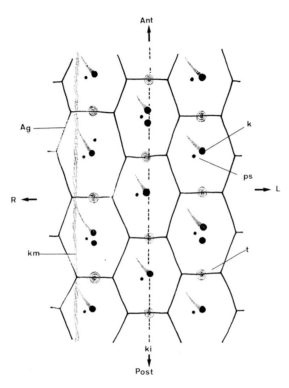

Fig. 11. Diagram showing the organization of the surface of *Paramecium*. The kinetodesmal fibers are represented in the left row only. Ag = argyrome; km = kinetodesme (kinetodesmal fibrils); ki = kinety; k = kinetosome; ps = parasomal sac; t = trichocyst; Ant = anterior side; Post = posterior side; R = right; L = left.

   i. The kineties, or longitudinal rows of kinetosomes (ciliary striae). The kineties, real anatomical units, are arranged according to the meridian of the ciliate. Along each kinety, and on its right, there is a sub-pellicular fibrillar system: the kinetodesmal fibers. Those definitions and the rule of desmodexy (constant disposition of the kinetosomes according to kinetodesmal fibers in the ciliates) were established by Chatton and Lwoff (42). The kineties are self-contained and all the kineties constitute the kinetome.

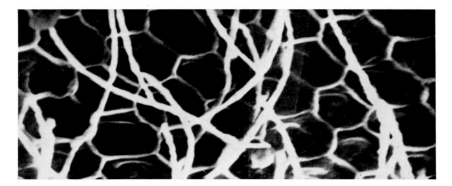

Fig. 12. View showing "argyrome" hexagons. Scanning microscopy, × 670.

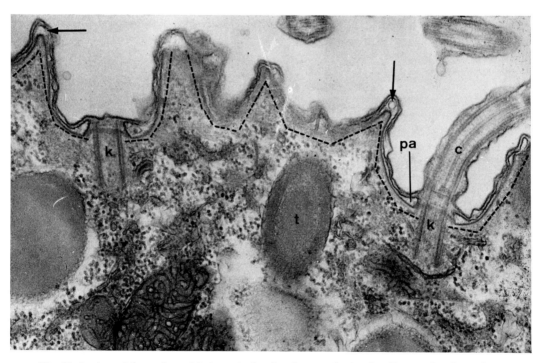

Fig. 13. Section of the surface of *P. bursaria* showing the organization in adjacent microdepressions of the pellicle. Notice (arrows) the passage of the sub-pellicular vesicles (pa) from one depression to another, as well as the location of the kinetosomes and the cilia in the center of the microdepressions.
k = kinetosome; t = trichocyst; c = cilia. × 35,000.

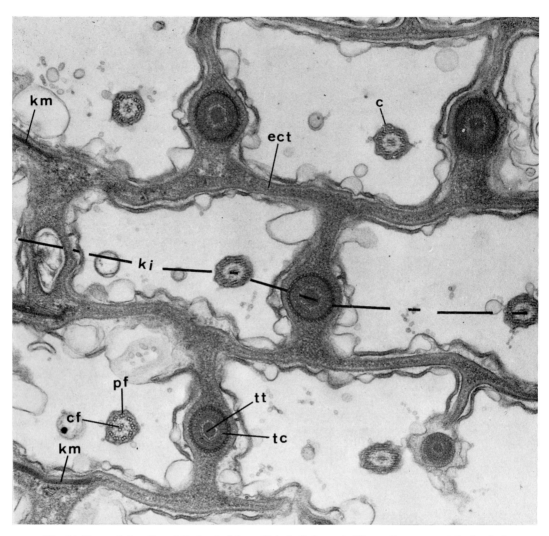

Fig. 14. Tangential section at the level of the pellicle in *P. bursaria*. The section passes at the level of the ectoplasmic crests that appear in a network around the depressions of the pellicle. Note the presence of cilia in the center of the depressions and the alignment of the trichocysts with the kineties (ki) at the level of the ectoplasmic crests. On the left of the picture, the kinetodesmal fibers (km) can be seen. c = cilia; ect = ectoplasm; ki = kinety; tt = trichocyst tip; tc = trichocyst cap; cf = central fibrils; pf = peripheric fibrils. × 35,000.

ii. The silverline system (Silberlinien system) *of Klein**, which corresponds to a network that can be impregnated with silver. In *Paramecium*, that network is regular;

---

* Chatton and Lwoff (43) had defined the silverline system as a system distinct from the kinetome in its substance, structure and topography. The definition of the Silberlinien system of Klein (192) agrees with this definition; yet Klein, prejudging the possible conduction role played by the silverline system, differentiated the direct conducting system (direkt Verbindungs-System) and the indirect conducting system (indirekt Verbindungs-System); the latter is the only homologue of the silverline system and plays no nervous or conducting part whatsoever (see below).

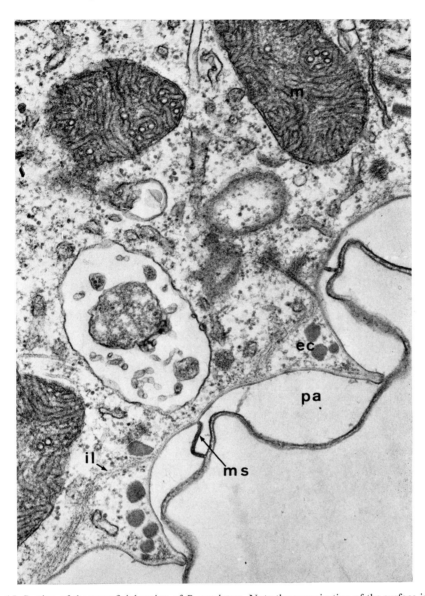

Fig. 15. Section of the superficial region of *P. caudatum*. Note the organization of the surface in the depressions. The section passes at the side of the cilia and makes it possible to see the median septum (ms) separating the sub-pellicular alveoli (pa) at the level of the meridian ciliary lines. Note, as well, the presence of microfibrils (il, infraciliary lattice) visible at the level of the ectoplasmic crests (ec). × 44,800. (from Estève, unpublished).

its meshes have an hexagonal shape over the whole of the body except in the vestibulum area where they are more rectangular.

This network spreads its meshes between the cilia so that each mesh is centered by a cilium (or a group of two cilia), together with, on its right, a small structure that can also be stained: the parasomal sac. On this silver-lined network, following the line of the kineties, are the heads of the trichocysts.

Studies with electron microscopy have clarified the organization of the whole system (Figs. 12–15).

The meshes of the silverline system correspond to a series of slight, superficial depressions from the center of which the cilia emerge, either alone or in groups of two. Underneath, there exist, under the cytoplasmic membrane, flattened sub-pellicular alveoles which make way for the fibrils of the kinetosome and of the cilium, as well as the parasomal sac constituted by an invagination of the superficial membrane. At the level of the meshes of the silverline system, the ectoplasm is flush with the cytoplasmic membrane and separates the sub-pellicular alveoles more or less completely.

A thick, granular epiplasmic coat lies under the alveoles and joins with the cilia at their limits with the kinetosome.

The questions that can be asked about the superficial structures essentially concern physiology. What meaning must be given to the subpellicular alveoles; do they play a part in keeping the shape of the organism? Or are they part of an osmotic boundary between the inside and the outside medium?

In fact, those problems are not particular to *Paramecium*, nor even to many ciliates, since such vesicles can be seen in other protozoa, namely, Sporozoa (307).

b. Sub-pellicular structures, including the fibrillar system and the trichocysts

i. The sub-pellicular fibrillar system includes different types of structures: (1) The kinetodesmal fibers, which are composed of fibers of a periodical structure, starting from each kinetosome; those fibers proceed toward the forepart of the ciliate, moving on the right of the kinety, and stretch parallel to the surface, above the vesicles, at a distance that has not been certainly determined, but that would approximately be equal to the distance between 6 kinetosomes. (2) Microtubular fibers that also start at the level of each kinetosome and move cross-wise under the surface. They are probably homologous to the transverse fibrilla described in other ciliates (116). (3) Various microfibrilles, more or less gathered in bunches, and more or less anastomosed.

What function can be attributed to those different fibrillar systems? A coordination of ciliary movements? A part in keeping the shape? A part in relation with the properties of elasticity and contractibility? If we can argue in favor of some of these hypotheses, nothing has yet been demonstrated and the problem is still to be solved.

ii. The trichocysts (Figs. 16 and 20) are comparatively voluminous organelles, fusiform, endowed with explosive properties, which are situated perpendicularly to the cellular membrane and are flush with the ectoplasmic crests at the precise places stated above. Their morphology is known thanks to numerous investigations; the most important and recent are those of Jakus (151), Jakus and Hall (152), Beyersdorfer and Dragesco (12), Yusa (330), Ehret and De Haller (77), Stockem and Wohlfarth-Bottermann (292). Those organelles are constituted of two parts: an ovoid body, topped by a head, in the shape of the tip of an arrow protected by a cap; the complexity of the structure of that cap does not seem to have been entirely revealed in previous investigations. If the mechanism of their extrusion is known, their functions are still obscure; their burst does represent a reaction to an excitation (mechanical, physical or chemical), but it can hardly be considered to play a defensive role.

On the whole surface of the body, including the oral groove and the vestibulum, the ciliature is of a somatic type, corresponding to what has been described above; the mouth marks the limit of that ciliature.

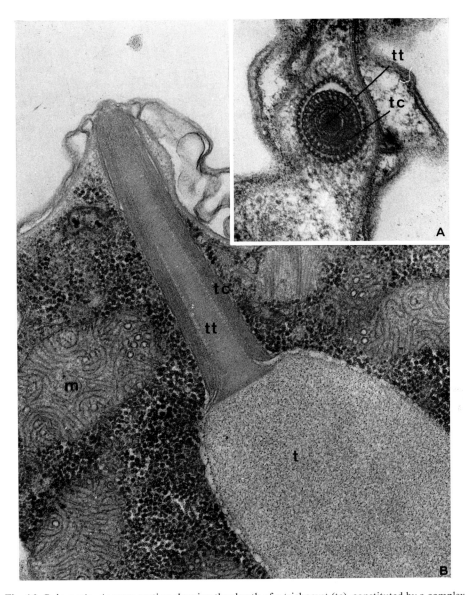

Fig. 16. *P. bursaria. A:* cross-section showing the sheath of a trichocyst (tc), constituted by a complex system of 36 longitudinal units, wrapping the point (tt) of a roughly polygonal section. × 52,500. *B:* section showing the different parts of a trichocyst in its place in the ectoplasm. The body (t) of the trichocyst is topped by the point (tt) wrapped with a sheath (tc = trichocyst cap). m = mitochondria. × 35,000.

c. Buccal structure. Starting from the mouth and inside the buccal cavity, there exist particular ciliary structures that already had been studied exactly by von Gelei; describing them in *P. nephridiatum* and *P. caudatum* in a series of investigations from 1925 to 1954. More precise complementary descriptions have been added later, thanks to the research of Lund (200), Mast (203), Fauré-Fremiet (86), Yusa (328) and Roque (255).

Whereas the right side of the buccal cavity remains naked, kineties, some of which constitute very special membranelles, exist on the right edge, on the dorsal side and on the left side; they are shown in Figs. 17 and 18.

i. The paroral kinety, also called endoral kinety; this kinety is single and auto-nomous. It runs along the internal right edge of the buccal cavity, but on a certain distance only; it is therefore comparatively short and is constituted of 15 to 20 kinetosomes.

ii. The quadrulus. It consists of 4 kineties, starting at the dorsal edge of the mouth.

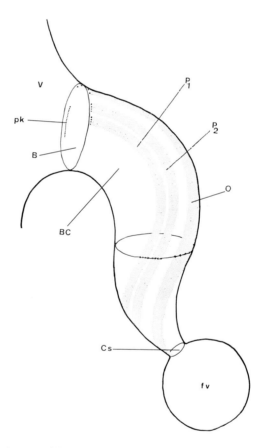

Fig. 17. Diagram showing the different buccal structures. V = Vestibulum; B = buccal overture; Cs = cytostome; fv = food vacuole; pk = paroral kinety; P_1 = ventral peniculus; P_2 = dorsal peniculus; Q = quadrulus.

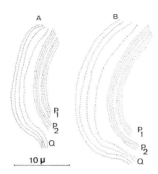

Fig. 18. Buccal kineties of *Paramecium:* Peniculus (P₁ et P₂) and Quadrulus (Q), *A:* in *P. aurelia; B:* in *P. caudatum.* (from Dragesco, 69).

They run along the dorsal side parallel to each other toward the bottom of the buccal cavity. Here they execute a small spiral turn which brings them from the dorsal side to the left at the ventral side at the level of the cytostome.

iii. The peniculi. The two peniculi are two polykineties, which generally consist of 4 parallel rows of kinetosomes. They are located on the left side of the buccal cavity, juxtaposed side by side. They are shaped in a helix in such a way that the farthest tip reaches a ventral position. In general they are described according to their particular location as the dorsal peniculus and the ventral peniculus. The ventral peniculus is slightly shorter than the dorsal one.

3. Internal organization

The cytoplasm has a classical cellular structure and it is relatively difficult to differentiate the ectoplasm from the endoplasm in *Paramecium*, in contrast to what has been found in other ciliates. Electron microscopy reveals a very dense cytoplasm containing an abundance of ribosomes intermixed with glycogen particles. The very numerous mitochondria are tubular as found frequently in the protozoa. These structures are necessary for the aerobic mode of living and the motor activity of this ciliate (Vivier, 305).

The Golgi apparatus was not observed for a long time and the first images of dictyosomes in paramecium were made by Pitelka (246). Dictyosomes are small and composed of only a very low number of saccules. They seem always to be connected with the ergastoplasmic membranes through coated vesicles (Estève, 85, 85a) (Fig. 19).

Some electron microscopic studies of the contractile apparatus (Schneider, 262) revealed the existence, around the nephridial canals, of nephridial tubules that are in close relation with the endoplasmic reticulum; other studies have also dealt with the part played by this apparatus (251) and its function (233, 234).

Aside from the classical cytoplasmic components, the existence of symbiotic organisms permanently present in some species must be pointed out. These are the symbiotic zoochlorella in *P. bursaria* (Fig. 20) whose ultrastructural study was recently investigated by Vivier *et al.* (310) and by Karakashian *et al.* (183). Symbiotic

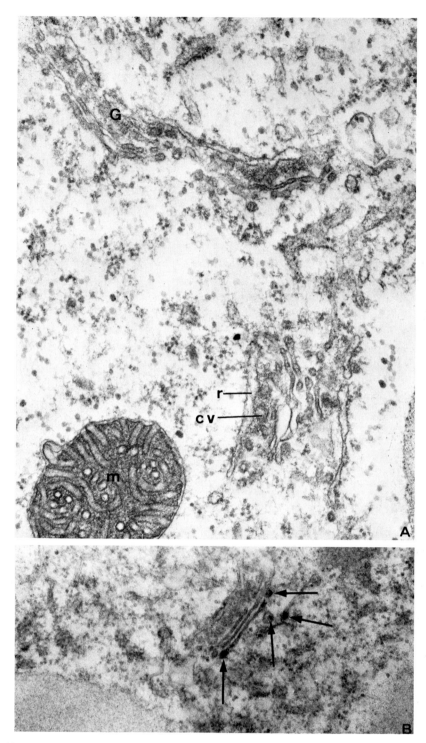

Fig. 19. *P. caudatum*. Golgi apparatus. (from Estève, unpublished). *A:* dictyosomes connected with some saccules of the endoplasmic reticulum (r) through "coated vesicles" (cv). × 60,000; *B:* revealing the acid phosphatase (arrows) in the saccules and the vesicles of a dictyosome. × 44,800.

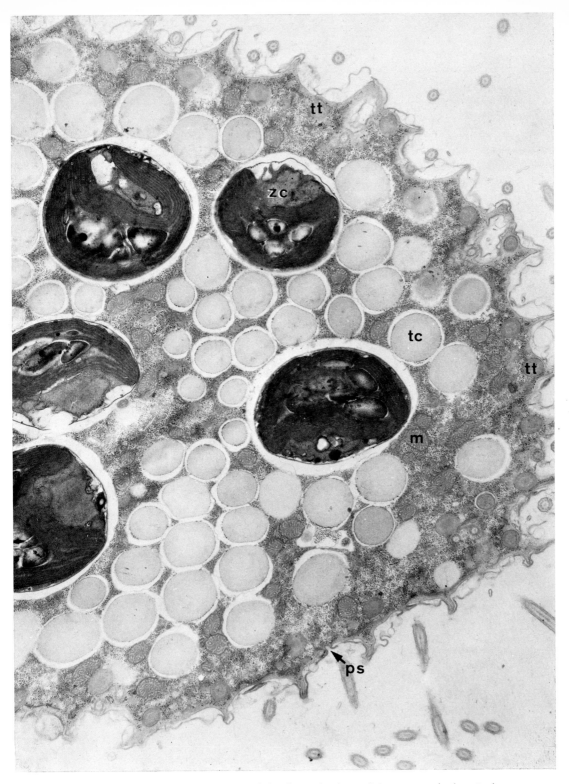

Fig. 20. *P. bursaria*. Section showing the peripheral organization and the presence, in the cytoplasm, of trichocysts and of symbiotic zoochlorella. tc = trichocyst cap; tt = trichocyst tip; m = mito-chondria; zc = zoochlorella; ps = parasomal sac. × 7,500.

References p. 74–86

bacteria have been found in numerous stocks of *P. aurelia*. The ultrastructural studies (Dippell, 66; Hamilton and Gettner, 121; Beale and Jurand, 6, 7; Jurand and Preer, 176; Beale *et al.*, 8; Preer, 249) made it possible to solve the problem of the nature of the particles discovered by Sonneborn (280) in "killer" *Paramecium*.

The nuclear system (Figs. 21 and 22) consists typically of a macronucleus and, depending upon the species, of one or several micronuclei. The composition of the nuclear envelope conforms to that of the nuclei of any cell, i.e. two membranes with pores. On the other hand, the contents are not identical with those of the nuclei of the cells of metazoa, but they correspond to a typical structure of a ciliate.

The macronucleus shows in its nucleoplasm very abundant and very small masses of chromatin, that are either isolated *(P. caudatum)* or more or less connected with each other *(P. bursaria)*, the nucleoli are abundant, much larger than the masses of chromatin and show a plain polymorphism (303) and a heterogeneity of composition which, though precise cytochemical studies have not been made, must be similar to that of *Tetrahymena* (30).

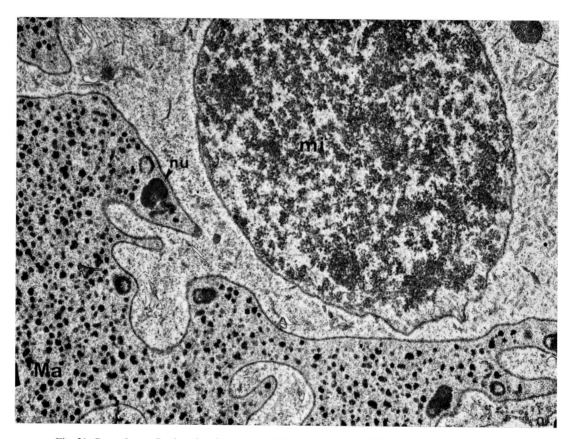

Fig. 21. *P. caudatum.* Section showing a part of the macronucleus (Ma) and the micronucleus (mi). × 12,000. (from Estève, unpublished).

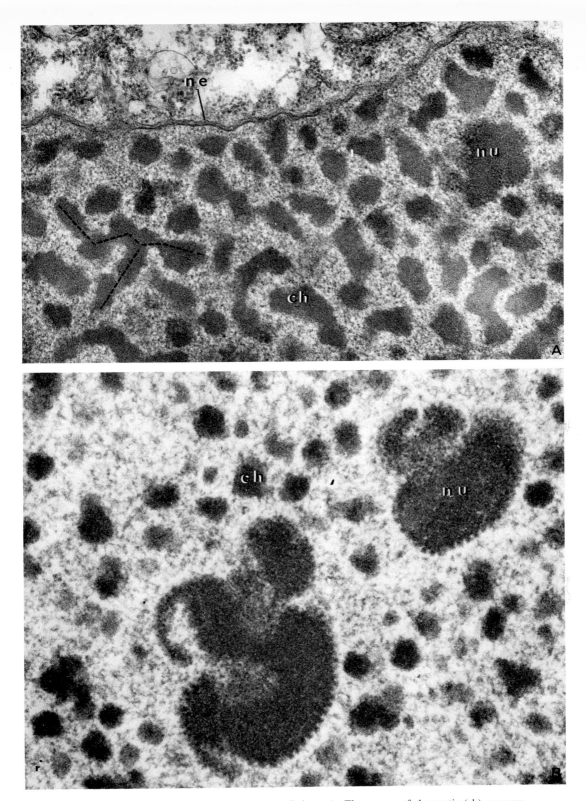

Fig. 22. Ultrastructure of the macronucleus. *A: P. bursaria*. The masses of chromatin (ch) are more or less coalescent (dotted line). ne = nuclear envelope; nu = nucleolus. × 35,000. *B: P. caudatum*. The chromatin (ch) makes clearly separated masses; the nucleoles (nu) have a complex structure. × 60,000. (from Estève, unpublished).

References p. 74–86

Besides those normal nuclear components, other structures have been found, whose significance is not as yet clear; for example, a bundle of protein fibers (306). It must be noticed that the ultrastructural studies were not able to reveal any sub-units whose existence would have supported some hypotheses about the genetic part played by this macronucleus.

The micronucleus also has an original composition in that only some accumulations of chromatin, very small and dense, can be detected in the absence of any typical nucleolus.

The micronucleus is generally situated close to the macronucleus; it is often situated in a groove of the latter, which makes it difficult to be detected "*in vivo*".

The chromosomes can be seen in the micronucleus only during the division cycle. According to the studies that have been made so far, their number does not seem to be permanent in a species or even for a given pure line; for instance, they can vary from 36 to 165 in *P. caudatum* (61), from 80 to several hundred in *P. bursaria* (45) and in *P. aurelia* in the same syngen from 86 to 126 (184). The estimates of quantities of DNA by cytophotometry yield similar results. This inconsistency cannot be interpreted by varying degrees of polyploidy, because their numbers do not seem to be regular multiples of a basic value (haploid or diploid). Therefore, we believe that there are complex chromosomes, or that variations of a polysomic or aneuploid type are perfectly consistent with a normal life of the organism. This hypothesis is the more plausible since we have known for a long time that the micronucleus does not play any part in the vegetative life of the ciliate and that the micronuclei can even be completely absent. In fact, we know that, in several species, amicronucleate strains can live perfectly and normally and can multiply.

While the micronucleus, of a spherical or ovoid shape, has a diameter smaller than 10μ (Table 1), the macronucleus can have a size as large as several tens of microns. The size of the latter show a precise correlation with the size of the cell; the correlation between the length of the nucleus and that of the cell formed the subject of a precise

TABLE 1

Characteristics of the micronuclei in different species of Paramecium. *(after Wichterman).*

Species	Number	Size	Shape
P. caudatum	1	8	ellipsoidal
P. aurelia	2	3–5	spherical
P. bursaria	1	7	ellipsoidal
P. multimicronucleatum	3–4 (up to 9)	0.7–2.5	spherical
P. trichium	1	4–7	ellipsoidal
P. calkinsi	1–2	3–5	spherical
P. polycaryum	4 (3–8)	3–4	spherical
P. woodruffi	3–4 (up to 8)	4–5	spherical

statistical study in *P. caudatum*, taking into account the conditions of culture and of the strains (231). The correlation generally varies from 0.18 to 0.28 and it seems to be a more reliable characteristic of the species and the strains, better than the size of the cell and of the nucleus when they are considered separately.

IV. TAXONOMY

Although different species of *Paramecium* have been known for many years, their present place in the systematics of the ciliates dates from 1950. This is due to the contributions of Fauré-Fremiet (87, 88). His observations were confirmed by Corliss (56, 57) and Roque (255). The study of the buccal structures and ciliatures, and their comparison with those of other ciliates, has been the basis for the taxonomic change in the classification of the genus *Paramecium* and some other closely related species.

A. General Characteristics of the Genus and its Place in the Systematics of the Ciliates

The genus *Paramecium* had been classified for a long time among the trichostome holotrichous ciliates. But von Gelei (101), after his studies of the buccal structures, the observation of the "vestibulum" and of the membrane which he called peniculus, had already understood that this classification was not correct and he had created, for *Paramecium*, a new sub-order: the *Trichohymenostomata*. In 1952, the same author (105), acknowledging a certain morphological and phylogenetical importance to the vestibulum, placed *Paramecium* among the hymenostomes *Vestibulata* and a posthumous work showed that von Gelei (105) considered the Parameciidae as real hymenostomes that must be placed next to the Frontoniidae.

Meanwhile, some studies of the comparative morphology of the buccal structures revealed to Fauré-Fremiet the existence of the dorsal and the ventral peniculi. This author (87), finding comparable structures in the *Frontonia* and *Disematostoma* ciliates, proposed to range those species in a group of the "Hyménostomes péniculiens".

The present systematics of the genus *Paramecium*, after Corliss (57) and Honigberg *et al.* (136), can be established as follows:

Phylum: *Protozoa* Goldfuss, 1818, emend. van Siebold, 1845
Subphylum: *Ciliophora* Doflein, 1901
Class: *Ciliatea* Perty, 1852
Subclass: *Holotrichia* Stein, 1859
Order: *Hymenostomatida* Delage and Hérouard, 1896
Suborder: *Peniculina*, Fauré-Fremiet (in Corliss, 56)
Family: *Parameciidae* Dujardin, 1841.

Paramecium shows typical features of:
(1) Somatic ciliature regularly distributed, with meridional kineties.
(2) A well-delimited mouth, situated on the ventral side, preceded by a vestibulum

with somatic ciliature and followed by a buccal cavity where typical ciliated structures are to be found.

(3) A differentiated buccal ciliature, in a precise area, constituted by 4 structures:

(a) a right paroral (or endoral) membranelle, also called "undulating membrane".

(b) three left adoral membranelles, or peniculi P_1, P_2, P_3.

The Parameciidae family can be defined as follows: Medium-sized ciliates (120–300μ) with a complete somatic ciliature. A well-marked ventral groove is present as far as the middle part where the vestibulum gives access to the mouth. The ciliature of the buccal cavity shows a differentiated peniculus called dorsal "quadrulus" (synonym: quadripartite membrane, "Vierermembran", "vierteilige Membran") which is an important feature of the family. Generally, two contractile vacuoles are present. The organism possesses trichocysts. The family includes only one genus: *Paramecium*.

It is obvious that the buccal ciliature makes *Paramecium* genuinely tetrahymenal: the paroral kinety corresponds to the undulating membrane (UM) of the group *Colpidium-Glaucoma-Leucophrys-Tetrahymena* (Corliss, 53–55). The peniculi P_1, P_2, P_3 are the homologues of the adoral zone of membranelles (AZM) of the tetrahymenal prototype (Furgason, 95). Though the correspondence between the membranelles of the tetrahymenal type and those of the peniculine type is delicate and presents some problems (Roque, 255), it seems that the quadrulus of *Paramecium* corresponds to P_3; that particular peniculus would reach, in *Paramecium*, a very high differentiation that would infer an evolution from the Frontonia type, with intermediary stages that would be represented by the P_3 of *Stokesia* and *Neobursaridium*.

B. Different Species

The genus *Paramecium* (O. F. Müller, 219) includes about 15 species, presently recorded, but fewer than 10 can be considered to be frequently occurring. Only these can be considered to be true paramecia.

In 1928 Wenrich (312) defined the characteristics of 8 species in precise terms: *P. aurelia, P. caudatum, P. multimicronucleatum, P. trichium, P. bursaria, P. calkinsi, P. polycaryum, P. woodruffi*.

Kahl (180) reported 16 species. He added the description of *P. traunsteineri, P. nephridiatum, P. chilodonides, P. putrinum, P. glaucum, P. pyriforme, P. chlorelligerum* and *P. pseudoputrinum*.

Wichterman (317) made a detailed critical study of the 16 species, but 2 species recorded before by Kahl (180) *(P. glaucum* and *P. pyriforme)* were not studied again while 2 other species *(P. ficarium* and *P. duboscqui)* were mentioned. According to this author, only the 8 species described by Wenrich (313) are well-defined; all the others are so ill-defined that they cannot be considered to belong in this genus.

Some species were later added to the genus: *P. silesiacum* reported by Šrámek-Hušek (289), *P. jenningsi* described by Diller and Earl (65), *P. porculus* recorded by Fauré-Fremiet (87), quoted by Beyersdorfer and Dragesco (12), *P. arcticum* described by

TABLE 2

New proposed taxonomy of the genus Paramecium. *(after Jankowski)*.

Groups	Sub-genera	Species
"putrinum"	Helianter	*P. putrinum*[a]
		P. bursaria
"woodruffi"	Cypreostoma	*P. woodruffi*
		P. calkinsi
		P. polycaryum
		P. arcticum
"aurelia"	Paramecium	*P. aurelia*
	s. str.	*P. caudatum*
		P. multimicronucleolata[b]
		P. jenningsi

[a] Synonym of *P. trichium*.
[b] This species is often named *P. multimicronucleolatum*.

Doroszewski (67a), and finally, *P. africanum* and *P. pseudotrichium* described by Dragesco (69).

All those species have been established on the basis of morphological criteria that are the only one that can be used effectively and easily. Yet, the discovery (see further on) of mating-types and varieties (or syngens) poses the problem of genetic and inbreeding criteria. If we take into account those distinctions in taxonomy, however valuable they may be in theory, the task of the systematist would be seriously complicated. It is therefore not possible, for the present, to classify the classical Linnean species in a different way.

Recently Jankowski (157a, 160a) proposed a new classification and a new taxonomy of the genus *Paramecium*. This author distinguishes (Table 2) three groups corresponding to three sub-genera: *Helianter, Cypreostoma* and *Paramecium s. str.* These subdivisions are based on their kinetome, body shape, nuclear structure, degree of cytostomal shift and invagination, outline of a prevestibular zone and complexity of its adesmokinety system, in mating type system and stomatogenetic patterns. Although this study is very interesting, we shall examine the various species according to the simpler system.

1. Well-defined species (Fig. 23)

It is possible to classify the well-defined species correctly into 2 groups according to the shape of the body, as Woodruff (323) and Wichterman (317) have stated: the "aurelia" group and the "bursaria" group.

The "aurelia" group: spindle or cigar-shaped body, with circular or subcircular sections, and with a more or less sharp posterior extremity. Moreover, as Wenrich (312) stated, the cytopyge is situated on the ventral side halfway between the posterior extremity and the mouth. This group includes: *P. aurelia, P. caudatum, P. multimicronucleatum* and *P. jenningsi*.

The "bursaria" group: a shorter and wider body, dorsoventrally flattened, rounded

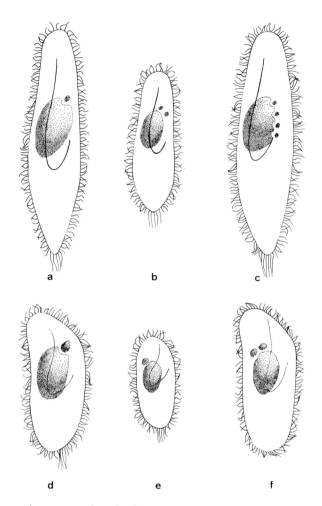

Fig. 23. Diagrammatic representation of a few common species of *Paramecium* (all are sketched at the same scale). a = *P. caudatum;* b = *P. aurelia;* c = *P. multimicronucleatum;* d = *P. bursaria;* e = *P. trichium;* f = *P. calkinsi.*

in its posterior, and obliquely truncated in its anterior part. The cytopyge is situated far in the posterior end and is sometimes at the very end of the body. This group includes: *P. bursaria, P. calkinsi, P. woodruffi, P. polycaryum, P. trichium** and *P. putrinum**.

In each group species exist with only one micronucleus, others with 2 or several micronuclei. The appearance of the micronucleus, after staining and when viewed with the light microscope, enables us to distinguish 2 types in the 2 groups: a compact type *(P. caudatum)*, in which the chromatin fills the whole nuclear space, and a vesicular type *(P. aurelia)*, in which the chromatin appears concentrated in one small central mass.

---

* See note on *Paramecium trichium*, p. 35 and note on *Paramecium putrinum*, p. 36.

a. Paramecium caudatum Ehrenberg, 1838

Description: Rounded anterior end. Conical posterior end whose lateral edges meet at an angle of 45 to 60° and whose blunt top wears a tuft of long cilia.

Length: 180 to 280μ.

One micronucleus of a compact type, generally situated in a concavity of the macronucleus. Two contractile vacuoles.

Note: A very frequent species, very easily recognized. Cosmopolitan species. It lives in fresh stagnant or slowly moving water at the level of fermenting organic substances. It feeds essentially on bacteria.

b. Paramecium aurelia O. F. Müller, 1773

Description: Rounded anterior end; posterior end forming a wide cone (angle: 90° (approximately).

Length: 80 to 170μ.

Two micronuclei, small, vesicular, near the macronucleus. Two contractile vacuoles with short radial canals.

Note: See *P. caudatum.* Feeds in nature on bacteria, but also on yeast and algae.

c. Paramecium multimicronucleatum Powers and Mitchell, 1910

This name is a modified one; the original name is *P. multimicronucleolata.*

Description: It looks like *P. caudatum* in its general aspects, but its anterior end is more blunted and its posterior end less pointed.

Length: 180 to 310μ.

Several vesicular micronuclei (3 to 12 but generally 4), very small, often situated at the level of the concavities of the macronucleus. Voluminous macronucleus. Two contractile vacuoles (sometimes 3 or more); the extra vacuoles are situated, when they exist, together in the posterior end.

Note: This is the largest species of the "aurelia" group. It can be mistaken for *P. caudatum* from which it is differentiated by the greater number of micronuclei (but this feature is visible only after staining).

The species is widespread all over the world, but it is found less frequently than *P. caudatum.* It lives in the same biotopes.

d. Paramecium bursaria (Ehrenberg, 1831) Focke, 1836

Description: Stocky and flat-shaped, clearly truncated anteriorly at the level of the wide oral groove. Broadly rounded posteriorly.

Length: 80 to 150μ.

One micronucleus, of a compact type, comparatively large. Two contractile vacuoles. Cytoplasm tightly packed with unicellular symbiont algae of the zoochlorella type.

Note: This is the "green" *Paramecium,* that can be easily recognized because of its color. It lives in fresh, clear waters.

e. Paramecium trichium Stokes, 1885

Description: General morphology of the "bursaria" type, but more rounded extremities. The vestibulum lies slightly anterior to the middle of the body on its right side.

Length: 70 to 90μ.

One micronucleus of a compact type. Two contractile vacuoles without any radial canals, filling by coalescence of small vesicles, connecting to the exterior by a long convoluted tube.

Note: It is the smallest species of the genus. The morphology of its contractile vacuoles is unique. According to Jankowski (157a, 160a), this species name is a synonym of *P. putrinum.* In this case, according to the law of priority, it is *P. putrinum,* the valid name.

f. *Paramecium calkinsi* Woodruff, 1921

Description: General morphology of the "bursaria" type, with the broadest part anteriorly.

Length: 110 to 140μ.

Two micronuclei of a vesicular type (occasionally 1 to 5). Two contractile vacuoles.

Note: Its way of swimming is different from that of the other species. *P. calkinsi* is characterized by its spiralling to the right on its long axis. Lives in fresh, brackish and even sea water.

g. *Paramecium polycaryum* Woodruff and Spencer, 1923

Description: General morphology very similar to *P. calkinsi,* but smaller and wider.

Length: 70 to 110μ.

Generally four micronuclei (occasionally 3 to 8), of a vesicular type. Two contractile vacuoles.

Note: It lives in fresh water. This species represents, in the "bursaria" group, the equivalent of *P. multimicronucleatum* in the "aurelia" group. According to Opton (232), the peniculi would be constituted by 15 kineties instead of the usual 8. But supplementary cytological studies would be necessary to verify this.

h. *Paramecium woodruffi* Wenrich, 1928

Description: General morphology of the "bursaria" type. Unusually long ventral oral groove, extending posteriorly to about two-thirds of the body length.

Length: 120 to 210μ.

Three or four vesicular micronuclei, often far from the macronucleus. Two contractile vacuoles.

Note: Lives in brackish water, but can also live in fresh water. It is the largest species of the "bursaria" group.

i. *Paramecium putrinum* Claparède and Lachmann, 1858.

Description: Morphology of the "bursaria" type, but with an extreme anterior location of buccal cavity and feeble development of a prevestibular cavity. The body also is more narrow and ovoid than that of *P. bursaria.*

Length: 80 to 140μ.

Only one micronucleus. Two contractile vacuoles. (After Jankowski, personal communication).

Note: Although *P. putrinum* was considered as a valid species in the treaties of Protozoology by Calkins (26) and Kudo (194), it was not considered by Wenrich (313), and it was also set aside by Wichterman (317). Those critics insist on the fact that *P. putrinum* are individuals of *P. bursaria* that have lost their zoochlorellae. Yet, important and recent biological studies (Jankowski, 155) have been carried out on this species and it seems that, though the descriptions are not complete, they must be acknowledged as valid. It is to be noticed that, in particular, *P. putrinum* has 4 Ma-anlagen after conjugation and there are only 2 in *P. bursaria*. Thus they do not belong to the same species.

2. *Rare, ill-known or doubtful species*

Some of the species described in this section have been the subject of important cytological or biological studies; unfortunately, their taxonomic basis has often not been sufficiently established beforehand. Other species have been described with comparative precision, but they have been spotted only locally and do not seem to be widespread. Lastly, some of them have not been correctly defined, or the criteria stated by their authors do not appear as very convincing and some confusion with abnormal types of other species or some mistakes in observation have been possible.

a. Paramecium nephridiatum von Gelei, 1925

Description: Looks like *P. caudatum*, but more narrow anteriorly and more widely rounded posteriorly.

Length: about 150μ.

Lives in brackish waters.

Note: Though reference was made to this species by Kahl (180) and Kalmus (181), it was not taken into account by Calkins (25) nor by Kudo (194), and it was not acknowledged as being a valid species by Wichterman (317).

Though it was studied by von Gelei in his important publication, it is not sufficiently defined and it is desirable that studies should be made to verify its existence and to define its specific features more precise.

b. Paramecium porculus Fauré-Fremiet in Beyersdorfer and Dragesco, 1953

Description: This species, which was not described by the author, has the following features (personal communication by Fauré-Fremiet): looks like *P. putrinum*, but is different from it by the existence of 2 to 3 compact micronuclei. Lives in putrid waters. Was found in France (near Paris).

c. Paramecium jenningsi Diller and Earl, 1958

Description: Looks like *P. aurelia*, but is larger.

Length: 115 to 220μ.

Larger micronuclei. Contractile vacuoles with very long radial canals.

Note: This species, found in India only, belongs to the "aurelia" group. It is desirable that studies should be made on its aptitudes for conjugation (mating types system) in order to establish more precisely its possible relationship with *P. aurelia*.

d. Paramecium africanum Dragesco, 1970

Description: Shape of the "aurelia" type, looks like *P. multimicronucleatum* because of its size (250 to 420µ) and its numerous micronuclei (4 to 9), but is mainly differentiated by the existence of an abnormal quadrulus.

Note: Species found only in the Cameroons where it is rather abundant.

e. Paramecium pseudotrichium Dragesco, 1970

Description: Shape of the "bursaria" type, but absence of anterior truncation. General aspects recall those of *P. putrinum* or *P. polycaryum*, except for its small size (90–100µ). It is mainly distinguished by a preoral suture, hardly noticeable and rectilinear, by an equatorial buccal overture, and by 2–4 micronuclei.

Note: Species found in the Cameroons where it seems to be rather numerous.

f. Other species

With regard to the other recorded species, i.e., *P. traunsteineri*, *P. chilodonides*, *P. ficarium*, *P. pseudoputrinum* (see Kahl, 180), *P. glaucum* (see Claparède and Lachmann, 49), *P. chlorelligeum* (see 180), *P. pyriforme* (see Gourret and Roeser, 115), *P. duboscqui* (see Chatton and Brachon, 32), *P. arcticum* (see Doroszewski, 67a), and *P. silesiacum* (see Šrámek-Hušek, 289), which are forms that are seldom encountered, described too superficially, and have not been studied in important morphological and biological contributions, there is no reason why they should be described again here. The reader who might be interested in them is referred to the original publications and to subsequent reviews (Kahl, 180; Wichterman, 317).

V. GENERAL BIOLOGY

The biology of *Paramecium* is a biology of protozoa and particularly of ciliates.

It is that of a cell which at the same time is an organism which must satisfy all its functions such as conjugation, nutrition, and reproduction with elementary, yet complex structures, whose role is still poorly understood. The biology of paramecium integrates in one exceedingly condensed organic unit the general and cellular biology of multicellular animals.

The main lines of the biology are known, but in reality the essentials and the fundamentals are still ignored or very unprecise. The following sections deal with the problems that are linked with some aspects of this biology, but have been artificially isolated from each other. In order to grasp its importance and to understand its significance, one must be familiar with the fundamentals of the biology of the organism and integrate these into the framework of the vital phenomena. For example, paramecia are cells which do conjugate, but one must not forget that they are also organisms which feed themselves, move, multiply and are sensitive to many outside factors as they are conditioned by their own hereditary traits.

We endeavor in this section to examine this simple basis of biology and will try to demonstrate the connection between the different facets and to examine a few of the problems which are relevant.

A. *Movements and Locomotion*

Paramecia move with their cilia. But *in vivo* observation shows that they can alter the shape of their body, they can twist, flatten or bend. Besides the movements of the cilia which have mostly caught the eye of the observer, there are movements of the body which must not be neglected.

1. *Movements of the cilia*

The posthumous publication of Párducz (242) which constitutes both a remarkable assessment and an original document on ciliary movement, makes an abundant use of paramecia in its examples. The metachronal wave system is perfectly analyzed thanks to the use of a technique perfected by the author, which fixes the movement of the cilia thus enabling one to observe this in great detail. Scanning microscopy can but bring supplementary confirmation without attempting to provide new information at present (Figs. 6 and 7).

The problems involved are not particular to paramecia, but are related to all the ciliates: the origin of ciliary movement, and the control and coordination of the ciliary beat.

As far as the origins of ciliary movement are concerned, the problem remains one of the fundamental problems of cellular biology and is encountered in all cells with cilia and flagella. In spite of many investigations, the problem has not yet been solved. Is there a contraction of the fibrils of the axoneme or do the fibrils slide over each other? Studies in this field have not used paramecia as material, but studies made on other organisms (for example by Satir, 258–260) seem to favor the sliding theory.

The determinism of ciliary movements constitutes another very complex problem. The phenomena have been seriously investigated only during the last ten years. The offered explanations indicate the overall excitability of cellular membranes and the dependence on membrane depolarization phenomena and the role of salt in the process of excitation. The research of Jahn (146–150), of the Japanese school (Kinosita *et al.*, 188), the Polish school (Dryl, Grebęcki *et al.*), naming but the main publications dealing with these problems, have established the following point: there is a relationship between depolarization of the membrane and the direction of ciliary beat.

The negative charge plays a role in the variation of the polarization and the Ca^{++} content of the membrane plays an essential role in the process.

The possibility exists that the environment has a physical effect upon the direction of the ciliary beat.

All these observations can be interpreted by the electronics theory of Ludloff, as was suggested by Jahn (146) and noted by Andrivon (4) in a recent study. But numerous aspects of the problem require further investigation.

The biochemical aspect of ciliary movements is also a concern for cellular biology, and various types of investigations have been carried out using Tetrahymena. It has been shown that ciliary movements require an ATP-ase activity, a need for Ca^{++} and contractile proteins. These problems are still far from solved and the relationship between structures and functions are still not too well defined.

An important point as yet unsolved concerns the possibility of a motor center at the level of the oral structures. Indeed Rees (254), Bozler (22), von Gelei (99) and especially Lund (200, 201) described a "neuromotorium", located behind the buccal cavity, from which fibrils radiate into the endoplasm and become organized in various patterns extending into the endoplasm. The presence of this neuromotor center, or nerve center, which has been claimed by these authors has never been proven, and the functions of this structure remain unknown, as are those of all subpellicular fiber systems (microtubules and microfibrils).

2. Movements of the body

Until now few authors have shown interest in the body movements of paramecia. However these movements do exist, but the question arises whether they are active or passive. In other words are the bending, the twitching and the apparent contraction initiated by the environment (obstacles met during course of movement) or are they governed by a mechanism within the cell itself?

These movements imply the presence of structures that maintain the shape, perhaps elastic structures, and in the case of active phenomena, contractile structures. In the latter the fibrillar structures may be of importance. There are various systems which could be responsible, but no definite results have been obtained concerning *Paramecium*. Electron microscopy, dealing with other types of ciliates—peritrichs (Fauré-Fremiet *et al.*, 90), spirotrichs (Vivier *et al.*, 308; Legrand, 197)—have established that genuine myonemes were present as bundles of microfibrils. Microfibrils are present in paramecia in the ectoplasm where they are called the "infraciliary lattice" (Fig. 15). This resemblance and their role in contraction has been noticed by Pitelka (246). In the case of the contractile peritrichs and spirotrichs, the myonemes are clearly linked with the vesicles of the endoplasmic reticulum as are the contractile structures in the cells of metazoa; this association is not so clear in the case of *Paramecium*, in spite of the abundance of vesicles throughout the ectoplasm.

Paramecia are not usually considered as contractile animals, although some physiological experiments (Kamoda and Kinosita, 182; Hisada, 127) have revealed some reaction to electronic and ionic stimuli. The problem is not yet settled; a weak degree of contraction is very much possible.

Whatever the contractile structures are, the cytoplasmic inclusions of *Paramecium* correspond to those of an active ciliate which uses a considerable amount of energy (Vivier, 305) as indicated by the presence of mitochondria and an abundant reserve of glycogen (Estève, 83).

B. Nutrition

The formation and the circulation of food vacuoles has been the subject of many studies including those of Bozler (21), Lund (201) and especially Mast (203) who gave a precise outline of the essential characteristics of the phenomena.

A few electron microscopical investigations during the last few years have yielded some additional information (Jurand, 175; Schneider, 265; Estève, 85).

Paramecia feed naturally on bacteria which are transported into the bottom of the buccal cavity by the currents which are caused by the movements of the membranelles (quadrulus and peniculus). Eventually, other bodies also can be drawn into the cavity. These particles may be digestible (starch for example), but it is also possible to introduce experimentally any inert particles. *Paramecium* not only ingests inert matter of a size comparable to bacteria (Indian ink particles of about 0.7μ in diameter, $CaCO_3$ particles of about 1.4μ) but also much larger particles (granules of vinyl resin from 11 to 16μ) (Estève, 81). The ingested bacteria obviously belong to different species, but all the observations that have been made show that the food value of the bacteria depends on the species: some seem to be digested more fully and enable a better rate of growth than others.

Paramecia can also ingest other living cells of a small size, particularly flagellates or yeast. Observations have been made showing that the nutritive value of these cells is very low or worthless. An unusual case is presented by the ingestion of chlorellae which in *P. bursaria* are brought into the cytoplasm as symbionts (Hirson, 126).

The selection of the food particles seems to be done mechanically, mainly in relation to the size, the larger ones being rejected by the mouth. But various observations also support the possibility of a qualitative selection (Lozina-Lozinsky, 199; Bragg, 23). More recently Hirson (126) has observed that *P. bursaria* ingests 6 out of 7 species of unicellular algae without any difficulty due to their size. It is therefore possible that a real and effective choice is made between the food particles made up of living cells. This suggests that paramecia are able to screen the particles for size. We remain completely ignorant about this process of discrimination.

The filling of the food vacuole is by way of phagocytosis. The vacuole is formed within a cone of the cytoplasm delimited by microtubular fibrils which originate from the cortex of the buccal cavity.

The separation of the vacuole from the cytopharynx and its circulation within the cytoplasm are well-known since the investigations of Mast (203), so it is needless to recall them here. The influence of various factors has been put forward in order to explain the separation of the vacuole (influence of the particles, the action of the post-esophageal fibrils; etc.). However, it seems that there is no case for seeking special influences other than the separation which favors the passage into the cytoplasm of any phagocytotic vacuole (Roth, 257). It is probable that the vacuole detaches itself as soon as it has reached a particular size, as does a soap bubble from its base; the increase of the size of this vacuole is due to the liquid brought forward by the ciliary beats. Further observations show that the filling and separation occur even when there are few particles in suspension. However, numerous observations made by several investigators are seemingly contradictory on some finer points and require further investigation. As was noted by Mast (203), considerable variations can be observed in the intracellular course of the food vacuole. Waste products are eliminated after digestion quite irregularly by way of the cytopyge and at time intervals that are far greater than those for the formation of the vacuoles at the cytopharynx. This suggests that the food vacuoles fuse more or less towards the end of their cycle.

Aside from the method of feeding by the formation of vacuoles at the bottom of the buccal cavity, one must point out the alimentary role of the parasomal sacs. Although their presence was established at the time of the first electron microscopical observations, their role as a site of pinocytosis has only recently been pointed out by the use of throrothrast in another ciliate (Noirot-Timothée, 228) and it is possible that a similar situation prevails in *Paramecium*.

The details of the phenomenon of digestion are still not well known. Many erroneous interpretations have been made. The morphological aspects of the vacuoles and

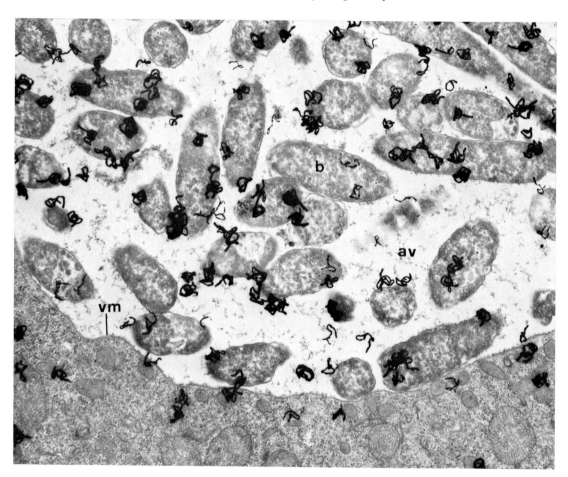

Fig. 24. *P. caudatum*. Alimentary vacuole (av) containing bacteria (b) marked with tritiated leucine; vm = vacuolar membrane. × 12,800. (from Estève, unpublished).

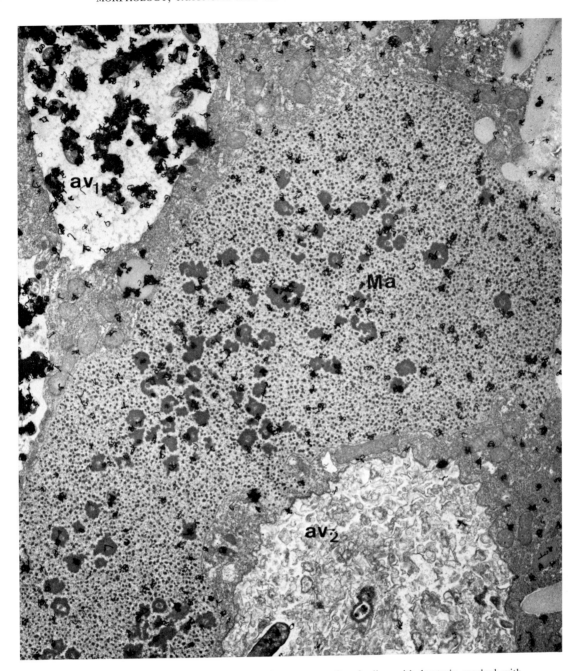

Fig. 25. *P. caudatum*. Cytoplasmic and nuclear aspects after feeding with bacteria marked with tritiated leucine. On the left and on top, a vacuole (av₁) containing un-digested and heavily labeled bacteria, on the bottom, another vacuole (av₂) whose contents have been digested; the labeled products have passed into the cytoplasm as well as into the macronucleus (Ma) where they are particularly numerous at the level of the nucleoles. × 6,400. (from Estève, unpublished).

their contents have been well described by Mast (203), but the enzymatic aspects of the digestion (the origin, nature and contribution of the enzymes, their action in relation to the food supply, the absorption of the ingested matter in the cytoplasm, etc.) require particular attention and the use of modern techniques. Such research has begun (Figs. 24, 25, 27 and 28) on a variety of species of ciliates (Favard and Carasso, 91, 92; Goldfischer *et al.*, 112, 113; Hunter, 141, 142; Seaman, 269). Only fragmentary data have been obtained using *P. multimicronucleatum* (Müller and Törö, 221) and *P. caudatum* (Rosenbaum and Wittner, 256; Estève, 82, 85) and these deal essentially with the distribution of acid phosphatase. This enzyme, which appears at the level of

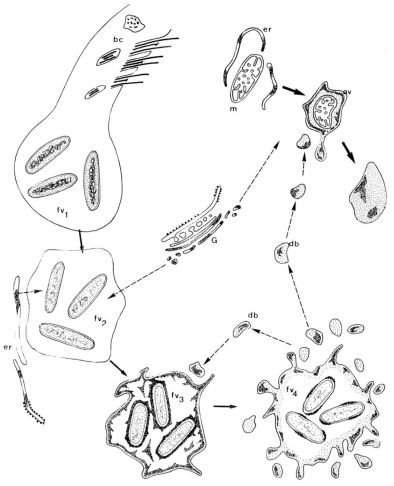

Fig. 26. Schematic drawings for the possible relationship between the acid-phosphatase-positive structures. The food vacuole likely gains its enzymatic activity from the endoplasmic reticulum (er) and, perhaps, from some small Golgi vesicles. In later stages which show micropinocytotic activity dense granules may arise from the food vacuole and may be storage sites for the enzyme. These dense bodies (db) may fuse with young food vacuoles and autophagic vesicles (av). bc = buccal cavity; fv$_1$, fv$_2$, fv$_3$, fv$_4$ = main successive stages of the evolution of the alimentary vacuoles; G = Golgi body; m = mitochondria (from Estève, 85).

the dictyosomes, seems to move into the endoplasmic reticulum, and is then diverted into the food vacuoles where it contributes to the degradation of the digested matter (Figs. 26 and 27). The digested matter seems to conglomerate against the membrane of the vacuole (Estève, 82, 85) and pass into the cytoplasm by a phenomenon of internal pinocytosis. The use of autoradiography (Figs. 24 and 25) also promises interesting results (Estève, unpublished). Metabolism is extremely active, as is shown by the structure and the great quantity of mitochondria and the abundance of ribosomes. Reserves seem to be made up mostly of glycogen. Grobicka and Wasilewska (120) had already assessed the quantity of glycogen as constituting 15% of the dry weight of *Paramecium*; Estève (84), using ultrastructural cytochemical methods, showed that glycogen is present in the form of β particles (Fig. 28). Using the terminology of Drochmans (70), Estève has defined the site and the mode of synthesis.

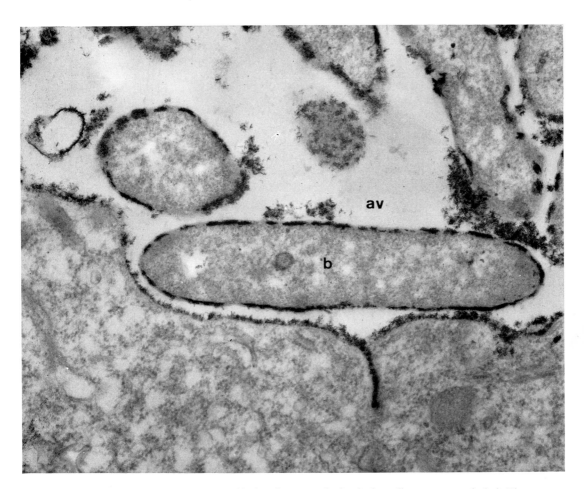

Fig. 27. *P. caudatum*. Showing the acid phosphatase at the level of an alimentary vacuole (av). The enzyme is abundant at the periphery of the bacteria (b) and inside the vacuolar membrane. $\times$ 44,800. (from Estève, 85).

References p. 74–86

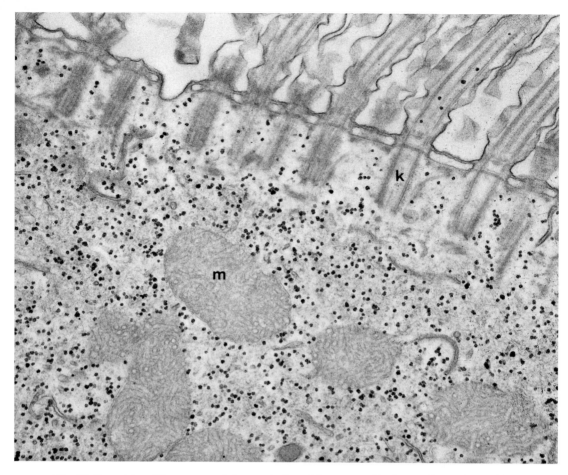

Fig. 28. *P. caudatum.* The presence of glycogen in β-particles scattered in the cytoplasm. The particles are contrasted by the specific reaction of Thiery. k = kinetosome; m = mitochondria. × 32,000. (from Estève, 84).

C. Reproduction, Sexual Phenomena and Life Cycle

A vast amount of research has been carried out in this area, especially with regard to sexual phenomena. In fact, the discovery of mating types by Sonneborn (278) in *Paramecium aurelia*, and by Jennings (162) in *P. bursaria*, brought about a considerable number of investigations on these problems in the various species. New problems, brought to light, have in turn initiated research in related areas.

While the division phenomenon has practically not benefited from the expansion in this field of research, the old Maupasian theory (Maupas, 206) of a life cycle has received renewed vigor and importance. Aside from the biological phenomenon of conjugation, observations have brought forward new facts concerning nuclear evolution as well as some physiological and biochemical phenomena. Electron microscopy has also made its own contribution which makes this subject certainly one of the richest on the biology of *Paramecium* over the last twenty or thirty years.

1. Binary fission

Binary fission is the normal process of reproduction of the species, as it is for most protozoa. This division occurs after a certain number of organelles and cellular structures have already divided or have been reproduced.

The division phenomenon lasts approximately half an hour; obviously variations occur depending on the species and the ambient temperature. The sequence of the fission process is as follows:

(a) The division of the micronucleus or the micronuclei by mitosis.

(b) The appearance of a kinetosomal field on the right side of the mouth, a field which will migrate toward a buccal bud from which the new mouth will develop.

(c) The appearance of new contractile vacuoles beside the old ones, and in an anterior position to each of them.

(d) The evolution of the buccal bud which moves posteriorly and will develop by invagination to form a buccal cavity, and will form the kineties proper to this organ.

(e) The stretching of the macronucleus.

(f) The appearance of an equatorial line of fission which will separate the old mouth and the new one. This line of fission will deepen and will finally separate the two new individuals: the former (proter) will inherit, with the old mouth, one, or a group of new micronuclei, 2 contractile vacuoles (the old one which will become the posterior vacuole, and the new one which will move anteriorly) and half of the macronucleus; the latter (opisthe) will have the new mouth and the groups of remaining organelles.

Although all these phenomena have been known for a long time, a certain number of details have been specified only during the last few years, and certain other ones are still very imperfectly known.

Thus, the origins of the kinetosomal field and the new buccal structures have been the object of the investigations of Roque (255). According to this investigator (Fig. 29), the kinetosomes of the paroral kinety, multiply and constitute the source from which the ciliated elements of the mouth of the opisthe develop. The kinetosomes multiply first *in situ* and then move anteriorly backwards. This results in an elongation of the paroral kinety. Subsequently, they form a field which stretches to the right of this kinety and which develops later into several parallel lines. During this time, the buccal wall forms an evagination in its right posterior part (the bud) where the infraciliar outline will develop. The bud lengthens into a crescent shape in the posterior part of the vestibule. At this stage the kinetosomes arrange themselves in order to reconstitute the characteristic quadripartite ciliar structures. The periphery moves to the right of the posterior suture towards the posterior region, then it enlarges, becomes hollow and gradually separates from the original mouth. The kinetosomal rows which cover the walls of the new buccal invagination first create eight penicular kineties. The quadrulus forms, and finally the upper field produces the new paroral kinety while the supplementary kinetosomes disappear.

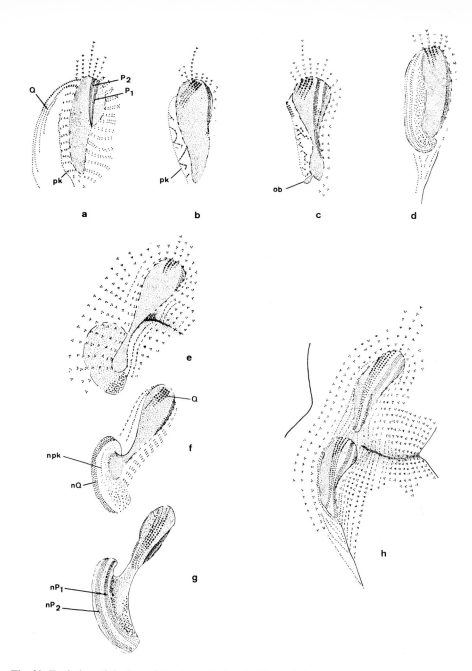

Fig. 29. Evolution of the buccal structures during the binary fission in *P. aurelia* (from Roque, 254). a = start: elongation of the paroral kinety (pk). Q = quadrulus; P_1 = ventral peniculus; P_2 = dorsal peniculus; b = multiplication of the kinetosomes on the right of the paroral kinety and the preliminary shape of the stomatogenesis area; c = growth of the kinetosomal field and appearance of an invagination of the wall of the buccal cavity (ob = oral bud); d = development of the buccal evagination and of the kinetosomal stomatogenesis field. e, f and g = preliminary buccal shape seen at 3 different levels: e: at the surface, f: medium level, g: deep level *in e*, the new buccal cavity is seen under the vestibulary and somatic kineties; *in f*, there appears the upper part of the new buccal cavity with the new quadrulus (nQ) and the new paroral kinety (npk); *in g*, the new peniculus (nP_1 and nP_2) can be seen; h = picture of the two mouths, the old one (at the top) and the new one (at the bottom); between the two the line of fission appears.

According to the studies of Roque (255), the paroral kinety is the stomatogenous kinety, and the buccal outline forms to the right of this kinety. These observations show the origin of the peniculine group, for in neighboring groups (in particular, the tetrahymenine hymenostomes) the kinetosomal field forms on the left of the stomatogenous kinety. This is a very interesting characteristic from the phylogenetic point of view. It would seem to contradict certain hypotheses, perhaps rather hastily formed, which try to show that the peniculines derive from the tetrahymenines. More complete data on the stomatogenic processes have been provided by various later electron microscopical studies, such as those by Ehret and De Haller (77) or by experimental work, such as that by Wille (318), Rasmussen (253, 253a) and Chen-Shan and Whittle (48). Wille (318) for example, blocked at different times, either the division and stomatogenesis together, or the stomatogenesis alone, or the formation of the cortical unit (ciliary unit) by the application of phenylalcohol which halts the synthesis of DNA. He observed that in order for the pharynx to form, the existence of the old endoral membranelle (paroral kinety) is necessary.

The question of the neoformation of the ciliated structures at the time of the division

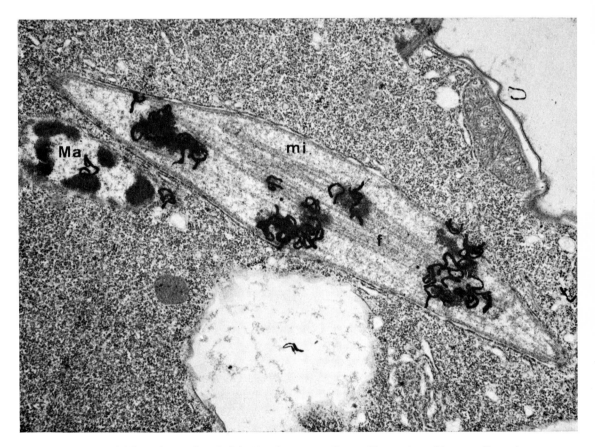

Fig. 30. Dividing micronucleus (mi) in *Tetrahymena pyriformis*. Observation with autoradiography, after incorporation of tritiated thymidine. × 20,000. (from Charret, unpublished).

is closely linked with the problems put forward in far more general terms by the duplication of the kinetosomes. Little research has been carried out on this subject with *Paramecium* (Dippell, 67). Far more numerous are the studies made with other neighboring species, in particular *Tetrahymena* (Allen, 1); these investigations have led to a new idea of the theory of genetic continuity of kinetosomes first established by Chatton and Lwoff (Lwoff, 202). A review of this question has recently been made by Fauré-Fremiet (89).

The question of the autonomous duplication of kinetosomes is also linked with the presence of nucleic acids and especially DNA. There, studies on *Paramecium* (Smith-Sonneborn and Plaut, 277; Hufnagel, 139), have also supported the results obtained with other species, in particular Tetrahymena (Seaman, 268; Hoffman, 135), results which agree with the presence of DNA in the cortex and especially in the kinetosomes.

The problems pertinent to stomatogenesis are not the only ones arising from division, since many other phenomena are concomitant, especially with regard to nuclear divisions. Here arises the question of DNA and RNA synthesis during the mitotic cycle both in the micronucleus and the macronucleus. This problem has been approached by various investigators, including Kimball and Perdue (185), Narasimha *et al.* (223), Pieri *et al.* (245), but many points are still unanswered.

The division of the micronucleus is a mitosis of the "orthomitosis" type; however, it presents, as with other ciliates (Fig. 30), some original aspects. It is acentriolar, the microtubular spindle is within the nucleus, the nuclear envelope remains until the completion of the division, and at anaphase the median part of the membrane stretches itself into a long tubule in which the spindle microtubules are concentrated. It has always been accepted that the macronucleus divides "amitotically"; however, Raikov (252) disputes this assumption.

Another aspect of the division of the macronucleus concerns the speed of this division, the way this influences the size, the growth, and the variations related to the various external or internal, natural or experimental factors. The studies produced in this field during the past ten years are extremely numerous and varied (Kimball *et al.*, 186, 187; Blanc *et al.*, 14; Nobili, 224; Cheissin *et al.*, 44; Whitson, 314; Beisson and Sonneborn, 11; Miyake, 212; Oger, 230; Oger and Vivier, 231; Siegel, 273; Borchsemius and Ossipov, 20; Jurand and Selman, 178). It is neither possible to cite nor to analyze all these studies. However, they have not resolved the many questions, even though every investigator has made some contributions with extremely varied means, such as morphological, cytochemical, biochemical, statistical and other techniques.

2. Conjugation

Although the conjugation phenomenon has been known for a long time (Müller, 220), it was not until 1889 before the process was correctly described and understood independently by Maupas and Hertwig. The morphological phenomena were then described with such exactitude and detail that nothing of importance could follow. Furthermore, the extensive contributions of Maupas also included the interpretation of the biological conditions of conjugation which were later shown to be exact in its

Fig. 31. Normal conjugation in *P. caudatum*.

general outline and from which, without a doubt, have originated all the great modern contributions to the physiology and the genetics of the phenomenon. The rise of interest in this line of research followed the discovery of mating types by Sonneborn (278) and by Jennings (162).

a. Morphological aspects. The cytological details of conjugation have been described in a great number of studies. All the familiar species have been studied: *P. aurelia, P. caudatum, P. bursaria, P. multimicronucleatum, P. trichium, P. calkinsi* (317); since Wichterman's review, other species have been added to the list: *P. polycarium* (Diller, 64), *P. woodruffi* (Jankowsky, 154), *P. putrinum* (Jankowsky, 155–160).

The conjugation of *Paramecium* (Fig. 31), as with nearly all ciliates, is characterized by: (1) A cell mating; (2) Nuclear phenomena with a double meiosis and a double amphimixis (prezygotic and zygotic phenomena); (3) Restoration of the vegetative nuclear state (postzygotic phenomena).

Although phenomena (1) and (2) are identical, aside from a few details in the different species (due to the various numbers of micronuclei), the nuclear regulation is different from one species to another.

i. Cell mating. The mating of paramecia is preceded by a clumping reaction, during

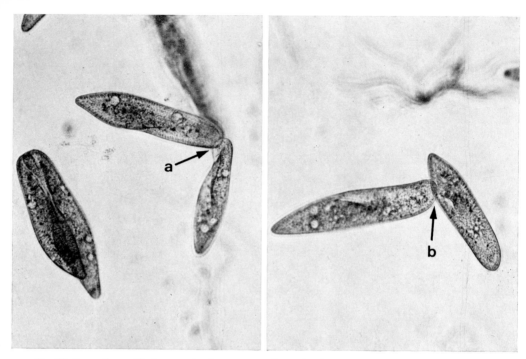

Fig. 32. *P. caudatum*. Mating reaction between anterior ends of two animals (a) and between anterior and posterior ends of two other ones (b).

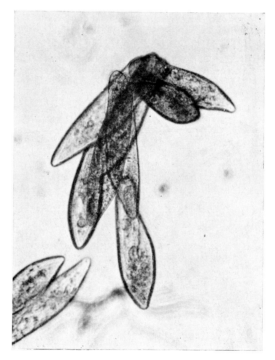

Fig. 33. *P. caudatum*. Mating reaction between several animals.

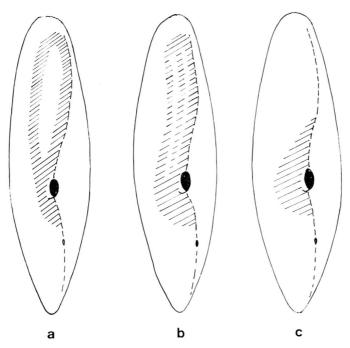

| a | b | c |

Fig. 34. Evolution of the area of coupling (hatched area) during conjugation. a = beginning of the conjugation; b = interim period; c = end of conjugation. (from Vivier and André, 306).

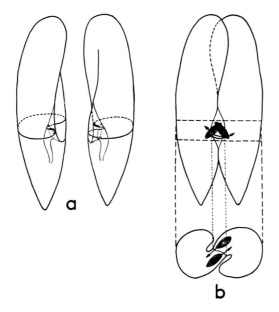

Fig. 35. Analytical diagrammatic representation of animals during conjugation. a = respective positions of the paired animals: the correspondance of joining area is shown by a thicker line at the level of the mouth; b = plane representation, and section, of a couple seen at the moment when they exchange the pronuclei. (from Vivier and André, 306).

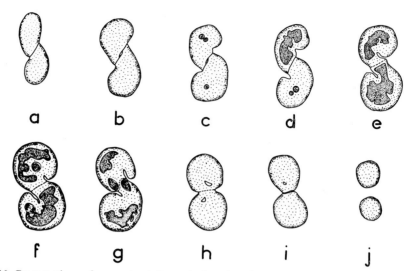

Fig. 36. Cross-sections of a couple at the end of conjugation, from the forepart (a) down to the hindpart (j). In c and d, we notice the presence of pycnotic micronuclei; in f, the stationary pronuclei; in g, the migrating pronuclei. (from Vivier and André, 306).

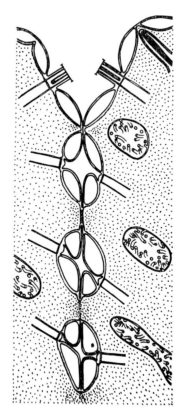

Fig. 37. Diagrammatic representation of the evolution of the surface structures of two conjugating *Paramecia*. The succession alterations are represented from the top to the bottom of the picture. (from Vivier and André, 306).

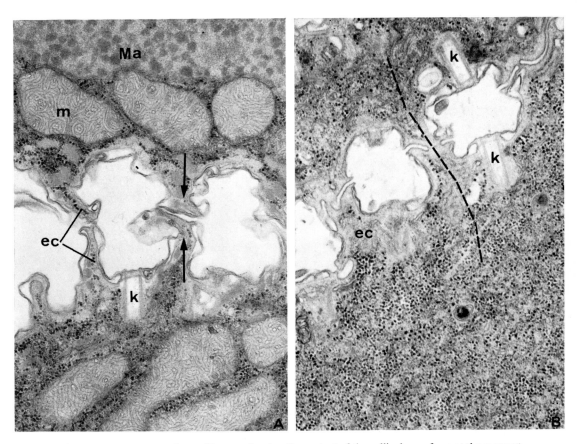

Fig. 38. *Paramecium caudatum*. Pictures showing the contact of the pellicular surfaces at the moment of conjugation. *A:* undulations at the level of the cytoplasmic crests; *B:* fusion of the membranes and formation of a cytoplasmic continuity (dotted line) between the two conjugating animals. Ma = macronucleus; ec = ectoplasmic cristae; k = kinetosomes. × 22,750.

which the individuals agglutinate into masses of various size (up to 50 units and even more in some cases). From these masses emerge pairs of individuals.

This clumping reaction is characterized by the units making contact, apparently, at the level of the cilia. The cilia on the ventral side, especially those at the anterior end, seem to have agglutinating properties. The animals agglutinate thus in various positions (Figs. 32 and 33); quite frequently, they make initial contact at the anterior region, but many other positions are possible (Jennings, 164, 165; Vivier, 302; Hiwatashi, 132). The clumped animals move in groups, modifying their positions in relation to each other in such a way that their union affects the greater part of the ventral side. It seems that the agglomerates disassociate when this position (ventral side against ventral side) has been attained.

Recent studies, mostly by electron microscopy, on the position and the relations of coupling animals (Vivier and André, 306; André and Vivier, 3; Schneider, 262, 264) have given interesting information about the nature and the evolution of the contacts.

References p. 74–86

Paramecia, placed opposite to each other, fuse over a large zone which extends from the tip of the anterior side to the postoral zone (Figs. 34–38) and includes nearly the whole of the zone to the right of the oral groove and the right side of the vestibulum (except perhaps for its deepest zone). An examination of the areas common to two paramecia during conjugation (306, 3) shows that the fusion occurs in such a way that the homologous structures of the conjugants are superimposed; the domed interciliary areas come into contact, the cilia break from their base while the kinetosomes remain; the cytoplasmic crests which are pointing at the level of the "argyrome" come into contact with the outside membrane by spacing of subpellicular alveoli; finally the cytoplasmic membrane of the conjugants disappears in the contact points while the membranes fuse together on their periphery. Continuity between the cytoplasmic membranes of the two conjugants and an intercytoplasmic continuity around the structures which correspond to the place of the "argyrome" are formed. Sections viewed with the electron microscope show only openings of 0.5μ at the most, at the points of junction between the two animals, but it is possible that these openings are slits of various length.

These apertures of intercytoplasmic communication are present on nearly the whole of the fused zone both in front and behind the paroral zone where the migration pronuclei move. Such a process obviously presents numerous problems which are without a solution for the moment. One does not know why it is that this zone has such properties, different from the other regions of the *Paramecium*; also one does not know what it is that determines the loss of cilia, the recognition of the competent areas, the mechanism of membrane lysis in the fusion zones. One must admit that a special pellicular sensitivity or an extremely delicate enzymatic or humorous mechanism about which we are completely ignorant must exist.

The fusion lasts several hours. In general it varies between 12 and 24 hours, according to the species and the temperature. The separation of the conjugants occurs very quickly; it starts at the anterior side, progressively reaches the rear and the paroral region of the right side; the zone of the exchange of nuclei is the last to separate by the opposite efforts of the two conjugants. Sometimes a fusion point can persist for an extended period of time, which can entail considerable cytoplasmic exchange.

ii. Prezygotic and zygotic nuclear phenomena. The nuclear phenomena in coupled individuals have been well known since the research of Maupas (206) and Hertwig (124). They usually comprise three successive divisions of micronuclei (Fig. 39), said to be prezygotic, which end in the formation of two functional pronuclei in each ex-conjugant; this is followed by the zygotic phenomena, which comprise the exchange between the conjugants of one of the formed pronuclei and the fusion in each individual of the two pronuclei introduced into each other (one said to be stationary, having been formed there, with the other, called migratory, having come from the other party). During this time the macronucleus apparently remains unchanged. However, Blanc (13) has shown that during these first phases of conjugation, the macronucleus starts degenerating. It becomes more and more irregular and eventually disintegrates. The start of this process varies according to the species: while it begins at the stage of the second prezygotic division of the micronuclei in *P. aurelia*, it is

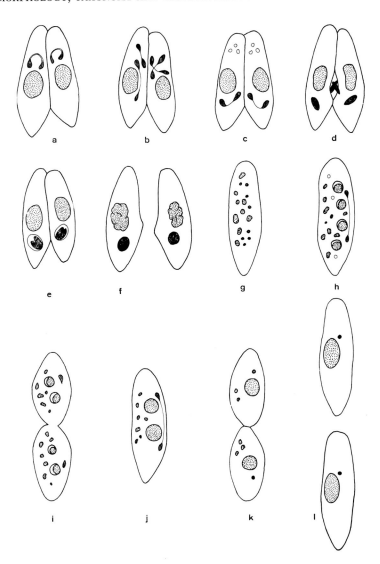

Fig. 39. Diagram showing the main stages of conjugation in *P. caudatum*. a = first prezygotic division; b = second prezygotic division; c = third prezygotic division; d = exchange of the migrating pronuclei; e = formation of the syncaryons; f = separation of the conjugating animals; g = early exconjugant; h = formation of 4 new macronuclei from the products of division of the syncaryon and disintegration of the old macronucleus; i = separation of the ex-conjugants soon after mitosis of the remaining micronucleus; j = ex-conjugants with 2 macronuclei and a micronucleus in mitosis; k = separation of the ex-conjugants; l = ex-conjugants that have become individuals with a normal vegetative nuclear system.

practically non-apparent when the syncaryon is formed in *P. caudatum* and *P. bursaria*.

The first two micronuclear divisions correspond to a meiosis, and the products of the second division are therefore haploid. If such a division of reductions had been postulated by Maupas and Hertwig, the proof was only given much later. Sonneborn (284) was the first to provide an indirect demonstration of this, thanks to genetic

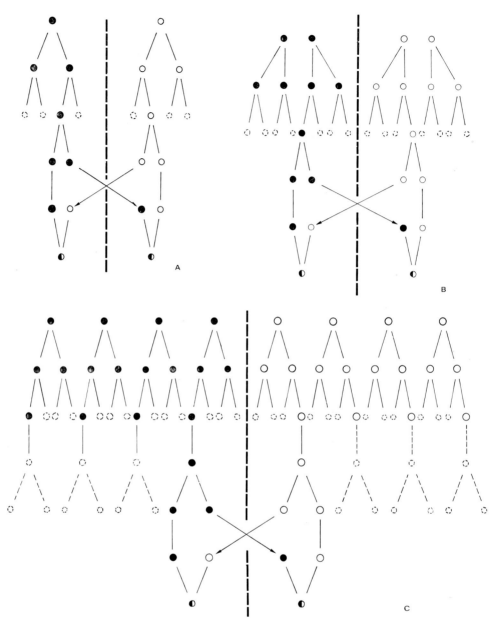

Fig. 40. The prezygotic and zygotic evolution of micronuclei. *A:* example of a species with single micronucleus *(P. caudatum)*; *B:* example of a species with two micronuclei *(P. aurelia)*; *C:* example of a species with several micronuclei *(P. multimicronucleatum)*. (The nuclei which degenerate are shown by a dotted line).

analyses. But we had to wait for cytophotometric techniques for a direct proof of the phenomenon; this was given by Dupy-Blanc (73), who thus confirmed in *P. caudatum* the results which Pieri (244) had obtained by studying *Stylonychia pustulata:* i.e., the micronuclei formed after the 2nd prezygotic division contain a quantity of DNA that equals half that present in a vegetative micronucleus. The 3rd division is thus equatorial and maintains the haploid nuclei.

As in every meiotic division, the prophase of the first division is indeed the longest phase in *Paramecium*. The size of the micronuclei then increases considerably; the different phases of the division are carried through without the nuclear membranes having disappeared; and, while the anaphase is taking place, the nuclei undergoing the process of division become tubular, in a characteristic crescent shape.

Of the haploid nuclei produced at the end of the second prezygotic division, only one remains in each conjugant; the others degenerate (Figs. 40A, B, C). This nucleus then undergoes a 3rd division which provides the pronuclei or gametic nuclei. One of those 2 pronuclei will then normally move towards the membrane that separates the 2 conjugants, and a simultaneous and mutual exchange of the pronuclei will take place. They then fuse with the stationary pronuclei to form the synkaryon.

Studies of this process with electron microscopy have only very rarely and in a most incomplete way witnessed these very short-lived phases. André and Vivier (3) and Vivier (304) have shown that the pronucleus migrating in *P. caudatum* shaped itself in a long spearhead-looking structure and formed some sort of pseudopodia over the entire surface. Inaba *et al.* (145) have succeeded in observing in *P. multimicronucleatum*, the passage of the pronucleus through the openings, and they have shown how it had to compress and take on the shape of a double-mallet in order to get through the narrow passages of the separating membrane.

Several hypotheses have been suggested to explain the different behavior of the 2 pronuclei that are formed, but up to now no proof has been given. The most common hypothesis, which had already been suggested by Maupas, is that the pronucleus which becomes migratory is the nucleus which, after the 3rd division, is in an area quite near to the partner and to the exchange area. The arguments put forward to sustain this hypothesis are the anomalies that can happen in the behavior of the pronuclei. These seem, indeed, to be numerous although they have rarely been studied precisely (Diller, 62; Jankowski, 156, 157). It seems, indeed, that it is occasionally possible for the 2 pronuclei coming from one and the same partner to migrate into the other one, or, conversely, neither of them. Both remain stationary and finally fuse (cytogamy).

One wonders about the nature of the stimulus which the migratory pronucleus obeys, also about the active or passive nature of its movement. Several authors have been trying to find an explanation of these questions (Schwartz, 267; Moldenhauer, 217). André and Vivier (3) have suggested the hypothesis, based on the examination of electron microscopical pictures, that the migration could be the result of an active movement related to a chemotactic phenomenon, but a direct proof of this is still missing.

On the other hand, it is convenient to endow these pronuclei with sexuality, the migratory one being considered as the male and the stationary one as the female. But a rational examination of the problem (Vivier, 304) shows that this interpretation is not founded on any valid argument; consequently, this terminology is to be abandoned since it can only lead to wrong ideas.

References p. 74–86

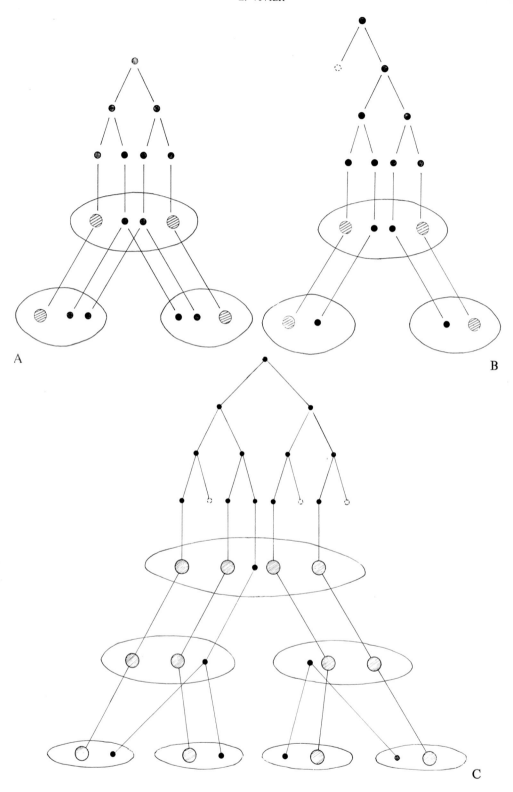

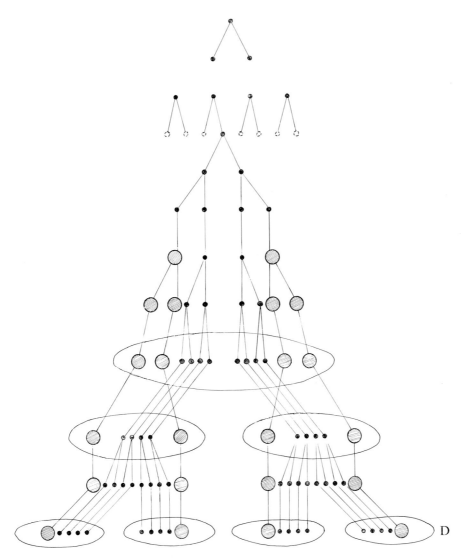

Fig. 41. The evolution of the syncaryon up to nuclear organization (according to different writers). *A:* in *P. aurelia; B:* in *P. bursaria; C:* in *P. caudatum; D:* in *P. multimicronucleatum.* (The products of the division, which degenerate, are shown by dotted lines the macronuclear outlines are indicated by hatching.)

The forming of the syncaryon corresponds to the return of the diploid state of the micronucleus (amphimixis). There is thus, normally, a double fertilization which brings about the mixing and the association of two different genomes, but the association is identical in each of the partners, which explains why, after conjugation, the individuals will be endowed with the same nuclear hereditary heritage.

iii. Postzygotic nuclear phenomena. The nuclear phenomena that follow the formation of the syncaryon are: (1) the gradual disintegration of the old macronucleus and its disappearance into the cytoplasm, and (2) the division of the syncaryon, a variable number of times according to the species. This division will create the new micronucleus and macronucleus. At the same time, both partners which have separated and are referred to as "ex-conjugants" will divide once or twice according to the species, which will ensure a normal distribution of the newly formed nuclei in the individuals.

The steps vary according to the species. Figs. 41A, B, C, D show the main types of the processes of some species. One of the simplest cases is provided by *P. aurelia* where the syncaryon divides twice, the 4 nuclei thus obtained provide 2 new macronuclei and 2 new micronuclei; only one division of the "ex-conjugant" individual is sufficient to obtain normal vegetative individuals again, which by convention are considered to be the origin of the caryonides (see above and Fig. 1).

In *P. bursaria* and *P. caudatum* the syncaryon undergoes 3 divisions in succession; but in *P. bursaria*, one of the products of the 1st division is abortive, which means that only 4 nuclei are obtained and the regulation is then similar to that of *P. aurelia;* an "ex-conjugant" provides only 2 caryonides. In *P. caudatum*, out of the 8 nuclei that are formed after the 3rd postzygotic division, 4 will form new macronuclei, one will form the new micronucleus, and the last 3 will degenerate; 2 divisions of the ex-conjugant in succession will then be necessary to re-establish the normal nuclear equipment. In *P. multimicronucleatum* the phenomena are still more complex because of the increase in number of postzygotic divisions; but regulation is obtained, as in *P. caudatum*, through 2 divisions of the "ex-conjugant". However, since the 4 new macronuclei that are formed in an "ex-conjugant" come from the division of 2 "anlagen" there would, theoretically, be only 2 caryonides out of the 4 individuals (which would then correspond to sub-caryonides).

The most interesting development during these phenomena of nuclear regulation is undoubtedly the formation of a new macronucleus from a product of the division of the syncaryon. Few precise data on this process in *Paramecium* are available; a few studies (Egelhaaf, 74; Ehret and Powers, 77; de Puytorac and Blanc, 250; Dupy-Blanc, 73), however, show the main lines of this development, a development which is characterized by a considerable increase in size and endomitosis that leads to an important polyploidization, a dispersion of DNA and an accumulation of RNA into the numerous nucleoli.

iv. Other cytological phenomena of conjugation. Besides the very characteristic nuclear process of conjugation, there happen or may happen other phenomena concerning the cellular organelles.

Thus, there is a correlation between the cessation of feeding and the regression of the buccal structures; these structures are reconstituted after the separation of the coupled cells when the mouth again becomes functional.

The cilia that had dropped off the whole of the ventral side are immediately formed again from the kinetosomes that remained.

We have very few data concerning the possible formation of the other cytoplasmic structures and we do not precisely know that happens to the large quantities of DNA that are liberated in the cytoplasm after the multiple nuclear degenerations.

On the other hand, we do know that cytoplasmic exchanges can take place between coupled individuals through the numerous openings that exist in the separating membranes. These exchanges have been proven by electron microscopy (Vivier, 302a) for the ergastoplasm and, as far as RNA is concerned, by autoradiography (McDonald, 202a); it may be possible that, at least in some special cases, more important organelles may pass through, such as mitochondria. It seems logical to believe that such exchanges can influence the physiology of the receiver but we are in total darkness on this point.

b. Physiological phenomena and sexuality. Maupas (206) was the first to outline the main aspects of the conditions that are necessary for the initiation of conjugation, to wit: (1) lack of food; (2) reaching a caryogamical maturity related to the existence of a life cycle; (3) the necessity of a cross-fertilization resulting from placing together different lines of descendants.

These 3 great principles of Maupas' theory, although greatly criticized in the following years, were to be used as a basis for all the research on the physiology and the conditions of conjugation (recent research has confirmed these views) while making them more precise.

Historically, it is possible to consider that research on this question can be divided into three periods:

(1) The research and hypotheses of Maupas.

(2) The research that was carried out between 1889 and 1937 with the following orientations: (a) a criticism of Maupas' principles, particularly his first 2 conditions (the 3rd condition being generally completely overlooked); (b) stressing and making clear the influence of certain external conditions.

(3) Research carried out from 1937 on, following the discovery of "mating types" by Sonneborn and corresponding in fact to the demonstration of the 3rd condition set forth by Maupas and to the development of research caused by the numerous physiological and genetic problems that became evident.

Unquestionably the publications, both numerous and diversified, that illustrate the second period are those of Jennings (161), Calkins (24, 25), Woodruff (321–325) supported by those of Galadjieff and Metalnikow (96), Hopkins (137), and lastly, those of Chatton and Chatton (40) which confirmed in *P. caudatum* the conclusions they had drawn from their experiments with *Glaucoma scintillans* (33–37).

No essential result has really emerged from all these publications. Contradictory conclusions were suggested both for the influence of fasting on the existence of a life cycle and its importance on the appearance of conjugation, and on the reality of the influence of external factors. All these points will consequently be taken up again, confirmed or questioned, sometimes made more precise during the 3rd period in relation to the notion of mating types.

i. Mating type systems and conjugation. Complementary sexual tendencies in individuals belonging to a same species have been stressed and made clear by Sonneborn (278) for *P. aurelia*. This was confirmed by Jennings (162) for *P. bursaria* the following year. This was the start of a modern era, that of research on conjugation. All the problems then had to be re-examined in the light of this new notion.

The technique that clarified the sexual tendencies which induce conjugation is that of the mixing of clones in every possible way. These complementary sexual tendencies have been defined as mating types and were found in all of the common species of *Paramecium:*

Paramecium aurelia (Sonneborn, 278–280. 284, 288; Beale and Schneller, 9).

Paramecium bursaria (Jennings, 162–167; Bomford, 18).

Paramecium caudatum (Gilman, 110, 111; Giese and Arkoosh, 109; Chen, 46; Hiwatashi, 128, 129; Vivier, 301, 302; Vivier *et al.*, 311; Ossipov, 236).

Paramecium multimicronucleatum (Giese and Arkoosh, 109; Giese, 107, 108; Sonneborn, 279; Boell and Woodruff, 17).

Paramecium calkinsi (Wichterman, 316).

Paramecium polycaryum (Hayashi and Takayanagi, 123).

Paramecium trichium (Sonneborn, 279, 281; Ammermann, 2).

Paramecium woodruffi (Ammerman, 2).

Paramecium putrinum (Jankowski, 155).

The studies have shown that each species is composed of several sets of complementary (or opposed) mating types which are sexually separated. Each of these sets has been originally referred to as a variety and, later on (Sonneborn, 288), as a syngen. The number of mating types by syngen is variable and depends on the species under study. In *P. aurelia* and *P. caudatum* there are always 2 complementary mating types in each syngen; these are species of a binary type; in *P. bursaria* and *P. trichium* there are several mating types for each variety (up to 8); these are species of a multiple type.

Conjugation normally happens as soon as the individuals belonging to complementary mating types of the same variety are put together under appropriate conditions. However, some conjugations may take place within one and the same type (intraclonal conjugations or selfing); they may also take place on the occasion of a mixing of types belonging to different varieties (intervarietal conjugations) or belonging sometimes even to different species (example of the crossing between *P. aurelia* and *P. multimicronucleatum*).

The conjugations which may appear, either naturally or experimentally, between mating types not normally complementary, present numerous biological, physiological and genetic problems that all are far from being solved. Thus, for example, can in some cases the appearance of intraclonal conjugations be the consequence of a change in the mating type of certain individuals of the clone? The causes of such a change have not always been made clear with any certainty. It may be due to nuclear alterations (autogamy in *P. aurelia*, for example), but also to variations in the speed of division [Hiwatashi (131) in *P. caudatum*]; in other cases (Vivier, 302; Vivier *et al.*, 311) no change in the mating type could be detected.

The modalities occasionally observed are such (Vivier, 301) that it is difficult to state whether one is confronted with a morphological, physiological or genetic variation and, in some cases (Oger and Vivier, 231), the appearance of intraclonal conjugations has been shown to be correlated with the existence of a dimensional bi-modality in the population of the clone.

From an examination of all the research carried out regarding this question, it appears with some degree of certainty that conjugation is not solely determined by genetic factors but that a physiological conditioning has a certain part to play. Several hypotheses have been suggested to attempt the explanation of the modalities and of the variations in the appearance of conjugation, but it seems that none of them can yet be applied to all the phenomena observed in the different species.

There exists also a practical problem relative to the classification of mating types in each species by interested scientists. Whereas the morphological criteria can always easily be used to identify the species, no criteria other than the method of conjugation can be used to recognize the mating types. Consequently, inasmuch as the lines of the descendants already identified as belonging to one particular mating type cannot be preserved, transported, exchanged and compared, it becomes impossible to take a valid census of the existing mating types and to discuss the results that were obtained at different times and in different places. At present these facts are particularly striking in the case of *P. caudatum* where it has not been possible to compare the varieties and the mating types discovered and identified separately by different scientists. Furthermore, this can never be done on account of the disappearance of some stocks or lines of descendants. It cannot be known whether the lines of descendants have identical or different sexual characteristics and this aspect of the problem can seriously impair the understanding of the results, results that may possibly be contradictory since they were arrived at by different laboratories.

The last point worth mentioning with regard to the existence of such varieties and mating types within one species is that of their evolution in relation to ecology and biogeography. Indeed, one and the same variety is very frequently discovered in one single natural source; however, identical varieties are to be found in different places, sometimes very far apart.

What mechanisms made these varieties become different, what are the factors that have influenced the evolution, which criteria are to be used to consider the present state as a stable one? These are some of the questions concerning this problem that cross the mind. For the present only hypotheses, among which the influence of geographical isolation is the most common, can be suggested.

ii. Conditions and physiology of conjugation. Apart from belonging to complementary mating types, varied conditions are necessary for mating to take place. Furthermore, conjugation is correlated with a certain number of physiological modifications, one of the most interesting being certainly the elaboration of sexual substances.

As far as the conditions of conjugation are concerned, it would be advisable to take into account the influence of many factors: starvation or influence of nutrition, maturity or influence of a generation cycle, speed of division and density of populations, external factors (light, pH, temperature, presence of chemical substances, etc.).

All these points cannot be examined in detail here and we shall only discuss a few problems.

The influence of previous fasting, a factor that has long been a subject of controversy, seems to have been explained. Indeed, after numerous observations made by different authors and experiments carried out on *P. caudatum* (Vivier, 302) it seems that

References p. 74–86

stopping nutrition after a period of plenty does favor the start of sexual reactions; these appear during the period corresponding to the metabolic transformations that follow a period when plenty of food is given; real starvation always corresponds to complete absence of conjugation.

A heavy density of population, following a very high fission rate, is also a favorable factor; and certain observations (Vivier and Mallinger, 309) have shown that, in this respect, conjugation appeared, the vegetative activity slowed down, obviously within certain limits. If one refers to the graphs concerning the development of the population (Figs. 3 and 4), conjugation appears normally possible and easy in the periods of stabilization and equilibrium and with a greater frequency at the beginning of the decline of the vegetative activity.

These conditions are intimately linked with those that concern the acquisition of maturity. The existence of a period of immaturity can be conceived only in relation to that of a life cycle, and it presents the problem of the origin of the descendant line. The existence of a possible life cycle will be examined further on, but it is necessary to stress the fact that the origin of a new line of descendants is traditionally the formation of a caryonide. Indeed, conditions of maturity, revealed by the existence of a delay after the isolation of a caryonide, have been considered as proof that the delay of maturity was genetically determined. It may vary according to the lines of the descendants, but it is often difficult to isolate it from other factors, depending upon the culture conditions in the laboratory.

Indeed the conditions of nutrition, the fission rate, the density of the population and the attainment of maturity are intimately linked, and it is afterwards very difficult to know the influence of each.

External factors do not appear as essential but are rather variable factors that favor or inhibit conjugation. The external factors which influence the appearance of sexual reactions are obviously more limited in their action than those that simply allow a vegetative life. This is particularly clear with regard to pH and temperature in certain species. Some lines of descendants show particular requirements—e.g., optimum conjugation at certain temperatures—whereas some other species require specific light conditions (alternation of day and night periods, or optimum conjugation at certain times of the day). The influence exerted by chemical substances is far more interesting. Prior to the discovery of mating types, Zweibaum (331), and Chatton and Chatton (33–40) had already shown the influence of certain salts on conjugation. But, more recently, the experiments of Miyake (212–216) have yielded very interesting data concerning the action of different chemical substances, data which can possibly throw a new light on the mechanism of the initiation to sexual reproduction.

The physiological determinism of conjugation appears to be related to the production of sexual substances. Metz's research (207–209) has shown the existence of a sexual attraction between individuals belonging to complementary types either dead or alive and initiating in the living animals the processes of conjugation. These investigations have pointed to the conditions and characteristics of this attraction that could only be explained by the action of substances produced by the individuals which can have a post-mortem action. Those results have been confirmed and later extended (Hiwatashi, 128, 129; Vivier, 302; Cohen and Siegel, 52; Cohen, 50, 51) to different species: *P. aurelia*, *P. calkinsi*, *P. bursaria* and *P. caudatum*. From these investigations and from others more recent, it appears that these "sexual substances" are possible proteinaceous in nature and that they are specific, and not normally diffused in the medium, and that they reside mainly at the level of the pellicle and of the cilia. It has also been shown that they are only formed during periods of reactivity and that there actually exist several active substances that are elaborated in succession or at the same time during conjugation or autogamy [among others, the research of Nobili and collaborators (224–226) and Beisson and Capdeville (10) have shown that the inhibition of RNA or protein synthesis with antimetabolites provided valuable information on this mechanism]. A great number of questions are still unanswered but we now certainly have a new and extremely fertile field of research. It is indeed necessary not only to obtain a better biochemical knowledge of the effect of these active substances on the initiation and the development of the sexual processes but also to link their genesis both to the genome of the individuals and to the conditions of the conjugation: nutrition, modalities of the vegetative life, external factors, etc.

iii. Heredity of mating type. The transmission of the mating type character through vegetative generations, conjugation and autogamy has been investigated ever since its discovery by Sonneborn and Jennings. One could have hoped that the experiments would demonstrate a simple system of Mendelian inheritance. In fact, this has not been the case. Such a mechanism has been proven only in certain species, and, even then, always with numerous exceptions that could only be explained with the greatest difficulty. The existence of the selfing process, of changes of mating types during vegetative life, and of numerous anomalies have really complicated the problem. However, if one disregards these special cases, it is possible to consider: (1) that the mating type is stable during normal vegetative live, and (2) that the transmission of mating type through conjugation and autogamy follows one of the following mechanisms:

(α) Synclonal heredity. All the descendants of the 2 individuals of a pair belong to the same mating type. As a consequence of the exchange of the migratory pronuclei, the syncaryons are genetically identical and it is therefore logical to accept a strict genotypical determinism.

(β) Clonal heredity. The descendants of the individuals of one pair possess a mating type that is identical with that of the conjugant from which they originated. In spite of the formation of the syncaryons, the features of sexuality have not been altered; a genotypical determinism has therefore to be rejected and one must accept another mechanism, because a continuity in the mating type prevails just as there is a cytoplasmic continuity.

(γ) Caryonidal heredity. The mating type of the descendants varies according to the caryonide; there may be caryonides of a different mating type among the descendants of one and the same individual of the pair. One has, therefore, to reject the predominating influence of the syncaryon, as well as the influence of the cytoplasm. A direct influence of the new macronuclei is probable.

Our knowledge in this direction has been increased mostly by the research of

Sonneborn and his school (Sonneborn, 284, 288; Nanney, 222). The subjects of these investigations have been mostly *P. aurelia* and *P. bursaria* species. Even if the situation is somewhat clearer in these 2 species, many problems still remain unsolved (Siegel, 272; Taub, 294, 295; Bomford, 18). The situation seems far more complex in *P. caudatum* (Hiwatashi, 130, 133, 134; Vivier *et al.*, 311; Vivier, 304). Finally, the mating types have been rarely studied in other species.

No hypothesis can at present give an adequate and general account for all the observations made concerning the heredity of mating types. Much research is still necessary in order to establish whether the complex inheritance of mating types is due to determinism, dependence on differentiation, genetic or epigenetic factors, or even simple physiological variations.

iv. Sexuality and sexes. It is incontestable that conjugation which shows a cellular coupling and nuclear phenomena with meiosis and amphimixis is a sexual act. For these characters are common to all sexual acts and to these alone.

Therefore, if there is a sexual phenomenon, there is sexuality. Where should we place sexualization and what are the sexes? These are problems that are still debated. Several authors have discussed them, among which Schwartz (267), Canella (27, 28, 29), and Vivier (304) be mentioned. The basic problem is really a choice between 2 conceptions of sexuality.

The first one, outdated and rigid, suggested by Hartmann (122), maintains that the fundamental manifestation of sexuality is the existence of a male or female bipotentiality which constitutes a fundamental property of living matter; the other one, more flexible, proposed by Vivier (304), accepts the fact that sexuality must be the whole of the characteristics which, within the sphere of one species or one variety, corresponds to the different individuals between which copulation can take place. Using this latter conception, one has to disregard any idea of masculinity and feminity, and, as it becomes possible to imagine more than two sexes.

It is obvious that the problem of sexes and of sexuality is very difficult to conceive clearly for *Paramecium* if one is to follow Hartmann's hypothesis. On the other hand, this problem becomes easy if one is to accept the second hypothesis since sexuality is considered at the level of the individuals and since the mating types now correspond to sexes or, rather, to sexual types. It also becomes much easier to interpret the variations of the mating types and the phenomena of so-called relative sexuality, because the sexuality of one category of individuals appears only in relation to another category. This is not a fundamental, absolute, and intrinsic property. One individual will only have the possibility to conjugate with another one if there is a large enough difference between them, whether this difference be conditioned by genetic, physiological or other factors.

3. Other nuclear phenomena

Apart from the nuclear changes that take place during vegetative binary fission or conjugation, there are other less classical, but very important phenomena.

Here four types of changes that have been described by several authors, endomixis, autogamy, macronuclear regeneration and hemixis, will be examined.

a. Endomixis. This nuclear reorganization has been described by Woodruff and Erdmann (326) taking place in *P. aurelia* as a periodical occurrence in isolated individuals, during the decline of the population. According to these authors, this process would entail the formation of a new macronucleus from one of the products of the division of the micronucleus. The old macronucleus degenerates into the cytoplasm and is lost from view. The micronucleus divides mitotically and no meiosis or amphimixis like that during conjugation was observed. Although in the following

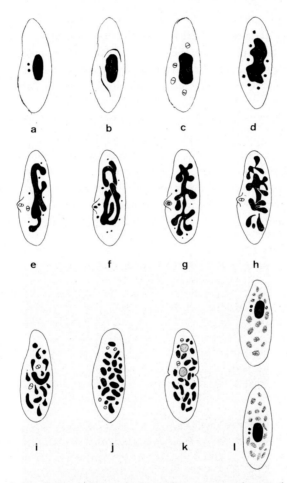

Fig. 42. Nuclear evolution during autogamy in *P. aurelia*. a = normal vegetative animal; b = first division of the micronuclei (diamond-shaped); c = second prezygotic division; d = micronuclei (8) produced by the second division. The macronucleus becomes irregular in shape. e = some micronuclei (2 of them are functional) divide a third time, and one of them migrates to a protuberance near the mouth (paroral cone); f = pronuclei gametes (4), formed after the third division. Other degenerating micronuclei. The macronucleus is dividing into fragments; g = fusion of 2 pronuclei to form a syncaryon; h = first division of the syncaryon; i = second division of the syncaryon; j = from the 4 nuclei produced by the division of the syncaryon, two will give rise to new macronuclei, 2 will develop into the new micronuclei; k = the two new macronuclei are formed. The micronuclei and the cell are dividing; l = the formed macronuclei are separated from the daughter-cells. The fragments of the old macronucleus are disintegrating.

years such a process was described in other species by different authors, the most recent investigations, cytological as well as genetic, have not confirmed the existence of endomixis. Diller (60) described the periodical nuclear reorganization in *P. aurelia* differently, under the name of autogamy, and Sonneborn has carried out a genetic study which conforms Diller's conceptions. There is, consequently, no apparent reason to show an interest in endomixis other than historical.

b. Autogamy. According to Diller, autogamy appears as a cytological process identical to that of conjugation, but occurring in single animals. The micronuclei undergo (Fig. 42) three prezygotic divisions, the first 2 of them being reductional, and the 3rd producing the 2 pronuclei; the only difference with conjugation is that there is no exchange of pronuclei but the syncaryon is formed by the fusion of two gamete nuclei that have been formed. The postzygotic nuclear changes are similar to those of conjugation.

Autogamy was later confirmed by others and is now generally accepted and even used to obtain clones that are homozygotic for all their characteristics; indeed, since the syncaryon is the result of the union of two pronuclei resulting from the division of an already haploid nucleus, the organism becomes homozygous.

Autogamy has been found in other species: *P. caudatum* (Vivier, 306; Ossipov, 237; Skoblo and Ossipov, 275; Ossipov and Skoblo, 239), *P. putrinum* (Jankowski, 156, 157), *P. polycaryum* (Diller, 63), *P. jenningsi* (Mitchell, 211), *P. bursaria* (Chen, 47).

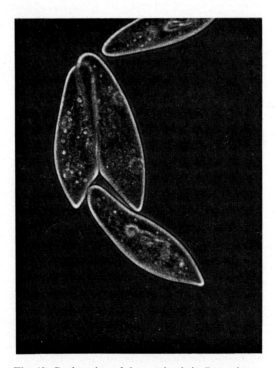

Fig. 43. Conjugation of three animals in *P. caudatum.*

It is necessary to specify that autogamy, in some of these species, does not normally appear in isolated individuals as in *P. caudatum* and in *P. bursaria*. But it can happen in coupled individuals, the two members of the couple not exchanging any pronuclei. The process is then known as cytogamy (Wichterman, 315). It can also happen in conjugation between three or four individuals (Fig. 43). To the present extent of our knowledge, autogamy appears to be a regular periodical and normal process only in certain lines of *P. aurelia;* in other species, it seems that autogamy is only a phenomenon provoked by an abortive conjugation. Thus it does take place in individuals of *P. caudatum* that have clumped during the mixing of clones belonging to complementary mating types, but have not undergone conjugation (Vivier, 302). In any case, it appears that autogamy is a process that is intimately linked with conjugation and whose determinism also presents interesting problems.

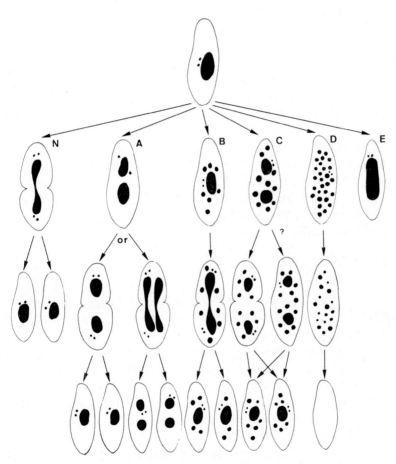

Fig. 44. Diagram showing the possible ways of the behavior of the macronucleus (excluding conjugation and autogamy). The inter-relations are indicated by arrows. N represents the normal evolution in the case of binary fission; A, B, C and D are animals undergoing hemixis, E represents a case of macronuclear hypertrophy. (from Diller, 60).

c. *Macronuclear regeneration*. This cytological process was discovered by Sonneborn in *P. aurelia* (282, 283, 284). It takes place during the postzygotic nuclear stage. The old macronucleus whose fragmentation has started is reconstituted instead of degenerating and, consequently, certain descendants of ex-conjugants possess a macronucleus that is not formed by one of the products of the division of the syncaryon, but is directly derived from the old one. In *P. aurelia*, this process can be induced experimentally by bringing the two coupled individuals, when they have reached an appropriate stage, to a temperature of 38°C (Sonneborn, 284). This process of macronuclear regeneration has a genetic and cytological significance that has been widely used by Sonneborn to specify certain modalities of the heredity of the mating type. Macronuclear regeneration caused by raising the temperature to 36°C when the 2nd division of the syncaryon is taking place has also been obtained in *P. caudatum* by Ossipov (238).

d. *Hemixis*. Hemixis has been originally described by Diller (60) in *P. aurelia*. It is a macronuclear change occurring during the vegetative life cycle and has nothing to do with any cellular or micronuclear division. Hemixis (Fig. 44) is mainly characterized by a fragmentation of the macronucleus which extrudes macronuclear fragments of varying size and number. These pieces then degenerate in the cytoplasm; in the extreme case, the macronucleus undergoes fragmentation and total disintegration. Such cytological manifestations of hemixic phenomena together with cases of macronuclear hypertrophy were also found by Vivier (302) in *P. caudatum*, particularly in old lines no longer able to conjugate. The proportion frequently reached 5% of the individuals that were examined. Diller has considered hemixis as the reflection of a degenerating condition of the cell; if it indeed seems to be "a sign of old age", it may also be a way of elimination of the macronucleus, aiming at restoring balance in the nucleo-plasmic relations. This could then be a process of rejuvenation enabling old lines to keep some vitality and therefore play a part in the life cycle.

D. *The Life Cycle*

Maupas (206) was, again, the first to have mentioned the existence of a life cycle in the ciliates. He thought that the successive generations which arise from each other by division go through several stages:

(1) A period of youth which is an agamous period. The individuals divide actively but, even when placed in conditions favorable for conjugation, they never unite; we can thus consider them immature.

(2) A period of maturity which is a eugamous period. It is during this period, and only this period, that fertile coupling can take place. After conjugation the animals enter a new period of youth.

(3) A period of senility which is a period of degeneration. It only exists in individuals which have not conjugated during the preceding period. Conjugation can still take place more or less easily but it is infertile, and this period ends inevitably in death.

This notion of the cycle of generations which questioned the principle of the potential immortality of unicellular beings was, in the following years, criticized by several investigators. Woodruff seems to have given the "coup de grace" in 1921 when he published the results of a long experiment on *P. aurelia*. By a technique of daily transfers to a fresh medium, he succeeded in preserving the same line of descendants for thirteen and a half years. During this period more than 8000 generations followed each other without there being any conjugation and without any appearance of the characteristics of senility.

These results which seemed a severe criticism of Maupas' theory did, however, reveal the existence of nuclear modifications, then called endomixis but in reality showing autogamy (see above). This phenomenon, leading to the elaboration of a young macronucleus by changing it for an old one, took the place of conjugation and brought about, under another form, the necessary rejuvenation.

An identical experiment was described a few years later by Galadjieff and Metalnikow (96) with *P. caudatum;* they cultivated this species for twenty-two years and also obtained more than 8000 generations without conjugation taking place. But the problem is a different one here because no one has ever noticed nuclear modifications in this species during its ordinary vegetative life. So, it must be admitted that either there is no aging or else that rejuvenation can take in the absence of conjugation by a process which is not autogamy, or else that the aging is extremely slow (much longer than the considered length of time).

The research which has been carried out, since the discovery of mating types on different species, particularly *P. aurelia* (Sonneborn, 286–288), *P. bursaria* (Jennings, 168–172) and *P. caudatum* (Vivier, 302) strongly upholds the theory of the existence of a life cycle, resembling in its main concepts the ideas of Maupas.

The period of immaturity seems to vary exceedingly in length according to the species and the lines. Sometimes it is very short (a few days) and practically non-existent, sometimes it is very long (some months or even some years). The period of maturity can also vary considerably in length, and so can the period of decline which seems, at least in some species such as *P. caudatum* (unpublished observations), to last a very long time (several years, and, doubtless, several decades).

It would seem that these phenomena of clonal evolution during the course of time must be in relation to a nuclear evolution, as Maupas had already suggested in 1889. Recent experiments of Siegel, using a technique analogous to nuclear grafts, tend to show that a differentiation of the macronucleus influences the different stages of the life cycle of *P. aurelia*.

Many points are still uncertain, and much work is still necessary in order to find out the modalities and the conditions of such a life cycle. Is aging a natural phenomenon, or is it solely caused by the laboratory breeding and culture conditions? Vivier (302) provided arguments which tend to show that aging also exists under natural conditions; it is only accelerated by the methods of culture, and particularly by feeding the paramecia on a single species of bacteria. And, also, hemixis has shown itself to be (see above) a relatively more frequent anomaly in the old clones than in the others; is this a sign of degeneration, is it a reaction of regulation and a means of struggling against this degeneration? The problem remains unsolved. Finally, what do the nuclear modifications, differentiations, or alterations which provoke the lack of ability to conjugate and the vegetative decline exactly consist of? According to Fauré-Fremiet (88) senescence has its origin in the hyperploidal characteristic of the macronucleus, a characteristic which comprises a risk of irreversible impotence, particularly in the highly differentiated species, as in the case of the *Paramecium*.

VI. PERSPECTIVES AND CONCLUSIONS

The essential points of morphology, taxonomy and biology of *Paramecium* have just been described. Nevertheless, many interesting aspects and problems have been passed over. But the reader can be sure that many points need to be looked at in more detail, proceeding from our traditional knowledge.

We can already see that taxonomy needs to be revised completely and the facts stated precisely in a systematic and comparative form, with the aid of modern techniques. And it is also necessary, in particular, to verify clearly and definitively the existence and possibly the characteristics of the numerous species which still remain undefined.

Doubtless, it would also be useful to take into consideration, besides the morphological data, the ecological, biogeographical and physiological observations that have been made.

But it is undeniably in the field of biology that much remains to be done. The multiple aspects of physiology, nutrition and reproduction can still provide very interesting research subjects. Many of the problems have been brought up in the preceding pages, in connection with the different points that have been described. Many others have not been discussed.

But it is certain that *Paramecium*, thanks to easy cultivation and the many possibilities that it offers to research, will long remain a privileged animal in laboratories and that it will provide protozoology and cellular biology with a great deal of valuable knowledge and much satisfaction to those who work with it.

VII. REFERENCES

(Abstracts are indicated by *)

1 Allen, R. D. 1969. The morphogenesis of basal bodies and accessory structures of the cortex of the ciliated protozoan *Tetrahymena pyriformis*. *J. Cell Biol.* **40**, 716–733.

2 Ammermann, D. von, 1966. Das Paarungssystem der Ciliaten *Paramecium* woodruffi und *Paramecium trichium*. *Arch. Protistenk.* **109**, 139–146.

3 André, J. & Vivier, E. 1962. Quelques aspects ultrastructuraux de l'échange micronucléaire lors de la conjugaison chez *Paramecium caudatum*. *J. Ultrastruct. Res.* **6**, 390–406.

4 Andrivon, C. 1969. Le phénomène du renversement ciliaire chez les Protozoaires (Ciliés et Opalines): Ses développements récents et son intérêt dans l'étude des mouvements ciliaires. *Ann. Biol.* **8**, 99–114.

5 Beale, G. H. 1953. Adaptations in *Paramecium*. In Gale, E. F. and Davies, R. *Adaptation in Micro-organisms*, Cambridge University Press, New York, N.Y., 294–305.

6 — & Jurand, A. 1960. Structure of the mate-killer (mu) particles in *Paramecium aurelia*, stock 540. *J. Gen. Microbiol.* **23**, 243–252.

7 — & — 1966. Three different types of mate-killer (mu) particles in *Paramecium aurelia* (syngen 1). *J. Cell Sci.* **1**, 31–34.

8 —, — & Preer, Jr., J. R. 1969. The classes of endosymbionts of *Paramecium aurelia*. *J. Cell Sci.* **5**, 65–91.

9 — & Schneller, M. 1954. A ninth variety of *Paramecium aurelia*. *J. Gen. Microbiol.* **11**, 57–58.

10 Beisson, J. & Capdeville, Y. 1966. Sur la nature possible des étapes de différenciation conduisant à l'autogamie chez *Paramecium aurelia*. *Compt. Rend.* **263**, 1258–1261.

11 — & Sonneborn, T. M. 1965. Cytoplasmic inheritance of the organization of the cell cortex in *Paramecium aurelia*. *Proc. Nat. Acad. Sci. U.S.* **53**, 275–282.

12 Beyersdorfer, K. & Dragesco, J. 1953. Etude comparative des trichocystes de sept espèces de paramécies. *Proc. 1st Int. Congr. Electr. Microsc., Paris, 1950*, Mémoire hors-série, no. 1, Rev. opt. theor. instrum. edit. 661–671.

13 Blanc, J. 1968. Détermination par cytophotométrie des teneurs en ADN des macronucleus de *Paramecium caudatum* au cours des premières phases de la conjugaison. *Protistologica* **4**, 415–418.

14 —, Vivier, E. & Puytorac, P. de, 1960. Action comparée de la colchicine et de la trypaflavine sur la vitesse de multiplication de *Paramecium caudatum*. *Bull. Biol.* **93**, 1–11.

15 *Bodian, D. 1936. A new method for staining nerve fibers and nerve endings in mounted paraffin sections. *Anat. Rec.* **65**, 89.

16 — 1937. The staining of paraffin sections of nervous tissues with activated protargol. The role of fixatives. *Anat. Rec.* **69**, 153–162.

17 Boell, E. J. & Woodruff, L. L. 1941. Respiratory metabolism of mating types in *Paramecium calkinsi*. *J. Exp. Zool.* **87**, 385–402.

18 Bomford, R. 1966. The syngens of *Paramecium bursaria:* new mating types and intersyngenic mating reactions. *J. Protozool.* **13**, 497–501.

19 — 1967. Stable changes of mating type after abortive conjugation in *Paramecium bursaria*. *Exp. Cell Res.* **47**, 30–41.

20 Borchsemius, O. N. & Ossipov, D. V. 1968. Polymorphism of micronuclei of *Paramecium caudatum*. II. Mitotical cycles of micronuclei of different morphological types. *Acta Protozool. (Warsaw)* **6**, 161–167.

21 Bozler, E. 1924. Über die Morphologie der Ernährung-organelle und die Physiologie der Nahrungsaufnahme von *Paramecium caudatum* Ehrb. *Arch. Protistenk.* **49**, 163–215.

22 — 1924. Über die physikalische Erklärung der Schlundfadenströmungen, ein Beitrag zur Theorie der Protoplasmaströmungen. *Z. Vergleich. Physiol.* **2**, 82–90.

23 Bragg, A. N. 1936. Selection of food in *Paramecium trichium*. *Physiol. Zool.* **9**, 433–442.

24 Calkins, G. N. 1902. Studies on the life history of Protozoa. I. The life cycle of *Paramecium caudatum*. *Arch. Entwicklungsmech. Organ.* **15**, 139–186.

25 — 1915. Cycles and rhythms and the problem of immortality in *Paramecium*. *Amer. Naturalist* **49**, 65–75.

26 — 1933. *The Biology of the Protozoa*, Lea and Febiger, Philadelphia, Pa.

27 Canella, M. F. 1958. Biologie degli infusori e ipotetici raffronti con i metazoi. I. *Monit. Zool. Ital.* **65**, 164–183.

28 — 1959. Biologie degli infusori e ipotetici raffronti con i metazoi. II. *Monit. Zool. Ital.* **66**, 198–228.

29 — 1960. Biologie degli infusori e ipotetici raffronti con i metazoi. III. *Monit. Zool. Ital.* **67**, 143–189.

30 Charret, R. 1969. L'ADN nucléolaire chez *Tetrahymena pyriformis:* Chronologie de sa replication. *Exp. Cell Res.* **54**, 353–361.

31 — & Fauré-Fremiet, E. 1967. Technique de rassemblement de micro-organismes: Préinclusion dans un caillot de fibrine. *J. Microscopie* **6**, 1063–1066.

32 Chatton, E. & Brachon, S. 1933. Sur une Paramécie à 2 races: *Paramecium duboscqui*, n. sp. *Compt. Rend. Soc. Biol.* **114**, 988.

33 — & Chatton, M. 1923. La sexualité provoquée expérimental chez un Infusoire: *Glaucoma scintillans*. *Compt. Rend.* **176**, 1091–1093.

34 — & — 1923. L'influence des facteurs bactériens sur la nutrition, la multiplication et la sexualité des Infusoires. *Compt. Rend.* **176**, 1262–1265.

35   — & — 1925. L'action des facteurs externes sur la sexualité des Infusoires. *Compt. Rend. Soc. Biol.* **93**, 675–678.

36   — & — 1925. L'action des facteurs externes sur les Infusoires. *Compt. Rend.* **180**, 1137–1139.

37   — & — 1927. Sur les conditions nécessaires pour déterminer expérimentalement la conjugaison de *Glaucoma scintillans. Compt. Rend.* **185**, 400–402.

38   — & — 1929. Les conditions de la conjugaison de *Glaucoma scintillans* en cultures letho-bactériennes. Action directe et spécifique de certains agents zygogènes. *Compt. Rend.* **188**, 1315–1317.

39   — & — 1929. L'état de jeûne, condition nécessaire mais non suffisante de la conjugaison expérimentale de *Glaucoma scintillans. Compt. Rend.* **189**, 59–62.

40   — & — 1931. La conjugaison de *Paramecium caudatum* déterminée expérimentalement par modification de la flore bactérienne associée. Races dites conjugantes et non conjugantes. *Compt. Rend.* **193**, 206–209.

41   — & Lwoff, A. 1930. Imprégnation, par diffusion argentique, de l'infraciliature des Ciliés marins et d'eau douce, après fixation cytologique et sans dessication. *Compt. Rend. Soc. Biol.* **104**, 834–836.

42   — & — 1935. La constitution primitive de la strie ciliaire des Infusoires. La desmodexie. *Compt. Rend. Soc. Biol.* **118**, 1068–1072.

43   — & — 1936. Techniques pour l'étude des Protozoaires, spécialement de leurs structures superficielles (cinétome et argyrome). *Bull. Soc. Microscopie* **5**, 25–39.

44   Cheissin, E. M., Ovchinnikova, L. P., Selivanova, G. V. & Buze, E. G. 1963. Changes of the DNA content in the macronucleus of *Paramecium caudatum* in the interdivisional period (in Russian). *Acta Protozool. (Warsaw)* **1**, 63–69.

45   Chen, T. T. 1940. Conjugation in *Paramecium bursaria* between animals with diverse nuclear constitution. *J. Heredity* **31**, 186–196.

46   — 1944. Mating types in *Paramecium caudatum. Amer. Naturalist* **78**, 334–340.

47   — 1946. Conjugation in *Paramecium bursaria.* I. Conjugation of three animals. *J. Morphol.* **78**, 353–395.

48   Chen-Shan, L. & Whittle, J. R. S. 1968. The effect of mitomycin C on the cortex of *P. aurelia. Symp. Ciliate Genet., Oak Ridge, Tenn., U.S.A.*

49   Claparède, E. & Lachmann, J. 1858–1861. *Etudes sur les Infusoires et les Rhizopodes,* Vols. *1, 2,* Genève.

50   Cohen, L. W. 1964. Diurnal intracellular differentiation in *Paramecium bursaria. Exp. Cell Res.* **36**, 398–406.

51   — 1965. The basis for the circadian rhythm of mating in *Paramecium bursaria. Exp. Cell Res.* **37**, 360–367.

52   — & Siegel, R. W. 1963. The mating-type substances in *Paramecium bursaria. Genet. Res.* **4**, 143–150.

53   Corliss, J. O. 1952. Comparative studies on holotrichous Ciliates in the *Colpidium-Glaucoma-Leucophrys-Tetrahymena Group.* I. General consideration and history of strains in pure culture. *Trans. Amer. Microscop. Soc.* **71**, 159–184.

54   — 1953. Comparative studies on holotrichous Ciliates in the *Colpidium-Glaucoma-Leucophrys-Tetrahymena Group.* II. Morphology, life cycles and systematic status of strains in pure culture. *Parasitology* **43**, 49–87.

55   — 1954. The buccal apparatus and systematics status of *Glaucoma frontaia. J. Morphol.* **94**, 199–219.

56   — 1956. On the evolution and systematics of ciliated protozoa. *Syst. Zool.* **5**, 68–91, 121–140.

57   — 1961. *The Ciliated Protozoa* (International Series of Monographs on Pure and Applied Biology, Vol. 7), Pergamon Press, London.

58   Delage, Y. & Hérouard, E. 1896. *Traité de Zoologie Concrète. La Cellule et les Protozoaires,* Vol. *1,* Schleicher Frères, Paris.

59   Dembowski, J. 1929. Die Vertikalbewegung von *Paramecium caudatum.* II. Einfluss einiger Faktoren. *Arch. Protistenk.* **68**, 215–261.

60   Diller, W. F. 1936. Nuclear reorganization processes in *Paramecium aurelia,* with descriptions of autogamy and "hemixis". *J. Morphol.* **59**, 11–67.

61   — 1940. Nuclear variation in *Paramecium caudatum. J. Morphol.* **66**, 605–633.

62   — 1948. Nuclear behavior of *Paramecium trichium* during conjugation. *J. Morphol.* **82**, 1–52.

63   — 1954. Autogamy in *Paramecium polycaryum. J. Protozool.* **1**, 60–70.

64 — 1958. Studies on conjugation in *Paramecium polycaryum. J. Protozool.* **5**, 282–292.

65 — & Earl, P. R. 1958. *Paramecium jenningsi*, n. sp. *J. Protozool.* **5**, 155–158.

66 Dippell, R. V. 1958. The fine structure of kappa in killer stock 51 of *Paramecium aurelia*. Preliminary observations. *J. Biophys. Biochem. Cytol.* **4**, 125–128.

67 — 1968. The development of basal bodies in *Paramecium. Proc. Nat. Acad. Sci. U.S.* **61**, 461–468.

67a Doroszewski, M. 1959. *Paramecium arcticum*, sp. nov. *Bull. Acad. Polon. Sci. Classe II* **7**, 73–78.

68 Dragesco, J. 1962. L'orientation actuelle de la systématique des Ciliés et la technique d'imprégnation au protéinate d'argent. *Bull. Micros. Appl.* **11**, 49–58.

69 — 1970. Ciliés libres du Cameroun. *Ann. Fac. Sci. Univ. Fed. Cameroun*, no. hors-série, 1–141.

70 Drochmans, P. 1962. Morphologie du glycogène. Etude au microscope électronique de colorations négatives du glycogène particulaire. *J. Ultrastruct. Res.* **6**, 141–163.

71 Dryl, S. & Grebęcki, A. 1966. Progress in the study of excitation and response in Ciliates. *Protoplasma* **62**, 255–284.

72 Dujardin, F. 1841. *Histoire Naturelle des Zoöphytes: Infusoires*, Paris.

73 Dupy-Blanc, J. 1969. Etude cytophotométrique des teneurs en ADN des micronucleus de *Paramecium caudatum* au cours de la conjugaison et pendant la différenciation des "anlages" en macronucleus. *Protistologica* **5**, 239–248.

74 Egelhaaf, A. 1955. Cytologisch-entwicklungsphysiologische Untersuchungen zur Konjugation von *Paramecium bursaria* Focke. *Arch. Protistenk.* **100**, 447–514.

75 Ehrenberg, C. G. 1833. *Abhandl. Akad. Wissensch. Berlin*, Druckerei der Königlichen Akademie der Wissenschaften, Berlin.

76 — 1838. *Die Infusionsthierchen als vollkommene Organismen*, Leipzig.

77 Ehret, C. F. & De Haller, G. 1963. Origin, development and maturation of organelles and organelle systems of the cell surface in *Paramecium. J. Ultrastruct. Res.*, Suppl. **6**, 3–42.

78 — & Powers, E. L. 1955. Macronuclear and nucleolar development in *Paramecium bursaria. Exp. Cell Res.* **9**, 241–257.

79 — & — 1959. The cell surface of Paramecium. *Int. Rev. Cytol.* **8**, 97–133.

80 Essner, E. & Novikoff, A. B. 1962. Cytological studies on two functional hepatomas. Interrelations of endoplasmic reticulum, Golgi apparatus and lysosomes. *J. Cell Biol.* **15**, 289–312.

81 Estève, J. C. 1966. Facteurs de groupement en anneau chez *Paramecium. Protistologica* **2**, 95–100.

82 — 1967. Observations ultrastructurales sur quelques aspects de l'évolution des vacuoles alimentaires chez *Paramecium caudatum. Compt. Rend.* **265**, 1991–1994.

83 — 1968. Données complémentaires sur le déterminisme du groupement en anneau chez les Paramécies. *Protistologica* **4**, 243–249.

84 — 1969. Observations sur l'ultrastructure et le métabolisme du glycogène de *Paramecium caudatum. Arch. Protistenk.* **111**, 195–203.

85 — 1970. Distribution of acid phosphatase in *Paramecium caudatum;* its relations with the process of digestion. *J. Protozool.* **17**, 24–35.

85a — 1972. L'appareil de Golgi des Ciliés. Ultrastructure, particulièrement chez *Paramecium. J. Protozool.* **19**(4), 609–618.

86 Fauré-Fremiet, E. 1949. Morphologie comparée des Ciliés holotriches *Trichostomata* et *Hymenostomata. Compt. Rend. XIIIe Congr. Int. Zool., Paris.* 1948, 215–216.

87 — 1950. Morphologie comparée et systématique des Ciliés. *Bull. Soc. Zool. Franç.* **75**, 109–122.

88 — 1953. L'hypothèse de la sénescence et les cycles de réorganisation nucléaire chez les Ciliés. *Rev. Suisse Zool.* **60**, 426–438.

89 — 1970. Microtubules et mécanismes morpho-poiétiques. *Ann. Biol.* **9**, 1–61.

90 —, Favard, P. & Carasso, N. 1962. Etude au microscope électronique des ultrastructures d'*Epistylis anastica* (Cilié péritriche). *J. Microscopie* **1**, 287–312.

91 Favard, P. & Carasso, N. 1963. Mise en évidence d'un processus de micro-pinocytose interne au niveau des vacuoles digestives d'*Epistylis anastatica. J. Microscopie* **2**, 495–498.

92 — & — 1964. Etude de la pinocytose au niveau des vacuoles digestives des Ciliés. *J. Microscopie* **3**, 671–696.

93 Focke, G. W. 1836. Über einige Organisations-Verhältnisse bei polygastrischen Infusorien und Rädertieren. *Isis.*

94 Frisch, J. A. 1939. The experimental adaptation of *Paramecium* to sea water. *Arch. Protistenk.* **93**, 38–71.

78   E. VIVIER

95   Furgason, W. H. 1940. The significant cytostomal pattern of the *Glaucoma-Colpidium Group*
     and a proposed new genres and species. *Arch. Protistenk.* **94**, 224–226.

96   Galadjieff, M. & Matalnikow, S. 1933. L'immortalité de la cellule. Vingt-deux ans de culture
     d'Infusoires sans conjugaison. *Arch. Zool. Exp. Gén.* **75**, 331–352.

97   Gelei, J. von, 1925. Uj *Paramecium* szeged Környékerol. *Paramecium nephridiatum nov. sp.*
     Allattani Közlemenyck Bud. **22**, 121–162.

98   — 1925. Nephridialapparat bei den Protozoen. *Biol. Zentralbl.* **45**, 676–683.

99   — 1929. A Veglenyck Indegrendszere. *Allattani Közlemenyck Bud.* **26**, 164–190.

100 — 1932. Die reizleitenden Elemente der Ciliaten in nass hergestellten Silber-bzw. Gold Präpa-
     raten. *Arch. Protistenk.* **77**, 152–174.

101 — 1934. Der feinere Bau des Cytopharynx von *Paramecium* und seine systematische Bedeutung.
     Arch. Protistenk. **82**, 331–362.

102 — 1938. Beiträge zur Ciliatenfauna der Umgebung von Szeged. VII. *Paramecium nephridiatum.*
     Arch. Protistenk. **91**, 343–356.

103 — 1939. Das äussere Stützgerüstsystem des Parameciumkörpers. *Arch. Protistenk.* **92**, 245–272.

104 — 1952. Nekany sze acsillosok *Trichostoma* alrendjenek rendszerfahoz. *Ann. Biol. Univ. Hung.*
     1, 350–360.

105 — 1954. Ueber die Lebensgemeinschaft einiger temporärer Tüpel. III. Ciliaten. *Arch. Biol.*
     Hung. **5**, 259–343.

106 Génermont, J. 1966. *Recherches sur les Modifications Durables et le Déterminisme Génétique*
     de Certains Caractères Quantitatifs chez Paramecium aurelia. Thèse Fac. Sc. Paris, Expansion
     Scientifique Française, Paris.

107 Giese, A. C. 1939. Studies on conjugation in *Paramecium multimicronucleatum. Amer. Naturalist*
     73, 432–444.

108 — 1957. Mating types in *Paramecium multimicronucleatum. J. Protozool.* **4**, 120–124.

109 — & Arkoosh, M. A. 1939. Tests for sexual differentiation in *Paramecium multimicronucleatum*
     and *Paramecium caudatum. Physiol. Zool.* **12**, 70–75.

110 Gilman, L. C. 1939. Mating types in *Paramecium caudatum. Amer. Naturalist* **73**, 445–450.

111 — 1941. Mating types in diverse races of *Paramecium caudatum. Biol. Bull.* **80**, 384–402.

112 Goldfischer, S., Carasso, N. & Favard, P. 1963. The demonstration of acid phosphatase
     activity by electron microscopy in the ergastoplasm of the Ciliate *Campanella umbellaria. J.*
     Microscopie **2**, 621–628.

113 —, Favard, P. & Carasso, N. 1967. The demonstration of acid hydrolase activities in digestive
     vacuoles of Peritrich Ciliates by hexazonium parasosanilin procedures. *J. Microscopie* **6**,
     867–872.

114 Goldfuss, G. A. 1782–1848. *Handbuch der Zoologie*, Nürnberg.

115 Gourret, P. & Roeser, P. 1886. Les Protozoaires du vieux port de Marseille. *Arch. Zool. Exp.* **4**,
     443–534.

116 Grain, J. 1969. Le cinétosome et ses dérivés chez les Ciliés. *Année Biol.* **8**, 53–97.

117 Grębecki, A., Kuźnicki, L. & Mikolajczyke, E. 1966. Some observations on the inversion of
     spiralling in *Paramecium caudatum. Acta Protozool. (Warsaw)* **4**, 383–388.

118 —, Kuznicki, L. & Mikolajczyke, E. 1966. Right spiralling induced in *Paramecium* by Ni++
     ions and the hydrodynamics of the spiral movement. *Acta Protozool. (Warsaw)* **4**, 389–408.

119 — & Mikolajczyke, E. 1968. Ciliary reversal and renormalization in *Paramecium caudatum*
     immobilized by Ni++ ions. *Acta Protozool. (Warsaw)* **5**, 297–303.

120 Grobicka, J. & Wasilewska, J. 1925. Essai d'analyse chimique quantitative de l'Infusoire
     Paramecium caudatum. Trav. Inst. Nencki Warszawa **3**, 1–23.

121 Hamilton, L. D. & Gettner, M. E. 1958. Fine structure of kappa in *Paramecium aurelia. J.*
     Biophys. Biochem. Cytol. **4**, 122–124.

122 Hartmann, M. 1943. *Die Sexualität*, Fischer, Jena.

123 Hayashi, S. & Takayanagi, T. 1962. Cytological and cytogenetical studies on *Paramecium*
     polycaryum. IV. Determination of the mating system based on some experimental and cytological
     observations. *Japan. J. Zool.* **13**, 357–364.

124 Hertwig, R. 1889. Ueber die Conjugation der Infusorien. *Abhandl. Bayer. Akad. Wiss.* **17**,
     150–233.

125 Hill, J. 1752. *General Natural History*, Vol. 3 (History of Animals), London.

126 Hirson, J. B. 1969. The response of *Paramecium bursaria* to potential endocellular symbionts.
     Biol. Bull. **136**, 33–42.

127 Hisada, M. 1952. Induction of contraction in *Paramecium* by electric current. *Annot. Zool. Japon* **25**, 415–419.

128 Hiwatashi, K. 1949. Studies of the conjugation of *Paramecium caudatum*. I. Mating types and groups in the races obtained in Japan. *Sci. Rep. Tôhoku Univ.* **18**, 137–140.

129 — 1949. Studies on the conjugation of *Paramecium caudatum*. II. Induction of pseudoselfing pairs by formalin killed animals. *Sci. Rep. Tôhoku Univ.* **18**, 141–143.

130 — 1958. Inheritance of mating types in variety 12 of *Paramecium caudatum*. *Sci. Rep. Tôhoku Univ., 4e Sér. Biol.* **24**, 119–129.

131 — 1960. Analysis of the change of mating type during vegetative reproduction in *Paramecium caudatum*. *Japan J. Genet.* **35**, 213–221.

132 — 1961. Locality of mating reactivity on the surface of *Paramecium caudatum*. *Sci. Rep. Tôhoku Univ., 4e Sér. Biol.* **27**, 93–99.

133 — 1968. Determination and inheritance of mating type in *Paramecium caudatum*. *Genetics* **58**, 373–386.

134 *— 1969. Genetic and epigenetic control of mating type in *Paramecium caudatum*. *Proc. 12th Int. Cong. Genet.* **2**, 259–260.

135 Hoffman, E. J. 1965. The nucleic acids of basal bodies isolated from *Tetrahymena pyriformis*. *J. Cell Biol.* **25**, 217–228.

136 Honigberg, B. M. *et al.* (Committee on Taxonomy and Taxonomic problems of Soc. Protozool.), 1964. A revised classification of the Phylum Protozoa. *J. Protozool.* **11**, 7–20.

137 Hopkins, H. S. 1921. The conditions for conjugation in *Paramecium*. *J. Exp. Zool.* **9**, 279–298.

138 Hubert, M. T., Carasso, N. & Favard, P. 1962. Méthode de rassemblement de petits organismes en suspension pour l'inclusion dans des milieux visqueux. *J. Microscopie* **1**, 163–166.

139 *Hufnagel, L. 1966. Fine structure and DNA of pellicles isolated from *Paramecium aurelia*. *Electron Microscopy, 6th Int. Congr. Electr. Micr. Kyoto, Vol. 2*, Ryozi Hyeda Edit. Maruzen Co. LTD, 239–240.

140 — 1969. Cortical ultrastructure of *Paramecium aurelia* studies on isolated pellicles. *J. Cell Biol.* **40**, 379–801.

141 Hunter, N. W. 1959. Enzyme systems of *Stylonichia pustulata*. II. Miscellaneous system (hydrases, hydrolases and dehydrogenases). *J. Protozool.* **6**, 100–104.

142 — 1961. Enzyme systems in *Colpoda cuculus*. II. Intracellular activity of some enzymes as determined by histochemistry. *Trans. Amer. Microsc. Soc.* **80**, 38–43.

143 Hyman, L. H. 1925. Methods of securing and cultivating Protozoa. I. General statements and methods. *Trans. Amer. Microsc. Soc.* **44**, 216–221.

144 — 1931. Methods of securing and cultivating Protozoa. II. *Paramecium* and other Ciliates. *Trans. Amer. Microsc. Soc.* **50**, 50–57.

145 Inaba, F., Imamoto, K. & Suganuma, Y. 1966. Electron microscopic observations on nuclear exchange during conjugation in *Paramecium multimicronucleatum*. *Proc. Japan. Acad.* **42**, 394–398.

146 Jahn, T. L. 1961. The mechanism of ciliary movement. I. Ciliary reversal and activation by electric current; the Ludloff phenomenon in terms of core and volume conductors. *J. Protozool.* **8**, 369–380.

147 — 1962. The mechanism of ciliary movement. II. Ion antagonism and ciliary reversal. *J. Cell. Comp. Physiol.* **60**, 217–228.

148 — 1967. The mechanism of ciliary movement. III. Theory of suppression of reversal by electrical potential of cilia reversed by barium ions. *J. Cell Physiol.* **70**, 79–89.

149 — & Bove, E. C. 1965. Movement and locomotion of micro-organisms. *Ann. Rev. Microbiol.* **19**, 21–58.

150 — & — 1967. Motile behavior of Protozoa. In Chen, T. T., *Research in Protozoology*, Pergamon Press, London, **1**, 40–198.

151 Jakus, M. A. 1945. The structure and properties of the trichocysts of *Paramecium*. *J. Exp. Zool.* **100**, 457–485.

152 — & Hall, E. E. 1946. Electron microscope observations of the trichocysts and cilia of *Paramecium*. *Biol. Bull.* **91**, 141–144.

153 Jankowski, A. W. 1960. Conjugation processes in *Paramecium trichium*. I. Amphimixis and autogamy (in Russian). *TSytologia* **2**, 581–588.

154 — 1961. Conjugation process in a *Paramecium* of unfrequent occurrence: *Paramecium woodruffi* (in Russian). *Compt. Rend. Acad. Sci. U.R.S.S.* **137**, 989–992.

155 — 1962. Conjugation processes in *Paramecium putrinum* Clap. Lachm. (in Russian). *TSytologia* **4**, 434–444.

156 — 1965. Conjugation processes in *Paramecium putrinum*. VII. Nuclear processes at autogamy in singles induced with a new technique: multiple mating (in Russian, summary in English). *Acta Protozool. (Warsaw)* **3**, 239–262.

157 — 1965. Conjugation processes in *Paramecium putrinum*. IV. The individual variability of the nuclear processes at apomictic conjugation (in Russian). *TSytologia*, **7**, 55–65.

157a — 1965. Processus de conjugaison de *Paramecium putrinum* Clap. et Lachm. III. Système pluralitaire des types d'accouplement chez *P. putrinum* (in Russian). *Rev. Biol. Gen. U.S.S.R.* **23**, 276–282.

158 — 1966. Conjugation processes in *Paramecium putrinum*. VI. The induction and cytological study of a triple conjugation (in Russian). *TSytologia* **8**, 70–80.

159 — 1966. Conjugation processes in *Paramecium putrinum*. VIII. The induction and cytological study of a triple conjugation (in Russian). *TSytologia* **8**, 70–80.

160 — 1966. Conjugation processes in *Paramecium putrinum*. IX. "Necrochromatine" and the functional significance of macronuclear fragmentation (in Russian). *TSytologia* **8**, 725–735.

160a — 1969. A proposed taxonomy of the genus *Paramecium* Hill 1752 *(Ciliophora)* (in Russian). *Rev. Zool. U.S.S.R.* **48**, 30–40.

161 Jennings, H. S. 1910. What conditions induce conjugation in *Paramecium?* *J. Exp. Zool.* **9**, 279–298.

162 — 1938. Sex reactions types and their interrelations in *Paramecium bursaria*. *Proc. Nat. Acad. Sci. U.S.* **24**, 112–120.

163 — 1939. Genetics of *Paramecium bursaria*. I. Mating types and groups, their interrelation and distribution; mating behavior and self sterility. *Genetics* **24**, 202–233.

164 — 1939. Mating types and their interaction in the Ciliate Infusoria. Introduction. *Amer. Naturalist* **73**, 385–389.

165 — 1939. *Paramecium bursaria:* Mating types and groups, mating behavior, self sterility; their development and inheritance. *Amer. Naturalist* **73**, 414–431.

166 — 1941. Genetics of *Paramecium bursaria*. II. Self-differentiation and self-fertilization of clones. *Proc. Amer. Phil. Soc.* **85**, 25–48.

167 — 1942. Genetics of *Paramecium bursaria*. III. Inheritance of mating types in crosses and in clonal self-fertilization. *Genetics* **27**, 193–211.

168 — 1944. *Paramecium bursaria*: life history. I. Immaturity, maturity and age. *Biol. Bull.* **86**, 131–141.

169 — 1944. *Paramecium bursaria*: life history. II. Age and death of clones in relation to the results of conjugation. *J. Exp. Zool.* **96**, 17–52.

170 — 1944. *Paramecium bursaria*: life history. III. Repeated conjugation in the same stock at different ages. *J. Exp. Zool.* **96**, 243–273.

171 — 1944. *Paramecium bursaria*: life history. IV. Relation of inbreeding to mortality of exconjugant clones. *J. Exp. Zool.* **97**, 165–197.

172 — 1945. *Paramecium bursaria*: life history. V. Some relations of external conditions post or present to aging and to mortality of exconjugants. *J. Exp. Zool.* **99**, 15–31.

173 Joblot, L. 1718. *Descriptions et Usages de Plusieurs Nouveaux Microscopes. . . avec de Nouvelles Observations* (etc.), Part 2, 2nd edition with addition by publisher, 1754.

174 Jollos, V. 1921. Experimentelle Protistenstudien. I. Untersuchungen über Variabilität und Vererbung bei Infusorien. *Arch. Protistenk.* **43**, 1–222.

175 Jurand, A. 1961. An electron microscope study of food vacuoles in *Paramecium aurelia*. *J. Protozool.* **8**, 125–130.

176 — & Preer, L. B. 1969. Ultrastructure of flagellated lambda symbionts in *Paramecium aurelia*. *J. Gen. Microbiol.* **54**, 359–364.

177 — & Selman, G. G. 1969. *The Anatomy of Paramecium aurelia*, MacMillan, London and St. Martin's Press, New York, N.Y.

178 — & — 1970. Ultrastructure of the nuclei and intranuclear microtubules of *Paramecium aurelia*. *J. Gen. Microbiol.* **60**, 357–364.

179 —, Beale, G. H. & Young, M. R. 1964. Studies on the macronucleus of *Paramecium aurelia*. II. Development of macronuclear anlagen. *J. Protozool.* **11**, 491–497.

180 Kahl, A. 1930. *Wimpertiere oder Ciliata (Infusoria)*, Dahl-Bischoff, Jena.

181 Kalmus, H. 1927. *Paramecium, das Pantoffeltierchen*, Fischer, Jena.

182 Kamoda, T. & Kinosita, H. 1945. Protoplasmic contraction of *Paramecium*. *Proc. Japan Acad.* **21**, 349–358.

183 Karakashian, S. J., Karakashian, M. W. & Rudzinska, M. A. 1968. Electron microscopic observations on the symbiosis of *Paramecium bursaria* and its intracellular algae. *J. Protozool.* **15**, 113–128.

184 Kent, W. S. 1880–1882. *A Manual of the Infusoria, 3 Vols.*, London.

185 Kimball, R. F. & Perdue, S. W. 1962. Quantitative cytochemical studies on *Paramecium*. V. Autoradiographic studies of nucleic acid syntheses. *Exp. Cell Res.* **27**, 405–415.

186 — & Vogt-Köhne, L. 1962. Effects of radiation on cell and nuclear growth in *Paramecium aurelia*. *Exp. Cell Res.* **28**, 228–238.

187 —, Caspersson, T. O., Svensson, G. & Carlson, L. 1959. Quantitative cytochemical studies on *Paramecium aurelia*. I. Growth in total dry weight measured by the scanning interference microscope and X-ray absorption methods. *Exp. Cell Res.* **17**, 160–172.

188 Kinosita, H. & Murakami, A. 1967. Control of ciliary motion. *Physiol. Rev.* **47**, 53–82.

189 Kirby, H. 1950. *Materials and Methods in the Study of Protozoa*, University of California Press, Berkeley, Calif.

190 Klein, B. M. 1926. Ergebnisse mit einer Silbermethode bei Ciliaten. *Arch. Protistenk.* **56**, 243–279.

191 — 1927. Über die Darstellung der Silberlinien-Systeme des Ciliatenkörpers. *Mikrokosmos* **20**, 233–235.

192 — 1927. Die Silberliniensysteme der Ciliaten. Ihr Verhalten während der Teilung und Konjugation, neue Silberbilder, Nachträge. *Arch. Protistenk.* **58**, 55–142.

193 Koscinszko, H. 1965. Karyologic and genetic investigations in syngen 1 of *Paramecium aurelia*. *Folia Biol.* **13**, 340–368.

194 Kudo, R. R. 1947. *Protozoology*, Thomas, Springfield, Ill.

195 Leeuwenhoek, A. van, 1674. *Phil. Trans. Royal Soc.* **9**, 178–182.

196 — 1677. *Phil. Trans. Royal Soc.* **12**, 821–831.

197 Legrand, B. 1970. Recherches expérimentales sur le déterminisme de la contraction et les structures contractiles chez le *Spirostome*. *Protistologica* **6** (3), 283–300.

198 Lozina-Lozinsky, L. 1929. Le choix de la nourriture chez *Paramecium caudatum*. *Compt. Rend. Soc. Biol.* **100**, 722–724.

199 — 1931. Zur Ernährungsphysiologie der Infusorien: Untersuchungen über die Nährungsauswahl und Vermehrung bei *Paramecium caudatum*. *Arch. Protistenk.* **74**, 18–120.

200 Lund, E. E. 1933. A correlation of the silverline and neuromotor systems of *Paramecium*. *Univ. Calif. (Berkeley) Publ. Zool.* **39**, 35–76.

201 — 1941. The feeding mechanisms of various ciliated Protozoa. *J. Morphol.* **69**, 563–573.

202 Lwoff, A. 1950. *Problems of Morphogenesis in Ciliates. The Kinetosomes in development, Reproduction and Evolution*. Wiley, New York, N.Y.

202a *McDonald, B. B. 1964. Exchange of cytoplasm during conjugation in *Tetrahymena*. *J. Protozool.* **11** (Suppl.), 11.

203 Mast, S. O. 1947. The food vacuole in *Paramecium*. *Biol. Bull.* **92**, 31–72.

204 Maupas, E. 1883. Contribution à l'étude morphologique et anatomique des Infusoires Ciliés. *Arch. Zool. Exp. Gén.* **1**, 427–664.

205 — 1888. Recherches expérimentales sur la multiplication des Infusoires ciliés. *Arch. Zool. Exp. Gén.* **6**, 165–277.

206 — 1889. Le rajeunissement caryogamique chez les Ciliés. *Arch. Zool. Exp. Gén.* **7**, 149–517.

207 Metz, C. B. 1947. Induction of "pseudo-selfing" and meiosis in *Paramecium aurelia* by formalin-killed animals of opposite mating types. *J. Exp. Zool.* **105**, 115–139.

208 — 1948. The nature and mode of action of the mating type substances. *Amer. Naturalist* **82**, 85–95.

209 — 1953. Mating substances and the physiology of fertilization in Ciliates. In Wenrich, D. H., *Sex in Micro-organisms* (Symposium of the American Association for the Advancement of Science, Washington, D.C.), 284–334.

210 —, Pitelka, D. R. & Westfall, J. A. 1953. The fibrillar systems of Ciliates as revealed by the electron microscope. I. *Paramecium*. *Biol. Bull.* **104**, 408–425.

211 *Mitchell, J. B. 1962. Nuclear reorganization in *Paramecium jenningsi*. *J. Protozool.* **9** (Suppl.), 26.

212 Miyake, A. 1955. The effect of urea on binary fission in *Paramecium caudatum*. *J. Inst. Polytech. Osaka, Ser. D*. **6**, 43–53.

213 — 1956. Physiological analysis of the life cycle of the Protozoa. III. Artificial induction of selfing conjugation by chemical agents in *Paramecium caudatum*. *Physiol. Ecol*. **7**, 14–23.

214 — 1958. Induction of conjugation by chemical agents in *Paramecium caudatum*. *J. Inst. Polytech*. **9**, 251–256.

215 — 1968. Induction of conjugation by chemical agents in *Paramecium*. *J. Exp. Zool*. **167**, 359–380.

216 — 1969. Mechanism of initiation of sexual reproduction in *Paramecium multimicronucleatum*. *Japan J. Genet*. **44** (Suppl.) 1, 388–395.

217 Moldenhauer, D. von, 1964. Zytologische Untersuchungen zum Austausch der Wanderkerne bei konjugierenden *Paramecium caudatum*. *Arch. Protistenk*. **107**, 163–178.

218 Moore, A. 1903. Some facts concerning geotropic gatherings of *Paramecia*. *Amer. J. Physiol*. **9**, 238–244.

219 Müller, O. F. 1773–1774. *Vermium Terrestrium et Fluviatilium. Historia*.

220 Müller, J. 1786. *Animalcula Infusoria Fluxiatilia et Marina*, Hauniae, Copenhagen.

221 Müller, M. & Törö, I. 1962. Studies on feeding and digestion in Protozoa. III. Acid phosphatase activity in food vacuoles of *Paramecium multimicronucleatum*. *J. Protozool*. **9**, 98–102.

222 Nanney, D. 1954. Mating type determination in *Paramecium aurelia*, a study in cellular heredity. In Wenrich, D. H., *Sex in Micro-organisms* (Symposium of the American Association for the Advancement of Science, Washington, D.C., 266–283.

223 Narasimha, Rao, M. V. & Prescott, D. M. 1967. Micronuclear RNA synthesis in *Paramecium caudatum*. *J. Cell Biol*. **33**, 281–285.

224 Nobili, R. 1961. Variazoni volumetriche del macronucleo e loro effetti nella riproduzione vegetativa in *Paramecium aurelia*. *Atti. Soc. Toscane Nat. Pisa* **B67**, 217–232.

225 — 1961. L'Azione del gene am sull'apparato necleare di *Paramecium aurelia* durante la riproduzione vegetativa e sessuale in relazione all' eta del elone ed alla temperatura di allevarnento degli animali. *Caryologia*, **14**, 43–58.

226 — & Agostini, G. 1964. Coniugazione e riproduzione vegetativa di *Paramecium aurelia* sotto l'azione del 6-azauracile e della *p*-fluorofenilalamina. *Atti. Ass. Genet. It. Pavia* **9**, 72–86.

227 — & Kotopolus de Angelis, F. 1963. Effetti degli antibiotici sulla riproduzione de *Paramecium aurelia*. *Atti. Ass. Genet. It. Pavia* **8**, 45–57.

228 Noirot-Timothée, C. 1968. Les sacs parasomaux sont des sites de pinocytose. Etude expérimentale à l'aide du thorothrast chez *Trichodinopsis paradoxa* (Cilié Péritriche). *Compt. Rend*. **267**, 2334–2336.

229 Novikoff, A. B. 1960. Biochemical and staining reactions of cytoplasmic constituents. In Rudnick, D., *Developing Cell Systems and their Control*, The Ronald Press Co., New York, N.Y., 167–203.

230 Oger, C. 1965. Analyse biométrique de la croissance de *Paramecium caudatum*. Intérêt pour l'étude de la conjugaison. *Protistologica* **1**, 71–80.

231 — & Vivier, E. 1965. Observations d'ordre biométrique sur quelques variétés françaises de *Paramecium caudatum* Ehrb. Intérêt pour l'étude de leur sexualité. *Arch. Zool. Exp. Gen*. **105**, 119–153.

232 *Opton, E. M. 1942. Demonstration of the cytosomic morphology of *Paramecium polycaryum*. *Anat. Rec*. **84**, 485.

233 Organ, A. E., Bovee, E. C. & Jahn, T. L. 1969. The mechanism of the nephridial apparatus of *Paramecium multimicronucleatum*. II. The filling of the vesicle by action of the ampullae. *J. Cell Biol*. **40**, 389–394.

234 —, —, —, Wigg, D. & Fonseca, J. R. 1968. The mechanism of the nephridial apparatus of *Paramecium multimicronucleatum*. I. Expulsion of water from the vesicle. *J. Cell Biol*. **37**, 139–145.

235 Orlova, A. F. 1941. Modifications durables chez *Paramecium caudatum* et *P. multimicronucleatum* (in Russian). *Zool. Zhur*. **20**, 341–370.

236 Ossipov, D. V. 1963. Varieties and mating types in *Paramecium caudatum* from U.S.S.R. (in Russian, summary in English). *Bull. Univ. Leningrad, Ser. Biologie* **21**, 106–116.

237 — 1966. Methods of obtaining homozygous *Paramecium caudatum* clones. *Genetics Moscow, Acad. U.S.S.R.* **2**, 41–48.

238 — 1966. On macronuclear regeneration in *Paramecium caudatum* (in Russian). *TSytologia* **8**, 108–110.

239 — & Skoblo, I. I. 1968. The autogamy during conjugation in *Paramecium caudatum* Ehrb. II. The ex-autogamous stages of nuclear reorganization. *Acta Protozool. (Warsaw)* **6**, 33–48.

240 Párducz, B. 1958. Das interziliäre Fasernsystem in seiner Beziehung zu gewissen Fibrillen-komplexen der Infusorien. *Acta Biol. Acad. Sci. Hung.* **8**, 191–218.

241 — 1962. On a new concept of cortical organization in *Paramecium. Acta Biol. Acad. Sci. Hung.* **13**, 299–322.

242 — 1967. Ciliary movement and coordination in Ciliates. *Int. Rev. Cytol.* **21**, 91–128.

243 Perty, M. 1852. *System der Infusorien*, Bern, pp. 57–67.

244 Pieri, J. 1965. Interprétation cytophotométrique des phénomènes nucléaires au cours de la conjugaison chez *Stylonichia pustulata. Compt. Rend.* **261**, 2742–2744.

245 —, Vaugien, C. & Trouillier, M. 1968. Interprétations cytophotométrique des phénomènes micronucléaires au cours de la division binaire et des divisions prégamiques chez *Paramecium trichium. J. Cell Biol.* **36**, 664–668.

246 Pitelka, D. R. 1965. New observations on cortical ultrastructure in *Paramecium. J. Microscopie* **4**, 373–394.

247 Poljansky, G. I. & Posnanskaya, T. M. 1964. A long-lasting culture of *Paramecium caudatum* at 0° (in Russian). *Acta Protozool. (Warsaw)* **2**, 271–278.

248 Powers, J. H. & Mitchell, C. 1910. A new species of Paramecium *(Paramecium multimicro-nucleatum)* experimentally determined. *Biol. Bull.* **19**, 324–332.

249 Preer, L. B. 1969. Alpha, an infectious macronuclear symbiont of *Paramecium aurelia. J. Protozool.* **16**, 570–578.

250 Puytorac, P. de & Blanc, J. 1967. Observations sur les modifications ultrastructurales des micronoyaux au cours de leur transformation en macronoyaux chez *Paramecium caudatum. Compt. Rend. Soc. Biol.* **161**, 297–299.

251 Raaze, C. & Schoffeniels, E. 1965. Rôle de la vacuole contractile chez *Paramecium caudatum* Ehrenberg. *Bull. Classe Sci. Acad. Roy. Belg.* 5e Sér. **51**, 1057–1073.

252 Raikov, I. B. 1969. The macronucleus of Ciliates. In Chen, T. T., *Research in Protozoology*, Pergamon Press, London, **3**, 1–128.

253 Rasmussen, L. 1965. Metabolic inhibitor analysis of events in the growth-duplication cycle of *Paramecium, Proc. 2nd Int. Conf. Protozool., London*, 85.

253a Rasmussen, L. 1967. Effects of metabolic inhibitors on *Paramecium aurelia* during the cell generation cycle. *Exp. Cell Res.* **48**, 132–139.

254 Rees, C. W. 1922. The neuromotor apparatus of *Paramecium. Univ. Calif. (Berkeley) Publ. Zool.* **20**, 333–364.

255 Roque, M. 1961. Recherches sur les Infusoires Ciliés: les Hyménostomes péniculiens. *Bull. Biol. France Belg.* **95**, 431–519.

256 Rosenbaum, R. M. & Wittner, M. L. 1962. The activity of intracytoplasmic enzymes associated with feeding and digestion in *Paramecium caudatum*. The possible relationship to neutral red granules. *Arch. Protistenk.* **106**, 223–240.

257 Roth, L. E. 1960. Electron microscopy of pinocytosis and food vacuoles in Pelomyxa. *J. Protozool.* **7**, 176–185.

258 Satir, P. 1963. Studies on Cilia. I. The fixation of metachronal wave. *J. Cell Biol.* **18**, 345–365.

259 — 1965. Studies on Cilia. II. Examination of the distal region of the ciliary shaft and the role of the filaments in motility. *J. Cell Biol.* **26**, 805–834.

260 — 1968. Studies on Cilia. III. Further studies on the cilium tip and a "sliding filament" model of ciliary motility. *J. Cell Biol.* **39**, 77–94.

261 Schneider, L. 1959. Neue Befunde über den Feinbau des Cytoplasmas von *Paramecium* nach Einbettung in Vestopal W. *Z. Zellforsch.* **50**, 61–77.

262 — 1960. Elektronenmikroskopische Untersuchungen über das Nephridial-System von *Para-mecium. J. Protozool.* **7**, 75–90.

263 — 1960. Die Auflösung und Neubildung der Zellmembran bei der Konjugation von *Paramecium. Naturwissenschaften* **47**, 543–544.

264 — 1963. Elektronenmikroskopische Untersuchungen der Konjugation von *Paramecium. Protoplasma* **56**, 109–140.

265 — 1964. Elektronenmikroskopische Untersuchungen an den Ernährungsorganellen von *Paramecium*. 2. Die Nährungsvakuolen und die Cytopyge. *Z. Zellforsch.* **62**, 225–245.

266 — & Wohlfarth-Bottermann, K. E. 1964. Grenzstrukturen und Hüllen bei Bakterien und Protisten. *Studium generale* **17**, 95–124.

267 Schwartz, V. 1952. Die sexualität der Infusorien. *Fortschr. der Zool.* **9**, 605–619.

268 Seaman, G. R. 1960. Large scale isolation of kinetosomes from the ciliated protozoan *Tetrahymena pyriformis*. *Exp. Cell Res.* **21**, 292–302.

269 — 1961. Acid phosphatase activity associated with phagotrophy in the ciliate *Tetrahymena*. *J. Biophys. Biochem. Cytol.* **9**, 243–245.

270 Sedar, A. W. & Porter, K. R. 1955. The fine structure of cortical components of *Paramecium multimicronucleatum*. *J. Biophys. Biochem. Cytol.* **1**, 583–604.

271 Siegel, R. W. 1961. Nuclear differentiation and transitional cellular phenotypes in the life cycle of *Paramecium*. *Exp. Cell Res.* **24**, 6–20.

272 — 1963. New results on the genetics of mating types in *Paramecium bursaria*. *Genet. Res.* **4**, 132–142.

273 — 1965. Hereditary factors controlling development in *Paramecium. Genetic Control of Differentiation*. (Brookhaven Symp. Biol.), 55–65.

274 Siebold, C. T. E. von, 1845. *Lehrbuch der vergleichenden Anatomie der wirbellosen Thiere*.

275 Skoblo, I. I. & Ossipov, D. V. 1968. The autogamy during conjugation in *Paramecium caudatum* Ehrb. I. Study on the nuclear reorganization up to stage of the third synkaryon division. *Acta Protozool. (Warsaw)* **5**, 274–290.

276 Small, E. B. & Marszalek, D. S. 1969. Scanning electron microscopy of fixed, frozen and dried Protozoa. *Science* **163**, 1064–1065.

277 Smith-Sonneborn, J. & Plaut, W. 1967. Evidence for the presence of DNA in the pellicle of *Paramecium*. *J. Cell Sci.* **2**, 225–234.

278 Sonneborn, T. M. 1937. Sex, sex inheritance and sex determination in *Paramecium aurelia*. *Proc. Nat. Acad. Sci. U.S.* **23**, 378–385.

279 — 1938. Mating types in *Paramecium aurelia*: diverse conditions for mating in different stocks, occurrence, number and interrelations of the type. *Proc. Amer. Phil. Soc. Philad.* **79**, 411–434.

280 *— 1938. Mating types, toxic interactions and heredity in *Paramecium aurelia*. *Science* **88**, 503.

281 — 1939. *Paramecium aurelia*: mating types and groups; lethal interaction, determination and inheritance. *Amer. Naturalist* **73**, 390–413.

282 *— 1940. The relation of macronuclear regeneration in *Paramecium aurelia* to macronuclear structure, amitosis and genetic determination. *Anat. Rec.* **78**, 53–54.

283 — 1941. Relation of macronuclear regeneration in *Paramecium aurelia* to macronuclear structure, amitosis and genetic determination. *Collecting Net*, **16**, 3–4.

284 — 1947. Recent advances in the genetics of *Paramecium* and *Euplotes*. *Advanc. Genetics* **1**, 263–358.

285 — 1950. Methods in the general biology and genetics of *Paramecium aurelia*. *J. Exp. Zool.* **113**, 87–147.

286 — 1954. The relation of autogamy to senescence and rejuvenescence in *Paramecium aurelia*. *J. Protozool.* **1**, 38–53.

287 — 1955. Heredity, development and evolution in *Paramecium*. *Nature* **175**, 1100–1103.

288 — 1957. Breeding systems, reproductive methods and species problems in Protozoa. In Mayer, E., *The Species Problem*, American Association for the Advancement of Science, Publication No. 50, 155–324.

289 Šrámek-Hušek, R. 1954. Ciliaten aus der Tschechoslowakei und ihre Stellung im Sapobiensystem. *Arch. Protistenk.* **100**, 246–267.

290 Stein, F. 1859–1878. *Der Organismus der Infusionsthiere*, Leipzig.

291 Stewart, J. M. & Muir, A. R. 1963. The fine structure of the cortical layers in *Paramecium aurelia*. *Quart. J. Microscop. Sci.* **104**, 129–134.

292 Stockem, W. & Wohlfarth-Bottermann, K. E. 1970. Zur Feinstruktur der Trichocysten von *Paramecium*. *Cytobiologie* **1**, 420–436.

293 Stokes, A. C. 1885. Some new Infusoria. *Amer. Naturalist* **19**, 433–443.

294 Taub, S. R. 1963. The genetic control of mating type differentiation in *Paramecium*. *Genetics* **48**, 815–834.

295 — 1966. Regular changes in mating type composition in selfing cultures and in mating type potentiality in selfing caryonides of *Paramecium aurelia*. *Genetics* **54**, 173–189.

296 — 1966. Unidirectional mating type changes in individual cells from selfing cultures of *Paramecium aurelia*. *J. Exp. Zool.* **163**, 141–150.

297 Taylor, C. V. 1941. Fibrillar systems in Ciliates. In Calkins, G. N. & Summers, F. M., *Protozoa in Biological Research*, Columbia University Press, New York, N.Y., 191–270.

298 Thiery, J. P. 1967. Mise en évidence des polysaccharides sur coupes fines en microscopie électronique. *J. Microscopie* **6**, 987–1017.

299 Tuffrau, M. 1964. Quelques variantes techniques de l'imprégnation des Ciliés par le protéinate d'argent. *Arch. Zool. Exp.* **104**, 186–190.

300 — 1967. Perfectionnements et pratique de la technique d'imprégnation au protargol des Infusoires Ciliés. *Protistologica* **3**, 91–98.

301 Vivier, E. 1955. Contribution à l'étude de la conjugaison de *Paramecium caudatum*. *Bull. Soc. Zool. France* **80**, 163–170.

302 — 1960. Contribution à l'étude de la conjugaison chez *Paramecium caudatum*. *Ann. Sci. Nat. Zool. Biol. Anim., 12e Sér.* **2**, 387–506.

302a — 1962. Démonstration, à l'aide de la microscopie électronique, des échanges cytoplasmiques lors de la conjugaison chez *P. caudatum*. *Compt. Rend. Soc. Biol.* **156**, 1115–1116.

303 — 1963. Etude, au microscope électronique, des nucléoles dans le macronucleus de *Paramecium caudatum*. *Proc. 1st Int. Congr. Protozool., Prague* 1961, 421.

304 — 1965. Sexualité et conjugaison chez la Paramécie. *Ann. Fac. Sci. Clermont-Ferrand*, **26**, 101–114.

305 *— 1966. Variations ultrastructurales du chondriome en relation avec le mode de vie chez les Protozoaires. *Proc. 6th Int. Congr. Electron. Microscop. Kyoto, Japan*, 247–248.

306 — & André, A. 1961. Données structurales et ultrastructurales nouvelles sur la conjugaison de *Paramecium caudatum*. *J. Protozool.* **8**, 416–426.

307 —, Devauchelle, G., Petitprez, A., Porchet-Henneré, E., Prensier, G., Schrevel, J. & Vinckier, D. 1970. Observations de cytologie comparée chez les Sporozoaires. I. Les structures superficielles chez les formes végétatives. *Protistologica* **6** (1), 127–150.

308 —, Legrand, B. & Petitprez, A. 1969. Recherches cytochimiques et ultrastructurales sur des inclusions polysaccharidiques et calciques du *Spirostome;* leurs relations avec la contractilité. *Protistologica* **5**, 145–159.

309 — & Mallinger, M. C. 1960. Variations de la réactivité sexuelle en fonction du rythme de multiplication chez *Paramecium caudatum*. *Compt. Rend. Soc. Biol.* **154**, 2071–2075.

310 —, Petitprez, A. & Chivé, A. F. 1967. Observations ultrastructurales sur les chlorelles symbiotes de *Paramecium bursaria*. *Protistologica* **3**, 325–333.

311 —, Schrevel-Debersee, G. & Oger, C. 1964. Observations sur les variétés et types sexuels de *Paramecium caudatum;* hérédité et changement de type sexuel. *Arch. Zool. Exp. Gen.* **104**, 49–67.

312 Wenrich, D. H. 1928. *Paramecium woodruffi* nov. spec. (Ciliata). *Trans. Amer. Microscop. Soc.* **47**, 256–261.

313 — 1928. Eight well-defined species of *Paramecium*. *Trans. Amer. Microscop. Soc.* **47**, 274–282.

314 Whitson, G. L. 1964. Temperature sensibility and its relations to changes in growth, control of cell division, and stability of morphogenesis in *Paramecium aurelia*, syngen 4, stock 51. *J. Cell. Comp. Physiol.* **64**, 455–464.

315 Wichterman, R. 1940. Cytogamy: a sexual process occurring in living joined pairs of *Paramecium caudatum* and its relation to other sexual phenomena. *J. Morphol.* **66**, 423–451.

316 — 1951. The ecology, cultivation, structural characteristics and mating types of *Paramecium calkinsi*. *Proc. Pennsylvania Acad. Sci.* **25**, 51–65.

317 — 1953. *The Biology of Paramecium*. The Blakiston Company, New York, Toronto.

318 Wille, Jr., J. J. 1966. Induction of altered patterns of cortical morphogenesis and inheritance in *Paramecium aurelia*. *J. Exp. Zool.* **163**, 191–214.

318a — 1966. Induction of altered patterns of cortical morphogenesis and inheritance in *P. aurelia*. *J. Exp. Zool.* **163**, 191–214.

319 Woodard, J., Woodard, M., Gelber, B. & Swift, H. 1966. Cytochemical studies of conjugation in *Paramecium aurelia*. *Exp. Cell Res.* **41**, 55–63.

320 Woodruff, L. L. 1911. Two thousand generations of *Paramecium*. *Arch. Protistenk.* **21**, 263–266.

321 — 1914. So-called conjugating and non-conjugating races of *Paramecium*. *J. Exp. Zool.* **16**, 237–240.

322 — 1917. Rhythms and endomixis in various races of *Paramecium aurelia*. *Biol. Bull.* **33**, 51–56.

323 — 1921. The present status of the long continued pedigree culture of *Paramecium aurelia* at Yale University. *Proc. Nat. Acad. Sci. U.S.* **7**, 41–44.

324 — 1921. The structure, life history and intrageneric relationship of *Paramecium calkinsi*, n. sp. *Biol. Bull.* **41**, 171–180.

325   — 1925. The physiological significance of conjugation and endomixis in the Infusoria. *Amer. Naturalist* **69**, 225–249.

326   — & Erdmann, R. 1914. A normal periodic reorganization process without cells fusion in *Paramecium. J. Exp. Zool.* **17**, 425–518.

327   — & Spencer, H. 1923. *Paramecium polycaryum*, sp. nov. *Proc. Soc. Exp. Biol. Med.* **20**, 338–339.

328   *Yusa, A. 1955. The systematic significance of the buccal organelles in *Paramecium. J. Protozool.* **2** (Suppl.), 6.

329   — 1957. The morphology and morphogenesis of the buccal organelles in *Paramecium* with particular reference to their systematic significance. *J. Protozool.* **4**, 128–142.

330   — 1963. An electron microscope study on regeneration of trichocysts in *Paramecium caudatum. J. Protozool.* **10**, 253–262.

331   Zweibaum, J. 1912. La conjugaison et la différenciation sexuelle chez les Infusoires. Les conditions nécessaires et suffisantes pour la conjugaison de *Paramecium caudatum. Arch. Protistenk.* **26**, 275–393.

VIII. ADDENDUM

(Additional references are indicated by the prefix Ad)

Many recent investigations have advanced our knowledge of both the morphology and the biology of *Paramecium*. At the same time new avenues of research have opened up in those areas that are now in the forefront of scientific interest.

Helmy Mohammed and Nashed Nawal (Ad 344) have described a new species, *Paramecium wichtermani*, which was found in fresh water ponds near Cairo, Egypt. It is characterized by: average dimensions: 240 × 38, a narrow and sharply pointed anterior part, one macronucleus, two large micronuclei of the vesicular type (diameter 5–7μ), and two contractile vacuoles.

This species is related to the "aurelia" group, which is borne out by a detailed comparison with the other species of the same group. Its dimensions are similar to those of *P. multimicronucleatum*, however, its body is more tapered and slimmer.

An important comparative morphological study by Didier (Ad 340) on the ultra-structure of the cortex in *Hymenostomatida peniculina* has also given insight into the various characteristic structures of *Paramecium*. These studies have established that:

(1) Only the kinetosomes of the right kinety of the peniculus have postciliary fibers, and that this peniculus is connected with a "complex" of microfibrillar lattices (superficial, medium, and deep) which are linked with the kinetosome.

(2) The "paroral kinety" is not a real kinety. It actually consists of two rows of kinetosomes. Their orientation indicates that the "paroral kinety" is in fact constituted of a sequence of dyads whose antero-posterior axis is more or less perpendicular to the general direction of the kinety. Consequently, the "paroral kinety" consists of a sequence of fragments of kineties, oriented perpendicular to the longitudinal axis of the cell.

Allen (Ad 333) has reported some interesting details of the morphology of the cortex. He established that the subpellicular alveoli are not completely separated, but are connected by pores. Furthermore, the cortex consists of two microfibrillar systems. One is made up from striated bands, composed of fibers, 60 Å in diameter, which connect the kinetosomes. The other system, generally called the infraciliary lattice, is an inner network formed by bundles of fibers, 30–40 Å in diameter, which do not seem to be linked to any cellular organelles. The author believes that these micro-fibrils, which persist after glycerization, might have a contractile function.

Concerning nuclear phenomena, there is a new and important contribution on *Paramecium putrinum* by Jankowski (Ad 346); this author has described, besides the amphimixis of the normal conjugation, many other nuclear evolutions with or without meiosis: automixis I (with second shift cariogamic phase) and automixis II (with loss of cariogamic phase), apomixis I (with block of chromosomal union) and apomixis II (with entire block of meiosis).

The problems related to the concept of mating type and to the various sexual reactions continue to be intensively studied. A few results are particularly noteworthy. Takahashi and Hiwatashi (Ad 356) were able to restore the sexual activity which was lost after repeated washing by adding the culture fluid from a clone in a stationary phase. Experiments by Takagi (Ad 355) confirmed the existence of a sequential expression of the sex traits during clonal development. Miwa and Hiwatashi (Ad 352) demonstrated that mitomycin shortens the period of immaturity. Comparative studies of the characteristics of clones belonging to different varieties, mating types, or conjugating and non-conjugating animals have been carried out, utilizing biometrical analysis (Ad 343) and cytophotometry (Ad 341). One notable observation showed that senescent stocks contained less macronuclear DNA than sexually active clones of the same mating type.

Among the many other results obtained in various areas such as ecology, adaptation to different conditions, endosymbionts and others the development of two avenues of research may be especially useful, e.g. the use of *Paramecium* as material for biochemical and genetic studies. Although *Paramecium* is not as important as *Tetrahymena* in biochemical research, it is frequently used for biochemical studies. In the area of genetics, and particularly in the field of non-mendelian genetics (cytoplasmic heridity) *Paramecium* is becoming a research organism of prime importance. Already well known for the existence of the killer trait which was remarkably analyzed in *P. aurelia*, the species is now being studied by Adoutte and Beisson (Ad 332) with regard to the resistance of certain mutants to various antibiotics (erythromycin and chloramphenicol). Other cytological observations and genetic analyses show the presence of mitochondrial mutations. These studies in the area of cellular biology deal with the mechanism of information and transcription at the level of nucleic acids, thus emphasizing the importance of *Paramecium* in this area of research.

REFERENCES

332 Adoutte, A. & Beisson, J. 1970. Cytoplasmic inheritance of erythromycin resistant mutations in *Paramecium aurelia. Mol. Gen. Genet.* **108**, 70–77.

333 Allen, R. D. 1971. Fine structure of membranous and microfibrillar systems in the cortex of *Paramecium caudatum. J. Cell Biol.* **49**, 1–20.

334 Allen, S. L. & Gibson, I. 1971. The purification of DNA from the genomes of *Paramecium aurelia* and *Tetrahymena pyriformis. J. Protozool.* **18**, 518–525.

335 Andrivon, C. 1970. Preuves de l'existence d'un transport actif de l'ion nickel à travers la membrane cellulaire de *Paramecium caudatum. Protistologica* **6**, 445–455.

336 Andrivon, C. 1970. Complément à l'étude de l'action des inhibiteurs du métabolisme sur la résistance aux sels de nickel chez *Paramecium caudatum* et chez quelques autres Protozoaires Ciliés. *Protistologica* **6**, 199–206.

337 Canella, M. F. 1972. Sur les organelles ciliaires de l'appareil buccal des Hyménostomes et autres ciliés. *Ann. dell' Univers. di Ferrara (Nuova Series), Sezione III, Biol. anim.* Suppl. 1–235.

338 Capdeville, Y. 1971. Allelic modulation in *Paramecium aurelia* heterozygotes. Study of G serotypes in syngen 1. *Mol. Gen. Genet.* **112**, 306–316.

339 Cummings, D. J. 1972. Isolation and partial characterization of macro- and micronuclei from *Paramecium aurelia. J. Cell Biol.* **53**, 110–115.

340 Didier, P. 1970. Contribution à l'étude comparée des ultrastructures corticales et buccales des Ciliés Hyménostomes péniculiens. *Ann. St. biol. Besse en Chadesse (Fr.)* **5**, 1–274.

341 Dupy-Blanc, J. 1969. Etude par cytophotométrie des teneurs en ADN nucléaire chez trois espèces de Paramécies, chez différentes variétés d'une même espèce et chez différents types sexuels d'une même variété. *Protistologica* **5**, 297–308.

342 Génermont, J. 1969. Quelques caractéristiques des populations de *Paramecium aurelia* adaptées au chlorure de calcium. *Protistologica* **5**, 101–108.

343 — & Dupy-Blanc, J. 1971. Recherches biométriques sur la sexualité des Paramécies *(P. aurelia et P. caudatum). Protistologica* 7, 197–212.

344 Helmy Mohammed, A. H. & Nashed Nawal, N. 1968–1969. *Paramecium wichtermani* n. sp., with notes on other species of *Paramecium* common in fresh-water bodies in the area of Cairo and its environs. *Zool. Soc. Egypt* **22**, 89–104.

345 Inaba, F. & Kudo, N. 1972. Electron microscopy of the nuclear events during binary fission in *Paramecium multimicronucleatum. J. Protozool.* **19**, 57–63.

346 Jankowski, A. W. 1972. Cytogenetics of *Paramecium putrinum* c and L. 1858. *Acta Protozool.* **10**(17), 289–394.

347 Jenkins, R. A. 1970. The fine structure of a nuclear envelope associated with an endosymbiont of *Paramecium. J. Gen. Microbiol.* **61**, 355–359.

348 Kurashvili, B. E., Kurashvili, T. B. & Gogebashvili, I. V. 1971. Study of ultrastructure of *Paramecium caudatum* (in Georgian). *Soobshch. Akad. Nauk. Gruz., U.S.S.R.* **62**, 460–464.

349 Kuźnicki, L., Jahn, T. L. & Fonseca, J. R. 1970. Helical nature of the ciliary beat of *Paramecium multimicronucleatum. J. Protozool.* **17**, 16–24.

350 Miwa, I. & Hiwatashi, K. 1970. Effect of mitomycin c on the expression of mating ability in *Paramecium caudatum. Japan J. Genet.* **45**, 269–275.

351 Selman, G. G. & Jurand, A. 1970. Trichocyst development during the fission cycle of *Paramecium. J. Gen. Microbiol.* **60**, 365–372.

352 Simon, E. M. 1971. *Paramecium aurelia:* recovery from —196 °C. *Cytobiology* **8**, 361–365.

353 Stevenson, I. 1970. Endosymbiosis in some stocks of *Paramecium aurelia* collected in Australia. *Cytobios* **7–8**, 207–224.

354 Takagi, Y. 1971. Sequential expression of sex-traits in the clonal development of *Paramecium multimicronucleatum. Japan J. Genet.* **46**, 83–91.

355 Takahashi, M. & Hiwatashi, K. 1970. Disappearance of mating reactivity in *Paramecium caudatum* upon repeated washing. *J. Protozool.* **17**, 667–670.

356 Tawada, K. & Oosawa, F. 1972. Responses of *Paramecium* to temperature change. *J. Protozool.* **19**, 53–57.

Mating Type Determination and Development in *Paramecium aurelia**

HENRY M. BUTZEL, JR.

Department of Biological Sciences, Union College, Schenectady, N.Y. 12308 (U.S.A.)

I. INTRODUCTION

The discovery of mating types in *Paramecium aurelia* by Sonneborn over 30 years ago (91) initiated modern genetic studies of the ciliated protozoa. The concept of mating type has been extended not only to other species of *Paramecium* (*P. bursaria, P. caudatum, P. multimicronucleatum* and *P. woodruffi*), but also to other genera of the ciliates such as *Tetrahymena, Euplotes* and *Uronychia*. The establishment of mating types has led to the concept of the syngen, or genetic isolate, within the species of *P. aurelia* as well as within the species of other genera of ciliates.

Despite the major importance of this discovery, little is as yet known of the basic physiology or biochemistry of mating type determination and development, and much remains to be accomplished. It is the purpose of this review to discuss recent findings in the light of the problems still to be solved. Although the review will be primarily concerned with *P. aurelia*, the close phylogenetic relationships and obvious physiological similarities of the various species of paramecia require that some data concerning their mating type systems be included. Many reviews dealing wholly or in part with this subject have been published (1, 6, 28, 29, 30, 52, 71, 75, 76, 80, 81, 86, 92, 93, 96, 97, 99); therefore this chapter will deal mainly with more recent research in this field.

Paramecium aurelia, as well as the other named species of *Paramecium*, is not a single genetic species but is made up of a series of genetic isolates earlier called varieties, but now termed syngens. Each syngen within the array is designated by an Arabic number, and is in turn composed of two complementary mating types designated by Roman numerals. Thus, syngen 1 contains mating types I and II, syngen 2 contains types III and IV, etc., through syngen 13 with its mating types XXV and XXVI. Syngens are also found in other species of *Paramecium:* those found in *P. bursaria* (12, 57, 85–89), *P. caudatum* (27, 33–44), and in *P. multimicronucleatum* (3–5, 100, 102) have been particularly well delineated. In *P. bursaria* multiple mating types are found within a single syngen; the others appear to follow the *P. aurelia*

---

* This chapter is dedicated to Dr. Tracy M. Sonneborn, on the 35th anniversary of the discovery of mating types in *Paramecium*.

References p. 118–122

pattern. The multiple mating types within syngens of *Tetrahymena* (10, 72, 74, 77, 78) are especially well known.

The genetic isolation of the syngens is demonstrated by several means. The main evidence is the almost absolute failure of the F_1 of the possible intersyngenic crosses to give rise to a viable F_2 either by autogamy, backcrossing, or crossing of F_1 to F_1 (14, 96). Second, while mixtures of sexually reactive mating types within a syngen may yield up to 90% conjugants, matings between syngens never yield more than 40% conjugation, and frequently yield only a few or no pairs. Third, the F_1 hybrids between the syngens in *P. aurelia* may exhibit a number of cytological abnormalities, including loss of micronuclei, abnormal macronuclei, and changes in size and shape (14). In *P. bursaria* even the F_1 between syngens is unable to survive (12).

By the observation of possible mating reactions between syngens it has been shown that many of the designated odd mating types of *P. aurelia* are homologous as are the even mating types (Fig. 1). The assignment of odd or even types in syngens 2, 6, 9, 11 and 13 is arbitrary as no mating reactions between any mating types within these syngens and any mating types of another syngen has been found.

The 13 syngens of *P. aurelia* may be further assigned to one of three major groups. Group A consists of syngens 1, 3, 5, 9 and 11. Group B contains all of the remaining syngens with the exception of syngen 13 which is placed in a separate group C. The A group of syngens is distinguished by caryonidal and random determination of mating type following nuclear reorganization either at autogamy or at conjugation. Group B syngens are characterized by the apparent cytoplasmic and clonal inheritance of mating type; the mating type of the exautogamous or exconjugant clone usually

GROUP	SYNGEN	MATING TYPE	A 1 · I	1 · II	3 · V	3 · VI	5 · X	5 · XI	B 4 · VII	4 · VIII	7 · XIII	7 · XIV	8 · XV	8 · XVI	10 · XIX	10 · XX	12 · XXIII	12 · XXIV	TYPE
A	1	I	0	95	0	40	0	0	0	0	0	0	0	t	0	0	0	0	ODD
	1	II		0	1	0	40	0	0	0	10	0	t	0	0	0	0	0	EVEN
	3	V			0	95	0	0	0	0	0	0	0	40	0	0	0	0	ODD
	3	VI				0	0	0	0	0	t	0	0	0	0	0	0	0	EVEN
	5	X					0	95	0	0	0	0	0	0	0	0	0	0	ODD
	5	XI						0	0	0	t	0	0	0	0	0	0	0	EVEN
B	4	VII							0	95	0	0	0	95	0	0	0	0	ODD
	4	VIII								0	0	0	60	0	t	0	t	0	EVEN
	7	XIII									0	95	0	t	0	0	0	0	ODD
	7	XIV										0	0	0	0	0	0	0	EVEN
	8	XV											0	95	0	0	0	0	ODD
	8	XVI												0	t	0	t	0	EVEN
	10	XIX													0	95	0	0	ODD
	10	XX														0	t	0	EVEN
	12	XXIII															0	95	ODD
	12	XXIV																0	EVEN

Fig. 1. The known mating reactions between syngens of *P. aurelia*. Only those cases where reactions have been reported are included. [Based on Sonneborn (89) and Grell (29)]. Numbers refer to percentage conjugants. *t* refers to tentative reaction which fails to yield conjugants.

remains the same as that of the cytoplasmic parent in which the syncaryon nucleus was formed. The single syngen in group C is noted by its direct genetic control of mating type by a single pair of alleles (98).

Two main problems in mating type studies can be clearly distinguished. First is the way in which mating type is *determined*, i.e., the establishment of the potential mating type of a cell. Second, mating type *development* refers to the physiological events leading to the expression of the potential mating type. It should also be understood that, at present, the sole criterion for the mating type of any cell is its ability to show a mating reaction with one, or a series of other complementary mating types. The term mating type therefore refers to the specific capabilities of the ciliary surface in relationship to the different specific capabilities of other ciliary surfaces. The term mating type refers only to the ability of the cilia to adhere to one another and is not synonymous with conjugation or the events leading up to conjugation following initial ciliary contact.

In addition to the importance of mating type studies for our understanding of the ciliated protozoa, the underlying problem of determination and development is directly related to basic biological questions. The method by which cells, despite identical genotypes, may differentiate and reproduce true to different phenotypes is basic to developmental biology. The ciliates, particularly in those syngens showing only two mating types, may offer a simple "either–or" system which is less complex to analyze than the multiple potentials of fertilized metazoan eggs. The complexity of the ciliate cell is in some ways comparable to the complexity of an entire metazoan animal, and an insight into the ways by which this complexity develops may lead to a more general understanding of development and differentiation in other organisms.

II. GENETIC FACTORS

A number of genes affect the determination of mating type in *P. aurelia*. Their action is different in each of the three major groups.

A. *Mating Type Determination in Group A*

The most widely studied group in the earlier work was Group A, specifically syngen 1. Stocks of this syngen are of two kinds. The more common is known as a two-type stock. Cells homozygous or heterozygous for the dominant allele, $+mt^I$ are capable of expressing either odd or even mating types following nuclear reorganization. The second kind consists of those stocks restricted to the production of odd mating types due to the presence of the homozygous recessive allele, mt^I. Three independent, and apparently allelic mutations to the single type have been reported (14, 15, 96).

In addition at least one modifying gene, IN^I, which acts in the presence of $+mt^I$ to increase the frequency in determination of the odd mating type has been analyzed (14, 15). The possibility that there may be additional modifying alleles which influence the relative frequency of the determination of mating types in a two-type stock has not been ruled out (14).

B. Mating Type Determination in Group B

The B syngens are, with one exception, characterized by only cytoplasmic mating type inheritance. This exception occurs in stocks of syngen 7. Here, a brief historical review of the position of this syngen is valuable. Originally only one stock of this syngen, stock 38, was known, and it was pure for the odd mating type XIII. Intersyngenal matings with mating type II of stock 90 of syngen 1 yielded up to 20% conjugant pairs. The resulting hybrids were completely viable. On this basis, it was believed that syngen 7 belonged to the A group (14, 96). Presumptive evidence that the basis for the one-type stock of syngen 7 was the same as that in syngen 1 was found by crossing heterozygotes for the mt^I allele (Table 1). The $1:1$ ratio of one-type to two-type synclones found in the F_1 of this cross, along with the failure to find any odd mating types in the synclones derived from the cross of the homozygote of syngen 1 to stock 38, was suggestive of the ratio to be expected if the allele controlling the one-type stock in syngen 7 were similar to that of syngen 1. A cross of a heterozygote in syngen 7 ($+/mt^{XIII}$) to stock 38 (108) yielded a $14:11$ ratio of one-type to two-type offspring, indicating that a single allele in stock 38, mt^{XIII}, restricted that stock to the one-type condition. Thus a similar basis for one-type stocks is found in both syngens 1 and 7 and there is suggestive evidence that this condition may be allelic in the two syngens. However, it must be emphasized that no viable F_2 clones from the intersyngenic crosses were obtained.

TABLE 1

The results of crossing homozygous and heterozygous syngen 1 stocks to a stock of unknown genotype in syngen 7.

Syngen	1	7	1	7
Stocks crossed	P × 90 F-1	× 38	90	× 38
Genotypes	$+mt^I/mt^I$	?	$+mt^I/+mt^I$	?
No. of pairs with both exconjugant synclones showing odd mating type	52		0	
No. of pairs with both exconjugant synclones showing even mating type	55		30	

The later discovery of stocks showing the even mating type and the finding that the mating type was determined cytoplasmically, lead to a different understanding of the genetic basis of mating type in syngen 7 and to its reclassification from the A to the B group of syngens.

Two independent mutations have been studied, that of the original stock 38, and a second unlinked mutation leading to the same phenotypic expression, e.g., inability to produce mating type XIV (104–110, 112).

The original mutation in stock 38 has been designated as mt^{XIII}; homozygotes for this allele are restricted to type XIII. The dominant allele, $mt^{XIII-XIV}$, permits expression of either type XIII or type XIV. However, a further complication is found in the activity of this set of alleles. Although mt^{XIII} is recessive in its action on mating type determination, it is dominant for the ability to determine the cytoplasm towards mating type XIV. Cells homozygous for this allele have their mating type determined as XIII, but are cytoplasmically able to control the determination of mating type XIV when the proper allele is introduced (104–110). The cytoplasm of cells of this genotype is thereby essentially controlling the expression of a mating type which the nucleus cannot express. This finding of cytoplasmic control is shown not only within the syngen, but was observed in intersyngenal hybrids with syngen 1 when all of the hybrids were of the even mating type after the $+mt$ allele was introduced from syngen 1 (Table 1).

A second allele, n, is also recessive and does not permit determination of mating type XIV in the homozygote. Taub (108) has shown that this gene is not located in the same cistron as mt^{XIII} and apparently is not linked to it. In addition, n does not produce the cytoplasmic factor which governs production of the even mating type.

The mating type in this syngen seems to be determined by at least two steps leading to the production of type XIV. Cells must have the dominant allele for both the mt and the n locus present if type XIV is to be determined. Two types of cells pure for type XIII exist, those containing the homozygous recessive for the mt locus, but containing cytoplasm determining type XIV, and those homozygous for the n allele whose cytoplasm does not determine type XIV.

The contradictory information contained in the mt^{XIII} locus is difficult to understand. Cells are limited by their nucleus to type XIII, but the cytoplasm is directed towards type XIV, a condition which the genotype cannot express. In addition, the two activities of this locus differ in their dominance-recessivity relationships. Although there is no evidence for this as yet, it might be suggested that more than one locus may be involved, and that very close linkage has resulted in the failure to differentiate the two loci.

In addition, the earlier work implying possible allelism between the mt locus of syngens 1 and 7 may need reinterpretation. Here, also a contradiction seems obvious; the n locus of syngen 7 appears to act more nearly like the mt locus of syngen 1 than does the mt locus of syngen 7. However, the great excess of even type hybrid cells found would imply that the mt locus, rather than the n locus, was involved in crosses to syngen 1. Until techniques are found whereby viable F_2 generations can be obtained from the intersyngenal hybrids, this contradiction must remain.

C. Mating Type Determination in Group C

The simplest case of genetic control of mating type is found in the syngen of Group C. Here, control of the alternative two mating types is under the command of a single gene locus (98). In this case the allele controlling the even mating type determination in dominant over that of the odd mating type allele. A similar control by a single pair of alleles has been reported by Hiwatashi (39, 40) in syngen 3 of *P. caudatum*. However, homozygous dominants of *P. caudatum*, unlike those of *P. aurelia*, can express the odd mating type during part of their life cycle. In addition, expression of the dominant allele is variable, and it is suggested that another set of alleles may be influencing its expression. The similarity of these alleles to a gene as IN^1 in syngen 1 of *P. aurelia* (15) may be suggested. Earlier data for syngen 12 of *P. caudatum* have been reinterpreted on a similar basis by Hiwatashi (38).

Other genes involved in the expression, or development, of mating type will be discussed in the sections dealing with immature periods and circadian rhythms.

In summary, two types of genetic factors appear to be involved in mating type expression. First are those genes which control the ability of the cell to express a given mating type, the *mt* or *n* loci. Second are those genes which control the frequency of expression of alternative mating types when it is possible for both mating types to be expressed. IN^1 would appear to be an example of the second type.

III. FACTORS AFFECTING MATING TYPE CHANGES DURING VEGETATIVE REPRODUCTION

A. Naturally Occurring Factors

1. Maternal lag

Kimball (49) reported that in stock S of syngen 1 of *P. aurelia* a period of several fissions following autogamy (then described as endomixis) might occur before the definitive mating type of the caryonide was established. Such a lag occurred only in one direction; some cells arising from a type II cell at autogamy remained mating type II for a short time before changing to type I. Following the change from mating type II to mating type I the caryonide continued as type I until the following autogamy. Butzel (14) suggested that this might be related to the time to dilute out or to deactivate the products of a locus controlling the expression of type II which would still be present in the cytoplasm of the type II cells following autogamy. The failure to find cells which were type I for a short time after nuclear reorganization before changing to type II, might be interpreted as evidence for the immediate function of the type II-controlling locus after its activation following autogamy.

2. Changes within a Caryonide

a. Selfing. A second form of mating type change was found in caryonides of stock S. This change occurred during the vegetative cycle and unlike the maternal lag, was not related to autogamy. Analysis of split pairs of such incipient conjugants demonstrated that a true change of mating type in one of the two cells had indeed occurred. These vegetative changes were relatively uncommon, occurring in only

1–3% of the caryonides studied (50). A later interpretation of these data (14) suggested that these changes might be due to changes in the expression of a single locus which would affect the rate of expression of type II cells. A change from mating type I to mating type II would be due to the activation of such a locus; the opposite change could be accounted for by a repression of this locus. Thus the vegetative change might be similar to the determination of mating type after autogamy, except that normally such determination does not occur during the vegetative reproduction of the caryonide. Only in a few cases is redetermination possible after initial mating type has been established.

Macronuclear control of selfing has also been suggested for stock 50 of syngen 1. The relationship between sister subcaryonides was determined by isolating these after the second post-zygotic fission and comparing their mating types. A high degree of correlation between the sister subcaryonides was found for the occurrence of selfing among the subcaryonides. Of the 25 selfing subcaryonides found in this study, 20 had a sister subcaryonide which was a selfer. The data were interpreted to suggest that some macronuclei were "differentiated" so as to control selfing (84).

A few selfing caryonides were also observed in the hybrids between syngens 1 and 7. These changes were of interest since split pair analyses demonstrated that a true change of mating type had occurred. However, this change was only temporary, since both members of the split pair gave rise to the same mating type when allowed to grow into clones (14).

It should also be pointed out that these hybrids are the only paramecia which are permanently sexually reactive regardless of the fission or the feeding cycle. In addition the hybrids show other indications of genetic imbalance. Changes in the activation of a locus controlling mating type expression may reflect such abnormality.

Some selfing is due to a sequential change of mating type within the caryonide. Taub (110) reported a unidirectional change from mating type XIII to mating type XIV in syngen 7. The reverse change, from type XIV to type XIII was not observed to occur in the 145 cases tested.

A similar unidirectional change from mating type IX to X occurred in stocks of syngen 4 (7, 9). Indeed, in syngen 5 caryonidal instability is the general rule, and caryonides pure for one mating type are uncommon. Evidence that this is under genetic control has been reported (7, 9). The change from mating type IX to mating type X may occur within a single cell without the occurrence of fission (9).

b. Immature periods. As is the case with most of the other factors affecting mating type, different stocks and syngens show differences in the pattern of their life cycles. Immature periods are known for many syngens of different species. They appear to be of two basic kinds, depending upon the pattern of the mating type system.

The life cycle of *P. aurelia* consists of two stages, the immature and the adult stage. Immature periods following conjugation are common, although no such periods occur after autogamy or after macronuclear regeneration. These immature periods are characterized by an inability to develop mating type substances to a functional level. As a consequence the cells are unable to initiate mating reactions. When mating type substances are functional, the cell is considered to be adult.

Genetic factors may play a role in the regulation of the duration of the immature period (86). Crosses of different stocks with different immature periods revealed a "complex system", possibly polygenes, controlling the duration of this stage. Further investigation of the role and numbers of such genes would seem worthwhile to elucidate the exact genetic mechanisms that are involved here.

In addition, the total interval of immaturity between periods of conjugation seems relatively constant. Cells mated early in their adult period showed long immature periods after this conjugation prior to the expression of the next adult stage, while cells mated late in the adult period tended to show a shortened immature period (86).

It is not clear why the immature stage should be found only after conjugation and not after autogamy. The major difference between the processes is that of cross fertilization vs. self-fertilization. Similar genetic results would be expected in the offspring after conjugation of two cells from the same isogenic stock as after autogamy of a single cell.

The second difference is the fact that during autogamy no cell-to-cell contact occurs. During conjugation there may be a more extensive change in the surface of the cells due to their fusion and the resulting structural changes in outer cellular components. The membranes of both cells also must be penetrated by the partner's pronucleus. Therefore more extensive restructuring of the outer parts of the cell might be necessary after conjugation. Sexual reactivity may be somewhat restricted to the anterior ventral portions of the cell (43), the region which is most strongly affected during conjugation. Perhaps the effect of the contact between cells is a causative factor. This cannot be the exclusive factor, however. Exconjugant cells which undergo macronuclear regeneration during which a piece of the old macronucleus is restored to the functional control of the cell, apparently have a short, or even no period of immaturity (92). These results might suggest that the old macronuclear fragments were still able to continue to control the production of mating type substances, whereas the exconjugant would normally contain an "immature" macronucleus which was not yet able to control the synthesis of the mating type substances. It remains difficult to understand how a cell can distinguish between a new macronucleus formed after conjugation from one formed after autogamy unless there is additional feedback from the cytoplasm caused by the contact during conjugation. This feedback might not suppress the ability to express mating type in those cells where the macronucleus already has this potential, as would the nucleus developed during macronuclear regeneration, but might be able to exert an influence on a macronucleus which as yet had not activated the mating type system.

A second type of life cycle is exemplified by *P. bursaria*. Siegel (83) has shown that a third stage, termed "adolescent", may occur in those syngens which have multiple mating types. During this stage, the cell manifested some, but not all, of its permanent or adult mating types. According to Siegel, the adolescent stage demonstrated a sequential gene activity whereby the gene controlling one set of the mating type substances became active before the gene controlling the second set of substances could express itself. Studies of mutations as to a permanent adolescent stage suggested that the activation of the second set of alleles was controlled by the first set. A sequence of immaturity, with neither set active, followed by adolescence with one set of alleles not functional, occurred. Active function of this first set in turn allowed expression in part of the potential mating types, and also controlled the onset of the function of the second, non-allelic, set of genes controlling the development of the remaining mating types.

Hiwatashi (41) reported that *P. caudatum* exhibited another type of life cycle, during which an early, transient maturity was manifested. This was followed by an immature period, after which the adult characteristics were regained. The incorporation of a fragment of the old macronucleus within the developing anlagen was postulated in order to explain that the cells continued to show mating reactivity after conjugation. Successive fissions would dilute out the effects of this fragment, leading to the usual immature period.

c. The effect of varying amounts of feeding. Hiwatashi (38) studied unstable caryonides in syngen 12 of *P. caudatum*. In stock 401a the caryonides at first were pure for mating type XXIV. During the course of successive fissions the culture began to self. Selfing continued to increase until nearly 100% selfing was observed in the daily transfers. Conjugation gradually decreased and eventually ceased. Upon the cessation of conjugation, the culture was found to contain only type XXIII cells. If such cells were now removed from the mass culture and allowed to resume rapid growth, their mating type reverted again to type XXIV. Hiwatashi interpreted these changes as being due to feeding conditions; animals undergoing rapid fission tend to be mating type XXIV, while those growing more slowly become type XXIII. These results may be similar to those found by Bleyman (9) for *P. aurelia*, syngen 5, where it was suggested that those cells maturing soonest, and therefore completing their last fission before sexual reactivity first, were of type X, while the cells which underwent their last fission later were of type IX.

Hiwatashi (33) further suggested that changes in the relative amounts of substances covering the cilia were taking place during the mating type changes. Studies of the

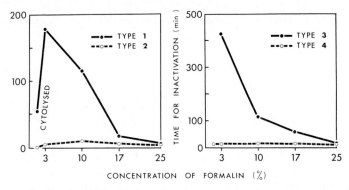

Fig. 2. Relations between concentration of formalin and time for inactivation of mating reactivity in complementary mating types of 2 syngens of *P. caudatum* (based on Hiwatashi, 33).

effects of the concentration of formalin upon retention of mating type ability by the killed cells showed a differential effect for different mating types in a syngen of *P. caudatum* (Fig. 2). The differential response is strongly indicative of a difference between the mating type substances found in the complementary reactive cells.

d. Circadian rhythms. Two distinct types of circadian ryhthms have been studied in the mating type systems of *Paramecium*. The first, reported in both *P. aurelia* and *P. bursaria*, deals with the cyclic ability to give a mating reaction. The second type, to date found only in *P. multimacronucleatum*, involves a cyclic shift of mating type expression within a single cell.

Ehret (24–26) showed that sexual reactivity in syngen 1 of *P. bursaria* normally occurs during the light period of the day. The time of reactivity, however, can be shifted readily by changing the period of illumination. Such new rhythmicity continued for at least 7 days in total darkness. Stocks deprived of their symbiotic green algae continued to show rhythmicity, proving that the algae were not acting as the photo-receptors. By studying the effects of various wave lengths of light upon the system, it was shown that the maximal effect for resetting the periodicity corresponded to the absorption range of flavins, suggesting that these compounds were the light receptors in paramecia.

Karakashian (46–48) studied a similar system in syngen 3 of *P. aurelia*. The main features were the same; the cycle continued for 4 days in the dark and could be shifted by an alteration of the light period. Complete loss of periodicity occurred when the 200 ft candles of illumination were applied continuously. No analysis of the effective wave length was attempted. It was also suggested that the diurnal mating reactivity long known for syngen 2 of this species may be a circadian pattern.

Reactive cells of syngen 3 continued to give a mating reaction for several days after becoming reactive without the addition of further food. Thus the loss and gain of reactivity cannot be attributed to changes in the feeding cycle. Both mating types V and VI showed the same patterns. Daily temperature cycles of 12 hour intervals at 17 °C (followed by 12 hours at 23 °C) under constant darkness or dim illumination also were found to be effective in resetting the periodicity. Variations in the effect of short periods of illumination were found to be dependent upon the time of day the illumi-nation was applied. Stock differences in response to the stimulus used for the setting of the periodicity suggested that genetic factors may be involved; however, no crossing analyses were carried out.

The second type of circadian rhythm seems more complex, for in this case the ability to mate did not cycle but the mating type itself did. Studies of syngen 2 of *P. multimicronucleatum* (95, 100, 102) showed that a circadian pattern exists for the alternative expression of mating types III and IV of this syngen. The pattern of ex-pression may be reset by varying photoperiods and will then continue for 3 cycles in total darkness. A novel finding of this work was the fact that a single dividing cell, which by itself cannot give a mating reaction, can be exposed to the *zeitgeber* and then give rise to a clone in which all cells show the response to entrainment. In this case, of course, the fission cycle is much shorter than the circadian cycle which persists in the

TABLE 4

Mating type of synclones arising at successive fissions from a cytoplasmically bridged pair (CBP).

	Synclone number						
	6	10	23	43	48	119	120
Mating type of synclone arising after 1st fission from CBP	VIII	S	S	VII	VIII	S	VIII
Mating type of synclone arising after 2nd fission from CBP	VII	S	VII	VII	S	S	VII
Mating type of synclone arising after 3rd fission from CBP	VII	VIII	D	VIII	—	—	—

S Selfing synclone.
D Synclone died before mating type could be tested.
— Not followed for this fission.

identical macronuclei in these cells. The data show that, contrary to expectation, a high frequency of selfing clones did arise from the two singlets given off from the CBP. A total of 32 of 109 clones followed for successive fissions showed opposite mating types in the two singlet clones arising from the original CBP. In addition, in some cases the second fission products are also of opposite mating type. A few cases occur where the mating type of the first fission clones was identical, but different from that of the second fission clones (Fig. 3, Table 4).

The formation of selfing synclones from the CBP may be due to several causes. First, it is possible that the macronuclear anlagen in two halves were not identically determined following conjugation so that the daughter cells received different types of macronuclei. In this case the cytoplasmic factor determining mating type would have acted only upon the macronuclear anlagen arising in the one half-cell rather than upon all the anlagen in the CBP. Second, it is possible that all macronuclei were identically determined, but that in some cases, one, or both, of the fission products arose by macronuclear regeneration and thereby retained their original mating types. This possibility is enhanced by the study of cells killed and stained prior to the first fission, and by time-lapse cinematographic studies.

Cytological observation of CBP cells prior to the first fission revealed that there are frequent abnormalities in nuclear behavior in such cells (Tables 5 and 6). Macronuclear distribution in these cells is not normal; only 1/3 of the CBP cells contained two anlagen on each side of the plane of fusion, indicating that anlagen had been shifted in these cells, or in some cases failed to form. Second, it can be seen that the number of macronuclear fragments in the CBP cells was strongly reduced so that the old macronucleus had not broken down completely as it would in a single cell.

TABLE 5

Distribution of anlagen within each half of a CBP after conjugation and prior to first fission. "Other" includes those with extra anlagen, or with less than 4 but at least 1 anlagen present. "No visible anlagen" implies that survival must have been by macronuclear regeneration.

Distribution of anlagen in the two halves of the CBP cells	No. observed	%
4:0	4	7.4
3:1	18	33.3
2:2	20	37.1
No visible anlagen	3	5.6
Other	9	16.6

TABLE 6

Distribution of macronuclear fragments in single and doublet cells following conjugation and before first fission.

No. of macronuclear fragments per cell*	No. of single cells	Cumulative (%)	No. of doublets	Cumulative (%)
0–2	0	0	0	0
3–4	0	0	1	2.2
5–6	0	0	4	11.1
7–8	0	0	3	17.8
9–10	0	0	3	24.4
11–12	0	0	3	31.1
13–14	0	0	5	42.2
15–16	1	4.0	12	68.8
17–18	2	12.0	6	82.2
19–20	5	32.0	5	93.3
21–22	2	40.0	2	97.7
23–24	3	52.0	1	100
25–26	6	76.0	0	
27–28	3	88.0	0	
29–30	3	100.0	0	

* Total number of fragments in doublets divided by 2 to give equivalent of single cells.

Fission from a cell with abnormally placed macronuclear anlagen might give rise to cells with extra or no macronuclei as the first fission usually results in a passive type. After a later fission, there must have been a redetermination of mating type within the CBP following first fission. Such redetermination is not unexpected since it is known that double cells are of the same mating type in all cases, and this may further indicate that determination of mating type within a doublet cell is such that only those cells with a single mating type are found.

Further confirmation of abnormal macronuclear assortment has been obtained from time-lapse cinematography of double cells prior to and after their first fission following conjugation. In addition the singlet cells from these fissions were recovered and the mating type of the clones arising from them was determined. In some, but not all, of the cases where mating type of the singlets had changed between the 1st and 2nd fissions, reduced numbers of macronuclear fragments and possible macronuclear regeneration was found, suggesting that the changes in the singlet synclones may be related directly to macronuclear regeneration in some instances.

The time-lapse studies also confirmed that macronuclear distribution was frequently abnormal. Only 8 of the 17 CBP's studied in this manner contained two anlagen in each half. Cytoplasmic fusion in these was seen to be complete and free flow of small endoplasmic particles was observed between the two halves. Thus it might be readily expected that if the cytoplasmic factor controlling mating type was readily diffusible, it could affect the developing anlagen on both sides of the CBP. However, in all cells studied there was little evidence of the exchange from side to side of particles as large as food vacuoles or macronuclear fragments. Such free flow was not observed until after the 2nd or 3rd fissions following conjugation. The cyclotic patterns were some-what separate, with each of the two gullets furnishing food vacuoles primarily to the half cell in which it was located. Where the two cyclotic patterns came into contact in the middle of the CBP a vortex-type pattern was seen with particles entrapped and rarely flowing from one pattern to the other. Thus, although a free flow of smaller particles could be seen, larger particles did not readily exchange sides, suggesting that abnormal distribution of macronuclei might be due to abnormal cytological events within the half-cell rather than by the flow of anlagen across the middle of the CBP.

A third hypothesis which cannot be eliminated by the data concerns the possible role of another cytoplasmic component, the basal bodies of the cilia. Dippell (23) has shown that the basal bodies are involved in the origin of new basal bodies. She pointed out: "Since the whole reaction system of the cell—swimming, mating, feeding, etc., involves cilia that develop from the basal bodies, position of the new by the old structures becomes a mechanism that doubtless would have positive selective advantage in evolution". Such a suggestion might offer another explanation for the origin of singlet cells of opposite mating types from each half of the cytoplasmically fused CBP. If the basal bodies of the original mating type could themselves carry some genetic information concerning the type of cilia to be produced, mating type might be maintained according to the original parental type through this mechanism. If this should be the case, a third component to the mating type system would have to be considered. The macronucleus would convey information to the cytoplasm concerning mating type determination, and this in turn might be further interpreted or modified by the basal bodies under certain conditions.

It is obvious that these hypotheses need further testing. Radioautographic studies would permit the following of macronuclear fragments and a determination of their role in these cases. Detailed cytological studies would also demonstrate whether distribution of macronuclear anlagen and macronuclear regeneration in themselves are sufficient to explain the origin of cells of opposite mating type from the common cytoplasm of the CBP.

References p. 118–122

C. Changes in Determination or Development induced by Physical or Chemical Agents

1. Temperature

In non-axenic cultures, the temperature at the time of the formation of the macro-nuclear anlagen has been found to affect the relative frequency with which the mating type is determined in many stocks of syngen 1 and in stock 51 of syngen 4 (73). The frequency of the even mating types increased in these stocks with increasing temperature. Such is not the case in axenic medium. Tables 7 and 8 show the data for the determination of mating type in stock 51 of syngen 4 and stock 90 of syngen 1 at two temperatures previously found to yield a different frequency of mating type. The data from syngen 4 are more conclusive; those from syngen 1 are of necessity derived from a smaller number of cells as survival following conjugation is usually low in this stock. Differential survival of one or the other mating type seems ruled out in both

TABLE 7

Mating type of exconjugant synclones and doublet cells of stock 51 from crosses made at 19 and 27 °C.

Experiment No.	Temperature									
	19 °C					27 °C				
	No. of synclones of mating type			No. of double clones of mating type		No. of synclones of mating type			No. of double clones of mating type	
	VII	selfer	VIII	VII	VIII	VII	selfer	VIII	VII	VIII
1	13	4	17	2	1	6	7	14	1	3
2	8	2	6	0	2	5	1	4	2	2
3	1	7	0	0	0	3	1	3	0	0
4	1	4	1	0	0	0	0	2	0	1
5	3	1	1	0	0	2	0	5	0	1
6	2	2	3	0	3	7	3	7	0	4
7	2	3	4	4	4	2	1	1	1	0
Totals	30	23	32	6	10	25	13	36	4	11
Per cent	35.3	27.1	37.6	37.5	62.5	33.8	17.6	48.6	26.7	73.3

TABLE 8

The effect of temperature upon mating type determination in stock 90, syngen 1.

Experiment No.	Temperature					
	20 °C			27 °C		
	No. surviving caryonides mating type		% survival	No. surviving caryonides mating type		% survival
	I	II		I	II	
1	3	13	33	6	17	41
2	13	37	26	8	12	11
Total	16	50		14	29	

sets of experiments since there was no correlation between survival rate and mating type frequency. It remains possible that the effect found in the non-axenic medium was due to effects upon the bacteria present in such systems. This could, if true, bring these data into the general category of feeding effects shown by Hiwatashi (41) for *P. caudatum* and in the axenic changes in stock 51.

2. Metabolic inhibitors

Butzel (14) reported that in bacterized medium the frequency of mating type 1 in stock 90 of syngen 1 rose in the presence of 2,4-dinitrophenol. Similar results were obtained with sodium azide (unpublished). No such effect was found for NaN_3 when axenically grown cells of syngen 4 were studied (Table 9). It is probable that these results were due to the action of these reagents upon the bacteria present, rather than on the paramecia themselves.

TABLE 9

Comparison of mating type from exconjugants grown with or without NaN_3 (16).

Number of survivors of each mating type found in

medium + NaN_3					medium only				
clones			double clones		clones			double clones	
VII	selfer	VIII	VII	VIII	VII	selfer	VIII	VII	VIII
8	0	10	1	3	17	4	16	1	3
44.4	00.0	55.6	25.0	75.0	45.9	10.8	43.3	25.0	75.0

3. Effect of the inhibitors of protein and RNA synthesis

The effect of these inhibitors upon circadian rhythms has already been mentioned. Other workers have found similar effects in bacterized systems.

Bleyman (8) found that puromycin and actinomycin D both inhibited reactivity when reactive cells of syngen 5 of *P. aurelia* were treated with concentrations of $50-200\mu g/ml$ for 1–2 hours. Ribonuclease had no effect in sublethal dosages but in high dosages gave a synergistic effect with actinomycin D. The data were interpreted to be indicative of the necessity for both RNA and protein synthesis for maintenance of mating reactivity.

Nobili (79) similarly studied 2 syngens of *P. aurelia*. Both were in the A group. The inhibitors used were actinomycin C, $0.05-1.6\mu g/ml$; *p*-fluorophenylalanine, $0.3-2.0-\mu g/ml$; 6-azauracil, $0.3-1.5mg/ml$. Each reduced clumping of reactive cells or totally inhibited the initiation of the mating reaction, dependent upon dosages. The effect was found only when totally non-reactive cells were treated; only a partial decrease of reactivity occurred when reactive cells were treated. It was suggested that there was *de novo* synthesis of RNA at the time sexually mature paramecia became reactive and that this required synthesis blocked by the inhibitors. Once reactivity began, synthesis of RNA ceased and the cells were no longer responsive to the inhibitors.

There seems to be a contradiction in these data derived in both cases from cells grown in bacterized media, one observer reporting loss of reactivity when reactive cells were treated and another reporting little loss. While the general evidence seems to point to some role of protein or RNA synthesis upon the development of mating type, confirmation must await the use of axenic systems, and data concerning the actual uptake of protein or nucleic acid precursors at these critical stages of the life cycle.

Kang and Taub (45) investigated the effects of the inhibition of both RNA and protein synthesis upon the mating type of live cells of syngen 7 of *P. aurelia* in a bacterized medium. They employed an unstable stock in which the caryonides change irreversibly from mating type XIII to type XIV during vegetative reproduction. Both actinomycin D and puromycin blocked the transformation in competent cells, suggesting that the "new m-RNA and protein synthesis" were both required for this transformation.

Longer treatment with actinomycin D resulted in the complete loss of mating activity in agreement with the data presented by Bleyman. It was suggested that there was a need for transcription of m-RNA for the development of the odd mating type also. Cells which had been treated with the inhibitors and had consequently lost all mating type ability were returned to normal bacterized culture fluid and their pattern of mating type reestablishment was followed. Such cells first regained mating type ability as cells of type XIII and then transformed to type XIV, suggesting a sequential change of mating type as found in *P. bursaria* (87). These authors also reported the ability of type XIV to transform back to type XIII in the presence of actinomycin D, a change never found spontaneously. It would appear that m-RNA is involved both in the development of the odd mating type and in the change to even mating type. Cells of type XIV treated with actinomycin D for a brief time reverted to type XIII: cells of type XIII were unable to continue to transform to type XIV. Longer treatment must have blocked the formation of m-RNA, required for the formation of any mating type substance and the cells lost reactivity.

Hiwatashi (40) used actinomycin and a variety of other compounds to test their effect upon mating type heterozygotes in syngen 3 of *P. caudatum*. Actinomycin D blocked the expression of the dominant gene for mating type V, and the treated cells remained mating type VI. Streptomycin, kanamycin, guanidine hydrochloride nitroso-guanine and acroflavin showed similar but less effective action, while puromycin was ineffective, possibly because it did not enter the cells. Hiwatashi concludes from these data that translation and transcription play a role in this syngen determining the expression of mating type alleles. Although the direction of this effect seems opposite to that found for *P. aurelia* by Kang and Taub (45), the similarity of blocking expression of one allele seems striking. Perhaps in *P. caudatum* the numbering system may be reversed from that of *P. aurelia* so that the odd and even mating types are reversed, or it remains possible that there is a sequential expression of mating type opposite to that of *P. aurelia*.

The experiments by Kang and Taub are critical for one of the theories of mating type development to be discussed later. It is hoped that they will be repeated in axenic medium to verify the direct effect of the antibiotic upon the paramecia and to eliminate any possible source of error due to the presence of bacteria.

4. *Miscellaneous effects*

Rudyansky (82) reported an effect of the homogenates of one mating type upon the development of mating type in syngen 2 of *P. aurelia*. The presence of the homogenates caused changes in the relative frequency of the two mating types following nuclear reorganization but the system was a crude one and no attempt to analyze the causative factors was made.

In addition, Bomford (13) has suggested that infective genetic units could change determination in *P. bursaria* following abortive conjugation. No nuclear reorganizations occurred during the change of phenotype, although micronuclear swelling indicated that "great physiological disturbances occur within the conjugating paramecia". No identification of the genetic unit was possible and further research should be undertaken in an attempt to isolate this factor.

D. *Chemical Nature of Mating Type Substances*

Although most workers today would agree with the hypothesis that proteins are involved in the mating type substances either as the substances themselves or as the matrix to which the mating type substances are bound, it should be emphasized that no isolation of any mating type substances has yet been successfully carried out. Indeed, little has been added to the knowledge of these substances since the work of Metz and his co-workers (58–64). The principal method has been either to attempt to preserve mating activity by killing cells with various agents or to study the effects of these agents upon the retention or loss of mating activity of the cilia of the dead cell. Metz (60) demonstrated that, of a series of enzymes employed, only proteolytic enzymes successfully destroyed mating activity. This had been confirmed by Cohen and Siegel (22) using homogenates of living cells instead of formalin-killed cells.

Metz (60) further ruled out the importance of S–S bonds by showing that formalin-killed cells treated with Hg-salts maintained the ability to give mating reaction. However, as these salts are lethal, the treated cells must have been washed prior to their being tested with live reactive cells. The possibility remains that some S–S linkages could be restored, because Hg reactions with sulfhydryl bonds are known to be reversible.

Additional evidence for the nature of the mating type substance was obtained by Klayman (55) using axenic systems. Paramecia were deciliated and preparations of cilia, "relatively free" of other cell components were obtained by two means. Of particular interest was the finding that when 15%(v/v) dilute cold ethanol was employed in deciliation, the cilia preparations lost all mating reactivity. When 12.5%(v/v) glycerol was used, ciliary mating reactivity was retained. Ethanol is known to cause denaturation of proteins, while glycerol is noted for its ability to stabilize proteins, again suggesting that protein may be involved directly in the mating type substances.

No qualitative difference between cilia of opposite mating type, or indeed from different syngens, was found when two-dimensional chromatograms of hydrolized cilia were prepared (18). Extensive comparisons were made between preparations from cilia, whole cell homogenates and culture media. Both whole cells and cilia from mating types of stock 90 of syngen 1, stock P of syngen 1, and stock 51 of syngen 4 were studied. Identical amino acid components were found in each, indicating no qualitative difference between animals, cilia, and culture media. Failure to utilize axenic systems may account for the similarity of all chromatograms although chromatograms of hydrolyzed bacteria at the same concentration as present in the preparations of paramecia did not contain sufficient protein to permit a positive identification of any amino acids. The failure to find differences may then indicate that qualitative amino acid differences in the cilia are not in themselves responsible for the different mating type substances. Further tests with purified reactive cilia should be carried out to determine whether these data are valid for axenic systems, and whether quantitative differences may yet be present in the various cilia.

A number of experimental approaches employed the study of differential loss of reactivity by different mating types following formalin treatment. While it is clear that in *P. caudatum* (33) differential effects of various concentrations of formalin are found (Fig. 2), the significance of these findings is not completely clear. An hypothesis of differential amounts of mating type substances upon the cell surfaces has been derived to account for the results (41).

No antigenic activity of mating type substances has as yet been reported, although a non-mating type specific inhibition by antiserum has been found (64). Since either whole cilia, or usually homogenates from whole cells, grown in bacterized media, were used as the antigen it is not surprising that mating type specific antisera are not available. Even in reactive cells, the mating substances must constitute only a small portion of the ciliary protein, and the action of this substance may be well masked by the overwhelming amount of non-mating type protein present. Again, the use of ciliary fractions from axenically grown cells might lead to the production of a specific antiserum and a proof of the protein nature of the mating type substances.

A reinvestigation of the role of antibodies in mating type reactions was carried out by Hiwatashi and Takahashi (44). A single serum, obtained from the injection of freeze-thawed cells, was able to block mating reactions nonspecifically. Its specificity did not depend upon the serotype or the mating type of the test cell. Indeed, this serum reacted with both *P. caudatum* and *P. multimicronucleatum*, inhibiting mating reactions in both species. Heat denaturation of the serum rendered it ineffective, suggesting that a protein is involved. It was suggested that this serum contained antibodies against a non-specific soluble co-factor necessary for mating. Only living cells are capable of secreting this co-factor, but attempts to isolate the factor for mating reactions from living cells and tests for its activity on formalin-killed cells have been unsuccessful, suggesting that this co-factor may be very unstable.

Only one stock of *P. caudatum*, and only one condition, i.e., freeze-thawing of the homogenate of cells prior to their being injected into a rabbit, produced this effect. Perhaps the freeze-thaw treatment is essential to break up the cells in such a manner as to free an active antigen and thereby to allow the production of the antibody against the soluble co-factor.

The possibility that compounds other than proteins may be found in the mating substances of *P. bursaria* has been suggested by Hayashi (31). RNA was revealed in the ectoplasm of cells of this species by staining with toluidine blue or Victoria blue. Formalin-killed reactive cells after treatment with RNAse had simultaneously lost the mating reactivity and the ability to react with the stains suggesting that RNA was involved. Hayashi himself urges caution, however, suggesting that the RNAse used may not have been completely specific. In the light of his caution and of the reported evidence that these stains color other structures such as elastic fibers (56), chondro-mucoproteins (2) or the "viscous thin gel amorphous ground substances of connective tissues" (11), it may be best to suggest that these data are possibly indicative, but not conclusive, evidence for compounds other than protein being involved in mating type substances.

Tartar and Chen (103) produced amacronucleate cells of *P. bursaria* by microsurgery. Such cells continued to demonstrate mating reactions for a short time. Although this may suggest that continued RNA and protein synthesis are not required for the maintenance of mating reactivity, a different explanation can be found based upon the data of Kimball and Prescott (54). The workers used stocks homozygous for the *am* gene (94) in whose presence fission may be abnormal, giving rise to cells lacking macronuclei. Amacronucleate cells continued to show protein synthesis at a relatively high rate, 30% of the control values. Surprisingly, RNA synthesis as measured by the uptake of radioactive nucleotides also continued, although at a lower rate than protein synthesis, 2% of the control values. The continued formation of RNA was interpreted by them as being possibly due to micronuclear synthesis or possibly to undetected endosymbionts. The data would seem to rule out the role of bacteria in the non-axenic medium used, but verification in axenic medium would be of value.

Although Kimball and Prescott (54) did not report upon the ability of the *am* cells to give mating reactions, it would seem logical that this could happen in the light of the results reported by Tartar and Chen (103). The maintenance of mating type reactivity may thus be related to the continuation of protein synthesis. It is also possible that *m*-RNA could continue to be present for a period of time and direct the synthesis of mating type proteins in the amacronucleate cell.

Hiwatashi (40) studied the effects of inhibition of protein and RNA synthesis in syngen 3 of *P. caudatum* in a non-axenic system. In this syngen, mating type is determined by a pair of alleles, with the allele for type VI dominant over that for type V, although some expression of type V can be found in the dominant homozygote. Treatment of type VI cells with actinomycin, RNAse, streptomycin, kanamycin, guanadine–HCl, nitrosoguanadine, and acriflavin resulted in a change from type VI to type V. The first two agents were more effective than the remaining compounds. Puromycin, in concentrations up to 200μg/ml had no effect. Treatment of type V with acromycin gave some temporary type VI cells. The change from type VI to type V was postulated to be caused by repression of the dominant allele by those agents known to interfere with the role of m-RNA. Failure to detect similar results with puromycin may have been due to the failure of this inhibitor of protein synthesis to enter the cell. No interpretation of the change from type V to type VI caused by acromycin was given. It should be pointed out that most of these agents could also have acted upon the bacteria present in the non-axenic medium. These potentially important findings should be verified in an axenic system.

Mention should also be made of the extensive work done upon the induction of conjugation by a variety of chemical agents (35, 42, 65–69). However, these studies were not directly related to mating type substances, but more likely to changes in the cortical structure allowing hold-fast unions which are formed after the mating type substances have drawn the paramecia close enough for this to occur. These data, while important in understanding the events in conjugation, do not seem to bear directly upon the problem of the nature of ciliary mating type substances.

Data from the ciliate *Euplotes* (32) may provide some indication of the role of one mating type in invoking reactivity in the cells of the opposite mating type. Here, contact with cells of the opposite types results in the development of reactivity within an hour. It has been postulated either that an exchange of some "foreign protein" results from an early contact, and that this stimulates the development of mating type substances to a level permitting cell union. Alternatively it is suggested that a low level of mating type substance is produced constitutively and the contact acts to permit the expression of mating type substances already present. Although this hypothesis was not favored by the authors, due to the 1-hour lag before expression, it might be reconsidered as a special case of immaturity with the need of cells of the opposite presumptive mating type needed to bring the affected cell to sexual maturity.

Mating type substances have been found to be remarkably resistant to irradiation. Wichterman (112) reported that dosages of X-irradiation up to 200,000r were insufficient to cause immediate loss of mating type reactivity of sexually reactive cells of *P. bursaria*. Although this work was carried out in a complex medium so that some of the irradiation must have been absorbed by the medium, it is clear that a dosage of this amount is surprisingly high not to cause loss of reactivity. The relatively high ploidy of the macronucleus (19, 51, 53) may have protected the cell from death and also may have allowed the continued control of protein synthesis by at least a few loci so that mating reactivity was unimpaired.

IV. DISCUSSION

Although most of the studies of mating type in paramecia have been conducted using bacterized media, certain genetic facts have emerged clearly, and a generalized hypothesis of mating type determination and development may be formulated. This hypothesis differs little from that first put forth by Butzel (15) but must be broadened to include more recent data.

The basic hypothesis, dealing with both determination and development, is diagrammed in Fig. 4. Mating type is determined by the presence or absence of specific genes. Cells homozygous for the recessive *mt* locus lack the necessary gene to produce

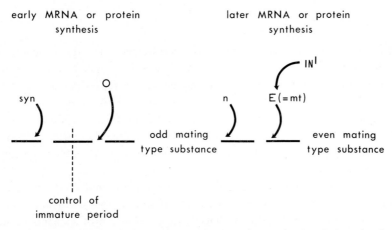

Fig. 4. Hypothetical scheme for mating type determination. See text for description of various genes.
Modified after Butzel (15).

even mating type, and are determined as the odd mating types. Thus there is first a simple either-or determination, and all else may be considered as subsequent steps in the determination of mating type.

The original hypothesis called for a number of genes leading to the development of the odd mating type substance, and a single locus involved in the determination of the even mating types. Taub's discovery of two non-allelic one-type stocks (108) indicates that the difference between the ability to produce two-type stocks and one-type stocks is dependent both upon the *n* and the $+^{mt}$ allele.

A second major feature of the hypothesis was the suggestion that the mating type substances are related in a sequential manner and the odd mating type substance was the basic substance and the activity of the *n* or *mt* locus changed the odd substance into the even mating type substance. None of the data presented here contradict this hypothesis and some seem to support it strongly. Of particular interest is the finding by Taub (110) that blocking of RNA synthesis prevented formation of even mating type substances even though the cell could express the odd mating type. In addition, cells producing the even mating type substance could be converted to cells producing odd mating type substance by interference with RNA synthesis. Thus the synthetic step between odd and even mating type substances seems to be one which requires RNA and protein synthesis, possibly the synthesis of an enzyme converting one mating type substance to another. Finally, it has been shown that strong doses of actinomycin may cause complete loss of mating type ability. When mating type substances reappear following cessation of treatment, the odd substance is formed first and the even type can only be formed at a later time, again implying a sequential relationship.

Another feature of the hypothesis explains whether two-type stocks express the even mating type or the odd mating type; the two-type stocks producing only the even mating type would have an active *mt* locus, while those producing only the odd mating type would have the *mt* locus in a repressed condition. Activation of this key locus could be controlled by genetic factors (IN^I), cytoplasmic control (the B system of syngens), or environmental factors.

The two types of circadian rhythm in *P. aurelia* also may be related to such a hypothetical activation of the *mt* locus. In those cells which change mating type during their periodicity, this locus would change from a repressed to an active state. It might be hypothesized that those cells which show a rhythmic gain and loss of mating ability have the entire train of events leading up to the production of mating type under circadian control, prior to the formation of the odd mating type as shown in Fig. 4. The *C* gene for ability to cycle would then be controlling the initiation of the *O* series of genes.

Similarly, the sequential development of mating types found for *P. bursaria* could be interpreted on the basis of the precursor activation hypothesis with additional steps under the control of additional loci. Siegel's suggestion (87) of sequential control is basically in accord with the general nature of the hypothesis.

Selfing stocks would also fit the hypothesis of sequential development dependent upon the activation of the *mt* locus. Such stocks may differ only in the timing of activation from those showing circadian changes of mating type.

Implicit also in the development of mating type is some control of the specified syngen pattern of mating type expression. While the odd and even mating types are known to be homologous in many of the syngens, they are not exactly similar as is shown by the different rates of reactivity between syngens (Fig. 1). There must be additional genes as yet unanalyzed, to permit this. It has been suggested earlier that genes confirming the exact syngenal mating type substances be termed "*var*" genes (14); with the change in name of the varieties to syngens it may be better to refer to these as yet unclearly depicted genes as "*syn*" genes. Possible evidence for the existence of such genes may be derived from hybrids between syngens. These cells show two mating types, one derived from each parental syngen. Thus the hybrid between syngens 1 and 7 showed mating types II + XIV, reacting equally well with either type I from syngen 1 or type XIII from syngen 7 (13). Similar results were reported for other intersyngenal hybrids of the A group (92). Such results might be expected if the hybrids were interpreted as being heterozygous for a "*syn*" gene from each parent. The number and exact point of activity of such genes remains unknown; they could work either before or after the *mt* locus.

Ehret (24) has also suggested that there is a precursor to mating type substances which is formed under the control of light in the circadian system of *P. bursaria*. Thus he suggested that a substance P^I gave rise to P which in turn gave rise to mating substance, with the reaction between P^I and P being under the control of the *zeitgeber*. This might correspond to the first series of reactions which lead to the odd mating type substance as suggested for *P. aurelia* (14, 15).

Some features of the system still remain unclear. A major problem would seem to be the highly polyploid nature of the macronucleus (19, 51, 53). A change in mating type may require the simultaneous repression of derepression of a large number of duplicate loci. Detection of the rate of such change would be difficult; a single active *mt* locus might be sufficient to permit development or the even mating type substances and convert an odd to an even mating type. The continued activity of only one *mt* site might be sufficient to continue production of the even mating type even though all other *mt* loci were repressed. It is possible that the cell would manifest an even mating type until all *mt* loci were repressed. If this is the case, the change from an odd to an even mating type might be a more sudden event than the opposite change. No specific experiments designed to test this hypothesis have been performed, although the rate of change in the maternal lag system studied by Kimball (49) is in accord with this.

The loss of reactivity following initiation of conjugation has been reported to be sudden; indeed, it was suggested that this might indicate enzymatic control and degradation of the mating type substance. It would appear that a more general problem is met with here, namely the relationship of mating type substance production to cell metabolism. Cells usually only express mating type substances under strictly defined physiological conditions; they must be neither too starved nor too well-fed. While it has been suggested in the past that the mating type substances must be synthesized at that time, it is possible that other events not related to synthesis may occur. Mating type substances might be an inherent part of the ciliary protein. Changes in molecular structure of such substances could result in a change from the non-reactive to the reactive condition. Such changes need not always require energy; folding or other configurational changes of a protein molecule may occur with little energy required.

Hiwatashi (38) has also suggested that both mating type substances are present at the time of reactivity and that there are two proteins involved. Changes in the relative amounts and positions of these proteins would determine which mating type is expressed. Weiss (112) pointed out that slight configurational changes of membrane protein can have vast effects on the surface of cells. Realignment of protein molecules would in effect present an entirely different surface. A somewhat similar suggestion to explain selfing in a stock of *P. bursaria* has been made (57).

The gaining of sexual reactivity might be interpreted as the exposure of the necessary protein in the ciliary membrane; the rapid loss during conjugation as either the rearrangement of the ciliary protein or of configurational changes of the molecule, neither of which would, in themselves, require enzymatically controlled processes. Mating type reactivity would then be envisaged as the specific stereotypic configuration and exposure of a particular protein inherent in the ciliary membrane itself.

Immature periods may also be interpreted as being caused by the failure of the entire system leading to formation of mating type substances to begin to function until a late time in the life of the caryonide. This could be due to the failure to form the specific mating type substance, or, in view of the present discussion, failure for specific configuration to occur.

V. PROSPECTS

It would now appear that the way to the solution of the mating type problem may be more clear. The development of axenic media whereby biochemical studies become feasible is a major breakthrough for further studies of mating type. The first clear need is for identification of the mating type substances themselves. A combination of the use of purified preparations of cilia derived from reactive, axenically-grown cells with modern methods of protein analysis should be undertaken. Once the nature of the substances, and their interrelationships, is determined, the biochemical events leading to their formation can be more fully grasped. Information concerning amino acid sequences and the shapes of various protein molecules themselves seem to be possible goals for the future. If the mating type substances are, as suggested, an integral part of the ciliary membrane, fractionation of ciliary protein may be required.

Another door may be opened with the refinement of the serological methods of Hiwatashi and Takahashi (44) combined with the use of purified reactive cilia (55). Analysis of the antibody reactions may give information concerning the number and type of mating type proteins. The obvious advantage of the axenic system, both with its purity, and with the high population of more than 20,000 organisms/ml should yield sufficient material for such analyses.

To date, most of the chemical studies have been carried out with factors which cause a loss of reactivity. Another approach might be to attempt to treat non-reactive pure cilia with various agents known to change protein structure, in an attempt to induce mating reactivity. If such results could be obtained, information regarding the structural relationships of the ciliary membrane protein to mating type substance protein would be possible.

Other useful information such as the chemical basis for circadian rhythms should be sought using axenic media. If, as has been suggested by Ehret (24) that flavins are generally involved, it might be possible to begin a study of the energetics of mating type development and the relationship of these compounds to the system.

Many stocks of paramecia now exist in axenic medium, and others are being introduced. The entire life cycles of both members of both the A and B group of syngens can be studied and it should be possible to look for biochemical changes in the mating type systems of either group. Genetic differences may be isolated and their biochemical expression analyzed. In this way much that is speculation may become fact, and the various groups of *Paramecia* will gain their rightful place in the realm of biochemical genetics.

VI. ACKNOWLEDGEMENTS

The author wishes to acknowledge the technical assistance of Mrs. Elizabeth Howard for her help in many of the experiments reported here.

This work was partially supported by an anonymous grant to Union College for Genetics Research.

VII. REFERENCES
(Abstracts are indicated by *)
(The literature search upon which this review is based was completed in September, 1969)

1   Allen, S. L. 1967. Chemical genetics of Protozoa. In Florkin, M. & Scheer, B. T., *Chemical Zoology*, Academic Press, New York, N.Y., Ch. 13, 617–694.

2   Barka, T. & Anderson, P. J. 1965. *Histochemistry. Theory, Practice and Bibliography*, Hoeber Medical Division, Harper & Row, New York, N.Y.

3   Barnett, A. 1961. The inheritance of mating type and cycling in *Paramecium multimicronucleatum*, syngen 2. *Amer. Zoologist* **1**, 341–342.

4   — 1965. A circadian rhythm of mating type reversal in *Paramecium multimicronucleatum*. In Aschoff, J., *Circadian Clocks*. North-Holland Publishing Comp., Amsterdam, 305–308.

5   — 1966. A circadian rhythm of mating type reversal in *Paramecium multimicronucleatum* syngen 2 and its genetic control. *J. Cell Physiol.* **67**, 239–270.

6   Beale, G. H. 1954. *The Genetics of Paramecium aurelia*. Cambridge University Press, New York, N.Y.

7   *Bleyman, L. 1963. Selfing studies in stock 210, syngen 5, *Paramecium aurelia*. *J. Protozool.* **10** (Suppl.), 21–22.

8   *— 1964. The inhibition of mating reactivity in *Paramecium aurelia* by inhibitors of protein and RNA synthesis. *Genetics* **50**, 236.

9   — 1967. Selfing in *Paramecium aurelia* syngen 5: persistent instability of mating type expression. *J. Exp. Zool.* **165**, 139–146.

10 — & Simon, E. M. 1968. Clonal analysis of nuclear differentiation in *Tetrahymena*. *Develop. Biol.* **18**, 217–231.

11 Bloom, W. & Fawcett, D. W. 1968. *A Textbook of Histology*, Saunders, Philadelphia, Pa.

12 Bomford, R. 1966. The syngens of *Paramecium bursaria:* new mating types and intersyngenetic mating reactions. *J. Protozool.* **13**, 497–501.

13 — 1967. Stable changes of mating type after abortive conjugation in *Paramecium aurelia*. *Exp. Cell Res.* **47**, 30–41.

14 Butzel, Jr., H. M. 1953. *The Genic Basis of Mating Type Determination and Development in the Varieties of Paramecium aurelia belonging to Group A*, Thesis, Indiana University, Bloomington, Ind.

15 — 1955. Mating type mutations in variety 1 of *Paramecium aurelia* and their bearing upon the problem of mating type determination. *Genetics* **40**, 321–330.

16 *— 1967. Conjugation and mating type determination of *Paramecium aurelia* in axenic medium. *J. Protozool.* **14** (Suppl.), 19.

17 — 1968. Mating type determination in stock 51, syngen 4, of *Paramecium aurelia* grown in axenic culture. *J. Protozool.* **15**, 284–290.

18 — & Martin, W. B. 1955. Studies of amino acid constituents of *Paramecium aurelia*. *Genetics* **40**, 565.

19 Cheissin, E. M. & Ouchinnikova, L. 1964. A photometric study of DNA content in macronuclei and micronuclei of different species of *Paramecium*. *Acta Protozool. (Warsaw)* **9**, 225–236.

20 Cohen, L. W. 1964. Diurnal intercellular differentiation in *Paramecium bursaria*. *Exp. Cell Res.* **36**, 398–406.

21 — 1965. The basis for the circadian rhythm of mating in *Paramecium bursaria*. *Exp. Cell Res.* **37**, 360–367.

22 — & Siegel, R. W. 1963. The mating type substance of *Paramecium bursaria*. *Genet. Res.* **4**, 143–150.

23 Dippell, R. V. 1968. The development of basal bodies in *Paramecium*. *Proc. Nat. Acad. Sci. U.S.* **61**, 461–468.

24 Ehret, C. F. 1953. An analysis of the role of electromagnetic radiation in the mating reaction of *Paramecium bursaria*. *Physiol. Zool.* **26**, 274–300.

25 *— 1955. The effects of pre- and post-illumination on the scotophilic recovery phase of the *Paramecium bursaria* mating reaction. *Anat. Rec.* **122**, 456–457.

26 *— 1955. The photoreactivability of sexual activity and rhythmicity in *Paramecium bursaria*. *Radiation Res.* **3**.

27 *Gilman, L. C. 1954. Distribution of the varieties of *Paramecium caudatum*. *J. Protozool.* **3**, (Suppl.), 4.

28 Grell, K. G. 1962. Morphologie und Fortpflanzung der Protozoen. *Fortschr. der Zool.* **14**, 1–85.

29 — 1967. Sexual reproduction in the protozoa. In Chen, T. T., *Research in Protozoology*, Pergamon Press, Oxford, V 2, 147–213.

30 — 1968. *Protozoologie*, 2nd edition, Springer-Verlag, Berlin.

31 Hayashi, S. 1959. On the relationship between the induction of pseudo-selfing pairing and RNA contents in *Paramecium bursaria*. *J. Fac. Sci. Hokkaido Univ., Ser. 6.* **14**, 129–133.

32   Heckman, K. & Siegel, R. W. 1964. Evidence for the induction of mating type substance by cell to cell contact. *Exp. Cell Res.* **36**, 688–691.

33   Hiwatashi, K. 1950. Studies on the conjugation of *Paramecium caudatum*. III. Some properties of the mating type substance. *Sci. Rep. Tôhoku Univ. Biol., Ser. 4.* **18**, 270–275.

34   — 1955. Studies on the nature of *Paramecium caudatum*. VI. The nature of the union of conjugation. *Sci. Rep. Tôhoku Univ. Biol., Ser. 4.* **21**, 207–218.

35   — 1959. Induction of conjugation by ethylenediamine tetraacetic acid (EDTA) in *Paramecium caudatum*. *Sci. Rep. Tôhoku Univ. Biol., Ser. 4.* **25**, 81–90.

36   *— 1960. Locality of mating reactivity on the surface of *Paramecium*. *J. Protozool.* **7** (Suppl.), 20.

37   *— 1960. An aberrant selfing strain of *Paramecium caudatum* which shows multiple unions of conjugation. *J. Protozool.* **7** (Suppl.), 20.

38   — 1960. Inheritance of differences in the life functions of *Paramecium caudatum* syngen 12. *Bull. Marine Biol. Sta. Assamushi Tôhoku Univ.* **10**, 157–159.

39   — 1968. Determination and inheritance of mating type in *Paramecium caudatum*. *Genetics* **58**, 373–386.

40   — 1968. Genetic and epigenetic control of mating types in *Paramecium caudatum*. *Proc. 12th Int. Congr. Genetics* 259–260.

41   — & Kasoga, T. 1960. Analysis of the change of mating type during vegetative reproduction in *Paramecium caudatum*. *Japan J. Genetics* **35**, 213–221.

42   *— & — 1960. Artificial induction of conjugation by manganese ion in *Paramecium caudatum*. *J. Protozool.* **7** (Suppl.), 20–21.

43   — & — 1961. Locality of mating reactivity on the surface of *Paramecium caudatum*. *Sci. Rep. Tôhoku Univ. Biol., Ser. 4.* **27**, 93–99.

44   — & Takahashi, M. 1967. Inhibition of mating reactions by antisera without ciliary immobilization in *Paramecium*. *Sci. Rep. Tôhoku Univ. Biol., Ser. 4.* **33**, 281–290.

45   *Kang, H. S. & Taub, S. R. 1968. Studies of the effects of actinomycin D and puromycin in cells of syngen 7, *Paramecium aurelia*. *J. Cell Biol.* **39**, 70a–71a.

46   Karakashian, M. W. 1961. *The Rhythm of Mating in Paramecium aurelia and its Significance for Genetic Analysis of Circadian Rhythm*, Dissertation, Univ. of California, Los Angeles, Calif.

47   — 1965. The circadian rhythm of sexual reactivity in *Paramecium aurelia*. In Aschoff, J., *Circadian Clocks*, North-Holland Publishing Comp., Amsterdam, 301–304.

48   — 1968. The rhythm of mating in *Paramecium aurelia* syngen 3. *J. Cell. Physiol.* **71**, 197–209.

49   Kimball, R. F. 1939. A delayed change of phenotype following a change of genotype in *Paramecium aurelia*. *Genetics* **24**, 49–58.

50   — 1939. Change of mating type during vegetative reproduction in *Paramecium aurelia*. *J. Exp. Zool.* **81**, 165–179.

51   — 1953. The structure of the macronucleus of *Paramecium aurelia*. *Proc. Nat. Acad. Sci. U.S.* **39**, 345–347.

52   — 1964. Physiological genetics of the ciliates. In Hutner, S. H., *Biochemistry and Physiology of Protozoa*, Academic Press, New York, N.Y., Ch. 3, 243–275.

53   — & Barka, R. 1959. Quantitative cytochemical studies on *Paramecium aurelia*. II. Feulgen microspectrophotometry of the macronucleus during exponential growth. *Exp. Cell Res.* **17**, 173–182.

54   — & Prescott, D. M. 1964. RNA and protein synthesis in amacronucleate *Paramecium aurelia*. *J. Cell Biol.* **21**, 496–497.

55   *Klayman, M. B. 1969. Clumping by isolated cilia of complementary mating types from axenically-grown *Paramecium aurelia*. *Yale Sci. Mag.* **43**, 32.

56   Langeron, M. 1949. *Précis de Microscopie*, Masson, Paris.

57   Larison, L. L. & Siegel, R. W. 1961. Illegitimate mating in *Paramecium bursaria* and the basis for cell union. *J. Gen. Microbiol.* **26**, 499–508.

58   *Metz, C. B. 1946. Effect of various agents on the mating type substance of *Paramecium aurelia*. *Anat. Rec.* **94**, 347.

59   — 1949. The nature and mode of action of the mating type substances. *Amer. Naturalist* **82**, 85–95.

60   *— 1953. Mating substance and the physiology of fertilization in Ciliates. In Wenrich, D. H., *Sex in Micro-organisms* (Symposium of the American Association for the Advancement of Science, Washington, D.C.), 284–334.

61   — & Butterfield, W. 1950. Extraction of a mating reaction inhibiting agent from *Paramecium calkinsi*. *Proc. Nat. Acad. Sci. U.S.* **36**, 268–271.

62   — & Foley, M. T. 1949. Fertilization studies on *Paramecium aurelia*: an experimental analysis of a non-conjugating stock. *J. Exp. Zool.* **112**, 505–528.

63   *— & Fusco, E. M. 1949. Mating reactions between living and lyophilized paramecia of opposite mating type. *Biol. Bull.* **97**, 245.

64   — & Fusco, E. M. 1948. Inhibition of the mating reaction in *Paramecium aurelia* with antiserum. *Anat. Record* **101**, 654–655.

65   Miyake, A. 1958. Induction of conjugation by chemical agents in *Paramecium caudatum*. *J. Inst. Polytech. Osaka City Univ.* **9**, 251–296.

66   — 1957. Aberrant conjugation induced by chemical agents in amicronucleate *Paramecium caudatum*. *J. Inst. Polytech. Osaka City Univ.* **8**, 1–10.

67   — 1959. Chemically induced mating without mating type differences in *Paramecium caudatum*. *Science* **130**, 1423.

68   *— 1960. Artificial induction of conjugation by chemical agents in *Paramecium aurelia, Paramecium multimicronucleatum, Paramecium caudatum*, and between them. *J. Protozool.* **7**, (Suppl.), 15.

69   *— 1961. Artificial induction of conjugation by chemical agents in *Paramecium* of the "aurelia group" and some of its applications to genetic work. *Amer. Zoologist* **1**, 373–374.

70   — 1968. Induction of conjugation by chemical agents in *Paramecium*. *J. Exp. Zool.* **167**, 359–380.

71   Nanney, D. L. 1954. Mating type determination in *Paramecium aurelia*, a study in cellular heredity. In Wenrich, D. H., *Sex in Micro-organisms* (Symposium of the American Association for the Advancement of Science, Washington, D.C.), 266–283.

72   — 1956. Caryonidal inheritance and nuclear differentiation. *Amer. Naturalist* **90**, 291–307.

73   — 1957. Mating type inheritance at conjugation in variety 4 of *Paramecium aurelia*. *J. Protozool.* **4**, 89–95.

74   — 1959. Genetic factors affecting mating type frequencies in variety 1 of *Tetrahymena pyriformis*. *Genetics* **44**, 1173–1184.

75   — 1964. Macronuclear differentiation and subnuclear assortment in ciliates. In Locke, M., *The Role of Chromosomes in Development*, Academic Press, New York, N.Y. 253–273.

76   — 1968. Ciliate genetics: patterns and programs of gene action. *Ann. Rev. Genetics* **2**, 121–140.

77   — & Caughey, P. A. 1953. Mating type determination in *Tetrahymena pyriformis*. *Proc. Nat. Acad. Sci. U.S.* **39**, 1057–1063.

78   —, — & Tefankjian, A. 1955. The genetic control of mating type potentialities in *Tetrahymena pyriformis*. *Genetics* **40**, 668–680.

79   *Nobili, R. 1963. Effects of antibiotics, base-and-aminoacid analogues on mating reactivity of *Paramecium aurelia*. *J. Protozool.* **10** (Suppl.), 24.

80   — 1965. La riproduzione sessuale nei ciliati. *Boll. Zool.* **32**, 93–131.

81   — 1957. Genetics of the Protozoa. In Locke, M. (Ed.), *Ann. Rev. Microbiol.* **11**, 419–438.

82   *Rudyansky, B. 1953. The effect of breis made from sexually reactive animals on mating type determination in *Paramecium aurelia*, variety 2. *Microbial Genet. Bull.* **7**.

83   Siegel, R. W. 1963. An analysis of the transformation from immaturity to maturity in *Paramecium aurelia*. *Genetics* **42**, 394–395.

84   *— 1962. A study of selfing caryonides in *Paramecium aurelia*. *J. Protozool.* **9** (Suppl.), 28.

85   — 1965. New results in the genetics of mating types of *Paramecium bursaria*. *Genetic Res.* **4**, 132–142.

86   — 1965. Heredity factors controlling development in *Paramecium*. In *Genetic Control of Differentiation (Brookhaven Symp. Biol.)*, **18**, 55–65.

87   — & Cohen, L. W. 1963. A temporal sequence for genic expression: cell differentiation in *Paramecium*. *Amer. Zoologist* **3**, 127–134.

88   — & Cole, J. 1967. The nature and origin of mutations which block a temporal sequence for genic expression in *Paramecium*. *Genetics* **55**, 607–617.

89   — & Larison, L. L. 1960. The genic control of mating type in *Paramecium bursaria*. *Proc. Nat. Acad. Sci. U.S.* **46**, 344–349.

90   Soldo, A. T., Godoy, G. A. & Van Wagtendonk, W. J. 1966. Growth of particle-bearing and particle-free *Paramecium aurelia* in axenic culture. *J. Protozool.* **13**, 492–497.

91   Sonneborn, T. M. 1938. Mating types in *Paramecium aurelia*: diverse conditions for mating in

different stocks: occurrence, number and interrelations of the types. *Proc. Amer. Phil. Soc.* **79**, 411–434.

92   — 1947. Recent advances in the genetics of *Paramecium* and *Euplotes*. *Advanc. Genetics* **1**, 264–358.

93   — 1949. Ciliated Protozoa: cytogenetics, genetics, and evolution. *Ann. Rev. Microbiol.* **3**, 55–80.

94   *— 1954. Gene-controlled, aberrant nuclear behavior in *Paramecium aurelia*. *Microbial Genet. Bull.* **11**, 24–25.

95   *— 1957. Diurnal changes of mating type in *Paramecium*. *Anat. Rec.* **128**, 626.

96   — 1957. Breeding systems, reproductive methods, and species problems in protozoa. In Mayr, E., *The Species Problem*, American Association for the Advancement of Science, Washington, D.C., 155–324.

97   — 1964. The differentiation of cells. *Proc. Nat. Acad. Sci. U.S.* **51**, 915–929.

98   — 1966. A non-conformist genetic system in *Paramecium aurelia*. *Amer. Zoologist* **6**, 589.

99   — 1967. Does preformed cell structure play an essential role in cell heredity? In Allen, J. M., *The Nature of Biological Diversity*, McGraw-Hill, New York, N.Y., 165–221.

100   *— & Barnett, A. 1958. The mating type system in syngen 2 of *Paramecium multimicronucleatum*. *J. Protozool.* **5** (Suppl.), 18.

101   *— & Dippell, R. V. 1957. The *Paramecium aurelia-micronucleatum* complex. *J. Protozool.* **4** (Suppl.), 21.

102   *— & Sonneborn, D. R. 1958. Some effects of light on the rhythm of mating type changes in stock 232–236 of syngen 2 of *Paramecium multimicronucleatum*. *Anat. Rec.* **131**, 601.

103   Tartar, V. & Chen, T. T. 1940. Preliminary studies on mating reactions of enucleate fragments of *Paramecium bursaria*. *Science* **91**, 246–247.

104   *Taub, S. R. 1958. Nucleo-cytoplasmic interactions in mating type determination in variety 7 of *Paramecium aurelia*. *Anat. Rec.* **134**, 646.

105   *— 1959. The breeding system of syngen 7 of *Paramecium aurelia*. *Anat. Rec.* **134**, 646.

106   — 1959. The genetics of mating type determination in syngen 7 of *Paramecium aurelia*. *Genetics* **44**, 541–542.

107   — 1960. *Genetic Studies on Syngen 7 of Paramecium aurelia*, Thesis, Indiana University, Bloomington, Ind.

108   — 1963. The genetic control of mating type differentiation in *Paramecium*. *Genetics* **48**, 815–834.

109   — 1966. Regular changes in mating type composition in selfing cultures and mating type potentiality in selfing caryonides of *Paramecium aurelia*. *Genetics* **54**, 173–189.

110   — 1966. Unidirectional mating type changes in individual cells from selfing cultures of *Paramecium aurelia*. *J. Exp. Zool.* **163**, 141–150.

111   Weiss, P. 1960. Molecular reorientation as unifying principle underlying cellular selectivity. *Proc. Nat. Acad. Sci. U.S.* **46**, 993–1000.

112   Wichterman, R. 1948. The biological effects of X-rays on mating type and conjugation in *Paramecium bursaria*. *Biol. Bull.* **94**, 113–127.

VIII. ADDENDUM

(Additional references are indicated by the prefix Ad)

Since the original literature search was completed in 1969, additional papers dealing directly or indirectly with mating type systems have appeared. These can be grouped into several categories and will be dealt with accordingly.

A. *Further Information Concerning Syngens*

1. *Extension of the syngen complex to other species*

The concept of the syngen continues to be applied to other ciliates, such as *Fabrina salina* (Ad 121) and *Glaucoma scintillans* (Ad 135). It would appear that as more forms are studied the concept of the syngen, with its mating types, may be almost universal in the ciliates, again emphasizing the importance of Sonneborn's original finding and definition of the syngen.

2. Further definition of the syngens

Careful enzyme studies using starch gel electrophoresis have been carried out with the syngens of *P. aurelia* (Ad 115, Ad 138) and *Tetrahymena pyriformis* (Ad 116). In paramecia 5 enzymes have been examined in each of the syngens. Differences in isoenzymes were found, enabling the positive identification of each syngen by its zymogram. Only syngens 1 and 5 (which show a strong intersyngenal mating reaction) could not be completely distinguished from one another. Some variation was reported for stocks within a single syngen, but there was sufficient intersyngenal difference to make positive identification in almost all cases. The discovery of these differences illustrates further the isolation of the syngens and the basic importance of mating type systems as isolating mechanisms.

Syngenal differences were not found, however, for *Tetrahymena*, and there may be some variation found from species to species among the ciliates. Nevertheless, the enzyme differences found in *P. aurelia* indicate that differences may occur in biological systems, and this finding might be extended to a study of systems leading to the production of mating type substances, and to the substances themselves.

3. Studies of DNA base content

The base composition of DNA of various syngens of *Paramecium* and *Tetrahymena* has also been studied (Ad 114). A value of 28 % for the C + G ratio for syngens 1, 2, 4, 5, and 9, with a slight variation for syngen 8 (29–30 %) has been found for the DNA of *P. aurelia*. These similarities indicate that the basis for mating type differences is not due to differences in DNA base content. Cummings (Ad 122) studied the base content of purified macronuclei, finding a slightly lower C + G content (21–26 %). Allen (Ad 114) reported that differences in base ratios were found between bacterized and axenic forms, and it was further found by Berger (Ad 118) that DNA from *E. coli* might be incorporated in the macronucleus. Since Cummings stocks were derived from bacterized cultures, it is possible that some of the differences found by him, compared to the figures given by Allen, may be in part related to the different culture methods used. In general, while these findings are not helpful for understanding syngenal or mating type differences, they do point to the need for standardized culture conditions when biochemical studies are carried out.

Although *Tetrahymena* showed no difference between syngens in their zymograms, it is of interest to note that syngenal differences were found for the C + G content of their DNA. Again, this emphasizes the possible dangers of generalizing from one group of ciliates to another when biochemical processes are involved.

4. Chromosomal studies

Within syngen 1 of *P. aurelia*, there appear to be great differences in the chromosome numbers of various stocks (Ad 133). Similar findings were reported earlier for syngen 4 (Ad 124). In syngen 1, a 2n number varying from 86–126 was reported, and it is apparent that chromosome numbers cannot be used to identify a syngen, particularly since these numbers overlap those found for syngen 4. However, the almost total infertility found for intersyngenal crosses, and the high F_2 mortality of crosses within a syngen may be related to a high degree of aneuploidy, and the peculiar features of the F_1 of intersyngenal hybrids (14) may also be due to this feature.

B. Use of Various Inhibitors of m-RNA or Protein Synthesis

1. Induction of autogamy

The effects of actinomycin D and puromycin upon the induction of autogamy suggests that m-RNA and protein synthesis are both required for normal nuclear processes during autogamy. Both these reagents blocked the nuclear events of autogamy in otherwise competent cells (Ad 117). The effect may be controlled by specific genes, inducible only under the proper conditions for autogamy (starvation, following a minimum number of fissions since the last nuclear reorganization) and blocked by either inhibitor from producing their effects. A role for the control of autogamy by the macronucleus had already been postulated by Sonneborn (Ad 137). Cells lacking a macronucleus were unable to undergo meiosis even though the micronuclei were present. Cells treated with actinomycin D might be considered to be chemically enucleated, leading to the same failure to be able to undergo autogamy.

As amacronucleate cells are viable for various periods of time following their origin, such cells must either contain all of the products required or be able to produce them. The latter was suggested earlier (54) when it was reported that amacronucleate cells showed some continuation of protein synthesis following the loss of the nucleus. Since amacronucleates arise at fission, and autogamy occurs upon starvation, it is possible that the amacronucleates did not contain transcribed information for autogamy and that the survival of the enucleate cell is dependent upon the presence at fission of m-RNA in the cytoplasm of the cell. New processes, not ordinarily transcribed at the time of fission, cannot be introduced.

Actinomycin D was again reported to decrease sexual reactivity. However, reactive cells, treated with puromycin, not only continued to mate with one another, but also carried out the normal nuclear changes found during conjugation. Thus, cells which could not undergo autogamy were able to complete normal conjugation, confirming that the two processes are different, at least, in the need for protein synthesis. This difference may reflect a biochemical basis for the earlier finding that immature periods never follow autogamy, but often follow conjugation in *P. aurelia*. Autogamy apparently is sensitive to inhibition of protein synthesis and does not yield immature periods, while conjugation is insensitive to the inhibition and may thus yield long immature periods.

2. Immature periods

In *P. caudatum*, syngen 3, the length of the immature period is shortened by the use of the inhibitor of DNA synthesis, mitomycin C (Ad 134). Treatment of exconjugant cells resulted in about a 12.5 % decrease in the number of fissions required for maturity, independent of time or temperature. Clones of mating type V reached maturity earlier than those of type VI. The difference in time to maturity between the two mating types was of the same order as that found for the mitomycin C treatment, e.g. approximately 12.5%. The blocking of DNA synthesis decreased the immature period. This suggests that the normal induction of reactivity, found only when the cells are relatively starved, may be due to the cessation of DNA synthesis at this time. It is possible that during rapid cell fissions in the immature period DNA synthesis is carried out at a high rate, and that when the cell reaches maturity cessation, or a slower rate, of DNA synthesis may permit transcription of the genes controlling the production of mating type substances so that sexual reactivity is possible. The finding that one type has a longer immature period may also suggest a sequential series of events as proposed earlier.

3. *Effects upon circadian rhythms*

Clark (Ad 120) demonstrated that the ability of cycler cells to change mating type is independent of the ability to produce or maintain mating type substance in syngen 2 of *P. multimicronucleatum*. Treatment with 10μg/ml of actinomycin D or 150μg of puromycin during various stages of the life cycle prevented attainment of mating type capability, but there was no effect upon the ability of the cell to undergo the cyclic change of mating type. The data suggest that attainment and maintenance of reactivity is dependent upon m-RNA and protein synthesis, but the cycling phenomenon is independent of such syntheses. The *cycler* gene controls which mating type will be expressed in normal cells, but is independent of the biochemical events leading to the development of the mating type substances.

A variation in sensitivity to puromycin was also reported for this syngen. Mating type IV cells were less sensitive to puromycin inhibition of reactivity than were animals of type III. A similar difference in recovery from the effects of the inhibitor were found. Based on these data, Clark proposed a detailed scheme interrelating cycling and mating type (Fig. 5). She suggests several hypotheses to account for the data, including differential transcription of the equivalent of the O and E genes (see Fig. 4, p. 115),

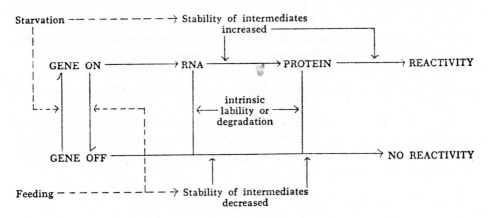

Fig. 5. Control of mating type expression (from Clark, 120).

differences in numbers of gene copies, difference in number of biosynthetic steps, or differences in the ability to detect mating type substance molecules.

The sequential model as proposed earlier, however, seems to be in accord with the newer data if one assumes that in this species the relative position of the odd and even mating types may be reversed; i.e., the even mating types in *P. aurelia* may be more homologous to the odd types in *P. caudatum*. As the data reported here for differential effects and recovery seem to be comparable to those previously reported by Kang and Taub for syngen 7 of *P. aurelia* (45) using the same inhibitors, such a suggestion does not seem improbable.

4. Effect of actinomycin upon a possible modifying allele in the mating type system

Based on work with actinomycin D, Hiwatashi (Ad 129) suggested that more than one gene is involved in the expression of the dominant mating type of syngen 3 of *P. caudatum*. The action of the antibiotic would be upon this gene controlling its expression rather than upon the genes responsible for the development of mating type substances. Such a gene seems to be an equivalent of the IN^1 locus found for syngen 1 of *P. aurelia* which also acted to control the frequency with which the dominant $+mt$ locus was expressed (15).

C. Other Biochemical Investigations

1. The effects of washing cells

Repeated washing of reactive cells of *P. caudatum* in isotonic salt solutions resulted in the loss of mating type reactivity. Cells washed in saturated $CaCO_3$ solutions did not lose reactivity (Ad 139). The effect of washing was interpreted as possible evidence that some factor in the culture medium was essential for continued maintenance of reactivity, a suggestion which was supported by the finding that cells replaced in exhausted culture medium after washing regained reactivity after two hours. These data were also interpreted to support the hypothesis put forth by Cohen (22) for the need of a continued synthesis of mating type substances during reactivity. The data were further supported by the finding that actinomycin D-treated cells of *P. caudatum* lost reactivity after six hours while formalin-killed cells exposed to the same antibiotic did not. A discrepancy in the postulate, that washing of non-reactive cells led to a high degree of synchronous gain of reactivity, was ameliorated by the additional hypothesis that the substances lost by repeated washings may not have been the mating type substances, but some precursors in the biosynthetic pathways towards the mating type substances.

The finding that Ca^{++} ions did not cause the loss of reactivity led to the hypothesis that these ions, known for their effect on the permeability of membranes as well as upon cortical structures may have changed these properties so that mating type substances could not be readily removed.

2. Effect of Ca^{2+} upon delayed determination of mating type

Hallet (Ad 128) found in syngen 1 of *P. aurelia* that treatment with Ca^{2+} resulted in the delay of mating type determination for several fissions following nuclear reorganization. Effects upon permeability and the cortex again seem to be indicated as playing a role in the timing of determination. Such effects may also be instrumental in the delayed determination of mating type reported previously for singlet clones given off from doublet cells in syngen 4 (Ad 119).

3. *Possible cytoplasmic factors involved in mating type determination*

Koizumi (Ad132) has begun a study of cytoplasmic factors which might play a role in mating type determination. Injection of conjugating or autogamous cells with cytoplasm from the opposite mating type resulted in a high frequency of mating type change in the recipient cells of syngen 4 of *P. aurelia*. Less frequent changes were found when vegetative cells were injected. Treatment of cells with RNAase or cyclohexamide also resulted in similar, but less frequent changes. Changes were observed in both directions; mating type VII cells injected with VIII cytoplasm changed to mating type VIII and vice versa. Streptomycin and mitomycin C were ineffective. The obvious suggestion is that protein synthesis and m-RNA synthesis are required to maintain the constancy of mating type in this member of the B syngens, while DNA synthesis is not.

4. *Identification of pellicular DNA*

An earlier report (Ad 136) that the pellicular fraction of *P. aurelia* homogenates contained DNA has been recently disputed (Ad 130, Ad 131). The DNA found in this fraction was found to have the same buoyant density as that of macronuclear DNA, and in addition it was found that DNA from bacteria may also bind with the pellicle. Although cortical structures are known to be involved in self-replication to some degree (23), suggesting that DNA may be involved in this process, positive identification of pellicular DNA seems in doubt as yet. Further precautions on the analysis of nuclear or cellular DNA are suggested in light of the work by Gibson (Ad 127) who found that DNA as large as 10^4 daltons was taken up by *Paramecium*. This DNA, while not completely localized, may be found in both nuclear and cytoplasmic fractions. Thus any DNA studies using paramecia grown in bacterized media have to be interpreted with extreme caution.

D. *Master-slave Relationship of Mating Type Determination*

Allen (Ad 113) has recently suggested that the determination of enzyme types in *Tetrahymena* takes place whether a gene is expressed or not. This is an irreversible process. She suggests there may be a "master" gene, while copies, or "slaves" are non-replicating and synthesized under the control of the master. Thus only one gene, the master, needs to be repressed in the macronucleus instead of the entire polyploid set of genes. Carrying this concept over to the mating type situation in *Paramecium* could account for the vegetative changes of mating type, as discussed previously (p. 117). If the *E* (or *mt*) locus is considered to be the master gene, then repression or activation of this one locus could account for the change in mating type, without attributing these changes to the simultaneous activation or inactivation of multiple duplicate genes for mating type determination. Once the master gene has been repressed, the slaves would be diluted out at subsequent fissions and a delayed change of mating type, one which was directed many fissions prior to its expression, would be detected.

E. Miscellaneous Reports dealing with Mating Type

1. Assortative mating

Studies with syngen 1 of *P. aurelia* and syngen 3 of *P. caudatum* indicate that assortative mating based on cell length occurs (Ad 123). In *P. aurelia* there is a tendency for mating between the smallest individuals of the population, while there is a biphasic distribution in *P. caudatum* whereby the smallest and the largest tend to mate assortatively with equally sized mates. In *P. caudatum* the length of the cell and the amount of nuclear DNA are also correlated. No similar data were given for *P. aurelia*. The authors conclude that the relative sizes represent various stages in the cell division cycle and that there exists a period following cell division during which conjugation is not possible, possibly the G-2 stage in *P. caudatum*. In this form individuals with high nuclear DNA do not conjugate. These data might be related to the suggestion that during periods of rapid increase of DNA the genes for production of mating type substances are not transcribed.

2. Obtaining purified, reactive cilia

Cilia which retain mating type substances as judged by their ability to induce conjugation in intact cells of the opposite mating type may be freed from centrifuged, reactive cells by treatment with 10mM MnCl$_2$. The cilia maintain their reactivity for 2–3 days when stored in the cold (Ad 125). The use of such cilia in attempts to identify the mating type substances should prove to be of value.

3. Mating type from doublet cells of P. aurelia

A brief abstract and the following paper dealing with the findings presented earlier in this chapter (p. 104ff.) has appeared (Ad 119 and Ad 119a).

F. Concluding Remarks

More attention is now being brought to bear upon the mating type system, particularly its biochemistry, and its relationship to gene activity. It would appear that this interest, plus the increasing use of defined culture conditions, should lead within the next few years to the identification of the mating type substances, their biochemical control, and thus to the solution of the problem of development and differentiation of these substances.

REFERENCES

(Abstracts are indicated by *)

113   Allen, S. L. 1971. A late-determined gene in *Tetrahymena* heterozygotes. *Genetics* **68**, 415–433.

114   — & Gibson, I. 1971. The purification of DNA from the genomes of *Paramecium aurelia* and *Tetrahymena pyriformis*. *J. Protozool.* **18**, 518–525.

115   — & — 1971. Intersyngenetic variations in the esterases of axenic stocks of *Paramecium aurelia*. *Biochem. Genet.* **5**, 161–181.

116   — & Weremiuk, S. L. 1971. Intersyngenetic variations in the esterases and acid phosphatases of *Tetrahymena pyriformis*. *Biochem. Genet.* **5**, 119–133.

117   Beisson, J. & Capdeville, Y. 1966. Sur la nature possible des étapes de différentiation conduisant à l'autogamie chez *Paramecium aurelia*. *Compt. Rend.* **263**, 1258–1261.

118   Berger, J. D. 1971. Kinetics of incorporation of DNA precursors from ingested bacteria into macronuclear DNA of *Paramecium aurelia*. *J. Protozool.* **18**, 419–429.

119   *Butzel, Jr., H. M. 1970. Abnormal cytological events and mating type determination in doublet cells of stock 51 syngen 4, of *Paramecium aurelia* following conjugation in axenic medium. *J. Protozool.* **17** (Suppl.), 8.

119a   — 1973. Abnormalities in nuclear behavior and mating type determination in cytoplasmically bridged exconjugants of doublet *Paramecium aurelia*. *J. Protozool.* **20**, 140–143.

120   Clark, M. A. 1972. Control of mating type expression in *Paramecium multimicronucleatum*, syngen 2. *J. Cell. Physiol.* **79**, 1–14.

121   Cho, P. L. 1970. The genetics of mating type in a syngen of *Glaucoma*. *Genetics* **67**, 377–390.

122   Cummings, J. D. 1972. Isolation and partial characterization of macro- and micronuclei from *Paramecium aurelia*. *J. Cell Biol.* **53**, 105–115.

123   Demar-Gervais, C. 1971. Quelques précisions sur le déterminisme de la conjugaison chez *Fabrea Salina* Henneguy. *Protistologica* **7**, 177–195.

124   Dippell, R. V. 1954. A preliminary report on the chromosomal constitution of certain variety 4 races of *Paramecium aurelia*. *Caryologia* **6** (Suppl.), 1109–1111.

125   *Fukushi, T. & Hiwatashi, K. 1970. Preparation of mating reactive cilia from *Paramecium caudatum* by $MnCl_2$. *J. Protozool.* **17** (Suppl.), 21.

126   Génermont, J. & Dupy-Blanc, J. 1971. Recherches biométriques sur la sexualité des Paramécies *(P. aurelia* et *P. caudatum)*. *Protistologica* **7**, 197–212.

127   Gibson, I. 1968. Studies on the incorporation of DNA into *Paramecium aurelia*. *J. Cell Sci.* **3**, 381–389.

128   *Hallet, M. M. 1971. Mise en évidence de l'action d'un facteur externe, le chlorure de calcium sur la différentiation du macronucléus chez *Paramecium aurelia* souche 60, variété I. *J. Protozool.* **18** (Suppl.), 52.

129   Hiwatashi, K. 1969. Genetic and epigenic control of mating type in *Paramecium caudatum*. *Japan J. Genetics* **44** (Suppl. 1), 383–387.

130   *Hufnagel, L. 1965. Structural and chemical observations on pellicles isolated from paramecia. *J. Cell Biol.* **27**, 46A.

131   — 1969. Properties of DNA associated with raffinose-isolated pellicles of *Paramecium aurelia*. *J. Cell Sci.* **5**, 561–573.

132   *Koizumi, S. 1971. The cytoplasmic factor that fixes macronuclear mating type determination in *Paramecium aurelia*, syngen 4. *Genetics* **68** (Suppl.), s34.

133   Kosciuszko, H. 1965. Karyologic and genetic investigations in syngen 1 of *Paramecium aurelia*. *Folia biol.* (Krakow) **13**, 340–386.

134   Miwa, I. & Hiwatashi, K. 1970. Effect of mitomycin c on the expression of mating ability in *Paramecium caudatum*. *Japan J. Genet.* **45**, 269–275.

135   Nataka, A. 1969. Mating types in *Glaucoma scintillans*. *J. Protozool.* **16**, 689–692.

136   Smith-Sonneborn, J. & Plaut, W. 1967. Evidence for the presence of DNA in the pellicle of *Paramecium*. *J. Cell Sci.* **2**, 225–234.

137   *Sonneborn, T. M. 1955. Macronuclear control of the initiation of meiosis and conjugation in *Paramecium aurelia*. *J. Protozool.* **2** (Suppl.), 12.

138   Tait, A. 1970. Enzyme variation between syngens in *Paramecium aurelia*. *Biochem. Genet.* **4**, 461–470.

139   Takahashi, M. & Hiwatashi, K. 1970. Disappearance of mating type reactivity in *Paramecium caudatum* upon repeated washing. *J. Protozool.* **17**, 667–670.

Note Added in Proof

Seven recessive mating type mutations in syngen 4 of *P. aurelia* have been found by Byrne (Ad 140); four of these have been studied in detail. Three of the 4 restrict the homozygous recessive to the odd mating type (VII). Unlike similar acting mutations of syngen 1 (14, 15) none of these is allelic. A somewhat similar picture of non-allelic mating type mutations restricting the homozygous recessive to the odd mating type was previously described for syngen 7 (104–110). As both syngen 4 and 7 belong to the B group of syngens while syngen 1 is a member of the A group (p. 92), this might suggest that there are more genes controlling the development of the even mating type in the B group. However, Byrne suggests the difference may not be real and that the relative ease in which mating type mutations can be detected in the B group with clonal inheritance of mating type makes the finding of mating type mutations more likely. Of further interest is the finding that double heterozygotes for these non-allelic mutations are expressed as selfing clones.

The 4th mutant studied is unique in that homozygotes for this gene are determined as selfer clones but only if the clone descended from an even mating type heterozygote which became homozygous at autogamy. The presence of E cytoplasm allows expression of the gene but the gene cannot be expressed in 0 cytoplasm. Byrne again compares this to the system in syngen 7 (104–110) where there is nuclear–cytoplasmic interaction in the expression of mating type.

While none of these data provide proof for the hypothesis of sequential development of mating type (p. 115), none is in disaccord. The hypothesis predicted the finding of mutations pure for the odd mating type and the failure to find those pure for the even mating type. In addition, Byrne interprets many of his data as being in accord with this hypothesis.

A brief report describing an antiserum specific against mating type Vi of *P. caudatum* and lacking immobilization activity has appeared. This may tend to substantiate further the protein nature of the mating type substances, and more importantly, may be the point of departure for an analysis of the mating type substances themselves if the antigen–antibody complex can be dissociated. If a number of such sera, each specific for a single mating type, can be obtained, it may become possible to analyse specific differences in mating type substances both within and between syngens.

Ad 140   Byrne, B. C. 1973. Mutational analysis of mating type inheritance in syngen 4 of *Paramecium aurelia. Genetics* **74**, 63–80.

Ad 141   *Sasaki, S. A., Ito, A., & Hiwatashi, K. 1972. Mating type specific antigen in *Paramecium caudatum. Genetics* **71**(Suppl), s55.

Surface Antigens of *Paramecium aurelia*

IRVING FINGER

Biology Department, Haverford College, Haverford, Pa. 19041 (U.S.A.)

I. INTRODUCTION

The earliest studies of the surface antigens of *Paramecium* by Rössle (96, 97), Jollos (52), Harrison and his coworkers (18, 48, 49), and Sonneborn and his coworkers (109, 110, 112, 121) were devoted chiefly to a description of the immobilization phenomenon and an enumeration of the spectrum of antigenic types, or serotypes, that a particular stock could express. Unlike reactions generally observed when bacteria or other microorganisms are introduced into specific antiserum, paramecia are not agglutinated to one another. Rather they are immobilized separately by tip-to-tip agglutination of patches of individual cilia (48, 49). Even in dilute serum (e.g., 1:400 or 1:800), the cilia are prevented from beating vigorously (8); sometimes in the space of a minute or two, although more often within ten or fifteen minutes, a cell becomes completely inert. Complement is not required for immobilization (18, 103). If allowed to remain in the serum for several hours, death usually ensues; in concentrated antiserum, immobilization and death may take place within a few minutes. The basis for the lethal effect is not known.

In some cultures of paramecia, cells may be present which swim freely in antiserum which has immobilized all of their sister cells. The descendants of these resistant cells in turn can be used to elicit specific antibodies which will only affect the initially unaffected cells (11, 110, 121). In this fashion, a battery of antisera can be built up which reflects the variety of immobilization antigens that a single clone is capable of making, which in some instances can be as many as a dozen (60). The remarkable specificity of most of these antisera illustrates two important features of the serotypes: (1) each serotype is generally antigenically unique, and (2) individual cells are not mosaics of antigens, but are uniformly of one type, the presence of one antigen excluding any others. To this day, although our knowledge of the chemistry, serology and genetics of the serotypes has grown enormously, the molecular basis for mutual exclusion is almost a complete mystery, even though the phenomenon was among the earliest uncovered.

The work carried out in the author's laboratory and reported in this review was supported by the National Science Foundation (GB 5926 and GB 8550) and the U.S. Public Health Service, National Institutes of Health (GM 12017).

References p. 158–162

II. INHERITANCE

A. General

When a number of stocks, derived from cells often isolated from geographically widely separated regions, were examined, it was noted that paramecia from one stock could be immobilized by serum prepared against another stock. However, to be effective more concentrated serum had to be used than was the case with the homologous paramecia of the original stock. Further studies demonstrated that not only were related but distinguishable antigens found in cells of different genetic backgrounds, but stocks also differed in the diversity of antigens available to them, with certain antigenic types apparently completely missing from some strains (9, 11, 112, 115).

B. Nuclear Control

Apparently the same locus which controls the ability to express an antigen also determines antigenic specificity (91). Appropriate crosses among these stocks showed that alleles generally determined the differences between related serotypes, with each set (or pair when only two stocks were being analyzed) of crossreacting serotypes under the control of a separate locus (30). Similarly, straightforward Mendelian inheritance was displayed in matings involving absence of a particular class of serotypes, with the capability for manufacturing a particular antigen being dominant (9, 112, 115). Of those that have been examined intensively, the loci specifying the various serotypes are unlinked. Indeed, there has only been one instance of linkage reported in *Paramecium aurelia*, although admittedly relatively few characters have been followed (18). Allen (1) has suggested that, if each antigen is composed of subunits, there should be a structural gene for each subunit and that, in theory, recombination within a "locus" should occur. The only attempt to obtain recombinants from exautogamous animals has yielded negative results (Beale, personal communication) although, in view of the difficulty in recognizing rare recombinants, it cannot be said that this hypothesis has been adequately tested.

The picture of the inheritance of the specificity of the serotypes outlined above, although correct in its more general aspects, is overly simplified. For example, stock 7 of syngen 2 has never shown serotype E, yet the F_1 from matings of 7 with 197 produces an E antigen which is not identical with 197E. Furthermore antisera against a heterozygous clone indicated that the F_1 could elicit antibodies directed against the F_1 alone. The interpretation given these data was that a hapten-like molecule was produced by the "E-deficient" strain, which apparently could be combined with portions of the E antigen contributed by the 197 parent (35). In other words, stock 7 had an E locus, but one that specifies an incomplete antigen. Another possible explanation for these results is that newly introduced genes from stock 197, other than the e gene, allowed the e^7 to be expressed.

Also, not all crossreacting serotypes are determined by alleles. There are a number of antigenic types within an individual strain that, although distinguishable, none-theless can be immobilized by a single antiserum in the appropriate concentrations. Since these strains are completely homozygous (having gone through autogamy), these serotypes cannot be determined by alleles and must be specified by at least two different loci. In other words, it is possible for non-allelic antigens to be structurally related.

Summarizing, a cell may have these options: (1) a particular antigen may be entirely absent; (2) it may be present as an incomplete antigen; (3) it may exist as one of several allelic forms; (4) it may be serologically unique or crossreact with another antigen under the control of a separate locus or (5) the same serotype may be present in all stocks with no stock-specific variants.

In light of what is now known about the molecular consequences of gene mutation on protein synthesis and structure, little of what has been described already is sur-prising except, perhaps, the *lack* of variability among some serotypes, e.g., G in syngen 2 (Finger, unpublished) and B in syngen 4 (85). By far the most significant phenomenon whose genetic basis has been studied is that of mutual exclusion. The earliest experiments dealing with this phenomenon involved crosses between cells expressing different serotypes. There seemed to be only three possible results. Either one serotype would be dominant; the exconjugants would be a mixture of the two parental serotypes; or the original serotypes would persist. However, it turned out that the answer depended, in part, on precisely how the crosses were performed. Surprisingly, the important variable was not necessarily the genotypic component, but often was the phenotypic one.

C. Cytoplasmic Control

Sonneborn and LeSuer (121) were the first to demonstrate the role the cytoplasm can play in determining which serotype a cell expresses. Intrastock matings between A and B cells of syngen 4 showed that the exconjugant deriving its cytoplasm from the A parent yielded an A clone, while the B parent gave rise to a clone of B cells. However, when large amounts of cytoplasm as well as nuclei were allowed to exchange, both exconjugants yielded B cultures. Other kinds of exconjugant clones could be formed depending on the volume of cytoplasm intermixed. Even after passage through autogamy, during which genes segregate, the serotype of the original parent (when no cytoplasmic exchange has occurred) was retained by the individual exautogamous clones. The conclusion drawn from these results was that the cytoplasm regulates the expression of the different serotypes, calling forth one antigenic type at a time. From other experiments it was concluded that the nucleus specifies the types of antigens a cell is capable of expressing. In *Tetrahymena* there is no evidence for a cytoplasmic role in the expression of serotypes (69). *Paramecium caudatum* appears to be more like *P. aurelia* in the genetic control of serotype expression and specificity (51).

Using clones of another syngen, 1, Beale (10, 11) arrived at the same conclusions, but through a somewhat different route. This time different stocks were used, each with antigenically distinguishable, but related, serotypes. When stock 90G was mated with stock 60D, the 90 parent produced a G expressing clone and the 60 parent a D clone. The significant feature about these clones was that the G cells were both 90G and 60G and the D cells were both 90D and 60D. Thus, apparently, the 90G cytoplasm allows the expression of both g^{60} and g^{90} genes and the D cytoplasm the d^{60} and d^{90} genes. The G clones can transform to D and the D to G by transferring the exconjugant clones to either 29° (for D) or 25° (for G). (Fig. 1) Since both exconjugants, as a result of the events preceding and accompanying reciprocal fertilization, come to possess the same genes it is apparent that the g^{60} gene was unexpressed in its original parent because somehow the cytoplasm was unsuitable; introduced into the 90G cytoplasm it now is expressed. The other copy of the g^{60} gene left behind in the

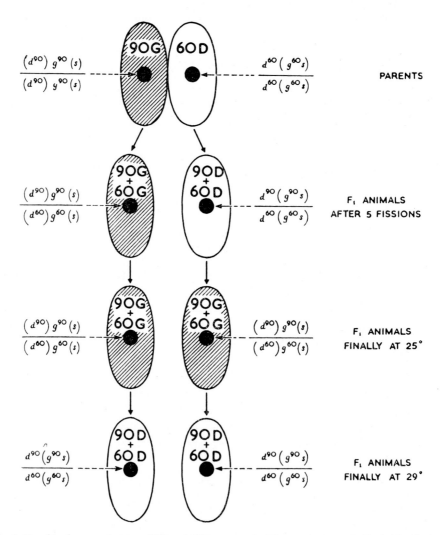

Fig. 1. Results of crosses between 90G and 60D paramecia. The serotypes are indicated by the large letters and numbers, and the genes by lower case italics. The general serotype expressed is represented by shading for G, and clear for D. (Beale, 10).

60 parent with D-sustaining cytoplasm remains unexpressed. A comparable situation exists with the *d* alleles. Thus, mutual exclusion appeared to be a reflection of the existence of a number of different cytoplasmic states, each state favoring the expression of one particular antigenic type.

D. *Allelic Repression*

More recently, in syngen 2, a somewhat different situation was found with respect to the simultaneous expression of different alleles in a heterozygote (37). For example, when 72E was mated with 197E, the majority of the exconjugant clones exhibited a mixed 72E + 197E serotype. However, when 72E was mated with 197G (or some other non-E serotype), among the E exconjugant clones, some of which were descended from the 197 parent, the majority expressed the 72E serotype. When the reciprocal cross was carried out, 197E × 72G, most of the F_1 clones were mixed 72E-197E or were 197E. (Table 1) When cytoplasm as well as nuclei were exchanged, the proportion of mixed E serotypes increased markedly. In *Tetrahymena*, also, some serotype variability in heterozygotes can exist but, unlike *Paramecium*, the particular parental serotype expressed becomes "fixed" with time (73). The difference in phenotypes of heterozygotes in syngens 1 and 2 of *P. aurelia* may be more apparent than real. Capdeville (22) has recently followed the appearance of several different G serotypes in syngen 1 [the same syngen and class of serotypes studied by Beale (10)], and found

TABLE 1

Examples of variations in influence of parental serotype on progeny serotype. (Finger and Heller, 37).

Serotypes of parents	Number of clones of each serotype among progeny		
	72E	mixed	197E
Set I			
197E × 72E	4	34	0
197non-E × 72E	26	4	0
197E × 72non-E	1	44	2
Set II			
197E × 72E	15	51	31
197non-E × 72E	7	13	14
197E × 72non-E	0	0	3
Set III			
197E × 72E	0	6	34
197non-E × 72E	0	3	66
197E × 72non-E	0	0	50
Set IV			
197E × 72E	0	39	46
197non-E × 72E	0	15	2
197E × 72non-E	0	0	21

that whether an F_1 clone expressed both parental types depended on the particular stocks employed. 156G and 90G were dominant to 33G and 168G, while crosses between 33G and 168G yielded heterozygotes which reacted with both parental antisera.

The experiments with the E and C serotypes in syngen 2 and some G serotypes in syngen 1 provide an example of *allelic* repression (as contrasted with mutual exclusion which is exclusion between *loci*) and illustrates how "dominance" can depend on the recent history of an allele. More detailed analyses of variability in expression in heterozygotes is discussed below.

Differential allele expression in a heterozygote is not restricted to *Paramecium*. Indeed, among the ciliates the most thoroughly analyzed examples are provided in another organism, *Tetrahymena*. Not only serotypes, but esterases and acid phosphatases have been shown to exhibit allelic repression (1, 70, 71). In mice (50) and rabbits (59) a similar phenomenon exists with regard to the allotypes of the immunoglobulins.

The E results suggest that the migratory micronucleus originating from the cell expressing E (as well as the stationary micronucleus) is different from a micronucleus from a non-E cell and, therefore, contrary to earlier views (119), the micronucleus is not inert. Rao and Prescott (90) using autoradiography noted in *P. caudatum* that the micronucleus synthesizes RNA. Support for the notion of an active micronucleus also comes from Pasternak's autoradiographic studies with transforming cells (76).

Pasternak reported that the micronuclei of amacronucleate cells were able to synthesize RNA and, further, that during serotype transformation (in normal cells) there was a burst of RNA synthesis in both micro- and macronucleus. This latter observation is puzzling if it is assumed that m-RNA synthesis is needed for both new and old antigen. The finding suggests that the total amount of all immobilization antigens a cell makes is not constant at all times. Also, since this increase occurs before any transformed cells can be detected by immobilization, one would expect that actinomycin D would block early antigen synthesis. Sommerville (106, 107) has reported otherwise using a somewhat different system (see below). In *Tetrahymena* Gorovsky and Woodward (47) were unable to detect RNA synthesis in the micronucleus during vegetative growth or reproduction. Nanney and Dubert (72) also showed that in crosses involving different H serotypes of *Tetrahymena*, the phenotype of the F_1 is independent of the serotypes expressed by the parental cells (72), in contrast to the situation in syngen 2 of *P. aurelia*.

E. Nucleo-cytoplasmic Interactions

That the nucleus is not simply a passive acceptor of cytoplasmic stimuli has also been shown by the experiments of Preer, Bray and Koizumi (86). Macronuclear regeneration was induced in exconjugants of matings between A and B cells. The lines from normally reorganizing exconjugants transformed with a very high frequency while the lines from exconjugants where macronuclei had regenerated from old anlagen were stable. Thus, the nucleus can exert control over the expression of sero-type, and may even play the dominant role, as suggested by Preer (85).

Recapitulating, then, at least during conjugation the cytoplasm may determine which structural genes in the nucleus will become expressed, while the ability of a cell to manufacture a member of a particular class of antigens and the specific kind of antigen it will make is determined by nuclear genes. The different serotypes are specified by separate unlinked loci. There is evidence that the nucleus regulates gene expression during most of the life of a cell, except during certain periods of nuclear reorganization, such as conjugation.

III. SEROTYPE TRANSFORMATION

A. Agents

Clearly, if several serotypes can be made by a cell and only one is, and must be, expressed at a particular time, there should exist an interval during which one serotype is transforming to a new type. Little is known about the events that transpire during transformation, although there are abundant data concerning agents which induce transformation, conditions for maintaining a particular serotype, etc. For example, the following factors are known to exert a pronounced effect: temperature (9), salt concentration (11), pH (3), stage in life cycle (26, 27), kind of culture medium (3, 6) quantity of food (11), homologous antiserum (8, 26), normal serum (43, 44), proteo-lytic enzymes (56, 128), UV- and X-irradiation (110, 118, 123), patulin (2, 6), actino-mycin D (4, 5, Finger, unpublished), puromycin (4, 5, 32, 34), chloramphenicol (4, 5), fluorophenylalanine (34), colchicine (Finger, unpublished), and acetamide (5). Doubt-less the list is incomplete, limited by the patience of the investigator and the particular aim of the experiment.

B. General Observations

Despite the tremendous variety of environmental conditions, stocks and serotypes examined, disappointingly few general conclusions can be made regarding the process of transformation. However, these conclusions are of considerable significance:

(a) The synthesis of the new antigen apparently does not require cell division (6), although complete conversion to a new serotype appears to take several fissions (7, 11, 107, 108).

(b) Transformation induced in a culture generally is not due to the selection of a few spontaneous transformants (109).

(c) Transformation is reversible and does not involve gene mutation (11).

(d) The original antigen generally decreases in quantity during transformation (but cf. 124). Its replacement increases nearly proportionately, although the rate of synthesis is exponential rather than linear (7). Preer (80) has pointed out that the autocatalytic kinetics of the induced synthesis are similar to those noted with "adaptive enzyme" formation (see discussion of hypotheses for mutual exclusion, p. 147). No molecules of intermediate or hybrid specificity are observed during transformation.

(e) Genetic and chemical analyses clearly indicate that the various antigens are not different combinations of the same subunits (as with some isozymes), or different configurations of the same polypeptide chains (54, 55). Each is a distinct molecule, but similar in certain overall chemical and physical properties (see below) (54, 55, 95, 125, 126, 127).

(f) Antigens determined by the same locus often respond similarly to a particular environment (9, 10, 12). For example, in syngen 1 among 37 stocks examined, serotype S is maintained at a lower temperature than is G; G, in turn, is formed at a temperature below that at which D is found (11). The optimal temperature favorable to a serotype will often vary from stock to stock, but the relative positions on a temperature scale will not vary (11, 88). This observation, however, does not necessarily mean that only the gene specifying the structure of the antigen being manifested determines the condition for maintenance (and thus for transformation) (9). Sonneborn *et al.*, (122) have shown that although temperature plays an important role in stabilizing the serotypes in syngen 4, certain types did not always have the same stability characteristics in different stocks. They also noted that genes other than the one determining the specificity of the serotype expressed play a role in stability. These "modifying" genes appear to be the determinants for the specificity of other serotypes.

(g) The *rate* at which a serotype is introduced into a new environment influences markedly the extent at which transformation occurs. Gradual transitions to higher temperatures or salt concentrations are not as effective in inducing transformation as is a change to the same temperature or concentration reached in a single step (11, 12, 78).

C. Influence of Cell's History

Several baffling observations merit special consideration, and an understanding of their underlying mechanisms seems to be especially critical to an understanding of the chemical basis of transformation and mutual exclusion. The oldest of these are Skaar's (103) experiments on the influence of the remote and recent history of a clone on transformation. Whether a culture is well-fed or partly starved for as long a time as 20 days was shown to determine its responsiveness to the induction of transformation by antiserum. Interpretations of the data assumed that the differences had a nutritional basis with other effects, such as autogamy, being indirectly related to this. In the light of recent (Ad 135) experiments by Finger and co-workers in which the frequency of spontaneous transformation in stable and unstable clones was followed (see below), Skaar's results may be viewed in other terms. Feeding paramecium involves the discarding of portions of the culture and replacing this amount with fresh culture fluid. Obviously, any cell products in the exhausted medium are removed by this procedure, and the remaining products are diluted by the added fresh medium. It has been shown that the conversion of unstable clones to stable clones by rapid growth is most likely due to this dilution effect rather than to changes in fission rate as a result of the added fresh nutrient. Cell products lost by dilution are thought to be either inhibitors or inducers of antigen synthesis. The level of these products may have been the controlling factors determining response to antiserum observed by Skaar.

D. *Effect of Normal and Immune Serum*

Although usually the consequence of exposure to dilute antiserum is transformation to a new serotype, this is not always the case. Sometimes the antiserum actually inhibited transformation when compared with the action of control heterologous antiserum (12, 43, 44). Since serum consists of many kinds of substances apart from antibodies, the variability in results could have been due to factors other than immunoglobulins. However, even after Finger *et al.* greatly increased the relative concentrations of antibodies through partial purification the variations were not eliminated (43). The most striking observations in these latter experiments were those in which pre-immune sera served as controls. These "normal" sera were not neutral; they also induced transformation. Equally surprising was the fact that the effect of serum from one rabbit was distinguishable from the effect of another rabbit's serum. Furthermore, sera prepared against culture medium in which paramecia had been living also were effective in transformation even though neither these nor the pre-immune sera immobilize paramecia.

The environmental factors that evoke transformation appear to owe their efficacy either to a non-specific traumatic physiological shock, or to any one of a host of specific interactions. The specific effects could be due to the binding, destruction or alteration of the antigen itself, or of messengers, repressors, or inducers. Possibly, interactions between substrates and effectors may have been interfered with. In some instances, one could envision a positive mode of action such as the added reagent serving as an inducer. Still another explanation for a specific effect would be that the transforming reagent may serve the same function as the antigen and by replacing it interfere in its synthesis.

IV. PARALLELS BETWEEN MUTUAL EXCLUSION AND CELL DIFFERENTIATION

Analogies between the expression of a single serotype, to the exclusion of all or most others, and differentiation of cell types within multicellular forms often have been drawn. In both instances, it would appear that only a portion of the potentialities of an individual cell are realized at any one time. However, there are major differences between the two situations whose significance and genuineness are still to be assessed.

First, although the function of the antigens is unknown, there is little doubt—from the similarities of their chemistry and their biological control—that all of them serve the same function. In contrast, phenotypic differentiation of tissues generally refers to differential activation or expression of genes which determine different sets of functions. If an analogy is to be made, it should be to isozymes such as lactic dehydrogenase, where one or several multiple forms of molecules with the same catalytic function are called into expression, or perhaps to the hemoglobins or immunoglobulins (62). In these molecules, where different loci specify subunits, one set of subunits becomes part of a complete molecule one time and another set part of a similar molecule some other time.

Second, the ready and complete reversibility of phenotype without any accompanying genotypic change is ordinarily not encountered in differentiated cells. Indeed, some definitions of differentiation include the characteristic of stability as a significant criterion. Whether this difference is more apparent than real may depend in part on finding the appropriate environments to induce and/or repress the activity of individual genes in cellular masses. With the serotypes the problem appears to exist at the opposite pole: the search for an environment which supports a particular antigenic type. It may, however, be that the more appropriate question is what prevents all the other serotypes from being expressed. Of significance is the fact that relatively minor physiological traumas often induce transformation of major proportions in cultures of paramecia.

And, lastly, in *Paramecium* the somatic nucleus (the macronucleus) is apparently composed of subunits (70, 111) assuring the presence of many copies of each locus, while in most differentiated cells a locus is represented by only two genes. Thus, in a sense, an individual paramecium is a population of genomes, and it is an open question whether a serotype is exhibited by only one gene or whether many copies of this gene are activated or expressed. If gene dosage need be taken into account, then the question that becomes particularly important is whether the detection (by cilia immobilization) of only a single antigen reflects the true situation with respect to gene activation. It is conceivable that, rather than a condition of complete mutual exclusion, several loci are derepressed but only the locus active in many "subnuclei" produces sufficient complete antigen to be recognized by agglutination. Thus, for some cells, the apparent expression of only a single gene at a time may be illusory and may merely be a consequence of quantitative differences in various gene products. Obviously, these considerations make any parallels between diploid or haploid cells and *Paramecium* exceedingly fragile.

Despite these provisos, the surface serotypes are of absorbing interest in themselves and may still serve as models for special situations in higher forms. The phenomena of mutual exclusion and transformation led Delbrück (24) to devise a scheme to account for mutual exclusion, invoking the notion of steady states. In several respects, his hypothesis foreshadowed some of the newer proposals of Monod and Jacob (63). Although, in the light of knowledge, since acquired, of the mechanisms of transcription and translation the proposed mechanism can be faulted, his general viewpoint is still the dominant one for theorists. Before considering his suggestion further, an account of the chemistry of the antigens would be especially appropriate at this point. This is because most hypotheses proposed in recent years are grounded on this knowledge and only make sense in the light of it and of what is now understood about the regulation of the synthesis of macromolecules in general.

V. CHEMISTRY

Early studies of the antigens led to data which could be constructed to indicate that the specificity resided in the carbohydrate, but that there was a significant protein moiety also. The ambiguity in the interpretation of these data was due in part to the heterogeneous nature of the material analyzed. A first step toward more direct chemical analysis was the detection of the antigens *in vitro*. Gel diffusion analysis of extracts of homogenates provided the necessary tool (28). Using a technique of Oakley and Fulthorpe (75), modified by Preer (77), it was shown that the antigens are easily soluble in saline or culture fluid and could be readily assayed in microtubes (81). Within a few years the most important single advance, the isolation of the antigens, was carried out by Preer, who extracted several of the antigens in a highly purified state as determined by ultracentrifugation, immunological techniques, and chemical analysis (79, 82, 83).

A. Size

From intensive physical, chemical, and immunological studies in four independent laboratories, a generally agreed upon picture of the antigens has emerged, although several areas of disagreement persist on significant details of structure. The antigens are primarily protein with a small variable carbohydrate content [0.9–2.0 (83)]. They are uniformly large with a molecular weight of 240,000–310,000 daltons (19, 54, 83, 95, 126, 127). (The significance of these variations in size will be discussed below). After treatment with trypsin and chymotrypsin at least 65 peptides are released, the exact number varying with the antigen (54, 56, 126).

B. Amino Acid Composition

Although there are considerable ambiguities involved in interpreting chromatographic and electrophoretic patterns (one of which is the fact that there is a core of protein resistant to digestion), one significant conclusion has emerged: each antigen probably possesses a unique primary amino acid sequence (Table 2). Thus the antigens cannot be considered as proteins which differ from each other solely in their folding. Nevertheless, the similarities among the antigens examined are striking indeed (32, 54, 95, 127). Cysteine is abundant (about 10 moles %), especially when contrasted with other proteins in which disulfide linkages play a prominent role, e.g., fibrinogen (2.3%). [Most proteins have less than 3% cystine plus cysteine (74)]. Another normally rare amino acid found in considerable quantities in hydrolyzates of the antigens is threonine (about 14 moles %). Arginine and lysine residues, those amino acids split from their neighbors by trypsin, make up 7 moles % of the antigens (Table 2).

C. Shape

Physically, the antigens have an axial ratio of about 11:1 and sediment at about

TABLE 2

Amino acid composition of six serotypic antigens from P. aurelia. (Reisner, Rowe, and Sleigh, 95). Values for "Steers" 51A obtained from Steers (127); for 90D, 178D, and 90G from Jones (54). The residue values are based on molecular weights of 301,500 (51A), 271,000 (51B), 259,000 (51D), 250,000 (90D, 178D, and 90G).

Amino acid	Steers' 51A (moles %)	51A moles %	51A residues	51B moles %	51B residues	51D moles %	51D residues	90D moles %	90D residues	178D moles %	178D residues	90G moles %	90G residues
Lysine	5.19	4.64	136	4.28	108	7.64	197	8.55	207	8.17	191	5.74	138
Histidine	0.71	0.60	18	0.69	17	0.35	9	0.45	11	0.38	9	0.87	21
Arginine	1.34	1.21	35	1.66	42	1.26	33	1.45	35	1.67	39	1.45	35
Aspartic acid	12.01	11.80	345	11.74	296	12.94	334	11.94	289	12.40	290	11.26	271
Threonine	15.31	16.42	481	16.53	417	13.94	360	12.27	297	13.39	313	15.34	369
Serine	8.61	9.94	291	9.99	252	7.46	193	7.64	185	7.53	176	7.02	169
Glutamic acid	6.27	6.12	179	5.33	134	6.03	156	6.36	154	6.89	161	5.74	138
Proline	2.42	1.87	55	2.19	55	2.72	70	2.48	60	2.14	50	1.91	46
Glycine	7.50	6.87	201	6.89	174	7.96	206	10.17	246	9.58	224	9.60	231
Alanine	12.57	12.11	354	12.73	321	11.33	293	9.79	237	9.07	212	12.72	306
½ Cystine	9.54	11.19	328	10.98	277	10.85	280	10.41	252	10.44	244	10.52	253
Valine	5.09	4.61	135	4.27	108	4.16	108	3.84	93	3.51	82	4.36	105
Methionine	0.42	0.50	15	0.30	8	0.23	6	0.29	7	0.38	9	0.50	12
Isoleucine	2.88	2.48	73	2.64	67	2.55	66	2.44	59	2.57	60	2.41	58
Leucine	4.09	3.86	113	3.98	100	3.59	93	4.30	104	4.49	105	4.78	115
Tyrosine	2.87	2.72	80	2.47	62	3.51	91	4.05	98	4.19	98	3.16	76
Phenylalanine	1.68	1.55	45	2.00	50	2.47	64	2.89	70	2.52	59	1.54	37
Tryptophan	1.50	1.49	44	1.32	33	1.02	26	0.66	16	0.68	16	1.08	26
Total	100.00	99.98	2928	99.99	2521	100.01	2585	99.98	2420	100.00	2338	100.00	2406

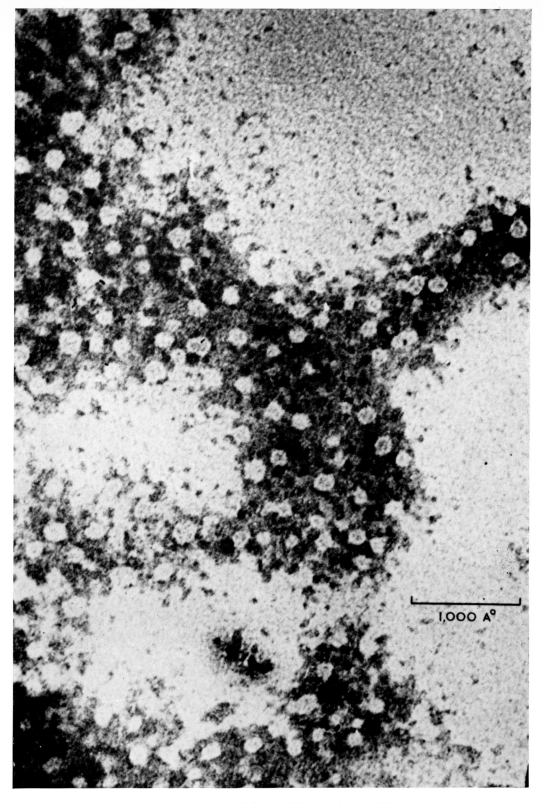

Fig. 2. Electron micrograph of purified immobilization antigen. (Mott, unpublished).

References p. 158–162

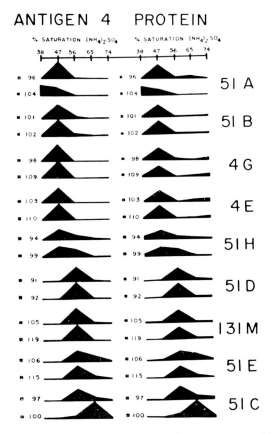

Fig. 3. Fractionation of immobilization antigen from different serotypes with ammonium sulfate. Each antigen was purified in duplicate (the duplicate preparations are indicated by the small numbers to the left) and the amount of antigen measured by gel diffusion (left hand column) and spectrophotometrically (right hand column). The crossreacting serotypes (A, B, G, and 4E; D and M) are precipitated by the same ammonium sulfate concentration. (Preer, 83).

8 S (83, 94, 95). Electron micrographs of antigen preparations indicate that the antigens are either clusters of subunits or tend to aggregate (67, 96). They could also be linear filaments which intertwine into an irregular sphere [Fig.3, (47a)]. In any case, the pictures of these particles present a structure somewhat different from that inferred from the hydrodynamic data, and Reisner *et al.*, have concluded that the antigens are probably oblate ellipsoids (95). Briefly, then, the antigens are large cigar-shaped globular proteins especially rich in disulfide-bonded amino acids.

D. Tertiary Structure

In some proteins this latter feature is responsible for the union of otherwise separate polypeptide chains (i.e., polypeptides with free amino and carboxy termini) into a single protein unit. Splitting the disulfide bond liberates the subunit, the scission being usually recognized by a lowering of molecular weight. With the immobilization antigens, Steers (127) detected a nine-fold decrease in molecular weight after reduction of the disulfide bonds and alkylation. An apparently uniform population of molecules with a molecular weight of 35,000 was observed in the ultracentrifuge. Added support for the existence of subunits came from an analysis of the fingerprints developed after tryptic digestion. From the number of arginine and lysine residues, assuming that all are accessible separately to enzymatic hydrolysis, 187 ninhydrin-positive spots should have appeared were the antigens not polymeric. Steers (126) actually observed 66, and concluded that there was a minimum of three unique primary sequences. By staining specifically for arginine, the same conclusion was reached. His model of the antigen (51A) was consistent with a molecule of nine chains, three of which were different in their amino acid sequence. Were his figure for the molecular weight of the native antigen to prove too high, he suggested that then the antigen would have only two identical subunits.

Jones (54), studying serotypes from a different syngen, also found that reduction and alkylation reduced the molecular weight of the antigen. From the number of spots after fingerprinting, he concluded that the antigen consists of two identical chains, with the possibility that these chains may in turn be composed of non-identical subunits. A more recent estimate places the number of chains at three (Jones, unpublished).

In marked contrast to these findings, Reisner *et al.*, working with some of the same antigens as Steers, were unable to detect any significant differences in molecular weight between reduced and native antigens (94). Their conclusion was that the immobilization antigens are "the largest known monomeric globular proteins", although they concede that the antigens Jones examined probably did have subunits. As Reisner *et al.* point out, if the antigens were composed of one uninterrupted polypeptide chain, they would be by far the largest such polypeptides in nature. Not only would they deserve attention on this score alone, but the interpretation of data demonstrating hybrid antigens in heterozygous cells would have to be drastically revised (see below). Nevertheless, in view of the several major points of agreement between Jones and Steers (allowing for differences in the estimates of the exact number of subunits) derived from the consideration of the fingerprint data and the immunological results obtained by Finger *et al.*, the weight of the evidence supports the view that the antigens are indeed composed of subunits.

The considerable range in molecular weights of some of the antigens has been interpreted as indicating that deletions or insertions may have occurred during the evolution of these proteins. Supporting this suggestion are the similarities in the relative abundance of amino acids in antigens of very different size (8.53 S vs 7.94 S and molecular weights of 301,500 vs. 259,000) (Table 2) (95). Similar reasoning could be used to support the view that there are large repeating sequences, such as might be expected if the antigens were polymers. Koizumi (58) has purified an immobilization antigen of *P. caudatum* and calculated its sedimentation coefficient to be 9.3 S and its molecular weight at 340,000, significantly larger than any of the *P. aurelia* antigens.

VI. RELATIONSHIPS AMONG ANTIGENS

A. Immunological Crossreactions

Comparisons of the various serotypes and of their purified antigens can either show marked similarities or equally striking differences, depending on the characteristic being measured (11, 35, 84). Immobilization tests reveal a number of crossreacting serotypes. Antigens specified by the same locus generally crossreact extensively, although there are some which do not at all. In many instances allelic antigens occurring in a number of stocks can barely be distinguished. Although antigens specified by different loci do not crossreact as often as do allelic antigens, nevertheless a large number are immobilized by heterologous antisera. Purified antigens studied by precipitation in gel present roughly the same picture: allelic antigens crossreact strongly, non-allelic antigens often not at all (31, 35, 42, 84).

B. Electrophoretic and Fingerprint Analyses

To purify the antigens Preer has employed ammonium sulfate to salt out the soluble proteins remaining after acidification of salt-alcohol extracted cells (79, 81, 82). Serologically related antigens are precipitated in the same ammonium sulfate cuts, non-crossreacting antigens in different cuts (Fig. 3). Jones modified this procedure and substituted for successive ammonium sulfate cuts gel filtration chromatography of the 75% ammonium sulfate precipitate (54). By further fractionation on hydroxyapatite columns he was able to separate two weakly crossreacting allelic antigens. Bishop and Beale have employed calcium phosphate columns to purify syngen 1 antigens (21).

When purified antigens are contrasted electrophoretically, non-crossreacting

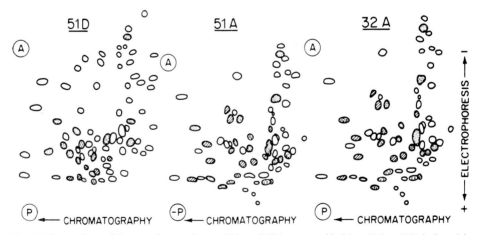

Fig. 4. Fingerprints of three surface antigens. 51A and 32A are specified by alleles. 51D is found in the same stock as 51A but is specified by a different locus. Tryptophan-positive peptides are indicated by cross-hatching while the stippled peptides represent tyrosine and/or histidine-positive peptides. (Steers, 126).

antigens usually can readily be distinguished; allelic antigens only with difficulty, if at all (21, 46, 84, 125). Fingerprints of some of these same allelic antigens generally show pronounced differences in patterns. The most distinctive patterns are found among serologically unrelated non-allelic antigens (Fig. 4). Non-crossreacting antigens (e.g., 51A and 51D) may have as few as 2 or 3 peptides of 75 in common (126). Crossreacting antigens may have somewhat similar patterns (51B and 51D), more nearly identical patterns (the allelic antigens 60G and 90G in syngen 1) or indistinguishable fingerprints (51A and 32A in syngen 4) (54, 126).

Reisner and Macindoe, using an adsorption technique, have reported that purified antigen and native antigen *in situ* do not have the same conformation. Surprisingly enough, the native antigen appears to have more exposed sites than the antigen in solution. Nevertheless, when antigens are compared by gel diffusion and adsorption, they crossreact more strongly than do these antigens as determined by immobilization tests, as though fewer sites were available in cells (20, 81, 84).

VII. THEORIES OF MUTUAL EXCLUSION

A. Delbrück Steady-state Model

Now that it has been firmly established that all immobilization antigens are very large proteins and share many physical and chemical properties, the various hypotheses proposed to account for mutual exclusion can be examined within this framework. Historically, the earliest, and still the most influential, scheme is that of Delbrück (24). According to his suggestion mutual exclusion of one antigen by another results from the inhibition, by an antigen itself, of an enzyme in the pathway leading to the synthesis of a second antigen. A steady state system is set up which, as long as one antigen is being made, determines that a second will be excluded. Should the original antigen be bound, destroyed, or its synthesis somehow interfered with, one of the antigens originally inhibited can now be made. In turn, this new antigen will assure the continued exclusion of the old antigen by inhibiting its synthesis. There are at least two objections to this theory in its specifics. First, it is difficult to imagine why the enzymes involved in the synthesis of the various antigens should differ. There is good evidence that the standard ribosomal system, which utilizes a common set of enzymes for all proteins, applies to these proteins also (93, 104). Of course, one could postulate that the specificity of inhibition resides in enzymes involved in final assembly or folding of the antigens, but there is no evidence for the existence of such a step. A more serious objection is how the specificity for exclusion can be extended to as many as a dozen antigens, which is the capacity of at least one strain that has been examined. Each antigen would have to be able to selectively inhibit eleven enzymes. Conceivably this objection could be overcome by postulating a rather complex system of highly polymeric enzymes, certain subunits of which are shared.

B. Modified Jacob–Monod Models

Preer favored a hypothesis according to which the antigens catalyze locus specific templates (presumably m-RNA's), and that there is competition for template precursors and for the amino acid pool (80). Another model, originating with Kimball (56, 57), also ascribes to the antigen a controlling role. According to this proposal the antigen acts as a self-inducer; the more antigen being made, the more that can be made. Finger (32), placing this same notion within a Jacob–Monod regulatory framework, has suggested that each antigen locus may have associated with it a regulatory gene producing a repressor specific for the antigen's operon. The loci for most antigens are normally repressed. One locus, that specifying the antigen currently expressed as the serotype, is not repressed. The serotype would be maintained through the inactivation, by the antigen, of the repressor for that operon.

C. Nanney Hypothesis

A third category of hypotheses is represented by one proposed by Nanney (68), in which the antigen itself plays no role in mutual exclusion. He suggests that "mutually exclusive elements" (DNA regions) have large regions of related sequences. Binding of homologous regions of an informational RNA molecule to DNA would occur with either its corresponding DNA, or with a related DNA. In the former case, messenger would continue to be produced normally; in the latter, messenger production would be impaired. One consequence of this model, presumably, is that crossreacting antigens, which might be expected to be specified by sequences of bases that have more regions of homology than would non-crossreacting antigens, should transform more often to each other than would non-crossreacting antigens. The data do not support this expectation. Also, judging by the fingerprints thus far published, many antigens appear to have relatively few regions in common (although admittedly the techniques of fingerprinting can magnify small differences in primary structure).

Even though mutual exclusion was recognized several decades ago and some of the theories are almost as ancient, the phenomenon is little understood and the hypotheses have not been very fruitful. No experiments have ruled out any of the models. Indeed, few experiments have been carried out with this in mind.

VIII. BIOCHEMICAL STUDIES IN TRANSFORMING SYSTEMS

Recently, Austin et al. (4, 5) have attempted to elucidate the sequence of events involved in transformation by the combined use of antigens and antibiotics or sequential exposure of cells to different combinations of antibiotics. The time at which the antibiotic (puromycin, chloramphenicol, or actinomycin D) was added relative to the time at which cells were exposed to antiserum, patulin, or acetamide determined the direction of transformation. When the inhibitors of protein synthesis were added at the start of transformation, a delay in transformation occurred; when added before antiserum treatment, transformation was promoted. Thus, it appeared that transformation occurred as a consequence of inhibition of antigen synthesis, either an antigen already on the surface or an antigen whose synthesis has just been induced.

Sommerville (106, 107, 108) has also studied the effect of actinomycin D on transformation, but his experiments differed in two major respects: (1) temperature change, rather than antiserum was used to induce transformation and (2) the appearance and disappearance of antigen was not measured by the immobilization reaction. Antigen synthesis was determined by following the incorporation of a pulse of radioisotope ^{35}S into antigen, using immunoelectrophoresis. Sommerville reported that, during transformation, synthesis of both old antigen and new antigen continued in both actinomycin D-treated and untreated cultures. He concluded that antigen synthesis possibly involved a pre-existing stable messenger RNA, and that the seeming contradiction between his and Austin *et al.*'s (4, 5) experiments could be due to the necessity for a large amount of antigen synthesis before the antigen would be recognized by the immobilization test. To reach this level new m-RNA might be needed. Finger (unpublished) has observed a pronounced effect of overnight exposure to actinomycin D on spontaneous transformation in unstable clones (see below). The results of Sommerville are of such importance that a confirmation of the reported differences in the effect of actinomycin D in short and long term synthesis is surely needed.

Direct evidence for the controlling role of antigen itself in its continuing synthesis is sparse. Paramecia fed purified antigen or antigen-antibody precipitates tend to become transformed in the direction of the kind of antigen introduced, especially if puromycin is added first (32). However, the variability in the responses was too great for undue significance to be attached to these results.

The studies by Beale's group (14, 15) on the molecular events occurring intracellularly during antigen synthesis and during transformation initially utilized fluorescent antibodies. While these experiments shed some light, the relatively poor resolution of the technique limited its usefulness. Sommerville (106, 107) has turned to radioactive isotopes to label *Paramecium* protein *in vivo*, using bacteria grown on ^{14}C-leucine or ^{35}S-magnesium sulfate. He has been able to demonstrate that within 30 minutes after feeding, labeled antigen is detectable in homogenates. Only 30–60 minutes later can the antigen be detected in extracts of the cell surface, although with ferritin-labeled antibody new surface antigen can be recognized from the time cells are placed at a new temperature [R. E. Sinden cited in Sommerville (108)]. Thus, it appears that newly synthesized antigen may remain internally for a considerable length of time before finding its way to the cilium and pellicle. Other pulse labeling experiments also indicate that a pre-existing pool of antigen does not exist (103). Since different techniques, having different sensitivities were used, it is possible that apparently discrepant conclusions may be due to this factor.

A perhaps equally significant step forward is the adaptation to *Paramecium* of a widely used cell-free preparation for amino acid incorporation: ribosomes, a soluble fraction, guanosine triphosphate, and adenosine triphosphate. Little if any *de novo* synthesis occurs. In this system apparently only already initiated chains are completed. The amount of labeled protein found free of ribosomes can be increased three-fold by the addition of puromycin after the original incubation (93). The chief sites for antigen synthesis *in vivo* are polyribosomes (108). Using the same components Sommerville (105) has reported that labeled immobilization antigen made in this system can be detected in immune precipitates. Unfortunately, the usefulness of this approach, while potentially great, is currently restricted by the low level of incorporation (about 40 counts/minute) and by the fact that complete molecules of new antigen may not be synthesized. In all of the *in vivo* experiments the paramecia were maintained on bacterized medium. Although this introduces a variable that may be of significance in certain experiments—e.g., those in which antibiotics were employed—internal controls often can take this into account. In some cases the possible role of bacteria has been explicitly examined. For example, in assaying the influence of cell products on sero-type transformation (see below) the bacteria from individual cultures were separated and introduced into various other cultures. The pattern of transformation induced was shown to be independent of the kind of bacterial flora present (Finger, un-published).

Many of the models suggested as explanations for the regulation of protein synthesis have implicit in them the premise that the protein is an enzyme. Often the catalytic function itself plays a direct role in the control of the protein's synthesis or an indirect role in the synthesis of another macromolecule. Undoubtedly different kinds of questions would be raised about mutual exclusion and transformation than have been dealt with heretofore if the antigens were enzymes. The truth of the matter, however, is that we do not know what function the antigens serve, we simply know how to recognize them.

IX. LOCATION AND POSSIBLE FUNCTION

Sometimes the location of a molecule offers a clue to its function. Since the earliest observations of the immobilization reaction it was assumed that the antigens were ciliary antigens. Using fluorescent antibody with the light microscope and, later, ferritin-labeled antibodies with the electron microscope, Beale and his collaborators (14, 15, 64, 65, 66) have been able to detect the antigen not only on the cilia, but also on the pellicle proper. None was detectable internally. During transformation Mott (66) showed that new antigen appeared before immobilization tests gave any indication of its presence, first at isolated pellicular sites and at the bases of cilia (Fig. 5). In time the entire pellicle became covered with the new antigen and, finally, entire cilia became coated, with the exception of those in the gullet. There was no evidence of antigen transport from the interior. Preer and Preer (87) confirmed the presence of the antigens in these locations when they were able to extract the antigen from isolated cilia and deciliated bodies. Seed *et al.* (100) were also able to adsorb ciliary antibody with de-ciliated bodies as well as with cilia.

Because originally the presence of the antigens was recognized by a cessation of

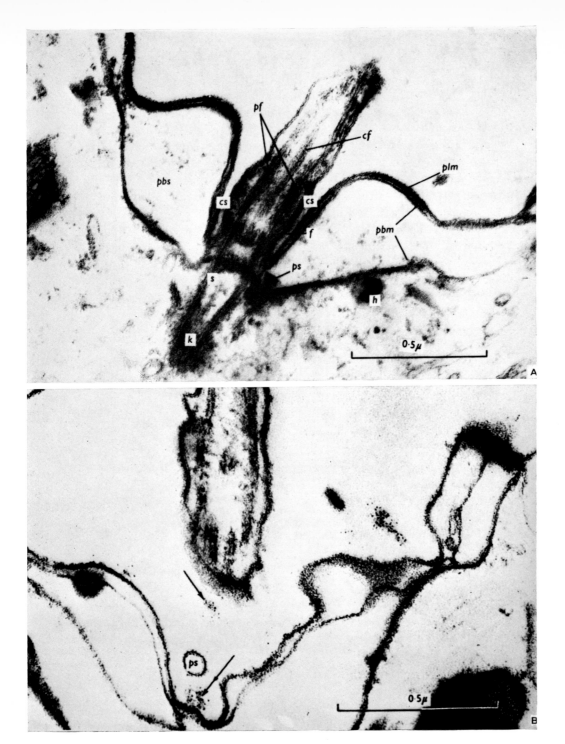

Fig. 5. Electron micrographs of sections of paramecia treated with ferritin-labeled antibody specific for the new serotype. *A:* Immediately preceding transformation. No ferritin granules visible. *B:* One hour after induction of transformation. Double arrow indicates ferritin as globulin "fuzz" on pellicle. Single arrow indicates fuzz on cilium.

Abbreviations: cf = central fibrils; cs = circumciliary space; csc = cross section of cilium; f = fold; k = kinetosome; pbm = peribasal membrane; pbs = peribasal sac; pf = peripheral fibrils; plm = pellicular membrane; ps = portion of parasomal sac; s = septum. (Mott, 66, 67).

swimming, it seemed probable that the antigens may be part of a paramecium's motor mechanism. From 51A cells, Van Wagtendonk and Vloedman (129) were able to isolate a preparation with ATPase activity that precipitated with 51A antiserum. Unfortunately, since neither the purity of the enzyme nor the specificity of the antibodies present in the antiserum were known, the data say little about the antigen as an ATPase. Reisner *et al.* (95), from a consideration of the amino acid distribution in the antigens, suggest that there are few exposed hydrophobic regions, a property which would enhance the antigen's ability to protect the cell's surface. Other possible functions would be to regulate the interaction of certain substances with the membrane or influence membrane permeability (85). Whatever the reason for the antigens' existence they are very likely important molecules. This is suggested by the large number of alternate forms available to a cell, their universal presence, and by the observation that when a cell's antigen is bound by antibody and protein synthesis is inhibited at the same time, the cell dies (Seed and Finger, unpublished). There is little evidence for an adaptive advantage of one serotype over another except under very special conditions (6, 88, 89).

X. HYBRID ANTIGENS AND ALLELIC EXCLUSION

A. Variations in Gene Expression in Heterozygotes

In recent years, immunological techniques have been employed to probe more deeply into the events occurring within hybrid cells. As already noted, heterozygotes generally do not display dominance; usually both alleles are expressed (10, 11, 25). To properly study in some detail the kinds of molecules produced, it was necessary to have a reasonable number of "markers" that could be used to distinguish two parental antigens. Through an analysis of gel diffusion patterns a number of allelic forms specified by the *c* locus, syngen 2, were assigned a mosaic of determinants (Table 3) (31, 42). Following a series of crosses, individual exconjugants were grown into mass cultures from which the immobilization antigen was extracted. The three major findings were (45, 46): (1) heterozygotes produced antigens that were truly hybrid molecules and not simply a mixture of individual parental molecules; (2) each

TABLE 3

Matrix of antigenic determinants for C serotypes as determined by gel diffusion. (Finger, 31).

Serotype	Determinants										
30C		F	G	H	I	J		L	M		O
7C	C	F	G		I	J			M		
72C		F		H		J	K	L	M		
83C				H	I				M	N	O
197C		F		H	I		K	L		N	O

kind of cross produced a spectrum of different hybrid molecules; and (3) from an individual exconjugant clone only one kind of hybrid antigen was detected. The methods used did not allow a firm conclusion as to whether free parental molecules were made by a heterozygote and whether relatively small amounts of several kinds of hybrid molecules were "contaminating" an apparently uniform population of antigen molecules. Jones (53), assaying antigen by the inhibition of the immobilization reaction, was able to clearly demonstrate parental antigen in heterozygotes. Indeed, in the stocks and serotypes employed by him, hybrid antigen was a minority component.

B. Model of Antigen Structure deduced from Hybrid Antigens

From the number and kind of hybrid molecules detected, Finger *et al.* (45, 46), concluded that the simplest model of the antigen that would account for their data was that of a dimer, each part of which was composed of three subunits. Hybrid molecules could be any one or more of twenty-five types, depending on the differences and kinds of subunits characterizing the two parents. This model is in agreement with the chemical data of Jones (54) and of Steers (126, 127). To reconcile the immunological findings with the report of Reisner *et al.* (95) that the antigen is a single uninterrupted chain, the assumption would have to be made that hybrid molecules result from either somatic recombination within a locus (which would have to occur with a high frequency), or recombination at the m-RNA level (cf. Schwartz) (98, 99).

Perhaps the most intriguing result from these experiments with heterozygotes was the clonal appearance of hybrid antigens. Even though each kind of heterozygote apparently possesses the capability for making several types of hybrid molecules, one class alone was preponderant within a clone; the other types were either completely or partly missing. In some respects this situation is not unlike mutual exclusion, in which a restricted amount of genetic information is expressed, with the difference that in heterozygotes the restriction is between alleles at a locus. Perhaps a better analogy would be to the phenomenon of "allelic repression", a form of which was described earlier with the E serotypes.

XI. CLONAL STABILITY

Equally important was the possibility that the exclusion might be related to conditions of growth in the flask or carboys prior to harvesting and extraction. Finger and co-workers were to suggest, in later publications, that clones with low densities of cells (32, 33, 41) (as in depression slides having few paramecia) appeared to be more heterogeneous in their serotypes than did clones in mass cultures (as in test tubes with many paramecia per ml). It was these observations that led to extended studies on the phenomenon of clonal stability, a condition related to but distinguishable from serotype stability. These new studies in turn were to lead full circle to a reconsideration of mutual exclusion.

Two test tube cultures, which can maintain the same serotype over an extended period of time, nonetheless may readily be told apart by either of two simple operations (Finger *et al.* Ad 135). The first class of clones will yield upon extraction a pure single antigen as determined by *in vitro* tests. Also, subclones originating from individual cells of this culture will consist exclusively of cells of the same serotype as the parental culture. A clone meeting these two criteria is considered stable. A second class of clones consists of cells which by the immobilization test do not differ in their serotype from stable clones. However, several kinds of immobilization antigen can be extracted from them. [These are the "secondary" antigens originally described by Sonneborn (110, 112) which are immunologically identical to primary antigens of other serotypes (43, 100)]. Moreover, subclones derived from these clones are not pure for the original serotype; many of them are a mixture of serotypes. A clone exhibiting these characteristics is termed unstable.

Fig. 6. Comparison of homologous (e.g., G recipient plus G donor's medium) and heterologous (e.g., G recipient plus C donor's medium) mixtures of conditioned medium. Each horizontal bar represents an individual mixture of a conditioned medium with the recipient clone listed below. The length of a bar represents the percentage of increase or decrease in the original serotype compared with the recipient clone resuspended in its own medium. In the heterologous mixtures 21 of 28 donors inhibited the expression of the original serotype, while in the homologous mixtures only 4 of 17 donors inhibited the recipient's initial serotype. (Finger, Heller and Larkin, 42).

The cells which constitute an unstable clone only retain their serotype if maintained *en masse*. When isolates are made, the derived subclones often display new serotypes. This observation led to the suggestion that an interaction may occur between cells, perhaps via the culture fluid, which affects the overall serotype of a culture. Experiments in which media from different cultures were added to a variety of "recipient" cells supported this idea (32, 39, 41) (Fig. 6). Only unstable clones were susceptible to such conditioned media and the kind of response evoked was independent of the serotype of the "donor" culture. The significant factors in determining the effect of the substituted medium appeared to be the kind of antigen that was being made most actively by a culture (Finger, unpublished). It was this antigen that was most likely to be replaced in the recipient clone after exposure to culture medium from an appropriate donor. Conversely, with respect to the donor providing a particular medium the effect of the medium was determined by the antigen being synthesized by the donor clone. In all of these experiments the effect of added medium could only be recognized after individual cells were isolated and the subclones derived from these cells tested; the parental tubes generally consisted of untransformed cells. These results could be explained in a general way by postulating that the effective agents in the medium were a collection of either inhibitors or of inducers. Native antigen as the active factor was excluded since none was usually detectable in conditioned medium (32).

Many stable and unstable clones, although capable of retaining their characteristic stabilities for months under a weekly feeding schedule, may spontaneously transform their serotype or stability pattern. If the culture medium of either a stable or an unstable culture is replaced with fresh culture fluid daily this conversion is vastly speeded up, with unstable clones converting to stable clones much more quickly than vice versa (34, Ad 132).

XII. PERSISTENT UNSTABLE CLONES AND SECONDARY ANTIGENS

One type of especially persistent unstable clone is unique in several respects. Cells from this clone are not immobilized by any of the antisera commonly used to diagnose serotypes and have not been successfully employed to elicit antibodies in rabbits. These "Z" cells also possess a remarkable array of other attributes. Secondary antigens can regularly be extracted from them (44) (Fig. 7); they are difficult to convert to stable clones by continuous replacement of medium; and their medium has little tendency to induce changes in recipient cells.

A. Characteristics of Secondary Antigens

Originally, the existence of secondary antigens was proposed by Sonneborn (110, 112, 113) to explain the crossreactivity of certain antisera prepared against a culture pure for a particular serotype. He suggested that there were several kinds of antibodies because these cells contained two kinds of antigens: primary antigens, which are on

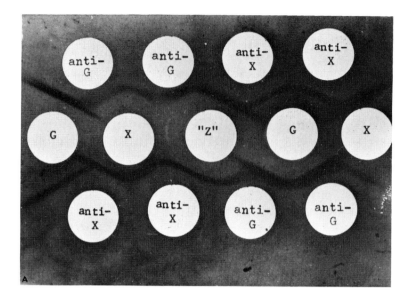

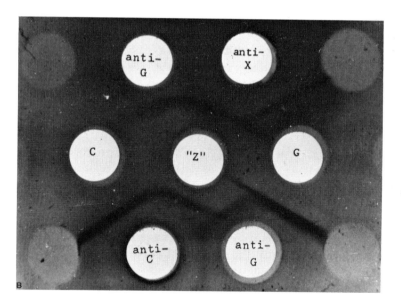

Fig. 7. Ouchterlony patterns of secondary antigens from "Z" clones. *A:* Antigens from a "Z" clone showing a reaction of identity with C antigen and partial identity with G antigen (with anti-G and anti-X sera). Since G and X antigens crossreact, this indicates that this "Z" clone has both C and X antigens. *B:* A direct demonstration that one of antigens from the "Z" clone is X, with the expected reaction of identity, using X antigen and anti-X serum. This particular "Z" clone contained no detectable G antigen. (Finger, Onorato, Heller and Dilworth, 44).

the surface and recognizable by the immobilization test, and secondary antigens, which may be internal. We now know that while some crossreactions are due to structural relatedness between immobilization antigens (29, 35, 54, 126), others are due to the presence of more than one antigen in the cells or homogenates injected, as Sonneborn suggested (40, 100).

Persistent multiple antigens were first demonstrated by Margolin (61) with serotypes M and D in syngen 4. The simultaneous presence of M and D, however, was restricted to the serotype M of stock 172 only. Because of the highly special nature of this situation it was difficult to assess the significance of this apparent exception to mutual exclusion. It was only after the antigens could be assayed *in vitro* that it was evident that secondary antigens could occur in many stocks in syngen 2 (40, 100), and that any primary antigen could appear as a secondary antigen. Thus, the seeming ubiquity of mutual exclusion need not be real, but may reflect the special conditions for recognizing and maintaining a serotype. In syngen 1, where unstable clones are apparently rare, the occurrence of secondary antigens has not been described.

B. *Localization of Secondary Antigens*

Of considerable relevance to any consideration of mutual exclusion is another observation having to do with secondary antigens. Adsorption of specific immobilizing antisera by cilia, deciliated bodies and other fractions indicates that the distribution of these antigens is the same as the primary antigens. Further evidence to support this conclusion comes from an experiment in which Z cells which were not immobilized after being exposed simultaneously to two different immobilizing antisera became immobilized when anti-rabbit gamma globulin serum was subsequently added. The explanation suggested for this result is that Z cells have several antigens on their surface but in such small quantities that bivalent antibody cannot find neighboring antigenic groups to bridge cilia, thus agglutinating them (Finger, unpublished). With the anti-gamma globulin serum this difficulty is overcome by the binding of these new antibodies to all antibody molecules without respect to the ciliary specificities of their combining sites.

These observations on Z clones, together with those of Seed *et al.* (100), on other more typical clones in which secondary antigens have been detected, dispel the notion that there may be something in the topology of a cell which plays a prominent role in mutual exclusion. More important, however, the very existence of Z clones and of Margolin's (61) M-D clones indicates that cells can regularly and continuously manufacture several antigens at the same time. In addition, from these studies and the quantitative data of Balbinder and Preer (7) on transformation, it appears likely that there exists a regulatory system which guarantees that the total amount of all antigens be kept relatively constant. How the cell manages this for a class of proteins that may not be enzymes is totally unknown. One could assume that all of the immobilization antigens, even those that appear to be serologically unrelated, possess a common subunit or sequence of amino acids which is the functional site of the molecule (cf. Allen for further discussion of several models involving shared subunits). The locus specifying this subunit need not be under the control of inducers or repressors (i.e., it is constitutive), although the loci determining the structure of the subunits distinguishing individual antigens are inducible. Such a system would probably account for mutual exclusion, where it exists, and the overall cell content of surface antigens, which would be determined by the rate of synthesis of the common subunit. One virtue of this suggestion is that it should lead to further experimentaion.

XIII. CODA

It is plain from this survey of the surface antigens that although much is known about their genetics, biology and chemistry, dismayingly little is understood about many aspects. Old phenomena have yet to be redescribed at a molecular level. The function of the antigens is unknown. The mechanisms of mutual exclusion and transformation are unknown. The mode of action of cellular products present in conditioned medium is unknown. Whether the antigens persist in some form in higher organisms is unknown. Nonetheless, it is encouraging that some of these questions are new and that the old questions are being recast within promising new frameworks.

A number of reviews have appeared in recent years, considerable portions of which have been devoted to the surface antigens. Especially useful are those of Preer (85), Allen (1), and Finger (32) and Sommerville (108). Others which touch upon this topic include articles by Nanney (71) and Child (23). Somewhat older but still valuable general reviews or articles are those by Beale and Wilkinson (16), Beale (12, 13), Nanney (70), Preer (78, 80) and Kimball (57). Beale's (11) classic monograph, Wichterman's (130) more general book, and Sonneborn's (111, 114, 115, 116) lengthy reviews of the methods and genetics of *P. aurelia* still contain much that is relevant and are needed to place the more recent information in proper perspective.

XIV. REFERENCES
(Abstracts are indicated by *)

1   Allen, S. L. 1967. The chemical genetics of the Protozoa. In Florkin, M. & Scheer, B., *Chemical Zoology*, Vol. 1, Academic Press, New York, N.Y. 617–694.
2   *Austin, M. L. 1963. The influence of an exchange of the genes determining the D antigen between stocks 51 and 32, syngen 4, *Paramecium aurelia*, on the transformation of D to B and A in patulin. *Genetics* **48**, 881.
3   *— 1963. Progress in control of the emergence and the maintenance of serotypes of stock 51, syngen 4, of *Paramecium aurelia*. J. Protozool. **10** (Suppl.), 21.
4   —, Pasternak, J. & Rudman, B. M. 1967. Studies on the mechanism of serotypic transformation in *Paramecium aurelia*. I. The effects of actinomycin D, puromycin, and chloramphenicol on an antiserum-induced transformation. *Exp. Cell Res.* **45**, 289–305.
5   —, — & — 1967. Studies on the mechanism of serotypic transformation in *Paramecium aurelia*. II. The effects of actinomycin D, puromycin, and chloramphenicol on transformations induced by patulin, acetamide, and antiserum. *Exp. Cell Res.* **45**, 306–322.
6   —, Widmayer, D. & Walker, L. M. 1956. Antigenic transformation as adaptive response of *Paramecium aurelia* to patulin; relation to cell division. *Physiol. Zool.* **29**, 261–287.

7 Balbinder, E. & Preer, Jr., J. R. 1959. Gel diffusion studies on serotype and serotype transformation in *Paramecium. J. Gen. Microbiol.* **21**, 156–167.

8 Beale, G. H. 1948. The process of transformation of antigenic type in *Paramecium aurelia*, variety 4. *Proc. Nat. Acad. Sci. U.S.* **34**, 418–423.

9 *— 1952. Genic control of gene expression in *Paramecium aurelia. Science* **115**, 480.

10 — 1952. Antigen variation in *Paramecium aurelia*, variety 1. *Genetics* **37**, 62–74.

11 — 1954. *The Genetics of Paramecium aurelia.* Cambridge University Press, New York, N.Y.

12 — 1957. The antigen system of *Paramecium aurelia. Intern. Rev. Cytol.* **6**, 1–23.

13 — 1966. The role of the cytoplasm in heredity. *Proc. Roy. Soc., B* **164**, 209–218.

14 — & Kacser, H. 1957. Studies on the antigens of *Paramecium aurelia* with the aid of fluorescent antibodies. *J. Gen. Microbiol.* **17**, 68–74.

15 — & Mott, M. R. 1962. Further studies on the antigens of *Paramecium aurelia* with the aid of fluorescent antibodies. *J. Gen. Microbiol.* **28**, 617–623.

16 — & Wilkinson, J. R. 1961. Antigenic variation in unicellular organisms. *Ann. Rev. Microbiol.* **15**, 263–296.

17 Beisson, J. & Rossignol, M. 1969. The first case of linkage in *Paramecium aurelia. Genet. Res.* **13**, 85–90.

18 Bernheimer, A. W. & Harrison, J. A. 1940. Antigen-antibody reactions in *Paramecium:* the *aurelia* group. *J. Immunol.* **39**, 73-83.

19 Bishop, J. O. 1961. Purification of an immobilization antigen of *Paramecium aurelia*, variety 1. *Biochim. Biophys. Acta* **50**, 471–477.

20 — 1963. Immunological assay of some immobilizing antigens of *Paramecium aurelia*, variety 1. *J. Gen. Microbiol.* **30**, 271–280.

21 — & Beale, G. H. 1960. Genetical and biochemical studies of the immobilization antigens of *Paramecium aurelia. Nature* **186**, 734.

22 Capdeville, Y. 1969. Sur les interactions entre allèles controlant le type antigénique G chez *Paramecium aurelia. Compt. Rend.* **269**, 1213–1215.

23 Child, F. M. 1967. The chemistry of protozoan cilia and flagella. In Florkin, M. & Scheer, B., *Chemical Zoology*, Vol. 1, Academic Press, New York, N.Y. 381–393.

24 Delbrück, M. 1949. See discussion to paper of Sonneborn, T. M. & Beale, G. H. in *Unités Biologiques Dousées de Continuité Génétique*, C.M.R.S., Paris **7**, 33–34.

25 *Dippel, R. V. 1953. Serotypic expression in heterozygotes of variety 4, *P. aurelia. Microbial Genet. Bull.* **7**, 12.

26 *Dryl, S. 1959. Antigenic transformation in *Paramecium aurelia* after homologous antiserum treatment during autogamy and conjugation. *J. Protozool.* **6** (Suppl.), 25.

27 — 1965. Antigenic transformation in relation to nutritional conditions and the interautogamous cycle in *Paramecium aurelia. Exp. Cell Res.* **37**, 569–581.

28 Finger, I. 1956. Immobilizing and precipitating antigens of *Paramecium. Biol. Bull.* **3**, 358–363.

29 — 1957. Immunological studies of the immobilization antigens of *Paramecium aurelia*, variety 2. *J. Gen. Microbiol.* **16**, 350–359.

30 — 1957. The inheritance of the immobilization antigens of *Paramecium aurelia*, variety 2. *J. Genet.* **55**, 361–374.

31 — 1964. Use of simple gel-diffusion techniques to assign antigenic markers to native proteins. *Nature* **203**, 1035–1039.

32 — 1967. The control of antigenic type in *Paramecium*. In Goldstein, L., *The Control of Nuclear Activity*, Prentice-Hall, Englewood Cliffs, N.J. 377–411.

33 — 1968. Gene activation by cell products. *Trans. N.Y. Acad. Sci.* **30**, 968–976.

34 —, Dilworth, L., Heller, C. & Fishbein, G. 1968. Transformation of antigenic types in *Paramecium* by conditioned medium. *Genetics* **60**, 177.

35 — & Heller, C. 1962. Immunogenetic analysis of proteins of *Paramecium*. I. Comparison of specificities controlled by alleles and by different loci. *Genetics* **47**, 223–239.

36 — & — 1963. Immunogenetic analysis of proteins of *Paramecium*. IV. Evidence for presence of hybrid antigens in heterozygotes. *J. Mol. Biol.* **6**, 190–202.

37 — & — 1964. Cytoplasmic control of gene expression in *Paramecium*, I. Preferential expression of a single allele in heterozygotes. *Genetics* **49**, 485–498.

38 — & — 1964. Immunogenetic analysis of proteins of *Paramecium*, V. Detection of specific determinants in strains lacking a surface antigen. *Genet. Res.* **5**, 127–136.

39 *— & — 1965. Induction of gene expression in *Paramecium* by cell-free culture fluid. *Amer. Zool.* **5**, 649.

40 —, — & Dilworth, L. 1969. Effects of immobilizing antiserum and normal serum on *Paramecium* surface antigen synthesis. *J. Protozool.* **16**, 12–18.

41 —, — & Green, A. 1962. Immunogenetic analysis of proteins of *Paramecium*, II. Coexistence of two immobilization antigens within animals of a single serotype. *Genetics* **47**, 241–253.

42 —, — & Larkin, D. 1967. Repression of gene expression by cell products in *Paramecium*. *Genetics* **56**, 793–800.

43 —, — & Smith, J. P. 1963. Immunogenetic analysis of proteins of *Paramecium*, III. A method for determining relationships among antigenic proteins. *J. Mol. Biol.* **6**, 182–189.

44 —, Onorato, F., Heller, C. & Dilworth, L. 1969. Role of non-surface antigens in controlling *Paramecium* surface antigen synthesis. *J. Protozool.* **16**, 18–25.

45 *—, —, — & Wilcox, III, H. B. 1965. Antigen structure and synthesis in *Paramecium*. *Progress in Protozoology* (Proc. 2nd Intern. Conf. Protozool., London), 244.

46 —, —, — & Wilcox, III, H. B. 1966. Biosynthesis and structure of *Paramecium* hybrid antigen. *J. Mol. Biol.* **17**, 86–101.

47 Gorovsky, M. A. & Woodward, J. 1969. Studies on nuclear structure and function in *Tetrahymena pyriformis*. I. RNA synthesis in macro- and micronuclei. *J. Cell Biol.* **42**, 673–682.

48 Harrison, J. A. 1955. General aspects of immunological reactions with bacteria and protozoa. In Butler, F. G., *Biological Specificity and Growth*, Princeton University Press, Princeton, N.J. 141–156.

49 — & Fowler, E. H. 1945. Antigenic variation in clones of *P. aurelia*. J. Immunol. **50**, 115–125.

50 Herzenberg, L. A., Minna, J. D. & Herzenberg, L. A. 1967. The chromosome region for immunoglobulin heavy chains in the mouse: allelic electrophoretic differences and allotype suppression. *Cold Spring Harbor Symp. Quant. Biol.* **32**, 181–186.

51 *Hiwatashi, K. 1963. Serotype inheritance in *Paramecium caudatum*. *Genetics* **48**, 892.

52 Jollos, V. 1921. Experimentelle Protistenstudien. I. Untersuchungen über Variabilität und Vererbung bei Infusorien. *Arch. Protistenk.* **43**, 1–222.

53 Jones, I. G. 1965. Immobilization antigen in heterozygous clones of *Paramecium aurelia*. *Nature* **207**, 769.

54 — 1965. Studies on the characterization and structure of the immobilization antigens of *Paramecium aurelia*. *Biochem. J.* **96**, 17–23.

55 — & Beale, G. H. 1963. Chemical and immunological comparisons of allelic immobilization antigens in *Paramecium aurelia*. *Nature* **197**, 205–206.

56 Kimball, R. F. 1947. The induction of inheritable modification in reaction to antiserum in *Paramecium aurelia*. *Genetics* **32**, 486–499.

57 — 1964. Physiological genetics of the ciliates. In Hutner, S. H., *Biochemistry and Physiology of Protozoa*, Vol. 3, Academic Press, New York, N.Y. 243–275.

58 Koizumi, S. 1966. Serotypes and immobilization antigens in *Paramecium caudatum*. *J. Protozool.* **13**, 73–76.

59 Mage, R. & Dray, S. 1965. Persistent altered phenotype expression of allelic γG immunoglobulin allotypes in heterozygous rabbits exposed to isoantibodies in fetal and neofetal life. *J. Immunol.* **95**, 525–535.

60 Margolin, P. 1956. The ciliary antigens of stock 172, *Paramecium aurelia*, variety 4. *J. Exp. Zool.* **133**, 345–387.

61 — 1956. An exception to mutual exclusion of the ciliary antigens in *Paramecium aurelia*. *Genetics* **41**, 685–699.

62 Markert, C. 1963. The origin of specific proteins. In Allen, J. M., *The Nature of Biological Diversity*, McGraw-Hill, New York, N.Y. 95–119.

63 Monod, J. & Jacob, F. 1961. General conclusions: Teleonomic mechanisms in cellular metabolism, growth and differentiation. *Cold Spring Harbor Symp. Quant. Biol.* **26**, 389–401.

64 Mott, M. R. 1963. Cytochemical localization of antigens of *Paramecium* by ferritin-conjugated antibody and by counterstaining the resultant absorbed globulin. *J. Roy. Microscop. Soc.* **81**, 159–162.

65 *— 1963. Identification of the sites of the antigens of *Paramecium aurelia* by means of electron microscopy. *J. Protozool.* **10** (Suppl.), 31.

66 — 1965. Electron microscopy studies on the immobilization antigens of *Paramecium aurelia*. *J. Gen. Microbiol.* **41**, 251–261.

67 *— 1965. Electron microscopy of the immobilization antigens of *Paramecium aurelia. Progress in Protozoology* (Proc. 2nd Intern. Conf. Protozool., London), 250.

68 Nanney, D. L. 1963. Aspects of mutual exclusion in Tetrahymena. In Harris, R. J. C., *Biological Organization at the Cellular and Supercellular Level*, Academic Press, London, 91–109.

69 — 1963. The inheritance of H-L serotype differences at conjugation in *Tetrahymena. J. Protozool.* **10**, 152–155.

70 — 1964. Macronuclear differentiation and subnuclear assortment in ciliates. In Locke, M., *Role of Chromosomes in Development*, Academic Press, New York, N.Y. 253–273.

71 — 1968. Ciliate genetics: patterns and programs of gene action. *Ann. Rev. Genet.* **2**, 121–140.

72 — & Dubert, J. M. 1960. The genetics of the H serotype system in variety of *Tetrahymena pyriformis. Genetics* **45**, 1335–1349.

73 —, Reeve, S. J., Nagel, J. & DePinto, S. 1963. H serotype differentiation in *Tetrahymena. Genetics* **48**, 803–813.

74 Needham, A. E. 1965. *The Uniqueness of Biological Materials*, Pergamon Press, Oxford.

75 Oakley, C. L. & Fulthorpe, A. J. 1953. Antigenic analysis by diffusion. *J. Pathol. Bacteriol.* **65**, 49–60.

76 Pasternak, J. 1967. Differential genic activity in *Paramecium aurelia. J. Exp. Zool.* **165**, 395–418.

77 Preer, Jr., J. R. 1956. A quantitative study of a technique of double diffusion in agar. *J. Immunol.* **77**, 52–60.

78 — 1957. Genetics of the protozoa. *Ann. Rev. Microbiol.* **11**, 419–438.

79 *— 1958. Isolation of the immobilization antigens of *Paramecium. Anat. Rec.* **131**, 591.

80 — 1959. Nuclear and cytoplasmic differentiation in the Protozoa. In Rudnick, D., *Developmental Cytology*, The Ronald Press, New York, N.Y. 3–18.

81 — 1959. Studies on the immobilization antigens of *Paramecium*, I. Assay methods. *J. Immunol.* **83**, 276–283.

82 — 1959. Studies on the immobilization antigens of *Paramecium*, II. Isolation. *J. Immunol.* **83**, 378–384.

83 — 1959. Studies on the immobilization antigens of *Paramecium*, III. Properties. *J. Immunol.* **83**, 385–391.

84 — 1959. Studies on the immobilization antigens of *Paramecium*, IV. Properties of the different antigens. *Genetics* **44**, 803–814.

85 — 1969. Genetics of the Protozoa. In Chen, T. T., *Research in Protozoology*, Vol. 3, Pergamon Press, Oxford, 129–278.

86 *—, Bray, M. & Koizumi, S. 1963. The role of cytoplasm and nucleus in the determination of serotype in *Paramecium. Proc. XIth Intern. Congr. Gen. The Hague* **1**, 189.

87 — & Preer, L. B. 1959. Gel diffusion studies on the antigens of isolated cellular components of *Paramecium. J. Protozool.* **6**, 88–100.

88 Pringle, C. R. 1956. Antigenic variation in *Paramecium aurelia*, variety 9. *Z. Indukt. Abstamm. Vererb.* **87**, 421–430.

89 — & Beale, G. H. 1960. Antigenic polymorphism in a wild population of *Paramecium aurelia. Genet. Res.* **1**, 62–68.

90 Rao, M. V. N. & Prescott, D. M. 1967. Micronuclear RNA synthesis in *Paramecium caudatum. J. Cell Biol.* **33**, 281–285.

91 Reisner, A. H. 1955. A method of obtaining specific serotype mutants in *Paramecium aurelia* stock 169, variety 4. *Genetics* **40**, 591–592.

92 — & Macindoe, H. 1967. Immunological evidence indicating conformational differences between surface-bound and solubilized serotypic antigen of *Paramecium. Biochim. Biophys. Acta* **140**, 529–531.

93 — & — 1967. Incorporation of amino acid into protein by utilizing a cell-free system from *Paramecium. J. Gen. Microbiol.* **47**, 1–15.

94 —, Rowe, J. & Macindoe, H. M. 1969. The largest known monomeric globular proteins. *Biochim. Biophys. Acta* **188**, 196–206.

95 —, — & Sleigh, R. W. 1969. Concerning the tertiary structure of the soluble surface proteins of *Paramecium. Biochemistry* **8**, 4637–4644.

96 Rössle, R. 1905. Spezifische Sera gegen Infusorien. *Arch. Hyg. Bakteriol.* **54**, 1–31.

97 — 1909. Zur Immunität einzelliger Organismen. *Verhandl. Deut. Pathol. Ges.* **13**, 158–162.

98 Schwartz, D. 1960. Genetic studies on mutant enzymes in maize: synthesis of hybrid enzymes by heterozygotes. *Proc. Nat. Acad. Sci. U.S.* **46**, 1210–1215.

99   — 1962. Genetic studies on mutant enzymes in maize. II. On the mode of synthesis of the hybrid enzymes. *Proc. Nat. Acad. Sci. U.S.* **48**, 750–756.

100 Seed, J. R., Shafer, S., Finger, I. & Heller, C. 1964. Immunogenetic analysis of proteins of *Paramecium*. VI. Additional evidence for the expression of several loci in animals of a single antigenic type. *Genet. Res.* **5**, 137–149.

101 Sinclair, I. J. B. 1958. The role of complement in the immune reactions of *Paramecium aurelia* and *Tetrahymena pyriformis*. *Immunology* **1**, 291–299.

102 *Sinden, R. E. 1969. Serotype transformation in *Paramecium aurelia*. 2. *J. Protozool.* **16** (Suppl.), 27.

103 Skaar, P. D. 1956. Past history and pattern of serotype transformation in *Paramecium aurelia*. *Exp. Cell Res.* **10**, 646–656.

104 Sommerville, J. 1967. Immobilization antigen synthesis in *Paramecium aurelia:* the detection of labelled antigen in a cell-free amino acid incorporating system. *Biochim. Biophys. Acta* **149**, 625–627.

105 — 1968. Immobilization antigen synthesis in *Paramecium aurelia*. *Exp. Cell Res.* **50**, 660–664.

106 *— 1969. Serotype transformation in *Paramecium aurelia*. 1. *J. Protozool.* **16** (Suppl.), 27

107 — 1969. Serotype transformation in *Paramecium aurelia*. Antigen synthesis after a temperature change. *Exp. Cell Res.* **57**, 443–446.

108 — 1970. Serotype expression in *Paramecium*. *Advanc. Microb. Physiol.* **4**, 131–178.

109 Sonneborn, T. M. 1943. Acquired immunity to specific antibodies and its inheritance in *P. aurelia*. *Proc. Indiana Acad. Sci.* **52**, 190–191.

110 — 1947. Developmental mechanisms in *Paramecium*. *Growth Symp.* **11**, 291–307.

111 — 1947. Recent advances in the genetics of *Paramecium* and *Euplotes*. *Advanc. Genet.* **1**, 263–358.

112 — 1948. The determination of hereditary antigenic differences in genically identical *Paramecium* cells. *Proc. Nat. Acad. Sci. U.S.* **34**, 413–418.

113 — 1949. Ciliated protozoa: cytogenetics, genetics, and evolution. *Ann. Rev. Microbiol.* **3**, 55–80.

114 — 1950. Cellular transformations. In *The Harvey Lectures*, Charles C. Thomas, Springfield, Ill. Series **44**, 145–164.

115 — 1950. The cytoplasm in heredity. *Heredity* **4**, 11–36.

116 — 1950. Methods in the general biology and genetics of *Paramecium aurelia*. *J. Exp. Zool.* **113**, 87–148.

117 — 1951. Some current problems of genetics in the light of investigations on *Chlamydomonas* and *Paramecium*. *Cold Spring Harbor Symp. Quant. Biol.* **16**, 483–503.

118 — 1951. The role of the genes in cytoplasmic inheritance. In Dunn, L. C., *Genetics in The Twentieth Century*, Macmillan, New York, N.Y. 291–314.

119 *— 1954. Is gene K active in the micronucleus of *Paramecium aurelia?* *Microbial Genet. Bull.* **11**, 25–26.

120 *— & Balbinder, E. 1953. The effect of temperature on the expression of allelic genes for serotypes in a heterozygote of *Paramecium aurelia*. *Microbial Genet. Bull.* **7**, 24–25.

121 — & LeSuer, A. 1948. Antigenic characters in *Paramecium aurelia* (Variety 4): Determination, inheritance and induced mutations. *Amer. Naturalist* **82**, 69–78.

122 *—, Ogasawara, F. & Balbinder, E. 1953. The temperature sequence of the antigenic types in variety 4 of *Paramecium aurelia* in relation to the stability and transformations of antigenic types. *Microbial Genet. Bull.* **7**, 27.

123 *— & Schneller, M. 1950. Transformations of serotype A, stock 51, variety 4, *Paramecium aurelia*, induced by ultraviolet light. *Microbial Genet. Bull.* **3**, 15.

124 *— & Whallon, J. 1950. Transformation of serotype A, stock 51, variety 4, *Paramecium aurelia*. *Microbial Genet. Bull.* **7**, 27–28.

125 Steers, Jr., E. 1961. Electrophoretic analysis of immobilization antigens of *Paramecium aurelia*. *Science* **133**, 2010–2011.

126 — 1962. A comparison of the tryptic peptides obtained from immobilization antigens of *Paramecium aurelia*. *Proc. Nat. Acad. Sci. U.S.* **48**, 867–874.

127 — 1965. Amino acid composition and quaternary structure of an immobilizing antigen from *Paramecium aurelia*. *Biochemistry* **4**, 1896-1901.

128 Van Wagtendonk, W. J. 1951. Antigenic transformations in *P. aurelia*, variety 4, stock 51, under the influence of trypsin and chymotrypsin. *Exp. Cell Res.* **2**, 615–629.

129 — & Vloedman, Jr., D. A. 1951. Evidence for the presence of a protein with ATP-ase and antigenic specificity in *Paramecium aurelia*, variety 4, stock 51. *Biochim. Biophys. Acta* **7**, 335–336.

130 Wichterman, R. 1953. *The Biology of Paramecium*, Blakiston, New York, N.Y.

XV. ADDENDUM

(Additional references are indicated by the prefix Ad)

Since this review was compiled, several papers have appeared that document significant new observations about the antigens. From studies on the effect of pH and salt concentration on transformation, both of which modify the extent of antiserum-induced transformation, Sikora (Ad 139) has suggested that transformation involves two phases. The first is an initiation stage, the second is the appearance of new antigen on the surface. By now, it is evident from these and other results (44, 108, Ad 135) that the immobilization reaction by itself is an inadequate indicator of the presence of an antigen. In addition, Macindoe and Reisner (Ad 138), using a technique involving adsorption of immobilizing antibodies by cell particulates and soluble antigen, claim that most of the antigen in the cell is not present in soluble form. Sodium deoxycholate, added to a 105,000 $\times$ *g* precipitate solubilizes antigen that is not released by the usual extraction procedures.

Finger *et al.* (Ad 132, Ad 133, Ad 135) have extended their studies on unstable clones. They find that: (a) stable clones are resistant to changes in culture medium and also are unaffected by most antisera; unstable clones are often markedly affected by these same agents; (b) the two kinds of clones are interconvertible when the medium from individual cultures is repeatedly and frequently replaced by fresh culture fluid, most likely due to the removal of cell products present in exhausted medium; (c) unlike stable clones, the original cultures of unstable clones which will give rise to subclones with the new serotypes possess several antigens, and (d) unstable clones fall into two classes: One is composed of cells, all of which tend to produce subclones with some cells which have transformed to new serotypes. Other unstable clones apparently are heterogeneous and are really composed of two or more kinds of cells, each of which tends to yield only subclones pure for either the original or a new serotype.

The selective expression of one allele in heterozygotes (i.e., allelic exclusion) has been re-examined by Capdeville (Ad 131) in syngen 1. Earlier, Beale (10) had shown that in this syngen, F_1 animals from interstock matings were hybrid in their serotypes, with both alleles apparently equally expressed. The newer work, based on a study of five stocks, shows that a hybrid cell can systematically exclude the expression of an allele, randomly exclude an allele, or express both alleles. Which dominance pattern is observed depends on which alleles are involved. Thus, the previously reported occurrences of allelic exclusion in *Paramecium* can no longer be thought to be exceptional (36, 60).

Interest has quickened in the molecular basis for transformation and mutual exclusion. Gibson (Ad 137) has employed the technique of DNA–RNA hybridization to demonstrate that, during the course of transformation, new RNA molecules are present shortly (1 hour) after transformation has been initiated. The interpretation, based on his data for the initial transformation and its reversal after various time intervals, is that many genes are switched on in the early stages. Some of the new gene products are missing at later stages. Thoughtful cautionary statements regarding this interpretation are included in the discussion.

References p. 164

These experiments, together with those of Pasternak (76) and Sommerville (106) suggest that synthesis of the surface antigens is vigorous. Direct evidence has now been accumulated that the surface antigens do indeed turn over more rapidly than other cell proteins. There is also evidence that pools of antigen precursors (perhaps subunits) are incorporated directly into antigen in the presence of puromycin. A new radioimmunoassay method for the recognition of serotype antigens has also been published (Ad 134).

A new review article dealing largely with the expression of the surface antigens (Ad140) and one covering mating types, killers, etc., in addition to the serotypes (Ad 137) have been published.

REFERENCES

131 Capdeville, Y. 1971. Allelic modulation in *Paramecium aurelia* heterozygotes: Study of G serotypes in syngen 1. *Mol. Gen. Genet.* **112**, 306–316.

132 Finger, I., Heller, C., Dilworth, L. & von Allmen, C. 1972. Clonal variation in *Paramecium*. I. Persistent unstable clones. *Genetics.* **72**, 17–33.

133 —, — & Magers, S. 1972. Clonal variation in *Paramecium*. III. Heterogeneity within clones of identical serotype. *Genetics.* **72**, 47–62.

134 —, Fishbein, G. P., Spray, T., White, R. & Dilworth, L. 1972. Radioimmunoassay of *Paramecium* surface antigens. *Immunology* **22**, 1051–1063.

135 —, Onorato, P., Heller, C. & Dilworth, L. 1972. Clonal variation in *Paramecium*. II. A comparison of stable and unstable clones of the same serotype. *Genetics.* **72**, 35–46.

136 —, Lavanchy, P. & Meany, L. 1973. Synthesis of *Paramecium* surface proteins. I. Puromycin insensitive amino acid incorporation. *J. Cell Biol.*, **56**, 434–440.

137 Gibson, I. 1970. Interacting genetic systems in *Paramecium*. *Advanc. Morphogenesis*, **8**, 159–208.

138 Macindoe, H. & Reisner, A. H. 1967. Adsorption titration as a specific semiquantitative assay for soluble and bound *Paramecium* serotypic antigen. *Austral. J. Biol. Sci.* **20**, 141–152.

139 Sikora, J. 1966. Immobilization by homologous antiserum and antigenic transformation in *Paramecium aurelia* in relation to the ionic composition of medium. *Acta Protozool. (Warsaw)*, **4**, 143–154.

140 Sommerville, J. 1970. Serotype expression in *Paramecium*. *Advanc. Microbiol. Physiol.*, **4**, 131–178.

Behavior and Motor Response of *Paramecium*

STANISLAW DRYL

Department of Biology, M. Nencki Institute of Experimental Biology, Polish Academy of Science, Warszawa (Poland)

I. INTRODUCTION

Motile behavior of *Paramecium* depends on the action of thousands of cilia which under normal conditions beat in a coordinated way obliquely backwards, so that the animal swims forward along a spiraling line. It should be noted that with the exception of *P. calkinsi* (29, 30) all other species of *Paramecium* show a typical forward left-spiraling movement which only sporadically or in special experimental conditions transforms to a forward right-spiraling one.

The forward movement may undergo a sudden interruption if the animal encounters a stimulus of sufficient intensity to evoke a change in the direction of locomotion. Jennings (144, 145, 147, 148) introduced the term "avoiding reaction" for this kind of motor response. The fully expressed avoiding reaction is characterized by three phases: (1) a short lasting backward movement, followed by (2) the pivoting and turning to the aboral side of the body and (3) a forward movement in the new direction

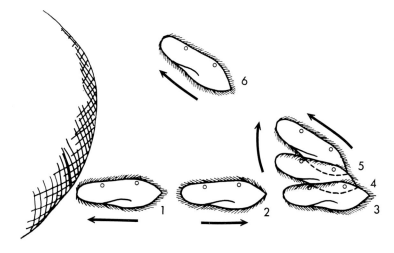

Fig. 1. Avoiding reaction of *Paramecium* towards external stimulus. (1–3) Short-lasting ciliary reversal; (3–5) aboral turning, pivoting or circling; (6) swimming in the new direction. (After Jennings, 154).

References p. 199–210

(Fig. 1). If the stimulus is weak, the avoiding reaction can be limited to the aboral turning only or to a slight change of direction of movement resulting in the forward swimming along an arc of a less or more accentuated curve. Many other motor reactions of *Paramecium* towards external stimuli were described by Alverdes (4, 6), Párducz (264, 265, 269) and Dryl (54, 56, 59). Longer lasting ciliary reversal or so-called "periodic ciliary reversal" can be induced in *Paramecia* exposed to an external medium of the appropriate ionic composition and concentration (K^+, Ca^{2+} or Ba^{2+}, Ca^{2+}). Besides, the swimming rate of the animal depends not only on the action of chemicals or drugs but also on pH, temperature, light intensity, oxygen and CO_2 tension, the stage of the life cycle and the general physiological state of the animal (74, 75, 104, 105).

There is little doubt, if any, that the ciliary activity plays an important role in the life of free-living ciliates like *Paramecium* because it enables the organism to move rather quickly from one environment to another. If necessary, the animals can slow down their forward movement or they can even stop for a longer period of time in media showing favorable living conditions (bacterial food in abundance, optimum pH and ionic composition, optimum temperature, etc.). They may avoid the unfavorable conditions of a medium by means of chemotactic, thermotactic, geotactic, thigmotactic and phototactic responses.

During the last decades our knowledge of the behavior of *Paramecium* has increased greatly due to the application of new techniques which gave a better insight in the morphology and ultrastructure (electron microscopy), and the physiology (new techniques for the recording of movement; the intracellular recording of membrane potential with glass capillaries). In spite of the great advances made, we still lack the appropriate theory of ciliary movement and we do not know much about the molecular basis of the contractile systems and the excitability of the protozoan cell and their various cellular components. Without this basic knowledge it would be difficult to attempt to explain some crucial questions like: (1) the mechanism of the normal ciliary beat and the coordination of ciliary movement; (2) the biological significance of motor response of protozoa to external stimuli, and (3) the evolutionary and comparative physiological aspects of the excitability and contraction phenomena in protozoa.

It is a generally accepted view that the physiological mechanism of ciliary movement and the response of cilia to external stimuli are similar in the whole living world, no matter whether we deal with metazoa, protozoa, animals or plants. This unity is also reflected in similarities of morphology and ultrastructure of cilia, flagella and sperm tails.

Jakus and Hall (143) showed in their pioneering electron microscopical studies on *Paramecium* that the cilium possesses two central and nine peripheral filaments. A number of investigators have confirmed these early findings and have extended our present knowledge of the morphology of the ciliary apparatus and the complicated fibrillar system within the protozoan cell. For details the reader is referred to monographs and review articles by Ehret (72), Ehret and Powers (73), Fawcett (78), Fawcett and Porter (79), Metz, Pitelka and Westfall (242), Párducz (266), Pitelka (280–282), Pitelka and Child (283).

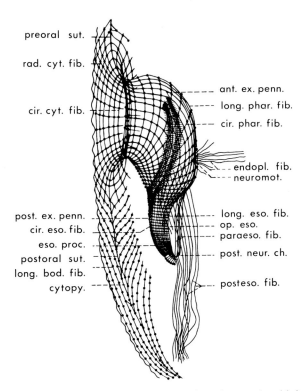

preoral sut.

rad. cyt. fib.

cir. cyt. fib.

ant. ex. penn.

long. phar. fib.

cir. phar. fib.

endopl. fib.

neuromot.

post. ex. penn.

cir. eso. fib.

eso. proc.

postoral sut.

long. bod. fib.

cytopy.

long. eso. fib.

op. eso.

paraeso. fib.

post. neur. ch.

posteso. fib.

Fig. 2. Diagrammatic representation of the pharyngo-esophageal network, with its associated fibrils of the body surface and the endoplasm, in *P. multimicronucleatum*. Ant. ex. penn. = anterior extremity of penniculus; cir. cyt. fib. = circular cytostomal fibril; cir. eso. fib. = circular esophageal fibril; cir. phar. fib. = circular pharyngeal fibril; cytopy. = cytopyge; endopl. fib. = endoplasmic fibril; eso. proc. = esophageal process; long. bod. fib. = longitudinal body fibril; long. eso. fib. = longitudinal esophageal fibril; long. phar. fib. = longitudinal pharyngeal fibril; neuromot. = neuromotorium; op. eso. = opening of esophagus; paraeso. fib. = paraesophageal fibril; posteso. fib. = postesophageal fibril; post. ex. penn. = posterior extremity of penniculus; post. neur. ch. = posterior neuromotor chain; postoral sut. = postoral suture; preoral sut. = preoral suture; rad. cyt. fib. = radial cytostomal fibril. (Lund, 220).

Cilia cover the whole body surface of *Paramecium* and are arranged in longitudinal or oblique rows, showing a more complicated pattern in the region of the cytopharynx and the mouth (219, 220) as shown in Fig. 2. In *Paramecium* there are at least three groups of cilia each serving various functions: (1) body or somatic cilia, (2) cilia of the peristomal area and (3) cilia of the gullet. The length of the somatic cilium in *Paramecium multimicronucleatum* is approximately 10–12μ and the diameter 0.27μ. The rate of the ciliary beat is between 12–28 cps (181, 319) and the swimming velocity between 700–1300μ/sec (10, 33, 60, 159, 182, 302, 308). The peristomal cilia are longer and beat stronger backward in a better coordination of movement than the somatic cilia. It is evident that they play a double role of locomotion and of directing the suspended food-particles from the medium (bacteria, etc.) towards the gullet and the cytostome where the food vacuole is formed. The gullet cilia show a stronger beat than the somatic cilia.

References p. 199–210

It is evident from the experimental data obtained with the rapid-camera movie technique utilizing *Paramecium multimicronucleatum* that the length of the meta-chronal wave in the peristomal groove is approximately $17-20\mu/\text{sec}$; the velocity of the metachronal wave is rather unstable with deviations between $455-1280\mu/\text{sec}$, while the number of appearing ciliary waves is $53.8\mu/\text{sec}$ on the average (205).

Although the early theories of ciliary movement were based on insufficient knowledge of the morphological and physiological basis of excitability of contractile systems in animals, it is interesting to note that Heidenhain (134) expressed the view that the effective stroke of a cilium appears as a result of a local contraction of the peripheral sheet of the stimulated organelle. In his brilliant monograph on ciliary movement (102), Gray pointed out the striking analogies and resemblances between the physiology of ciliary movement and the function of muscle cells of invertebrates and the heart muscle of vertebrates. He mentioned such processes as: rhythmic activity, lack of fatigue, dependence on external temperature and oxygen tension, etc. (103).

Bradfield (21) postulated the contractile role of the peripheral filaments of the cilium which were supposed to be stimulated by impulses produced rhythmically in the basal body. Harris (131) suggested that, in analogy to the function of striated muscle (138), hexagonal complexes are formed during the contraction of the cilium by means of changed configurations of peripheral and central fibers. This hypothesis was ruled out by Pitelka and Child (283) since there is no experimental evidence of a hexagonal arrangement of fibrils within the cilium. The sliding of fibrils is undoubtful since they never attach to each other at their distal terminations as shown by Gibbons and Grimston (94). Sleigh (316) modified the hypothesis of Bradfield by postulating that in 9 outer fibrils the localized and self-propagating contractions may occur which result in localized bending of the cilium. The contraction cycle starts at the base of one pair of outer fibrils and the wave of contraction spreads up the fibrils along the periphery of the cilium, being followed by a recovery phase. Sleigh suggests that the form of the beat of the cilia or flagella may depend on variations of such properties as frequency of stimulation at the basal body, the rate of propagation along the fibrils, the interval of the passage of a stimulus to the successive one and the duration of the contraction process within the fibrils.

The "contractile" component of all mentioned theories seems to be confirmed by experimental results achieved recently with glycerol- or saponin-extracted models of flagella (136), cilia of *Tetrahymena* (35) and whole-cell models of protozoa (253, 306) which respond with contraction to a solution containing ATP. The ATP-reactivated cilia and flagella have repeated contractions while the muscle models contract only once. According to Sleigh (318) this ability to show alternately contraction and relaxation may be related to the specific contractile mechanism of cilia and flagella. In connection with this, it is worthwhile to mention that cilia contain a complex similar to actinomycin (35), ATP and proteins with ATP-ase activity (34, 37, 95, 340). It is assumed that ATP is used as an energy source for the bending tension of the contractile proteins during the performance of ciliary movements in a similar way as it is utilized by other contractile elements of the cell.

sense and effector organs, which were later connected by a conductile system of neurons. Grundfest (120–125) expressed the view that in spite of the highly developed specialization during evolution, all kinds of excitable cells have preserved the primitive, ancestral type of receptor–effector system with eventual addition of the conductile portion which appeared at the more advanced level of development.

Grundfest called the three components of action: input, conductile and output. The input component is chemosensitive but electrically not excitable. It acts as a transducer converting the energy of the specific stimulus into a localized, non-propagated electrical response, which may have depolarizing or hyperpolarizing properties. The change of the resting potential of the cell membrane appears as a local graded response ("receptor potential") which may eventually as a so-called "generator potential" induce spike potentials in the conducting portion of the excitable cell.

The conductile component is electrically sensitive but does not respond to chemical stimulation. Its main function is to conduct the impulse by means of spike potentials which are typical, "all-or-none" responses, invariable in size and uniform in shape.

The output component is characterized by the secretion of the chemical transmitter which may act upon the input component of another excitable cell (Fig. 3).

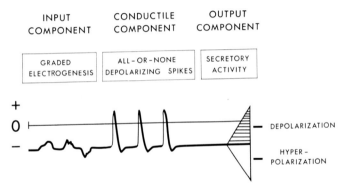

Fig. 3. Three components of excitable cells with corresponding functions and electrophysiological responses. (based on Grundfest, 121).

Parker considered the receptor-effector cells of sponges as a primitive 2-component type of the ancestral cell organization. However Duncan (70) indicated that an even more simple and better known receptor-effector system is found in *Amoeba*.

Does *Paramecium* possess an ancestral, 2-component system like *Amoeba* and presumably many other protozoa or does it represent a more advanced stage of cell organization? Before answering this question we must switch for a moment to experimental data obtained by Kinosita, Dryl and Naitoh (181, 183). These investigators provided strong evidence that graded responses of the cell membrane in *Paramecium* (Fig. 4) correspond to the local activity of cilia on the cell surface. Depolarizing spikes of the "all-or-none" character did accompany the Ba^{2+}/Ca^{2+}-induced cycles of reversed ciliary beat during the so-called "Periodic Ciliary Reversal" (PCR). The motor response during PCR is characterized by short-lasting (0.2–0.25

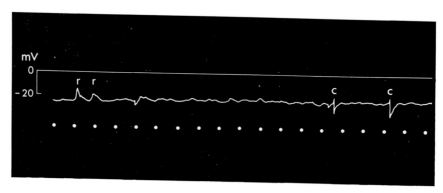

Fig. 4. Membrane potential of *P. caudatum* recorded by means of glass capillary intracellular elec-
trodes. Paramecia placed in 1 m*M* concn. of CaCl₂ diluted with 1 m*M* Tris-HCl solution of pH 7.2.
r = graded depolarization of cell membrane accompanied by spontaneous reversal of ciliary beat,
c = graded hyperpolarization of cell membrane associated with contraction of the ectoplasm.
(Kinosita, Dryl & Naitoh, 182).

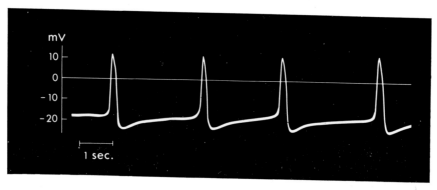

Fig. 5. Spontaneous action potentials appearing in *P. caudatum* during periodic ciliary reversal
evoked by the mixture of 2 m*M* solutions of BaCl₂ + 1 m*M* solution of CaCl₂ diluted in 1 m*M* soln.
of Tris-HCl of pH 7.2. (Kinosita, Dryl & Naitoh, 182).

sec) periods of ciliary reversal, followed by a normal forward movement. Each cycle
of movement appears at 1–2 sec intervals (Fig. 5). It is evident that paramecia possess
both input and conductile components of action; Dryl (69) suggested that the output
component is expressed by a transmission of impulses from the cell membrane to the
locomotor organelles, localized under the pellicle. In other words *Paramecium* could
be considered as a model unicellular organism possessing all three components of
action within the cell membrane. Thus far no specialized regions of the pellicle for
any one of the components could be distinguished. It is therefore tentatively assumed
that the whole cell surface of *Paramecium* possesses the polyvalent capacity of input,
conductile and output components of cation excitation. However, the underlying
physiological mechanism of this system remains to be elucidated by future experimental studies. It should be noted that such an organization of components within
a cell is reminiscent of the "primary type" sense organs (123) of invertebrates where

the sensory message is received in the receptor portion of the cell. After a graded generator potential has been produced, the message is encoded into single or multiple all-or-none spikes which appear in the conductile portion of the same cell.

If we assume in accordance with Parker, Grundfest and Duncan that the simple receptor-effector system of a cell is the most primitive system in the animal kingdom, then the three-component system in *Paramecium* represents a more advanced step in the evolution and organization of excitability on the cellular level. It remains an open question, whether the separation and strict localization of components within the cell (like in the case of the "primary type" invertebrate sense organs) constitute a more advanced evolutionary stage or a physiological adaptation to functions of the multicellular nervous system of metazoa.

III. COORDINATION OF CILIARY MOVEMENT

The mechanism of metachronal coordination of ciliary movement in protozoa is at present one of the most controversial topics of experimental protozoology in spite of the tremendous efforts of a number of investigators dealing with this problem. For more detailed studies of this aspect, the reader is referred to monographs and review articles by Gray (102), Jahn and Bovee (141), Párducz (273), Pitelka and Child (283), Rivera (293), Sleigh (317) and Smagina (323).

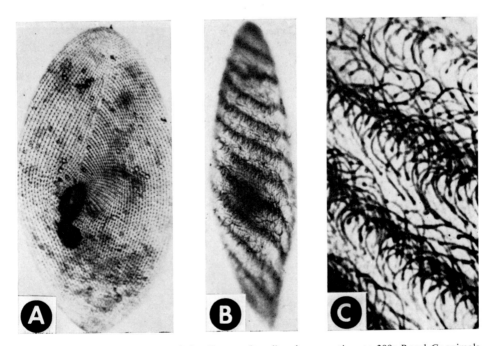

Fig. 6. *P. multimicronucleatum. A:* infraciliature after silver impregnation. × 300; *B* and *C:* animals fixed by osmium–hematoxylin rapid fixation technique during normal FLS movement, showing antiplectic metachronal waves. × 300 and × 1200 respectively. (Párducz, 263).

174 S. DRYL

The direct observation of ciliary activity in *Paramecium* is difficult even under high magnification because of the rather small size of the cilium, its high angular velocity and its specific optical properties. These difficulties led to an elaboration of instantaneous fixation with osmium–toluidine blue (92) and finally with the osmic acid–hematoxylin method (262, 113) for studying the metachronal ciliary pattern of *Paramecium* and other protozoa. Párducz (273) noticed that the fixed ciliary pattern in *Opalina* corresponded well with the metachronal wave pattern observed in living specimens. This may suggest that the fixed metachronal waves in *Paramecium* reflect the real conditions of ciliary coordination in this ciliate. Párducz indicated that under normal conditions the metachronal waves in *Paramecium* run obliquely in a NW–SE direction. Usually 10–13 metachronal waves are visible on both the ventral and dorsal side of the body (Fig. 6B). A more detailed study of the positions of single cilia indicated that the effective stroke is directed obliquely backwards (NW–SE) but at a greater angle to the longitudinal axis of the body than the crest of the corresponding metachronal wave. It can be concluded from the photograph of a specimen (Fig. 6C) and from the corresponding scheme in Fig. 7 that in normal left-spiraling movement the animal shows dexio-antiplectic metachronism of ciliary waves according to the

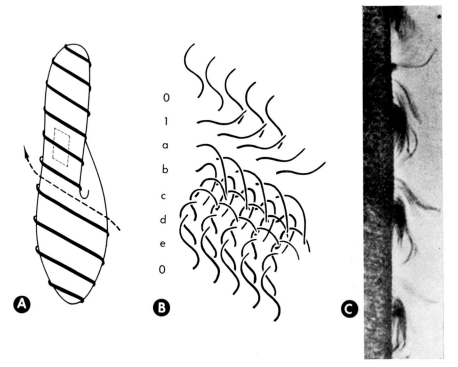

Fig. 7. *P. multimicronucleatum. A:* scheme of the metachronal waves. The arrow indicates the direction of movement. *B:* scheme of a single beating cycle. 0-1-a effective stroke; a-b-c-d-e counterclockwise recovery movement. *C:* the side view of three metachronal waves in *P. multimicronucleatum* (Photograph from preparation made with osmium–hematoxylin rapid fixation technique). Elevated, rigid cilia correspond to position No. 1 on scheme *B.* (Párducz, 263).

terminology introduced by Knight–Jones (189). As a consequence the effective stroke of the cilium runs in the opposite direction and to the right in relation to the progressive movement of the metachronal waves, which are moving obliquely forwards from the posterior to the anterior pole of the animal. It should be added that the crests of the metachronal waves (Fig. 6B) correspond to the stages of recovery (Fig. 7A, B) while the region between the crests corresponds to the stages of the effective stroke (0–1–a). It is clear from the scheme in Fig. 7B that the cilium is beating obliquely backwards during the effective stroke. A tip of the cilium is moving along 0–1 (Fig. 7B) whereas during the recovery phase the cilium starts a counter-clockwise rotation parallel to the surface. During the whole cycle of the beat the proximal curvature is moving towards the distal end of the cilium. Since the complete cycle of the beat is represented by seven cilia with only one involved in the effective stroke, it might be supposed that the angular velocity of the effective stroke is approximately 6 times faster than during the recovery phase. On this basis Párducz suggested that the metachronal waves are initiated at the posterior pole of the animal and transmitted to the anterior pole with the effective stroke of the cilia directed obliquely backwards and to the right, almost parallel to the crests of the metachronal waves.

It was noticed that the metachronal waves in the peristomal region show a similar orientation as in other parts of the animal, however the direction of their beat is perpendicular to the course of the waves, showing a pure antiplectic pattern (Fig. 8A, B, C). An interesting change of the ciliary metachronal wave pattern was observed during the avoiding reaction (AR) which begins with a short-lasting phase of ciliary

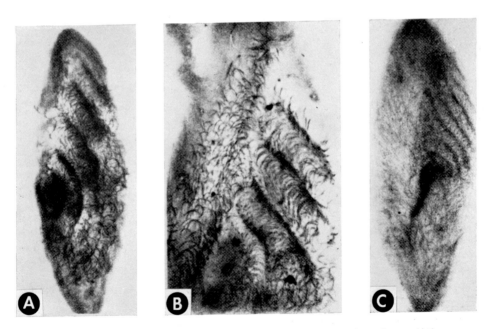

Fig. 8. *P. multimicronucleatum*. Ciliature of the oral groove. *A, B, C*: metachronal waves in the groove during the aboral turning of the anterior end of the body. Photographs of preparations made with osmium–hematoxylin rapid fixation technique. *A, C*, × 300, *B*, × 600. (Párducz, 264).

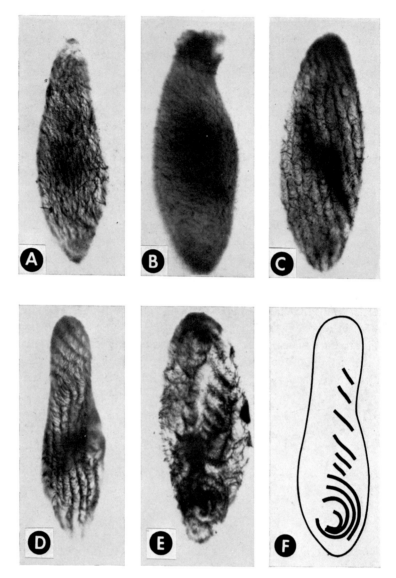

Fig. 9. *P. multimicronucleatum. A:* animal treated with vapors of chloroform. No metachrony present. *B, C, D:* various stages of avoiding reaction: simultaneous ciliary reversal *(B)*, swimming backward with spiralization to the right *(C)*; normal ciliary pattern reappearing at the anterior end *(D)*; *E: Paramecium* attacked by *Didinium nasutum*, concentric, metachronal waves around the point of local stimulation by *Didinia's* trichocysts at the posterior end of the body. Metachrony in other areas absent. *F:* scheme of metachronal waves corresponding to *E*. Photographs of preparations made by the osmium–hematoxylin rapid fixation technique; × 300. (Párducz, 273).

reversal followed by a pivoting, circling or spinning movement and finally by forward swimming in the new direction. Párducz established that the first phase of AR is characterized by the lack of any metachronism. This corresponds to the quick synchronized ciliary reversal (Fig. 9B). A transition to the coordinated reversal, accompanied by metachronal waves running parallel to the longitudinal axis of the body (Fig. 9C) follows later. It is assumed that during the appearance of the ciliary reversal the direction of the effective stroke is changed from normal NW–SE to SW–NE, reflected by the reversed right-spiraling movement. The step-wise renormalization of movement (pivoting, spinning and circling) results from the gradual recovery of the normal metachronal pattern at the anterior end of the animal (Fig. 9D) extending backwards to the posterior end. Finally the normal pattern of ciliary metachronal waves becomes again dominant and the animal starts to swim in the new direction along the left-wound spiral.

Sometimes the localized stimulus may evoke the new pattern of ciliary waves, concentrated around the stimulated or damaged surface of the pellicle as demonstrated by Párducz (264, 265) in the case of *Paramecium* being attacked by *Didinium* (Fig. 9E, F).

Paramecium exposed to a direct current (d.c.) shows during the galvanotactic response various steps of changed patterns of metachronal waves at the anterior end as a result of ciliary reversal due to the stimulating action of the current (Fig. 10A, B, C).

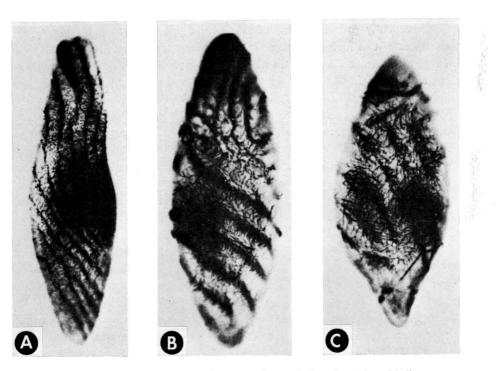

Fig. 10. Metachronal wave patterns of *P. multimicronucleatum* during stimulation with direct current. *A:* stimulation by weak current. *B:* stimulation by current of medium intensity. The pattern of the ciliary waves typical for ciliary reversal is visible at the anterior end of the animal. *C:* stimulation by strong current. The changed ciliary pattern at the anterior end of *Paramecium*. Normal ciliary pattern in the middle part of the body. Contraction of the posterior end of the animal. All photographs from preparations made by osmium–hematoxylin rapid fixation technique, × 350. (Párducz, 272).

Paramecium treated with chloroform vapor (Fig. 9A) or other narcotic agents may show after 15–20 minutes of exposure a complete suppression of the metachronal coordination. Under these conditions the animals settle down to the bottom of the container without any progressive or backward movement. The normal pattern of metachronal waves reappears again when the animals are allowed to recover slowly after being washed in normal medium devoid of narcotic agents.

The direct observation of isolated cilia at the periphery of blebs (produced by exposure of ciliates to ammonium vapors) brought evidence that they move counterclockwise in a wide angle at a constant angular velocity. The rotatory counter-clockwise movement of the cilium was also observed in dying animals.

All the above mentioned observations point out that uncoordinated cilia of *Paramecium* and other ciliates show apolar counter-clockwise movement along a cone-shaped path without any visible signs of an effective stroke. According to Párducz this kind of movement is probably a primitive, ancestral form which during the course of evolution underwent transformation to a coordinated movement of protozoan cilia with a rotatory component of the recovery phase. This development finally led to a coordinated pendular movement of metazoan cilia with the effective and recovery strokes occurring within the same plane.

On the basis of his observations, Párducz (262, 263) put forward the hypothesis that the metachronally coordinated movement of the cilia in protozoa emerges as a consequence of two factors: (1) a permanent apolar rotation of the cilium with an autonomous contractile mechanism within the organelle, and (2) waves of excitation of endogenous origin conducted through the undifferentiated ectoplasm.

Under normal conditions the system of metachronal waves of excitation is superimposed on the autonomous rotatory movement of the cilium. This induces an effective stroke of a determined direction: NW–SE. Párducz suggested that the site of transmission of the metachronal impulse is the cortical ectoplasm (264, 266, 271, 273) and that the subpellicular system of fibers does not have any relation to this process. Against the conducting role of these fibrils speak the observations which show that as a rule the pattern of metachronal waves changes very often and is quite different from the pattern of the subpellicular fiber system. Párducz (267, 269, 272, 273) postulated that the metachronal waves may pass invisibly through the areas of the denuded pellicle of *Paramecium* after treatment with chloroform.

Párducz (264, 265) also demonstrated that the various regions of the pellicle in *Paramecium* may initiate the coordinated ciliary movement. Thus each cilium may play the role of the facultative pace-maker.

Alverdes (4, 5) pointed out that the cilia of the anterior region of *Paramecium* are more sensitive to chemical, thermal and mechanical stimuli than those in the posterior end of the animal. Other authors (146, 148, 149, 151, 154, 265) suggested that the peristomal cilia might be more responsive to stimuli than the somatic ones.

A number of observed facts in other ciliated protozoa suggests the existence of the anterior–posterior polarization of excitability in *Opalina* (80), *Dileptus* (49–53) and *Spirostomum* (309, 310). Taking into account the anterior-posterior gradient of excitability and the apparent functional autonomy of the peristomal cilia in *Parame-*

cium, Jensen (157) postulated the existence of three pace-makers: posterior, anterior and cytostomal. Grębecki proposed the stomatocaudal gradient of excitability in *Paramecium*, basing his view on the observations that during ciliary reversal the metachronal waves originate not at the anterior pole of the animal but in the cytostomal area. These waves are first transmitted radially forwards along the peristome and then backwards on the dorsal side of the body (114).

Machemer (224) opposed the views of Jensen and Grębecki and denied the existence of a cytostomal pace-maker and favored the two-gradient hypothesis of Párducz and Kitching (186).

Sleigh (314–322) demonstrated that the conduction of metachronal impulses in ciliates might be a step-by-step process in which the movement of an undulipodium can excite the next one through "hydro-mechanical" or "neuroid" transmissions by means of internal cell structures. He suggested that both *Paramecium* and *Opalina* would utilize hydro-mechanical transmission, as evidenced by the change of velocity and direction of metachronal waves under conditions of increased viscosity of the medium (318). This is in accord with the observations of Pigon and Szarski (279) that the viscosity of the medium may play a significant role in the coordination and metachrony of the ciliary beat of *Paramecium*.

Recently Kuźnicki *et al.* (202–205) have questioned the validity of Párducz's theory concerning the coordination of ciliary movement and response to external stimuli as revealed by Párducz's osmium–hematoxylin instantaneous fixation technique.

On the basis of direct observations and analysis of motion-pictures achieved with the rapid-camera technique during the normal forward movement of *Paramecium multimicronucleatum* in media of increased viscosity (solutions of 0.8–1 % of methyl–cellulose) Kuźnicki, Jahn and Fonseca suggested that:

(a) The ciliary beat of body cilia is continuous with a helical wave travelling from the base to the tip; no power and recovery stroke could be distinguished.

(b) The cytopharyngeal cilia beat always posteriorly, independently of the motor response of other groups of cilia.

(c) The metachrony of the ciliary beat was observed in cilia of the oral groove, while it was scarcely visible or completely absent in somatic cilia.

(d) As a rule the metachrony of ciliary movement was lost during the start of forward motion in thigmotactic animals attached to the substrate and during the transition from forward to reversed beat of cilia.

(e) No significant differences in functions of cilia and flagella could be found in protozoa.

(f) The pattern of ciliary waves—as revealed by Párducz's osmium–hematoxylin rapid fixation technique—does not reflect the real activity of cilia in *Paramecium*.

The results obtained by Kuźnicki, Jahn and Fonseca are to some extent handicapped by the fact that they are based mainly on behavior of *Paramecium* exposed to methyl–cellulose solutions of high concentration (0.8–1 %) which may influence in a significant way the swimming rate and the pattern of ciliary coordination. For that reason it is at present not possible to give any decisive answer to the postulated mechanisms of movement and coordination of ciliary beat until more infallible information would be available about the metachrony of ciliary movement in *Paramecium* under conditions of a more or less normal activity of its locomotor apparatus. It should be mentioned that the coordination of ciliary movement in *Paramecium* may be inhibited by nickel ions (96, 118, 287, 308, 309), exposure to homologous antiserum (285, 296, 326, 327), or by the toxic action of the killing substance "Paramecin" on sensitive strains of *P. aurelia* (284, 326, 328). In this connection it is interesting to state that at the beginning of their specific action all the above mentioned agents cause some change in the state of excitability of the cell as revealed by spontaneous avoiding reactions, pivoting, circling and spinning movements or short-lasting ciliary reversals which are followed by the characteristic change from the FLS to the FRS pattern (207, 270, 326), before the movement is slowed down or completely stopped (Fig. 11).

IV. THE IONIC BACKGROUND OF EXCITATION AND MOTOR RESPONSE IN *PARAMECIUM*

There is no resting or really stationary state in *Paramecium* which could be characterized by a cessation of ciliary activity or by a complete lack of responsiveness. Even thigmotactic specimens, attached to the substrate, show a more or less normal beat of the peristomal and body cilia, except for those cilia which are immobilized as a consequence of the direct adhesion to the substrate. As mentioned before, the normal forward left-spiraling swimming is accompanied by the metachronal beat of the cilia backwards whereas the motor response to external stimuli appears in the form of AR or ciliary reversal, the duration of which depends on the intensity and exposure time of the stimulation.

Observations on the isolated cilia (102, 341) showed that these cell-organelles possess the capacity of contractile rhythmic activity which is preserved even in the non-living glycerinated or saponine models. It was demonstrated by Alexandrow and Arronet (3) and Seravin (308, 311) that in saponine models of protozoa cilia could be activated by exposure to ATP $+$ Mg^{2+} while Naitoh (253) brought evidence that the reversed position of non-beating cilia in glycerinated whole-cell models of *Paramecium* could be induced by ATP $+$ Ca^{2+} or ATP $+$ Zn^{2+}. It should be pointed out in this connection that the two above-mentioned states of ciliary activity in *Paramecium* are also reflected in the results obtained in electrophysiological studies which indicate that ciliary reversal is accompanied by depolarization of the cell membrane while no significant change of membrane polarity could be noticed during normal forward movement (182, 184).

The classical studies by Jennings (147, 150, 152, 154) on the behavior of *Paramecium* and other protozoa concentrated the attention of protozoologists on the important role of AR in the mechanism of response to external stimuli. On the other hand, the polar effects of d.c. on *Paramecium* could be recognized as a local ciliary reversal at the cathodal end of the animal while the augmentation of the ciliary beat in the normal direction was noticed at the anodal end (217, 342, 346, 332, 333, 335). A clear analogy between the observed cathodal ciliary reversal and the increased excitability

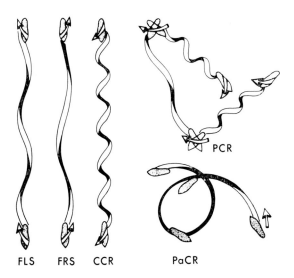

FLS   FRS   CCR   PaCR

Fig. 11. Schematic presentation of various kinds of response in *P. caudatum*. FLS = normal, forward left-spiraling movement, FRS = forward right-spiraling movement, CCR = continuous ciliary reversal, PaCR = partial ciliary reversal, PCR = periodic ciliary reversal. (Kuźnicki, 200).

(cath-electronus) of nerve and muscle cells suggested that ciliary reversal in protozoa is correlated with an increased excitability state of the organism (151, 153, 159, 169, 173, 192, 217). The effects of d.c. on the protozoan cell were explained by the change of concentrations of various cations at cell-regions affected by the electric current with special attention paid to the relations between calcium and monovalent cations e.g. potassium and sodium (139).

It is necessary to point out that at present we distinguish two patterns of normal forward motion in *Paramecium* (69):

(1) Forward Left Spiraling (FLS),

(2) Forward Right Spiraling (FRS),

and at least four patterns of reversed ciliary motion (Figs. 11, 12):

(1) Avoiding Reaction (AR),

(2) Continuous Ciliary Reversal (CCR),

(3) Periodic Ciliary Reversal (PCR),

(4) Partial Ciliary Reversal (PaCR).

In these definitions AR must be considered in a broader sense than usually, since it has been already pointed out in the introduction that Jennings' description of the classical "Avoiding Reaction" did not include all of the boundary reactions of *Paramecium* which may appear at the first contact with the external stimulus.

Jennings noticed as early as 1887 (144) that *Paramecia* may show long-lasting ciliary reversal in response to high concentrations of potassium chloride. Mast and Nadler (229) and Oliphant (259, 260) showed that monovalent cations induce ciliary reversal, whereas divalent or trivalent cations are indifferent in this respect.

The dependence of potassium-induced ciliary reversal in *P. caudatum* in the presence

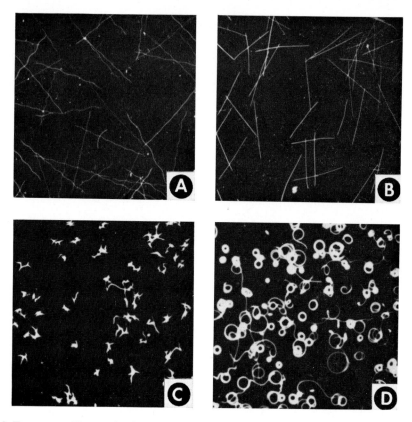

Fig. 12. Four types of locomotion in *P. caudatum. A:* FLS movement; *B:* CCR; *C:* PCR; *D:* PaCR. Recording of movement carried out by time-exposure dark-field macrophotographic technique. (Dryl & Grębecki, 69).

of calcium ions in an external medium was first emphasized by Kamada (162, 163) and Kamada and Kinosita (164). Kamada showed that intracellular injection of Ca-precipitating anions caused a ciliary reversal while injection of KCl into the cell interior was ineffective. On the basis of the above mentioned observations and taking into account the possibility of cationic exchange across the cell membrane (2, 161), Kamada expressed the view that an anion (X) exists within the cytoplasm of *Paramecium* which by binding calcium ions forms a stable compound Ca–X; the release of X caused by potassium ions or by an outward transmembrane current from this hypothetical substance would result in a reduction of the intracellular free calcium ions. This might be the initiating factor for ciliary reversal. This view was to some extent confirmed by more recent studies on the behavior of calcium ions in *Paramecium* as revealed by the use of radioactive isotopes and flame spectrophotometry techniques (358, 359). Kamada and Kinosita (164) noticed an antagonism between the external K^+ and Ca^{2+} in their effects on the excitation state of *Paramecium*, as expressed by the duration of induced ciliary reversal. Jahn (140) recalculated the data of Kamada and Kinosita and found that the maximal duration of ciliary reversal is constant when the K^+ concentration is proportional to the square root of the Ca^{2+} concentration in the external medium. This represents in fact the Gibbs–Donnan ratio $[K^{1+}]/[Ca^{2+}] = $ const. and it suggests that the ciliary reversal is caused by potassium ions which compete with calcium ions adsorbed to the cell membrane. The membrane will release Ca^{2+} ions from their binding sites within the postulated ion exchange system, localized on the cell surface.

The detailed studies on stimulating effects of external chelating (EDTA), calcium-binding (citrates) or calcium-precipitating (oxalates) agents in *Paramecium*, showed the following sequence of motor reactions depending on the concentration of the substance under study (112, 115): FLS → PCR → CCR → PaCR → FLS. This proved to be reversible if the recalcification procedure was started early enough so that the animals could recover from the transient damage of the cell membrane. Similar series of reactions were observed by Kuźnicki (197) in paramecia exposed to various cations (Cs^{1+}, Tl^{1+}, Na^{1+}, K^{1+}, Ba^{2+}, Mn^{2+}, Sr^{2+}) at different levels of external calcium. Kuźnicki suggested that all the above mentioned cations induce ciliary reversal by their competing action on the membrane-bound calcium. This assumption was confirmed by Naitoh and Yasumasu (257) who were able to show that ^{45}Ca-binding by paramecia was inhibited by monovalent cations in good agreement with the Gibbs–Donnan principle. The authors suggested that calcium ions, liberated from the anionic sites of the cell membrane by an exchange reaction with other ions, induce ciliary reversal.

A number of hypotheses concerning the possible mechanism of the excitation and ciliary reversal process and the role of calcium are available at present.

Jahn (140) postulated the mechanism of Ca-loss assuming that reversal of the ciliary beat appears when the bound calcium is depleted from its binding sites on the cell surface.

Grębecki (115) assumed that the degree of excitability depends on the amount of calcium remaining on the adsorption sites of the cell membrane and that no ciliary reversal occurs if the amount of membrane-bound calcium is below or above the postulated optimum for induction of PCR or CCR. In this way Grębecki denies the possible role of calcium as a coupling factor in linking excitation with contraction similar to the mechanism in muscle cells.

Naitoh (252–254) suggested that calcium ions activate the contractile system energized by ATP which is necessary for the induction of the reversal of the ciliary beat. According to this view the ciliary reversal would be analogous in its mechanism to the contraction of muscle cells. It is not yet clear where the postulated cation exchange system which releases calcium ions is located. The autoradiographic studies on calcium binding in *Paramecium* suggest that ^{45}Ca accumulates at the surface of the pellicle. The recent studies by Naitoh and Eckert (255) brought evidence that cationic permeability depends on the level of saturation of the anionic binding sites with Ca^{2+} and that an increase of the cation conductance is associated with a decrease of bound calcium. These findings point out that a calcium-binding mechanism plays perhaps an essential role both in the permeability of the cell membrane and in the ciliary response, coupling in this way the excitation and contractile systems in *Paramecium*.

Kuźnicki (202) postulates that the release of calcium ions from the ionic exchange system of cell membranes constitutes the trigger mechanism for the corresponding release of calcium from the inner surface of the cytoplasmic membrane and the intracellular membrane system. Because of this increased calcium level the contraction of the ciliary filaments is affected, which is manifested by a reversed beat of the cilia.

Recently Ebashi and Endo (71) reported a calcium-receptive role of troponin, a protein which together with tropomyosin is located within the thin filaments in muscle cells. It was suggested that troponin in the absence of calcium inhibits the interaction of myosin and actin. This repression can be released by addition of calcium ions. In this way the shuttle movement of calcium ions between the two hypothetical Ca-binding sites, i.e. between troponin and the proteins of the sarcoplasmic reticulum, could be the basic factor for the induction of the state of contraction and relaxation. There is no information available at present whether a calcium-receptive protein also plays a role in the excitation of the cell membrane and the contraction of undulipodia in protozoa. It seems however to be a promising idea to verify this possibility in further experimental studies on excitability and ciliary reversal in protozoa.

TABLE 1

The magnitude of the membrane potential and the ciliary activity of P. caudatum *in media of different ionic composition. (Kinosita, Dryl & Naitoh, 182).*

Cationic component and concentration (mM)	Membrane potential (Inside-negativity in mV)	Swimming velocity (μ/sec)	Beat frequency of cilia (cps)
Ca 2	19.7 ± 0.24	911 ± 6.5	13.4 ± 0.29
Ca 20	16.9 ± 0.22	751 ± 5.6	12.4 ± 0.27
Mg 2	14.3 ± 0.26	359 ± 4.2	11.9 ± 0.26
Mg 20*	8.3 ± 0.17	—	9.2 ± 0.17
Mg 2 + Ca 2	18.2 ± 0.37	844 ± 3.6	12.4 ± 0.07
Mg 20 + Ca 2	17.0 ± 0.73	719 ± 2.6	11.2 ± 0.24
Ba 2*	7.6 ± 0.07	—	6.1 ± 0.10
Ba 20*	4.6 ± 0.12	—	—
Ba 2 + Ca 2	15.1 ± 0.23	607 ± 5.6	10.5 ± 0.28
Ba 20 + Ca 2	8.3 ± 0.19	353 ± 4.8	8.0 ± 0.18
K 2	17.5 ± 0.59	407 ± 2.5	9.8 ± 0.09
K 20	4.7 ± 0.12	167 ± 1.5	5.1 ± 0.04
K 2 + Ca 2	18.7 ± 0.31	854 ± 7.4	12.8 ± 0.29
K 20 + Ca 2	5.2 ± 0.09	365 ± 3.0	6.7 ± 0.10
Na 2	14.1 ± 0.19	547 ± 2.8	12.4 ± 0.11
Na 20	10.0 ± 0.16	161 ± 2.9	12.2 ± 0.15
Na 2 + Ca 2	21.8 ± 0.83	822 ± 5.4	13.0 ± 0.13
Na 20 + Ca 2	13.8 ± 0.21	633 ± 4.6	14.8 ± 0.13
pH 4.63	11.7 ± 0.30	746 ± 4.3	12.6 ± 0.23
pH 5.15	12.9 ± 0.19	951 ± 3.7	13.0 ± 0.27
pH 5.98	15.8 ± 0.19	1060 ± 3.3	13.1 ± 0.26
pH 6.84	15.6 ± 0.28	962 ± 2.8	12.8 ± 0.20
pH 7.95	18.0 ± 0.21	869 ± 2.9	10.9 ± 0.27
pH 8.56	19.2 ± 0.22	806 ± 3.8	10.5 ± 0.21

A new approach to the ionic background of excitability in ciliate protozoa is reflected in experimental studies on the membrane potential. Kamada (161) in his pioneer study on the membrane potential of *P. caudatum* suggested that the negative intracellular potential in this species is caused by the specific permeability of the cell membrane to cations. It was shown by Yamaguchi (358) that CCR induced by K^{1+}/Ca^{2+} ions is accompanied by a depolarization of the cell membrane in *Paramecium*. Extensive studies by Kinosita, Dryl and Naitoh (181–183) on the changes in the membrane potential and its response to external stimuli in *Paramecium* showed that the spontaneous reversal of the ciliary beat was associated with a graded depolarizing response of the cell membrane whereas the contraction of the ectoplasm at the posterior end of the animal was related to a graded hyperpolarizing response. As it is indicated in Table 1, the decrease of the swimming rate and the beat of cilia induced by cations was correlated with a lower level of the membrane potential. The cations could be arranged in the following sequence with regard to the value of the induced intracellular potential: $Ca^{2+} > Na^{1+} > Mg^{2+} > K^{1+} > Ba^{2+}$, if these ions were added in concentrations of 20mM to the medium. Increase of the membrane potential was correlated with an increase of pH value in the external medium. However, the most interesting and rewarding results were recorded in the case of PCR induced by the simultaneous action of barium and calcium ions. Dryl (61, 64) demonstrated that paramecia exposed to the Ba^{2+}/Ca^{2+} or Sr^{2+}/Ca^{2+} factors show a short-lasting (0.2–1 sec) ciliary reversal which is followed by forward movement (FLS or FRS), each phase of movement appearing alternately at 1/2–2 sec long intervals.

Kinosita, Dryl and Naitoh (182) brought evidence to bear that during the Ba^{2+}/Ca^{2+}-induced PCR, each cycle of the reversal beat of the cilia was accompanied by "all-or-none" depolarizing spikes. These were uniform in shape and amplitude which is typical for the action potential (AP) phase of depolarization of the membrane overshooting to positive values of potential. This was followed by a repolarization with a transient period of hyperpolarization of the cell membrane (Fig. 5, p. 172). PCR response could last 24 hrs or longer, if a sufficient concentration of cations were present in the external medium and if the animal were not damaged by other external factors. The recorded potentials proved to be similar in form to the classical action potentials recorded in the nerve fibers although their rather long duration (approx. 350 msec) and rhythmicity made them similar to AP found in the cardiac muscle fibers or in the intestinal smooth muscle cells. Ba/Ca-induced AP in *Paramecium* occurred in the medium devoid of sodium ions so that the possible role of these ions in the formation of spikes can be ruled out. Some investigators showed that in sodium-free medium external calcium and barium ions increase the magnitude of AP in crustacean muscle fibers (77). The spike potentials from the giant barnacle muscle fibers (127) and from arthropod muscles (351, 352) could be induced in the medium devoid of sodium and the authors suggest that these APs are probably determined by the trans-membrane conductance of $[Ca^{2+}]$ and $[K^{1+}]$ only. Kinosita, Murakami and Yasuda (185) showed by the simultaneous recording of the ciliary response and the membrane potential that in *Paramecium* showing Ba/Ca-induced PCR, each reversal response of cilia appeared 22–36 msec after the beginning of the depolarizing spike

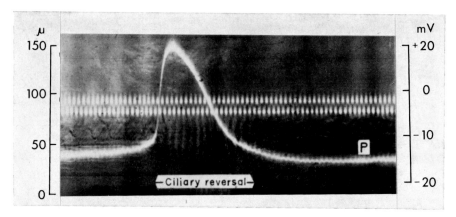

Fig. 13. Simultaneous recording of the action potential and ciliary response in *P. caudatum* during Ba^{2+}/Ca^{2+}-induced periodic ciliary reversal. (Kinosita, Murakami & Yasuda, 185).

(Fig. 13). A correlation was found between the values for the duration of the ciliary reversal and that of the depolarizing spike, supporting the view that the reversal of the ciliary beat in *Paramecium* is associated with the depolarization of the membrane potential.

Naitoh and Eckert using *P. caudatum* (256) established recently that addition of BaCl$_2$ in a final concentration of 0.25 mM to a 1 mM CaCl$_2$ solution was sufficient for the induction of an all-or-none transient reversal of the membrane potential by an outward current pulse of 10^{-10}A or greater. This reversal had a duration of about 40 msec. An increase of [Ba^{2+}] resulted in (1) a lower resting potential, (2) a positive shift in the critical firing level, (3) an increased overshoot of the action potential, (4) a decreased hyperpolarizing afterpotential, and (5) an increased duration of the action potential. When the concentration of Ca^{2+} was increased along with that of Ba^{2+} in order to keep the ratio [Ba^{2+}]/[Ca^{2+}] constant, similar results were obtained except that the duration of the action potential remained unchanged. These effects appeared to be related to [Ba^{2+}] and not to [Ca^{2+}] or [Cl$^-$]. The degree of overshoot in 1 mM Ca^{2+} was linearly related to log [Ba^{2+}]. The evidence indicated that prolongation of the action potential was due to a delayed onset of Ba inactivation. Other features of the action potential were absolute refractoriness during its rising and plateau phases, relative refractoriness lasting several seconds, and repetitive firing in response to steady current depolarization. The response was not affected by Tetrodotoxin and tetraethylammonium. Manganese prolonged the action potential.

The role of the acetylcholine–acetylcholine esterase system in the chemical regulation of excitation in ciliate protozoa is not clear. Some authors found ACH in *Paramecium* (15) and *Trypanosoma* (28) while others (275, 300, 301) reported the presence of ACH-ase in *Tetrahymena*. Müller and Toth (246) and Seravin (310) found that low concentrations of ACH and anti-ACH-ase agents lengthen K^{1+}/Ca^{2+}-induced ciliary reversal, whereas high concentrations shorten the response.

Similar results were reported by Grębecki (115) who noticed the antagonistic effects of external calcium on ACH-action. According to Seravin (310, 312) ACH evokes ciliary reversal and contraction of myonemes in *Spirostomum*, while Kosh-toyants and Kokina (195) observed that ACH and eserine in low concentrations raised cathodal excitability of *Paramecium* and decreased it when applied in higher con-centrations. The author (Dryl, unpublished) observed negative chemotactic response in *P. caudatum*, *P. aurelia*, *Stentor coeruleus* and *Spirostomum ambiguum* in response to low concentrations of ACH.

It looks as if the ACH-ACH-ase system may influence in some way the excitability of the protozoan cell but more experimental data are necessary in order to elucidate this interesting topic in terms of the physiology of ciliary movement and the electro-physiology of the cell membrane.

V. RESPONSE TO VARIOUS KINDS OF EXTERNAL STIMULI

This section deals with reactions of animals which are usually described as tropism (209), taxis (81, 208) or kinesis (126).

It would be beyond the scope of this review to go into details concerning the history of the problem and the development of hypotheses and theories on the mechanism of various responses of animals exposed to external stimuli. The reader interested in this field is referred to monographs and papers by Loeb (209), Jennings (154), Kuhn (208), Mast (227, 228), Rose (294), Russel (296) and more recently by Fraenkel and Gunn (81).

The author here uses the term "taxis" for responses of *Paramecium* to almost all external stimuli, being however aware that according to the classification by Kuhn some of these responses should be called phobotaxis (e.g. chemotaxis) or topotaxis (e.g. phototaxis) whereas according to the new classification by Fraenkel and Gunn the author should use the terms: klinotaxis or klinokinesis (126) or tropotaxis (e.g. geotaxis). The author decided to use only generally accepted simple terms for re-sponses of *Paramecium* and this decision seems to be justified by the fact, that new scientific terms concerning tactic phenomena in the animal kingdom are not yet accepted by all scientists.

A. Response to Mechanical Stimuli

Paramecium may react with a typical AR in response to mechanical stimuli or it may slow down the progressive movement remaining attached to the substrate. The last mentioned response is called thigmotaxis (288, 291) and it plays a very important role in the daily life of the animal because it helps to keep the animal in a region with rather favorable conditions. It was reported by Bullington (30) that *P. aurelia* and *P. calkinsi* are more sensitive to touch than *P. multimicronucleatum* or *P. trichium*. Jennings noticed that the anterior tip of *Paramecium* is much more sensitive to me-chanical stimulation than the remainder of the pellicle. This observation was recently confirmed by Naitoh and Echert (256) who found that *Paramecium* stimulated with a crystal-driven glass stylus (a tip diameter of 25μ) at the anterior end shows reversal of its ciliary beat which is accompanied by a transient increase of permeability of the anterior region of the cell membrane to calcium which may flow into the cell interior. On the contrary, *Paramecium* stimulated mechanically at the posterior end shows an increase of the frequency of the ciliary beat in the normal direction accompanied by an increased permeability of the posterior region of the cell membrane to potassium. The membrane becomes hyperpolarized.

During the thigmotactic response the cilia attached to the substrate are stiff, other somatic cilia are moving slowly and those behind the region of contact are usually at rest, whereas peristomal cilia seem to beat normally.

Sometimes *Paramecium* shows a so-called "peripheric reaction" which is manifested by fast movement of the animal at the periphery of the vessel with a more or less constant angle of rebounding from the marginal glass-edges (42, 100). It was noticed by the author (Dryl, unpublished) that such peripheral reactions are very often accompanying the positive rheotactic response, (i.e. against water current) which usually is associated with the negative geotactic response in *Paramecium*. Jennings found the positive rheotaxis to occur when at a certain velocity of the water current *Paramecia* place themselves with the anterior end upstream in line with the current. The physiological mechanism of these responses is unknown.

B. *Chemotaxis and Chemokineses*

The pioneer studies on chemotaxis were started by Pfeffer (276, 278) on bacteria, flagellates and fern sperm. Massart (226) considered the ciliate protozoa to be completely insensitive to chemical substances and he attributed the observed motor response of these animals to changes of osmotic pressure in the external medium. Massart proposed even to use the term "tonotaxis" instead of "chemotaxis" but his views were strongly opposed by Jennings (144) who was able to show that osmotic pressure does not play an essential role in the observed response of *Paramecium* to chemical stimuli and that ciliate protozoa may show positive or negative chemotactic responses depending on the chemical properties of the substance under study and on its concentration in the external medium (83, 84, 152, 154).

It is well known that *Paramecium* can live and reproduce under variable conditions of the external environment. They are typical cosmopolitan organisms which can be easily found in streams, rivers, lakes, ponds and bodies of stagnant water in every part of the world. The relatively high rate of swimming and the capacity to perceive the chemical stimulus in the environment ahead of it, gives additional opportunities for finding the favorable living conditions in the surrounding medium. There is no doubt that the chemotactic response of *Paramecium* towards chemical changes in the medium is most important in the life of this organism. As it was pointed out by Jennings (145, 154) in his classical studies on the response of free living protozoa to external stimuli, *Paramecium* responds positively to chemotactic stimuli which signalize the presence of bacteria or other kind of food. It was proposed by Jennings and Moore (156) and by Moore (244) that the positive chemotactic response of *Paramecium* to a slightly acidic medium has an adaptive character. Under natural conditions the chemotactic response may bring the animal to an environment abundant in food and bacteria. The latter produce large quantities of carbon dioxide in the process of respiration and thus bring the pH to an acidic range. Barrat (14) criticized Jennings' view; he suggested that the positive chemotactic response is only a transient phase of the negative one and that there is no reason for distinction between them. Positive chemotactic response of *Paramecium* to oxygen was observed by Verworn (341) only when the external medium was deficient in oxygen. Lozina–Lozinski (211–216) noticed that paramecia show a higher rate of food vacuole formation in suspensions of dyes causing positive chemotaxis, when compared with animals exposed to suspensions of dyes inducing negative chemotactic response or no response at all. The same author reported a correlation between positive chemotaxis and phagotrophic response of *Paramecium* towards suspensions of India Ink, Carmine and *B. subtilis*.

The application of more efficient long-exposure dark-field photographic techniques for recording the movement of protozoa (55, 80) and suspended particles in the medium (106) made it possible to conduct extensive, quantitative studies on chemotaxis and other motor responses of protozoa (Figs. 14 and 15).

It was shown by Dryl (54, 56, 59) that the optimum pH for chemotaxis to which *P. caudatum* responds is between 5.4–6.4. This corresponds well with the established optimum pH for the highest swimming rate (60) and confirms to some extent the data reported previously by Johnson (159) on the motor response of *Paramecium* to external pH. Grębecki (108, 109) found a similar value for the pH of the isoelectric point of the cell surface (5.25). At this pH pinocytosis in *Paramecium* fell abruptly. The cations

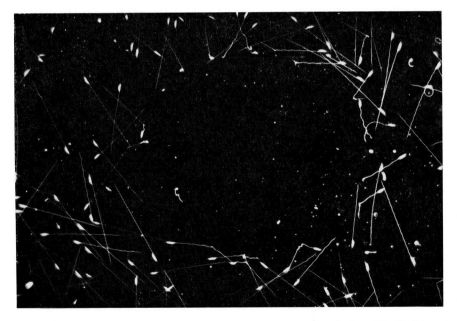

Fig. 14. Negative chemotactic response of *P. caudatum* towards a 20 mM solution of MgCl₂. Recording of movement carried out by Dryl's time-exposure dark-field photomacrographic technique.

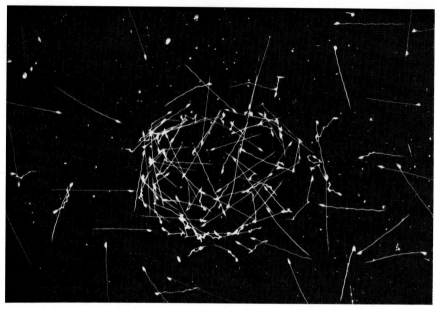

Fig. 15. Positive chemotactic response of *P. caudatum* towards the optimum pH range 5.2–6.4. The pH of the surrounding medium is 7.8. A number of avoiding reactions visible on the boundary between two solutions. Recording of movement carried out by Dryl's time-exposure dark-field macrophotographic technique.

can be arranged depending on the intensity of the negative chemotactic action as follows:

$$K^{1+} > Mg^{2+} > Ca^{2+} > Na^{1+}$$

This range of cations is similar to the results obtained in studies on the ciliary activity and intracellular potential of *Paramecium*, e.g.: $K^{1+} > Mg^{2+} > Na^{1+} > Ca^{2+}$. This may suggest that the negative chemotactic properties of cations are in some way related to their depolarizing effects on the cell membrane. The facts indicating a correlation between the lower rates of forward movement and the depolarization of the cell membrane (Table 1, Fig. 12) and the lower swimming rates associated with the negative chemotactic properties of cations are also in favor of this postulate (59). It is evident from the observed effects of pH and cations on chemotaxis that this response appears always at a much lower concentration than the corresponding LD_{50}. This may suggest an important biological role of the chemotactic response which prevents the animals to enter the area containing high concentrations of toxic substances.

In a separate series of experiments with ten lower alcohols, Dryl (57) found evidence that a stronger negative chemotactic response was correlated with the higher molecular weight of the alcohols and that normal compounds proved to be more effective than the isomers (Fig. 16). In this connection it should be emphasized that a similar range of alcohols was reported by other investigators studying smell sensitiveness in invertebrates (47) and vertebrates (249). This seems to point to a rather high degree of development of chemical sense in the ciliate protozoa.

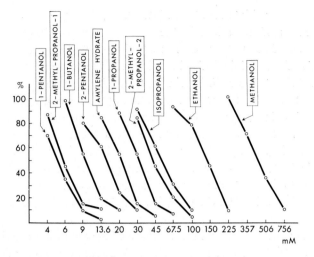

Fig. 16. Chemotactic effects of lower alcohols on *P. caudatum* expressed in percentage of animals entering the area of tested solution of alcohol in relation to the number of animals in the control area devoid of alcohol. (Dryl, 57).

Although various biological effects have been ascribed to alcohols (17, 95, 98, 210), it is known from studies by Bills (18) using *P. caudatum* that narcotic and toxic effects of lower alcohols are in direct proportion to the molecular weight of the compound. This was confirmed by the author (57).

We still lack the quantitative data on the action of gases of biological importance on the motile behavior of protozoa. Sears *et al.* (302–305) reported that a complete removal of carbon dioxide causes a gradual immobilization of *Paramecium* whereas the inert gases (helium, nitrogen, argon) had little influence on the locomotion, except xenon which at 250 lbs/sq. in. caused retardation of movement and narcosis.

Paramecia proved to be chemotactically insensitive to other organic compounds of biological interest such as urea and glucose (59). In contrast Adler reported (1) that *Escherichia coli* may show a positive chemotactic response towards glucose and galactose.

It is obvious that not only chemotactic, CCR, PaCR or PCR responses can be induced in *Paramecium*. Nagai (247), Andrejewa (10, 11) and Seravin (306, 307) reported that cations may affect the rate of forward swimming depending on the concentration and the presence of other cations in the medium. There is evidence that *Paramecium* responds by contraction of the ectoplasm when exposed to higher concentrations of calcium ions and to the transmembrane stimulation (135, 165, 243–250). Although some authors (12) stress the significant role of sodium ions in the external medium for the physiology of the cell membrane in ciliates, it should be pointed out that calcium ions play perhaps an even more important role in this respect not only in protozoa (115, 161, 181–183, 198, 206, 252) but also in muscle (71) and nerve fibers. Dryl provided evidence that some chemical and physical changes in the external medium may influence chemotactic and chemokinetic response in

Paramecium and other ciliated protozoa (54, 56, 59, 64–68). *Paramecium* exposed to high concentrations of potassium in the external medium (30–40 mM KCl in the presence of 1 mM CaCl$_2$) shows during the first 10–20 minutes a typical motile phenomenon for the physiological adaptation, i.e. the response changes from FLS through CCR, PaCR to slow FLS movement. In this physiological state the animals do not respond to chemical or mechanical stimuli; after being washed in a medium devoid of potassium ions, paramecia will recover their normal swimming rate almost immediately, while the normal response to chemotactic stimuli reappears 12–20 minutes after washing. The possibility of a changed sensitivity of *Paramecium* due to a physiological adaptation to an external medium should be taken into consideration in experimental studies on the motor response of protozoa and specially in the case of the analysis of experimental data obtained in studies on "learning" in unicellular organisms.

Although many authors suggest the existence of the anterior–posterior gradient of excitability in ciliates (4, 5, 114, 144, 154, 265, 309), Jennings and Jamieson (155) and Horton (137) demonstrated that fragments of paramecia possess the same capacity of response towards chemical stimuli as whole, intact animals.

C. Galvanotaxis

Paramecia swim towards the cathode, when exposed to a direct current of appropriate strength (Fig. 16). Verworn (341–347) noticed that at a threshold value of a d.c. only a few animals showed galvanotaxis, but that the number of paramecia responding to d.c. increased proportionally to the increase of the current. At higher intensities of d.c. paramecia were swimming very slowly forwards while at a very strong current they would even move backwards towards the anode, but with the anterior end of the body pointing towards the cathode. At a still higher, lethal strength of d.c. the animals bursted at the anodal end and perished.

Ludloff (217) observed that the endosmotic movement of the intracellular fluid in *Paramecium* towards the cathodal region of the ectoplasm reversed the beat of the cilia. The same author noticed that paramecia oriented obliquely towards the lines of the electrical field showed a symmetric ciliary reversal around all sides of the cathodal end with the larger portion of the cilia beating in a reversed direction on the surface of animals facing the cathode (Fig. 17A). The asymmetry of the ciliary response forces the animal to turn the anterior end towards the cathode and to swim further along the lines of the electric field with a symmetric action of cilia on both sides of the body (Fig. 17B).

Roesle (295) found the peristomal area of the cell surface of *Paramecium* to be more sensitive to d.c. stimulation than other parts of the body, while Statkewitsch (330–333) noticed that the contraction of the ectoplasm and the extrusion of trichocysts always occurred at the anodal end of the body.

The polar effects of a galvanic current result in a functional division of the protozoan cell into the cathodal and anodal part of the body as stressed by Verworn, Jennings, Statkewitsch (334) and Koehler (192). As a rule cilia from the anterior cathodal region

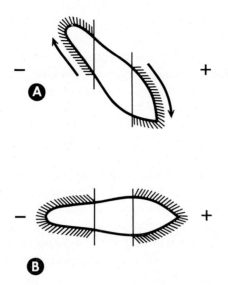

Fig. 17. Schematical presentation of *P. caudatum* exposed to direct current. Cilia affected by cathodal and anodal stimulation are shown in positions indicating the direction of their effective stroke. *A:* Ludloff's phenomenon. In the oblique position of the animal in relation to the lines of the electric field, the ciliary reversal at the cathodal end of the body covers a more extensive area on the surface facing the cathode than on the opposite one. The asymmetry of ciliary response forces the animal to turn and swim towards the cathode. *B: P. caudatum* oriented parallel to the lines of the electric field. The ciliary response is symmetrical both in the cathodal and the anodal region of the body. (Dryl, 63).

of *Paramecium* show ciliary reversal while the normal backward beat of the cilia appears at the posterior anodal region of the cell surface. There is much evidence that the galvanotactic response depends on the physiological state of the animal and on the chemical changes in the external medium (76, 230, 231, 349, 350).

On the basis of experiments with vital stains (Nile blue sulphate and neutral red), Kinosita (170, 172, 177, 178) reported that the cytoplasm of *Paramecium* becomes acid at the cathodal end of the body and alkaline at the anodal end. After some period of exposure to d.c. the alkaline reaction dominates over the entire cytoplasm of the ciliate.

It was noticed first by Bancroft (13) and was confirmed later by Kamada (159), Bonaventure, Viaud and Bonaventure (20, 348) and Grębecki (116) that in a medium devoid of calcium ions but containing low concentrations of potassium, barium or sodium, paramecia were swimming towards the anode instead of to the cathode. Under these conditions ciliary reversal did not appear during d.c. stimulation. Statkewitsch (330, 331, 334) made the observation that in response to frequent induction shocks, paramecia were more effective in their reaction to the "break" than to the "make" shock.

The region of ciliary reversal induced by the galvanic current spreads rapidly towards the anodal pole, immediately after the application of the electric stimulus, but shows a slow regression during longer-lasting stimulation (169, 173, 174). This finding and the observations of Kinosita (175, 177) on the behavior of *Paramecium* during two successive stimuli or with a linearly increasing current (179) indicated that the

References p. 199–210

galvanotactic response included both excitability and accommodation phenomena.

Verworn postulated that the galvanic current caused polar effects on the cytoplasm of *Paramecium*. This general idea was accepted by other authors although it did not explain the physiological mechanisms underlying the observed changes in the direction of the effective beat of cilia at the cathodal end of the body. Loeb and his students considered the electrolytic effects of d.c. in the external medium as a basic factor for the induction of galvanotaxis, which according to them should be treated as a special case of chemotropism. Coehn and Barrat (36) tried to explain galvanotaxis as an electrophoretic (cataphoretic) phenomenon while Bancroft (13) attributed galvanotactic phenomena to the local changes of the calcium content and that of other ions in the cytoplasm of the stimulated organism. Recently Butzel, Brown and Martin (31) revived the old concept of the electrophoretic mechanism of galvanotaxis, but these authors obviously did not take into account the occurrence of cathodal ciliary reversal. Grębecki (107, 116) expressed the interesting view that the electrokinetic factor may at least play a role in the galvanotactic responses of *Paramecium* induced by very weak electric fields.

In accordance with the new electrotonic theory of galvanotaxis by Jahn (139), the polar effects of direct currents on the protozoan cell are explained in a similar way as those reported for muscle and nerve cells. Jahn assumes that depolarization of the cell membrane on the cathodal end of the body induces ciliary reversal while hyperpolarization at the anodal end causes augmentation of the ciliary beat in the normal backward direction. It is suggested that the body of *Paramecium* should be considered as a core conductor immersed in a volume conductor (external medium) and that ciliary reversal appears when the appropriate degree of depolarization of the cell membrane is reached. The electrotonic theory of galvanotaxis provides a good explanation of Ludloff's phenomenon and of the extension of the cathodal ciliary reversal on the cell surface with an increase of current strength.

It was noticed by Dryl (63) that paramecia swimming close to the substrate (slide or cover glass) show an oblique orientation in relation to the lines of the electric field: SE–NW by animals moving closely to the slide and NE–SW by animals moving closely to the cover glass, provided that the cathode is on the left and that the observation is made from above. It should be pointed out that the same orientation of *Paramecium* towards the lines of the electric field was observed both in animals showing normal FLS movement as well as in those involved in PCR or CCR movement due to exposure to the Ba^{2+}/Ca^{2+} or K^{1+}/Ca^{2+} factors (Figs. 18B and 18C).

It is evident from motion-picture analyses (63) that the oblique galvanotactic response of *Paramecium* associated with the asymmetry of ciliary response at the anodal end of the body is in accordance with the electrotonic theory of galvanotaxis. On the basis of this fact and taking into consideration some new observations, Dryl assumes that the oblique galvanotactic response of *Paramecium* appears as a result of two forces acting simultaneously on the animal:

(1) A ciliary force which is induced by the polar effects of d.c. on the paramecium body. This force tends to orient the animal E–W, i.e. parallel to the lines of the electric field with the anterior end facing the cathode.

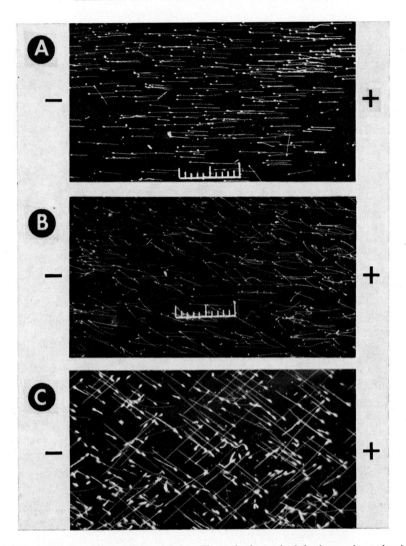

Fig. 18. Galvanotactic response of *P. caudatum*. The cathode on the left, the anode on the right. *A:* paramecia showing typical cathodal galvanotaxis E–W. The direction of swimming indicated by the small bright spot at the end of the recorded path by a single animal. *B:* normal cathodal E–W response of animals swimming some distance from the substrate and oblique SE–NW cathodal response of specimens swimming close to the slide. *C:* induced oblique galvanotactic response in paramecia showing Ba^{2+}/Ca^{2+}-induced periodic ciliary reversal. The animals are swimming obliquely backwards to the anode with the anterior end oriented towards the cathode. The orientation SE–NW was recorded in paramecia swimming close to the bottom slide while orientation NE–SW was observed in those swimming close to the cover-glass. The movement of animals and its direction were recorded by Dryl's time-exposure dark-field photomacrographic technique. (Dryl, 63).

    (2) A thigmotactic mechanical force which tends to turn the animal counter-clockwise. It was revealed by direct observation that this response appears as a consequence of the thigmotactic reaction of the anterior end of the animal and by a simultaneous tendency of counterclockwise rotation of the posterior end around the

less mobile anterior tip of *Paramecium* both during FLS or CCR depending upon the experimental conditions (Dryl, unpublished).

The ciliate protozoa show transversal orientation of the longitudinal axis towards the lines of the electric field, when exposed to an alternating current (133), while according to observations by Teixeira-Pinto, Nejelski, Cutler and Heller (335) they may show parallel or transversal orientation to the lines of force of the electric component in electromagnetic fields of high frequency. The physiological mechanism of the reported phenomena is obscure since the observations were not conducted with the known modes of behavior in ciliates in mind.

D. Thermotaxis

Although *Paramecium* can live within a wide range of external temperatures, an optimal temperature for the highest rate of swimming, food intake, food-vacuole formation and the action of the contractile vacuoles was found to be between 22°–26°C. It was shown by Mendelsohn (232–235) that paramecia placed in long glass tubes collect in the region nearest the optimum temperature range: 24°–28°C, responding with avoiding reactions towards higher and lower temperatures.

E. Geotaxis

It was noticed by a number of investigators that under suitable conditions paramecia show a negative geotactic response. This motor reaction can be easily induced by a sudden change of the external medium such as mixing of a culture medium containing large number of animals with Pringsheim's, Lozina–Lozinsky's or Dryl's salt solution (58), centrifugation, shaking of the container, addition of carbon dioxide to the culture medium etc. During the typical negative geotactic response the animals are swimming toward the top of the container with the anterior end of the body directed upwards. The animals usually stay for a short time near the surface of the fluid and show afterwards avoiding reactions in response to the deeper regions of the medium. Jensen (157) postulated that the difference in pressure between the upper and lower surface of the organism would induce geotaxis. Davenport (38) considered the differences of resistance in moving upwards or downwards as a causal factor for the induction of geotaxis. However, both views were ruled out by Lyon (221, 222) since he was able to show that paramecia were not able to detect such small differences in pressure or mechanical resistance of the surrounding medium. Sosnowski (329) and Moore (244) could induce positive geotaxis by exposure to low temperature and by shaking the culture medium. It was noticed by King and Beams (169) that mechanical stimulation (shaking of the container, etc.) decreased the rate of swimming and food-ingestion.

Lyon (221, 222) and after him Koehler (190–193) suggested the statocyst theory, which assumed that the cytoplasm of *Paramecium* contained various cell-organelles (macronucleus, micronuclei, food-vacuoles) and many inclusions which are of higher specific gravity than the remainder of the cell interior. These organelles could play the role of statoliths in an analogy to the function of statoliths in metazoa. Koehler suggested that these hypothetical statoliths could exert a pressure on the cytoplasm and on the ectoplasm of *Paramecium*, thus giving rise to an increased activity of the cilia within the limits of the stimulated area and initiate the geotactic response.

The mechanical theory of geotaxis was first suggested by Verworn (341) and Harper (128–130), and was more completely developed in a series of papers by Dembowski (43–46). According to Dembowski the posterior end of *Paramecium* is heavier than the anterior end. Due to this fact an eventual slackening or stopping of the forward movement during chemical or mechanical stimulation renders it possible for the animal to attain the vertical position with the heavier posterior end of the body directed downwards and with the anterior tip directed upwards. Dembowski was able to demonstrate that the so-called "centrotactic" response of paramecia induced by centrifuging (190, 191) is based on the geotactic orientation of the animal in space and that there are no reasons to treat this reaction as a specific response to centrifugal forces. This critical approach to the statocyst theory of Koehler finds support in recent studies on the physiology of the ciliary response of *Paramecium*, since the occurrence of "intrinsic" receptors for ciliary activity in the cytoplasm is excluded. On the other hand, Koehler (194) brought convincing evidence against the mechanistic concept of Dembowski by demonstrating that even fragments of *Paramecium* with an amputated posterior end still show the negative geotactic response in a similar way as normal specimens.

In recent studies on four species of *Paramecium* (*P. caudatum*, *P. aurelia*, *P. multimicronucleatum* and *P. bursaria*) Kuźnicki (199, 200) demonstrated that living animals immobilized with $NiCl_2$ showed various orientations of their longitudinal axis during the slow process of sinking thus casting doubt on the validity of Dembowski's "heavier posterior end" concept as a basic factor in the mechanistic theory of geotaxis.

F. Phototaxis and the Response to UV and X-irradiation

It is well known that all species of *Paramecium* are insensitive to the visible light, except *P. bursaria* which is positively phototactic due to the presence of zoochlorellae in the cytoplasm.

Wichterman (354) reported an acceleration of the forward movement in *Paramecium* exposed to sublethal doses of X-rays, while high doses of approximately 200,000 r caused retardation of the forward movement of *P. bursaria*.

Giese (96, 97) reported a negative response of *Paramecium* to UV-irradiation, an increase of ciliary activity in the beginning of the exposure, followed by a gradual decrease of the swimming rate during longer exposure. The wavelengths 245 mμ and 280.4 mμ proved to be most critical for *Paramecium*. This may be suggestive that the effects of UV-irradiation are limited mainly to the cytoplasmic proteins and RNA.

G. Response to Combined Stimuli and the Problem of "Learning"

There is a large body of evidence that stimulation of *Paramecium* by one stimulus

References p. 199–210

may change or even inhibit the response to the second one (144, 145, 289, 338). Jennings and Pütter noticed that a strong thigmotactic stimulus can prevent the normal reaction of the animal to electric or thermal stimuli. Exposure of *Paramecium* to a low temperature decreases or completely inhibits the normal response to chemotactic stimuli, while it is well known that potassium ions in the presence of low concentrations of calcium ions may completely abolish the response to mechanical, thermal and chemotactic stimuli (56, 59, 65, 68, 69). Many other examples of combined effects of stimuli have been discussed in section *C. Galvanotaxis*.

It was always a very tempting idea to find some primitive form of a learning process in unicellular organisms. Early investigators (Verworn and Jennings) believed that protozoa must possess the capacity of learning and this was one of the reasons why Jennings (152) was so eager to apply Morgan's theory of "trial and error" to the behavior of free-living protozoa.

Metalnikow (238–241) tried to obtain conditioned reflexes in *Paramecium* by coupling the food reaction with red light or alcohol as conditioned stimuli. His experimental techniques and final conclusions have been strongly criticized by Wladimirsky (355). Dembowski (40–42) and Alverdes (4, 7–9) described in *Paramecium* a direct orientation towards the food source instead of a "trial and error" activity which was suggested by Jennings as a basic motor response to external stimuli. Further studies along this line by Lozina–Lozinsky (211–216) and Bragg (22, 23) did not show any true conditioning in *Paramecium*. Smith (324) demonstrated that paramecia could learn to turn around in capillary tubes and the animals were able to retain this acquired feature for the next 20 minutes after being allowed to swim in an open dish. However, Buytendijk (32) showed that some physiological change resulting in the increased flexibility of the ciliate body was responsible for the observed reactions, and that no true learning was involved.

Bramstedt (24–26) tried to induce conditioned reflexes in *Paramecium* and *Stylonychia* using light, darkness, heat and water-shake as conditioned stimuli. According to him the conditioned reflexes in *Paramecium* could be preserved for at least 15 minutes after the "learning" procedure. Bramstedt suggested that after some training paramecia could "remember" the shape of a container, since after being replaced to another one, the animals continued to swim along a very similar pattern to the outline of the first container. In a similar way Tschakotine (339) suggested that *P. caudatum* can learn to avoid the location of a beam of UV for a half hour after the beam has been removed. Grabowski (101) showed that Bramstedt's and Tschakotine's results could be explained by the local chemical changes in the external medium induced by heat or by UV action on the cytoplasm of protozoan cell. Soest (325) did not succeed in elaborating the conditioned reflexes between light or darkness and cold.

Although Alverdes (8–9), Diebschlag (48) and French (82) claimed that a learning process may occur in *Paramecium* and other ciliate protozoa, Dembowski (46), Best (16), Dabrowska (39), Mirsky and Katz (243) and Kinastowski (167, 168) ruled out this possibility, indicating a number of errors in the techniques and the procedures of stimulation with electric current, light, heat or water shake.

Gelber (85–91) in a series of experiments with *P. aurelia* tried to reinforce the mechanical stimulus with bacterial food. She suggested that such training was successful and the positive conditioned reflexes could be elaborated. However, Katz and Deterline (166) repeated some of her experiments and were able to show that the observed phenomena by Gelber do not belong to the real learning process. They proved that the reported responses occurred in the presence of bacterial food while dipping of the platinum wire into the food for the induction of conditioning was without any effect on the "training". Jensen (158) came to the same conclusions.

It is clear that so far no really positive results could be achieved in the experimental studies on the process of learning in *Paramecium* and other ciliate protozoa.

VI. REFERENCES

(Abstracts are indicated by *)

1   Adler, J. 1966. Chemotaxis in bacteria. *Science* **153**, 708–716.
2   Akita, Y. K. 1941. Electrolytes in *Paramecium*. *Mem. Fac. Sci. Agr. Tôhoku Imp. Univ.* **23**, 99–120.
3   Alexandrov, V. Y. & Arronet, N. I. 1956. Motion caused by adenosinetriphosphate of cilia in ciliates killed by glycerol extraction (a cellular model). *Dokl. Akad. Nauk. SSSR* **110**, 457–460.
4   Alverdes, F. 1922. Zur Lokalisation des chemischen und thermischen Sinnes bei *Paramecien* und *Stentor*. *Zool. Anz.* **55**, 19–21.
5   — 1922. Zur Lehre von der Reaktionen der Organismen auf äussere Reize. *Biol. Zentralbl.* **42**, 218–222.
6   — 1923. Beobachtungen an *Paramecium putrinum* und *Spirostomum ambiguum*. *Zool. Anz.* **55**, 277–287.
7   — 1937. Das Lernvermögen der einzelligen Tiere. *Z. Tierpsychol.* **1**, 35–38.
8   — 1939. Zur Psychologie der niederen Tiere. *Z. Tierpsychol.* **2**, 258–264.
9   — 1943. Die Marburger Untersuchungen über das Lernvermögen niederer Tiere. *Forsch. Fortschr.* **5**, 60–62.
10 Andrejewa, E. W. 1930. Die elektrische Ladung und die Bewegungsgeschwindigkeit der Infusioren *Paramecium caudatum*. *Kolloid Z.* **51**, 348–356.
11 — 1931. Zur Frage der physikalisch-chemische Bestimmung der Korrelation einiger physiologischen Prozesse bei *Paramecium caudatum*. *Arch. Protistenk.* **73**, 346–360.
12 Andrus, W. W. & Giese, A. C. 1963. Mechanism of sodium and potassium regulation in *Tetrahymena pyriformis* Strain W J. *J. Cell. Comp. Physiol.* **61**, 17–30.
13 Bancroft, F. W. 1906. The control of galvanotropism in *Paramecium* by chemical substances. *Univ. Calif. Publ. (Berkeley) Physiol.* **3**, 21–31.
14 Barrat, J. O. W. 1905. Der Einfluss der Konzentration auf die Chemotaxis. *Z. Allg. Physiol.* **5**, 73–94.
15 Bayer, G. & Wense, T. 1936. Über den Nachweis von Hormonen in einzelligen Tieren, Teil 1 (Cholin und Acetylcholin im *Paramecium*). *Pflüger's Arch. Ges. Physiol.* **237**, 417–422.
16 Best, J. B. 1954. The photosensitization of *Paramecium aurelia* by temperature shock. A study of reported conditioned response in unicellular organisms. *J. Exp. Zool.* **126**, 87–99.
17 Bills, C. E. 1923. A pharmacological comparison of six alcohols, singly and in a mixture, on *Paramecium*. *J. Pharm. Exp. Therap.* **22**, 49–60.
18 — 1924. Some effects of the lower alcohols on *Paramecium*. *Biol. Bull.* **47**, 253–264.
19 Blättner, H. 1926. Beiträge zur Reizphysiologie von *Spirostomum ambiguum* Ehrbg. *Arch. Protistenk.* **53**, 253–311.
20 Bonaventure, N. 1956. L'électrocinèse dans le galvantropisme de *Paramécium caudatum*. Etude de l'action du courant galvanique sur les battements ciliaires. *Compt. Rend. Soc. Biol.* **149**, 2230–2232.
21 Bradfield, J. R. G. 1955. Fibre patterns in animal flagella and cilia. In *Fibres and their Biological Significance* (Symp. Soc. Exp. Biol.), **9**, 306–344.
22 Bragg, A. N. 1936. Selection of food in *Paramecium trichium*. *Physiol. Zool.* **9**, 433–442.

23   — 1939. Selection of food by Protozoa. *Turtox News* **17**, 41–44.

24   Bramstedt, F. 1935. Dressurversuche mit *Paramecium caudatum* und *Stylonychia mytilus. Z. Vergleich. Physiol.* **22**, 490–516.

25   — 1935. Die Lokalisation des chemischen und thermischen Sinnes bei *Paramecium. Zool. Anz.* **112**, 257, 262.

26   — 1939. Über die Dressurfähigkeit der Ciliaten. *Zool. Anz.* **12**, 111–132.

27   Brown, V. E. 1930. The neuromotor apparatus of *Paramecium. Arch. Zool. Exp. Gen.* **70**, 469–481.

28   Bulbring, E., Lourie, E. M. & Pardoe, A. U. 1949. The presence of acetylcholine in *Trypanosoma rhodiesiense* and its absence from *Plasmodium gallinaceum. Brit. J. Pharmacol.* **4**, 290–294.

29   Bullington, W. E. 1925. A study of spiral movement in the ciliata infusoria. *Arch. Protistenk.* **50**, 219–275.

30   — 1930. A further study of spiraling in the ciliate *Paramecium*, with a note on morphology and taxonomy. *J. Exp. Zool.* **56**, 423–449.

31   Butzel, Jr., H. M., Brown, L. H. & Martin, Jr., W. B. 1960. Effects of detergents upon electro-migration of *Paramecium. Physiol. Zool.* **33**, 39–41.

32   Buytendijk, F. J. 1919. Acquisition d'habitudes par des êtres unicellulaires. *Arch. Neerl. Physiol.* **3**, 455–468.

33   Chase, A. M. & Glaser, O. 1930. Forward movement of *Paramecium* as a function of the hydrogen ion concentration. *J. Gen. Physiol.* **13**, 627–636.

34   Child, F. M. 1959. The characterization of the cilia of *Tetrahymena pyriformis. Exp. Cell Res.* **18**, 258–267.

35   — 1961. Some aspects of the chemistry of cilia and flagella. *Exp. Cell Res.* **8** (Suppl.), 47-53.

36   Coehn, A. & Barrat, W. 1905. Über Galvanotaxis vom Standpunkte der physikalischen Chemie. *Z. Allg. Physiol.* **5**, 1–9.

37   *Culbertson, J. R. & Banerjee, S. D. 1965. Biochemistry of cilia isolated from *Tetrahymena. Progress in Protozoology* (Proc. 2nd Int. Conf. Protozool., London), 240.

38   Davenport, C. B. 1897. *Exp. Morphol.* **1**, 1–280.

39   Dabrowska, J. 1956. Tresura *Paramecium caudatum, Stentor coeruleus, Spirostomum ambiguum* na bodźce świetlne. *Folia Biol. Warsaw* **4**, 77–91.

40   Dembowski, J. 1921. Ueber die Nahrungsauswahl und die sogenannten Gedächtnisserschei-nungen bei *Paramecium caudatum. Trav. Lab. Biol. Nencki Varsovie* **1**, 1–16.

41   — 1922. Weitere Studien über die Nahrungsauswahl bei *Paramecium caudatum. Trav. Lab. Biol. Nencki Varsovie* **1**, 1–16.

42   — 1923. Ueber die Bewegungen von *Paramecium caudatum. Arch. Protistenk.* **47**, 25–54.

43   — 1929. Die Vertikal-Bewegungen von *Paramecium caudatum*. I. Die Lage des Gleichgewichts-zentrums im Körper des Infusors. *Arch. Protistenk.* **66**, 104–132.

44   — 1929. Die Vertikal-Bewegungen von *Paramecium caudatum*. II. Einfluss einiger Aussen-faktoren. *Arch. Protistenk*, **68**, 215–261.

45   — 1931. Die Vertikal-Bewegungen von *Paramecium caudatum*. III. Polemisches und Experi-mentelles. *Arch. Protistenk.* **73**, 153–187.

46   — 1950. On conditioned reactions of *Paramecium caudatum* towards light. *Acta Biol. Exp. Polish Acad. Sci.* **15**, 5–18.

47   Dethier, V. A. & Yost, M. T. 1952. Olfactory stimulation of blow flies by homologous alcohols. *J. Gen. Physiol.* **35**, 823–839.

48   Diebschlag, E. 1940. Über die Lernfähigkeit von *Paramecium caudatum. Zool. Anz.* **130**, 257–271.

49   Doroszewski, M. 1961. Reception areas and polarization of ciliary movement in the ciliate *Dileptus. Acta Biol. Exp. Polish Acad. Sci.* **21**, 15–34.

50   — 1962. The occurrence of the ciliary reversion in the *Dileptus* fragments. *Acta Biol. Exp. Polish Acad. Sci.* **22**, 3–9.

51   — 1963. Some features of the ciliary activity in *Dileptus. Acta Protozool. (Warsaw)* **1**, 189–192.

52   — 1963. The response of *Dileptus* and its fragments to the puncture. *Acta Protozool. (Warsaw)* **1**, 313–319.

53   — 1963. The response of the ciliate *Dileptus* and its fragments to the water shake. *Acta Biol. Exp. Polish Acad. Sci.* **23**, 3–10.

54   Dryl, S. 1952. The dependence of chemotropism in *Paramecium caudatum* on the chemical changes in the medium. *Acta Biol. Exp. Polish Acad. Sci.* **16**, 23–53.

175 —, 1936. Supernormal phase of electric response in *Paramecium*. *J. Fac. Sci. Imp. Univ. Tokyo, Sect. IV*, **4**, 171–184.

176 — 1936. Effect of change in orientation on the electrical excitability of *Paramecium*. *J. Fac. Sci. Imp. Univ. Tokyo, Sect. IV*, **4**, 189–194.

177 — 1936. Addition of subliminal stimuli in *Paramecium*. *J. Fac. Sci. Imp. Univ. Tokyo, Sect. IV*, **4**, 195.

178 — 1938. Electrical stimulation of *Paramecium* with two successive subliminal current pulses. *J. Cellular Comp. Physiol.* **12**, 103–117.

179 — 1939. Electrical stimulation of *Paramecium* with linearly increasing current. *J. Cellular Comp. Physiol.* **13**, 253–261.

180 — 1954. Electric potentials and ciliary response in *Opalina*. *J. Fac. Sci. Univ. Tokyo, Sect. IV*, **7**, 1–14.

181 —, Dryl, S. & Naitoh, Y. 1964. Changes in the membrane potential and the response to stimuli in *Paramecium*. *J. Fac. Sci. Univ. Tokyo, Sect. IV*, **10**, 291–301.

182 —, Dryl, S. & Naitoh, Y. 1964. Relation between the magnitude of membrane potential and ciliary activity in *Paramecium*. *J. Fac. Sci. Univ. Tokyo, Sect. IV*, **10**, 303–309.

183 —, — & — 1964. Spontaneous change in membrane potential of *Paramecium caudatum* induced by barium and calcium ions. *Bull. Acad. Polon. Sci. Ser. Sci. Biol. (Classe II)* **12**, 459–461.

184 — & Murakami, A. 1967. Control of ciliary motion. *Physiol. Rev.* **47**, 53–82.

185 —, — & Yasuda, M. 1965. Interval between membrane potential change and ciliary reversal in *Paramecium* immersed in Ba–Ca mixture. *J. Fac. Sci. Univ. Tokyo, Sect. IV*, **10**, 421–425.

186 Kitching, J. A. 1961. The physiological basis of behavior in Protozoa. In Ramsay, J. A. & Wigglesworth, V. B., *The Cell and the Organism*, Cambridge University Press, London, 60–78.

187 Klein, B. M. 1927. Die Silberliniensysteme der Ciliaten. Ihr Verhalten während der Teilung und Konjugation, neue Silberbilder, Nachträge. *Arch. Protistenk.* **58**, 55–142.

188 — 1928. Die Silberliniensysteme der Ciliaten. Weitere Resultate. *Arch. Protistenk.* **62**, 177–260.

189 Knight–Jones, E. W. 1954. Relations between metachronism and the direction of ciliary beat in metazoa. *Quart. J. Microscop. Sci.* **95**, 503–521.

190 Koehler, O. 1922. Über die Geotaxis von Paramecium. *Arch. Protistenk.* **45**, 1–94.

191 — 1922. Über die Geotaxis von *Paramecium. Verhandl. Zool. Botan. Ges. Wien* **26**, 69–71.

192 — 1928. Galvanotaxis. *Handb. norm. Pathol. Physiol.* **11**, 112–123.

193 — 1930. Über die Geotaxis von *Paramecium. Arch. Protistenk.* **70**, 279–307.

194 — 1939. Ein Filmprotokoll zum Reizverhalten querzertrennter *Paramecien. Zool. Anz. Suppl.* **12**, 132–142.

195 Koshtoyants, K. S. & Kokina, N. N. 1957. On the role of acetylcholine and cholinesterase systems in galvanotaxis and summation of stimuli in *Paramecium. Biofizika* **2**, 46–50.

196 Kuźnicki, L. 1963. Recovery in *Paramecium caudatum* immobilized by chloral hydrate treatment. *Acta Protozool. (Warsaw)* **1**, 177–185.

197 — 1963. Reversible immobilization of *Paramecium caudatum* evoked by nickel ions. *Acta Protozool. (Warsaw)* **1**, 301–312.

198 — 1966. Role of Ca^{2+} ions in the excitability of the protozoan cell. Calcium factor in the ciliary reversal induced by inorganic cations in *Paramecium caudatum. Acta Protozool. (Warsaw)* **4**, 241–256.

199 — 1968. Ciliary reversal in *Paramecium caudatum* in relation to external pH. *Acta Protozool. (Warsaw)* **4**, 257–261.

200 *— 1967. Studies on geotaxis of four species of *Paramecium. Amer. Zoologist* **7**, 310.

201 — 1968. Behavior of *Paramecium* in gravity fields. I. Sinking of immobilized specimens. *Acta Protozool. (Warsaw)* **6**, 109–117.

202 — 1970. Mechanisms of the motor responses of *Paramecium. Acta Protozool. (Warsaw)*, In press.

203 *—, Jahn, T. L. & Fonseca, J. R. 1968. The helical nature of the ciliary beat of *Paramecium. Amer. Zoologist* **8**, 164.

204 *—, — & — 1969. Cinematographic analysis of ciliary reversal in *Paramecium. Progress in Protozoology* (Proc. 3rd Int. Conf. Protozool., Leningrad), 171–172.

205 *—, — & — 1969. Ciliary activity of *Paramecium multimicronucleatum*. I. Evolution of techniques, II. Body cilia, III. Oral groove and cytopharynx, IV. Metachrony. *Progress in Protozoology* (Proc. 3rd Int. Conf. Protozool., Leningrad), 384.

206   *— & Mikolajczyk, E. 1969. Ciliary response of ciliates evoked by EGTA. *Progress in Proto-zoology* (3rd Int. Conf. Protozool., Leningrad), 172–173.

207   — & Sikora, J. 1966. Inversion of spiralling of *Paramecium aurelia* after homologous antiserum treatment. *Acta Protozool. (Warsaw)*, **4**, 263–268.

208   Kühn, A. 1919. *Die Orientierung der Tiere im Raum*, Fischer Verlag, Jena.

209   Loeb, J. 1918. *Forced Movements, Tropism and Animal Conduct*, Lippincott, Philadelphia and London.

210   Loefer, J. B. & Hall, R. P. 1936. Effect of ethyl alcohol on the growth of eight protozoan species in bacteria-free cultures. *Arch. Protistenk.* **87**, 123–130.

211   Lozina–Lozinsky, L. 1929. Le choix de la nourriture dans les différents milieux par le *Para-mecium caudatum*. *Compt. Rend. Soc. Biol.* **100**, 321–323.

212   — 1929. Zur Physiologie der Ernährung der Infusorien. I. Nahrungsauswahl und Vermehrung bei *Paramecium caudatum*. *Izv. Nauk. Issled. Inst.* **15**, 91–136.

213   — 1929. Chemotaxis in relation to the choice of food stuffs by Infusoria. *Compt. Rend. Acad. Sci. U.S.S.R. Ser. A*, **17**, 403–408.

214   — 1929. Physiology of nutrition of Infusoria. I. Food selection and multiplication of *Paramecium caudatum*. *Izv. Nauk. Issled. Inst.* **15**, 91–136.

215   — 1929. Le choix de la nourriture chez *Paramecium caudatum*. *Compt. Rend. Soc. Biol.* **100**, 722–724.

216   — 1931. Zur Ernährungsauswahl und Vermehrung bei *Paramecium caudatum*. *Arch. Protistenk.* **74**, 18–120.

217   Ludloff, K. 1895. Untersuchungen über den Galvanotropismus. *Pflügers Arch. Ges. Physiol.* **59**, 525–554.

218   Ludwig, W. 1927. Untersuchungen über Schraubenbahnen niederer Organismen. *Z. Vergleich. Physiol.* **9**, 734–801.

219   Lund, B. L. 1918. The toxic action of KCN and its relation to the state of nutrition and age of the cell as shown by *Paramecium* and *Didinium*. *Biol. Bull.* **35**, 211–231.

220   Lund, E. E. 1933. A correlation of the silverline and neuromotor systems of *Paramecium*. *Univ. Calif. (Berkeley) Pub. Zool.* **39**, 35–76.

221   Lyon, E. P. 1905. On the theory of geotropism in *Paramecium*. *Amer. J. Physiol.* **14**, 421–432.

222   *— 1918. Note on geotropism of *Paramecium*. *Biol. Bull.* **34**, 120.

223   Machemer, H. 1966. Versuche zur Frage nach der Dressierbarkeit hypotricher Ciliaten unter Einsatz hoher Individuenzahlen. *Z. Tierpsychol.* **6**, 641–654.

224   — 1969. Eine 2-Gradienten Hypothese für die Metachronieregulation bei Ciliaten. *Arch. Protistenk.* **111**, 100–128.

225   — 1969. Regulation der Cilienmetachronie bei der "Fluchtreaktion" im *Paramezium*. *J. Protozool.* **16**, 764–771.

226   Massart, J. 1889. Sensibilité et adaptation des organismes à la concentration de solutions salines. *Arch. Biol.* **9**, 515–570.

227   Mast, S. O. 1911. *Light and the Behavior of Organisms*, Wiley, New York, N.Y.

228   — 1938. Factors involved in the process orientation of lower organisms to light. *Biol. Rev.* **13**, 186–224.

229   — & Nadler, J. E. 1926. Reversal of ciliary action in *Paramecium caudatum*. *J. Morphol. Physiol.* **43**, 105–117.

230   Mayeda, T. 1928. De l'effet de la réaction (pH) du milieu sur le galvanotropisme de la paramécie. *Compt. Rend. Soc. Biol.* **99**, 108–110.

231   — & Date, S. 1929. De l'influence de la concentration en ions H sur le galvanotropisme de la paramécie. *Compt. Rend. Soc. Biol.* **101**, 633–635.

232   Mendelssohn, M. 1895. Über den Thermotropismus einzelliger Organismen. *Pflügers Arch. Ges. Physiol.* **60**, 1–27.

233   — 1902. Recherches sur la thermotaxie des organismes unicellulaires. *J. Physiol. Pathol. Gen.* **4**, 393–409.

234   — 1902. Recherches sur la interférence de la thermotaxie avec d'autres tactismes et sur le mécanisme du mouvement thermotactique. *J. Physiol. Pathol. Gen.* **4**, 475–488.

235   — 1902. Quelques considérations sur la nature et la rôle biologique de la thermotaxie. *J. Physiol. Pathol. Gen.* **4**, 489–496.

236   Mendelsohn, J. H. & Warmoth, M. 1965. Induced avoidance behavior of *Paramecium caudatum*. *Progress in Protozoology* (Proc. 2nd Int. Conf. Protozool., London), 241.

237 Merton, H. 1935. Versuche zur Geotaxis von *Paramecium*. *Arch. Protistenk*. **85**, 33–60.

238 Metalnikow, S. 1907. Über die Ernährung der Infusorien und deren Fähigkeit, ihre Nahrung zu wahlen. *Trav. Soc. Naturalistes, Petersburg* **38**, 181–187.

239 — 1912. Contributions à l'étude de la digestion intracellulaire chez les protozoaires. *Arch. Zool. Exp. Gen.* **10**, 373–497.

240 — 1914. Les Infusoires peuvent-ils apprendre à choisir leur nourriture? *Arch. Protistenk*. **34**, 60–78.

241 — 1917. On the question regarding the capability of Infusoria to "learn" to choose their food. *Russ. J. Zool., Petersburg* **2**, 397.

242 Metz, C. B., Pitelka, D. R. & Westfall, J. A. 1953. The fibrillar systems of ciliates as revealed by the electron microscope. I. *Paramecium*. *Biol. Bull.* **104**, 408–425.

243 Mirsky, A. F. & Katz, M. S. 1958. "Avoidance conditioning" in *Paramecium*. *Science* **127**, 1498–1499.

244 Moore, A. 1903. Some facts concerning geotropic gatherings of *Paramecia*. *Amer. J. Physiol.* **9**, 238–244.

245 Moulton, D. G. & Eayrs, J. T. 1960. Studies in olfactory acuity. II. Relative detectability of n-aliphatic alcohols by the rat. *Quart. J. Exp. Psychol.* **12**, 99–109.

246 Müller, M. & Toth, E. 1959. Effect of acetylcholine and eserine on the ciliary reversal in *Paramecium multimicronucleatum*. *J. Protozool.* **6** (Suppl.), 28.

247 Nagai, H. 1907. Der Einfluss verschiedener Narkotika, Gase und Salze auf die Schwimmgeschwindigkeit von *Paramaecien*. *Z. Allgem. Physiol.* **6**, 195–212.

248 Nagai, T. 1956. Elasticity and contraction of *Paramecium* ectoplasm. *Cytologia* **21**, 65–75.

249 — 1960. Contraction of *Paramecium* by transmembrane electrical stimulation. *J. Fac. Sci. Univ. Tokyo, Sect. IV*, **8**, 617–631.

250 — 1960. Local contraction of *Paramecium* ectoplasm by $CaCl_2$ stimulation. *J. Fac. Sci. Univ. Tokyo, Sect. IV*, **8**, 633–641.

251 Naitoh, Y. 1966. Reversal response elicited in nonbeating cilia of *Paramecium* by membrane depolarization. *Science* **154**, 660–662.

252 — 1968. Ionic control of the reversal response of cilia in *Paramecium caudatum*. A calcium hypothesis. *J. Cellular Comp. Physiol.* **51**, 85–103.

253 — 1969. Control of the orientation of cilia by adenosinetriphosphate, calcium, and zinc in glycerol extracted *Paramecium caudatum*. *J. Gen. Physiol.* **53**, 517–529.

254 — & Eckert, R. 1969. Ionic mechanisms controlling behavioral responses of *Paramecium* to mechanical stimulation. *Science* **164**, 963–965.

255 — & — 1968. Electrical properties of *Paramecium caudatum*: modification by bound and free cations. *Z. Vergleich. Physiol.* **61**, 427–452.

256 — & — 1968. Electrical properties of *Paramecium caudatum*: all-or-none electrogenesis. *Z. Vergleich. Physiol.* **61**, 453–472.

257 — & Yasumasu, I. 1969. Binding of Ca ions by *Paramecium caudatum*. *J. Gen. Physiol.* **50**, 1303–1310.

258 Okajima, A. & Kinosita, H. 1966. Ciliary activity and coordination in *Euplotes eurystomus*. I. Effect of microdissection of neuromotor fibers. *Comp. Biochem. Biophys.* **19**, 115–131.

259 Oliphant, J. F. 1938. The effect of chemicals and temperature on reversal in ciliary activity in *Paramecium*. *Physiol. Zool.* **12**, 19–30.

260 — 1942. Reversal of ciliary action in *Paramecium* induced by chemicals. *Physiol. Zool.* **15**, 443–452.

261 — 1943. Effects of some chemicals, which affect smooth muscle contraction on ciliary activity in *Paramecium*. *Proc. Soc. Exp. Biol. Med.* **54**, 62–64.

262 Párducz, B. 1953. Zur Mechanik der Zilienbewegung. *Acta Biol. Acad. Sci. Hung.* **4**, 177–220.

263 — 1954. Reizphysiologische Untersuchungen an Ziliaten. II. Neuere Beiträge zum Bewegungs- und Koordinationsmechanismus der Ziliaten. *Acta Biol. Acad. Sci. Hung.* **5**, 169–212.

264 — 1956. Reizphysiologische Untersuchungen an Ciliaten. IV. Über das Empfindungs- bzw. Reaktionsvermögen von *Paramecium*. *Acta Biol. Acad. Sci. Hung.* **6**, 289–316.

265 — 1956. Reizphysiologische Untersuchungen an Ziliaten. V. Zum physiologischen Mechanismus der sogenannten Fluchtreaktion und der Raumorientierung. *Acta. Biol. Acad. Sci. Hung.* **7**, 73–99.

266 — 1957. Über den feineren Bau des Neuronensystems der Ziliaten. *Ann. Hist. Natur. Mus. Nat. Hung.* **8**, 231–246.

267 — 1957. Das Problem der vorbestimmten Leitungsbahnen. *Acta Biol. Acad. Sci. Hung.* **8**, 219–251.

268 — 1958. Reizphysiologische Untersuchungen an Ziliaten. VII. Das Problem der vorbestimmten Leitungsbahnen. *Acta Biol. Acad. Sci. Hung.* **8**, 219–251.

269 — 1959. Reizphysiologische Untersuchungen an Ziliaten. Anlauf der Fluchtreaktion bei allseitiger und anhaltender Reizung. *Ann. Hist. Natur. Mus. Nat. Hung.* **51**, 227–246.

270 — 1962. Studies on reactions to stimuli in ciliates. IX. Ciliary coordination of right-spiraling paramecia. *Ann. Hist. Natur. Mus. Nat. Hung.* **54**, 221–230.

271 *— 1962. On the nature of metachronal ciliary control in *Paramecium*. *J. Protozool.* **9** (Suppl.), 27.

272 — 1963. Reizphysiologische Untersuchungen an Ziliaten. X. "Momentbilder" über galvanotaktisch frei schwimmenden *Paramecien*. *Acta Biol. Acad. Sci. Hung.* **13**, 421–429.

273 — 1967. Ciliary movement and coordination in ciliates. *Intern. Rev. Cytol.* **21**, 91–128.

274 Parker, G. H. 1919. *The Elementary Nervous System*, Lippincott, Philadelphia, London.

275 Pastor, E. P. & Fennel, R. A. 1959. Some observations on the esterase of *Tetrahymena pyriformis*. II. Some factors affecting aliesterase and choline-esterase activity. *J. Morphol.* **104**, 143–158.

276 Pfeffer, W. 1883. Lokomotorische Richtungsbewegungen durch chemische Reize. *Ber. Deut. Bot. Ges.* **1**, 524–533.

277 — 1884. Lokomotorische Richtungsbewegungen durch chemische Reize. *Untersuch. Bot. Inst., Tübingen* **1**, 363–482.

278 — 1888. Chemotaktische Bewegungen von Bakterien, Flagellaten und Volvocineen. *Untersuch. Bot. Inst., Tübingen* **2**, 582–661.

279 Pigoń, A. & Szarski, H. 1955. The velocity of the ciliary movement and the force of the ciliary beat in *Paramecium caudatum*. *Bull. Acad. Polon. Sci., Ser. Sci. Biol. (Classe II)* 99–102.

280 Pitelka, D. R. 1963. *Electron Microscopic Structure of Protozoa*, Pergamon Press, New York, N.Y.

281 — 1965. New observations on cortical ultrastructure in *Paramecium*. *J. Microscopie* **4**, 373–394.

282 — 1968. Fibrillar systems in protozoa. In Chen, T. T., *Research in Protozoology*, Pergamon Press, Oxford, 280–388.

283 — & Child, F. M. 1964. The locomotor apparatus of ciliates and flagellates: Relation between structure and function. In Hutner, S. H., *Biochemistry and Physiology of Protozoa*, Vol. 3, Academic Press, New York, N.Y. 131–198.

284 Preer, Jr., J. R. 1948. A study of some properties of the cytoplasmic factor "kappa" in *Paramecium aurelia*, variety 2. *Genetics* **33**, 349–404.

285 — 1959. Studies on the immobilization antigens of *Paramecium*. I. Assay methods. *J. Immunol.* **83**, 276–283.

286 — & Stark, P. 1953. Cytological observations on the cytoplasmic factor "kappa" in *Paramecium aurelia*. *Exp. Cell Res.* **5**, 478–491.

287 Puytorac, P. de, Andrivon, C. & Serre, F. 1963. Sur l'action cytonarcotique de sels de nickel chez *Paramecium caudatum* Ehrbg. *J. Protozool.* **10**, 10–19.

288 Pütter, A. 1900. Studien über die Thigmotaxis bei Protisten. *Arch. Anat. Physiol., Physiol. Abteil.* (Suppl.), 243–302.

289 — 1903. Die Flimmerbewegung. *Ergebn. Physiol.* **2**, 1–99.

290 — 1904. Die Reizbeantwortung der ciliaten Infusorien. *Z. Allg. Physiol.* **3**, 406–454.

291 — 1914. Die Anfänge der Sinnestätigkeit bei Protozoen. *Umschau*, 87–94.

292 Rees, C. W. 1922. The neuromotor apparatus of *Paramecium*. *Univ. Calif. (Berkeley) Publ. Zool.* **20**, 333–364.

293 Rivera, J. A. 1961. *Cilia, Ciliated Epithelium and Ciliary Activity*, Pergamon Press, New York, N.Y.

294 Rose, M. 1929. *La Question des Tropismes*, Presses Universitaires, Paris.

295 Roesle, E. 1902. Die Reaction einiger Infusorien auf einzelne Inductionsschläge. *Z. Allg. Physiol.* **2**, 139–168.

296 Roesle R. 1905. Spezifische Sera gegen Infusorien. *Arch. Hyg.* **54**, 1–31.

297 Russel, E. S. 1938. *The Behavior of Animals*, 2nd edition, Edward Arnold, London.

298 Schaefer, G. J. 1922. Studien über den Geotropismus von *Paramecium aurelia*. *Pflügers Arch. Ges. Physiol.* **195**, 227–244.

299 Schuberg, A. 1905. Über Cilien und Trichocysten einiger Infusorien. *Arch. Protistenk.* **6**, 61–110.

300 Seaman, G. R. 1951. Localization of acetylcholinesterase activity in the protozoan *Tetrahymena gelei* S. *Proc. Soc. Exp. Biol. Med.* **76**, 169–170.

301 — & Houlihan, R. K. 1951. Enzyme systems in *Tetrahymena gelei* S. II. Acetylcholinesterase activity. Its relation to motility and to coordinated ciliary action in general. *J. Cellular Comp. Physiol.* **37**, 309–321.

302 Sears, D. F. & Elveback, L. 1961. A quantitative study of the movement of *Paramecium caudatum* and *Paramecium multimicronucleatum*. *Tulane Stud. Zool.* **8**, 127–139.

303 — & Gittleson, S. M. 1960. Effects of CO_2 on *Paramecia*. *Physiologist* **3**, 140.

304 *— & — 1961. Narcosis of *Paramecia* with xenon. *Fed. Proc.* **20**, 142.

305 — & — 1964. Cellular narcosis of *Paramecium multimicronucleatum* by xenon and other chemically inert gases. *J. Protozool.* **11**, 538–546.

306 Seravin, L. N. 1959. Wzaimost' razlicznych funkcji *Paramecium caudatum* w processie priwykania k rastworam solej. *TSitologiya* **1**, 120–126.

307 — 1959. Wlijanie chimiczeskich agentow na podwiznost' *Paramecium caudatum*. *Zool. Z.* **38**, 626–632.

308 — 1961. Rol adenozintrifosfata w bijenii resniczek i infuzorij. *Biokhim.* **26**, 160–164.

309 — 1962. Fiziologicheskie gradienty infusorii *Spirostomum ambiguum*. *TSitologiya* **4**, 545–554.

310 — 1963. O roli sistemy atsetilkholin-kholinesteraza v koordinatsii bieniya resnichek i infuzorii. *S.B. Rabot Inst. TSitologiya Akad. Nauk. S.S.S.R.* **3**, 111–122.

311 — 1964. Ritmicheskaya aktivnost' sokratilelnoi sistemy odnokletochnych organizmov i usloviya ee vesniknovinaya. *TSitologiya* **6**, 516–520.

312 — 1969. Izuczenije wremiennych swiaziej uslownych refleksow u protiejszych organimow: kriticzeskij obzor. *TSitologiya* **11**, 659–680.

313 *— 1969. Some hydrodynamic aspects of movement in ciliates. *Progress in Protozoology* (Proc. 3rd Int. Conf. Protozool., Leningrad), 164–165.

314 Sleigh, M. A. 1956. Metachronism and frequency of beat in the peristomal cilia of *Stentor*. *J. Exp. Biol.* **33**, 15–28.

315 — 1957. Further observations on coordination and the determination of frequency in the peristomal cilia of *Stentor*. *J. Exp. Biol.* **34**, 106–115.

316 — 1960. The form of beat of cilia of *Stentor* and *Opalina*. *J. Exp. Biol.* **37**, 1–10.

317 — 1962. *The Biology of Cilia and Flagella*, Pergamon Press, London.

318 *— 1965. Ciliary coordination in Protozoa. *Progress in Protozoology* (Proc. 2nd Int. Conf. Protozool., London), 110–111.

319 — 1966. Some aspects of the comparative physiology of cilia. *Ann. Rev. Resp. Dis.* **93**, 16–31.

320 — 1966. The coordination and control of cilia. Nervous and hormonal mechanisms of integration. *Symp. Soc. Exp. Biol.* **20**, 11–31.

321 — 1968. Patterns of ciliary beating. Aspects of cell motility. *Symp. Soc. Exp. Biol.* **23**, 131–150.

322 — 1969. Coordination of the rhythm of beat in some ciliary systems. *Int. Rev. Cytol.* **25**, 31–54.

323 Smagina, A. P. 1948. *Mercatnelnogie Dvizenie*. Medgiz, Moscow.

324 Smith, S. 1908. Limits of educability of *Paramecium*. *J. Comp. Neurol. Psychol.* **18**, 499–510.

325 Soest, H. 1937. Dressurversuche mit Ciliaten und *Rhabdocoelen turbellarien*. *Z. Vergleich Physiol.* **24**, 720–748.

326 Sonneborn, T. M. 1943. Gene and cytoplasm. I. The determination and inheritance of the killer character in variety 4 of *P. aurelia;* II. The bearing of determination and inheritance of characters in *P. aurelia* on problems of cytoplasmic inheritance, pneumococcus transformations, mutations and development. *Proc. Nat. Acad. Sci. U.S.* **29**, 329–343.

327 — 1943. Acquired immunity to specific antibodies and its inheritance in *P. aurelia*. *Proc. Indiana Acad. Sci.* **52**, 190–191.

328 — 1959. Kappa and related particles in *Paramecium*. *Advanc. Virus Res.* **6**, 229–356.

329 Sosnowski, J. 1899. Untersuchungen über die Veränderungen des Geotropismus bei *Paramecium aurelia*. *Bull. Int. Acad. Cracoyie*.

330 Statkewitsch, P. 1903. Über die Wirkung der Induktionsschläge auf einige Ciliata. *Physiol. Russe* **3**, 1–55.

331 — 1903. Über die Wirkung von Induktionsschläge auf einige Ciliata: Galvanotropismus und Galvanotaxis von Organismen. *Physiol. Russe* **3**, 1–55.

332 — 1904. Galvanotropismus und Galvanotaxis der Ciliata. *Z. Allg. Physiol.* **4**, 296–332.

333 — 1905. Galvanotropismus und Galvanotaxis der Ciliata. *Z. Allg. Physiol.* **5**, 511–534.

334 — 1907. Galvanotropismus und Galvanotaxis der Ciliata. *Z. Allg. Physiol.* **6**, 13–43.

335 Teixeira-Pinto, A. A., Nejelski, L. L., Cutler, J. L., & Heller, J. H. 1960. The behavior of unicellular organisms in an electromagnetic field. *Exp. Cell Res.* **20**, 548–564.

336 Ten Kate, G. G. B. 1927. Über das Fibrillensystem der Ciliaten. *Arch. Protistenk.* **57**, 362–426.

337 Tobias, J. M. 1960. Further studies on the nature of the excitable system in nerve. I. Voltage-induced axoplasm movement in squid axons; II. Penetration of surviving, excitable axons by proteases; III. Effects of proteases and of phospholipases on lobster giant axon resistance and capacity. *J. Gen. Physiol.* **43**, 57–71.

338 Towle, E. W. 1904. A study on the effects of certain stimuli, single and combined, upon *Paramecium. Amer. J. Physiol.* **12**, 220–236.

339 Tschakhotine, S. 1938. Réactions "conditionées" par microponction ultraviolette dans le compartement d'une cellule isolée de *Paramecium caudatum. Arch. Inst. Prophylac.* **10**, 119–131.

340 Van Wagtendonk, W. J. & Vloedman, D. A. 1951. Evidence for the presence of a protein with ATP-ase and antigenic specificity in *Paramecium aurelia* var. 4 stock 51. *Biochim. Biophys. Acta* **7**, 335–336.

341 Verworn, M. 1889. *Psychophysiologische Protistenstudien. Experimentelle Untersuchungen,* Fischer, Jena.

342 — 1889. Die polare Erregung der Protisten durch den galvanischen Strom. *Pflügers Arch. Ges. Physiol.* **45**, 1–36.

343 — 1889. Die polare Erregung der Protisten durch den galvanischen Strom. *Pflügers Arch. Ges. Physiol.* **46**, 281–303.

344 — 1892. *Die Bewegung der lebendigen Substanz.* Fischer, Jena.

345 — 1896. Untersuchungen über die polare Erregung der lebendigen Substanz durch den constanten Strom. *Pflügers Arch. Ges. Physiol.* **62**, 415–450.

346 — 1896. Die polare Erregung der lebendigen Substanz durch den constanten Strom. *Pflügers Arch. Ges. Physiol.* **65**, 47–62.

347 — 1913. *Irritability,* Yale University Press, New Haven, Conn.

348 Viaud, G. & Bonaventure, N. 1956. Recherches expérimentales sur le galvanotropisme des Paramécies. *Bull. Biol. France Belg.* **90**, 287–319.

349 Wallengren, H. 1903. Zur Kenntnis der Galvanotaxis. I. Die anodische Galvanotaxis. *Z. Allg. Physiol.* **2**, 341–384.

350 — 1903. Zur Kenntnis der Galvanotaxis. II. Eine Analyse der Galvanotaxis bei *Spirostomum. Z. Allg. Physiol.* **2**, 517–555.

351 — 1903. Zur Kenntnis der Galvanotaxis. III. Die Einwirkung des konstanten Stromes auf die innere Protoplasmabewegung bei den Protozoen. *Z. Allg. Physiol.* **3**, 22–32.

352 Werman, R., McCann, F. V. & Grundfest, H. 1961. Graded and all-or-none electrogenesis in arthropod muscle. I. The effects of alkalie-earth cations on the neuromuscular system of *Romalea microptera. J. Gen. Physiol.* **44**, 979–995.

353 — & Grundfest, H. 1961. Graded and all-or-none electrogenesis in arthropod muscle. II. The effect of alkali earth and onium ions on lobster muscle fibers. *J. Gen. Physiol.* **44**, 997–1027.

354 Wichterman, R. 1953. *The Biology of Paramecium,* Blakiston, Philadelphia, Pa.

355 Wladimirsky, A. P. 1916. Are the Infusoria capable of "learning" to select their food? *Russ. J. Zool.* **44**, 4–21.

356 Worley, L. G. 1933. The intracellular fiber systems of *Paramecium. Proc. Nat. Acad. Sci. U.S.* **19**, 323–326.

357 — 1934. Ciliary metachronism and reversal in *Paramecium, Spirostomum* and *Stentor. J. Cellular Comp. Physiol.* **5**, 53–72.

358 Yamaguchi, T. 1960. Studies on the modes of ionic behavior across the ectoplasmic membrane of *Paramecium.* I. Electric potential difference measured by the intracellular micro-electrode. *J. Fac. Sci. Univ. Tokyo, Sect. IV,* **8**, 573–591.

359 — 1960. Studies on the modes of ionic behavior across the ectoplasmic membrane of *Paramecium.* II. In- and outfluxes of radioactive calcium. *J. Fac. Sci. Univ. Tokyo, Sect. IV,* **8**, 593–601.

360 — 1963. Time changes in Na, K and Ca contents of *Paramecium caudatum* after γ-irradiation. *Annot. Zool. Japon.* **36**, 55–65.

VII. ADDENDUM

(Additional references are indicated by the prefix Ad)

It was shown by Andrivon (Ad 361c, Ad 362) that nickel ions are actively transported throughout the cell membrane of *P. caudatum* and it is evident that the immobilizing effects of Ni^{2+} on ciliary movement depend on the concentrations of potassium and calcium in the external medium. Andrivon suggests that both ions may act as allosteric effectors on the enzymatic system allowing penetration of Ni^{2+} into the cell interior.

Machemer (Ad 397) found that the instantaneous fixation by the hematoxylin–osmium technique of Párducz of forward swimming paramecia revealed only 40 % of animals with preserved metachronal waves showing the characteristic pattern for forward movement while at least 50 % showed signs of early stages of ciliary reversal. In extensive studies on metachrony of *Paramecium* Machemer (Ad 398) brought evidence that the observed pattern of the metachronal beat of cilia in living animals corresponds well to that described by Párducz on the basis of an analysis of fixed preparations.

The considerable progress was achieved during the past years in the experimental approach to the mechanism of ciliary movement and to the role of the cell membrane in the bioelectric control of the ciliary activity of *Paramecium* (Ad 375–378, Ad 400, Ad 402). Naitoh and Kaneko (Ad 401) demonstrated the direct role of intracellular Ca^{2+} in the control of the ciliary beat with paramecia models extracted with the detergent Triton X-100 according to the method of Gibbons *et al.* (Ad 382). The extracted models showed forward motion and coordinated metachrony of ciliary movement in a mixture of 4 mM ATP, 4 mM $MgCl_2$, 3 mM EDTA, 50 mM KCl and 10 mM Tris-maleate (adjusted to pH 7.0 with NaOH). It is interesting that the reactivation of ciliary beating could occur in the presence of EDTA since this may suggest that calcium ions are not indispensable for reactivation of the ciliary beat in *Paramecium*. However, Mg^{2+} and ATP were required for reactivation and this is in agreement with results obtained previously on glycerinated models (252). The significance of Mg^{2+} seems to be confirmed by the recently discovered fact that the structural protein dynein in cilia of *Tetrahymena* (Ad 381) possessing ATPase activity was strongly activated by Mg^{2+} and much less so by Ca^{2+} and other divalent cations (Ad 380). The presence of dynein in the arms of the outer fibers of the cilium (Ad 380) may favor the sliding filament theory of ciliary movement proposed by Satir (Ad 404, Ad 405). On the basis of the calculated rate of turnover of dynein, Gibbons (Ad 380) made the interesting suggestion that a single dynein molecule may hydrolyze just one molecule of ATP during each cycle of the ciliary beat.

Returning to the interesting discovery of Naitoh, it is worth while to stress that the forward movement of cilia in detergent models was maintained when the concentration of external calcium was maintained between 10^{-8} to 10^{-7} mole/l. It should be emphasized however that Triton-extracted "skinned" paramecia show ciliary reversal when the concentration of calcium is raised to the level between 10^{-6} and 10^{-5} mole/l. It can be concluded from these data that in the living ciliate the reversal of the ciliary beat is also produced by an increase of intracellular Ca^{2+}. If we assume that the calcium sensitivity of extracted models and of the living paramecia is approximately the same, the concentration of Ca^{2+} within the normal cell should be in the range of 10^{-7} mole/l or slightly below this level. Consequently, the concentration of free intracellular Ca^{2+} during induced ciliary reversal should be higher than 10^{-6} mole/l and this may correspond to the case when the cell membrane is depolarized by the action of an external stimulus. The achieved results in experiments on Triton-extracted paramecia models support the view of Naitoh (250, 252) that the ciliary movement of *Paramecium* is under control of two separate physiological systems using ATP as the energy source: the first one causes cyclical bending of the cilium and requires Mg^{2+} as a cofactor for activation, while the second one requires Ca^{2+} and controls the direction of the ciliary movement.

Taking into consideration the recent advances in electrophysiological studies on the cell membrane and those observations which were carried out on Triton X-100-extracted models, Eckert (Ad 375) put forward the hypothesis that in paramecia the calcium influx current produced by an increase in calcium conductance of the cell membrane causes ciliary reversal. This resembles very similar mechanisms suggested by some authors for the striated muscles of the guinea pig (Ad 409) and the frog (Ad 364, Ad 406). According to this theory, calcium ions play an essential role in coupling the mechanical responses to the electrical responses of the cell membrane. Depolarization of the cell membrane causes an increase of calcium conductance (253) and in this way Ca^{2+} ions enter the region of the contractile fibers of the cilia, being driven by a calcium electro-chemical gradient. The hypothesis of a calcium current finds some support in previously reported observations that ciliary reversal does not occur at all, unless sufficient concentrations of external Ca^{2+} ions are present in the external medium (Ad 363, 179, 115, Ad 392, 251). It is obvious that in such cases there should be no net influx of calcium, because the electrochemical gradient for calcium ions between the external environment and the cell interior may approach zero. According to the new hypothesis the induced calcium current must stop as a consequence of the active removal of calcium from the cell. It is also conceivable that calcium ions may be sequestered inside the cell, causing in this way a termination of ciliary reversal.

Eckert and Naitoh summarized their results in a review article on the bioelectrical control of locomotion in ciliates (Ad 379). They gave a comprehensive description of the bioelectric phenomena within the cell membrane and their relation to well known patterns of behavior in *Paramecium* in response to external stimulations. Emphasis was placed on the fact that due to local stimulation of the anterior tip of the body the stimulus was transduced into a receptor potential which was accompanied by a local influx of Ca^{2+}. This receptor current would cause a depolarization of the whole membrane by an electronic spread. The resulting increase of the concentration of intracellular Ca^{2+} would cause activation of the locomotor organelles as expressed by ciliary reversal. This regenerative depolarization, spreading over the entire cell membrane is due to the electrical excitability of the membrane. There is evidence that both the nonregenerative receptor current and the regenerative current producing spike-like depolarizations are caused by Ca^{2+} (Ad 399, Ad 402). It should be added that active responses coupled with regenerative depolarization have a graded character. A weak mechanical stimulus evokes a small receptor potential and a small regenerative depolarization, whereas a stronger stimulus will produce a larger receptor potential, which in turn will evoke a larger regenerative depolarization of the cell membrane, accompanied by stronger motor responses of the ciliary apparatus. Eckert and Naitoh conclude that the gradation of the electrical response of the membrane provides fine gradations of control of ciliary activity of the protozoan cell and thus is sufficient for the passive spreading of the signal over the whole surface (length) of the organism. However, it is well known that under special environmental conditions (61, 180, 182) PCR responses, accompanied by all-or-none action potentials may appear with a signal propagation typical for this kind of response in excitable cells. It should be pointed out that all the observations and experimental results of Eckert, Naitoh and other protozooal physiologists seem to fit well into the general scheme of Grundfest (121, 123, 125) concerning excitable cells, with some modifications suggested by Dryl (68) for unicellular organisms.

Hildebrand suggested that some protein component of the cell membrane of ciliates may undergo conformational changes due to rupture of disulfide linkages. This might lead directly to an increased conductivity of the cell membrane for calcium (Ad 385). This could be connected with the activation of an ATPase-like enzyme which controls the transport of Ca^{2+} in accordance with the hypothesis of Bowler and Duncan (Ad 365). Since the enzymatic process must be involved both in the mechanism of forward and reversed ciliary beat it would be expected that both kinds of ciliary activity will depend on changes in external temperature. This is evident from early observations of Oliphant (259) who noticed a longer duration of cations-induced reversed swimming at lower temperatures. Dryl and Łukowicz (Ad 373) found that not only the velocity of the normal forward movement and of the K/Ca-induced ciliary reversal, but also the frequency of ciliary reversal cycles during RCP response induced by Ba/Ca depended on the external temperature. Paramecia adapted, for three hours to a constant temperature and then exposed for eight minutes to a medium containing 2 mM BaCl$_2$, 1 mM CaCl$_2$, and 1 mM Tris-HCl (pH 7.4) showed: (a) 28 $\pm$ 5.4 ciliary reversal cycles per min at 10°, (b) 44 $\pm$ 6.6 at 20° and (c) 62 $\pm$ 4.8 at 30°C. The rate of forward swimming and KCl induced backward swimming was increasing parallel to the increase of the external temperature. Tawada and Oasawa (Ad 407) reported that when paramecia cultured at 25° were transferred to higher or lower temperatures, their swimming velocity was increased immediately after the change of environment. During the first minute the velocity was decreasing exponentially with time to a new steady velocity. The optimum temperature for thermotaxis (thermotactic preferendum) was correlated with the temperature of the maximum steady velocity under given experimental conditions.

Positive phototactic responses towards chromatic light of 375–725 nm by *P. bursaria* (containing intracellular symbiontic algae of the genus Chlorella) were observed by Pado (Ad 403). At intensities of light, higher than 250 lux (with an optimum range of 3500–4200 lux), two phases of phototactic response could be distinguished: (a) the early phase, appearing during the first minutes of exposure, was induced by short wave lengths (436–450 nm) suggesting the role of riboflavin or carotinoids, and (b) the late phase, appearing after three hours, with an action spectrum at 680 nm which corresponds to the chlorophyll absorption spectrum.

A new promising field of investigation was started recently on behavioral mutants of *P. aurelia*, Stock 51 (Ad 368, Ad 389, Ad 390). Kung (Ad 389) was able to distinguish three mutant forms which were called by him: "Paranoiac", "Fast" and "Pawn". "Paranoiac" forms showed frequently occurring spontaneous ciliary reversal responses; "Fast" mutants showed an abnormally high speed of forward movement while "Pawn" individuals did not respond with ciliary reversal to depolarizing chemical and mechanical stimuli. The genetic mutants were characterized by molecular lesions within the excitable cell membrane. It was proven by genetic analysis (Ad 390) that the "Paranoiac" form is controlled by a pair of incompletely dominant alleles while the "Fast 1", "Fast 2" and "Pawn" forms are governed by unlinked recessive loci. Electrophysiological studies on normal strains of *P. aurelia d4-85* and the derived "Pawn" mutant d4-95 (originally designated 1-2-34) gave evidence that the behavioral deficiency of the "Pawn" mutant is due to an impaired electric excitability of the cell membrane (Ad 391). The cell membrane of the mutant did not show a depolarization-activated increase in calcium conductance, which normally plays an essential role in the regenerative depolarization of the wild type. Since the "Pawn" mutant differs from the wild type by a single gene, it was suggested that the impaired mechanism for calcium activation in the mutant strain might be due to a modification of single gene products.

After the pioneering study of Butzel *et al.* (31) on the inhibiting effects of the cationic detergent cetyl trimethyl ammonium bromide (CTAB) on the galvanotactic response of *Paramecium* it was shown recently by Bujwid-Ćwik and Dryl (Ad 366) that in the ciliate protozoa detergents may evoke a number of motor responses such as (a) various kinds of ciliary reversal, and (b) slackening of forward movement or contraction of the ectoplasm or the myonemes. It was found (Ad 370) that CTAB in toxic concentrations ($2-4 \times 10^{-6}$ g/ml) causes characteristic pear-shape deformations of the body with visible contractions of the posterior end, a lowering of the speed of forward movement accompanied by a decrease of the function of the contractile vacuoles and the abolishment of food-vacuole formation accompanied by a marked inhibition or almost complete cessation of cytoplasmic streaming (cyclosis) within the cell interior. In less toxic, sublethal concentrations ($0.5-2 \times 10^{-6}$ g/ml) CTAB caused short-lasting PCR responses, followed by a forward movement at a decreased rate. Within a wide range of concentrations (0.62×10^{-7} g/ml to 4×10^{-6} g/ml) CTAB gives rise to a marked decrease of sensitivity towards various external stimuli, with the strongest effect found at higher concentrations. Animals exposed to CTAB at $2-4 \times 10^{-6}$ g/ml did not respond with CCR to solutions with a high K/Ca ratio or to Ba/Ca ratios, while responses to strongly negative chemotactic agents were abolished and the thresholds for galvanotaxis were significantly increased. It was proven (Ad 371) that the observed changes of sensitivity were reversible phenomena since the animals were able to recover the normal level of sensitivity towards the K/Ca gradient after being washed for 16–32 minutes in a medium devoid of detergent. It was found recently that anionic detergents such as sodium dodecyl sulphate (SDS) and sodium dodecyl benzoate (SDB) evoke opposite effects. These detergents increase the sensitivity of *Paramecium*: the K/Ca-induced CCR lasts longer and the threshold values for chemotactic and galvanotactic responses are lowered (Ad 367). Similarly the changed state of cell excitability proved to be reversible and thus paramecia could recover the normal level of sensitivity after being washed for 16–32 minutes in a medium devoid of anionic detergents. It might be suggested that ionic detergents (CTAB, SDS, SDB) act on the cell membrane because of their charge and their affinity to the lipoprotein component of the membrane. CTAB may possess blocking effects on the chemoreception or conduction of impulses from the cell surface to the locomotor organelles. This view is confirmed by the fact that Ba/Ca-induced PCR responses and the accompanying all-or-none depolarizing spikes quickly disappear during the exposure of paramecia to the appropriate concentrations of CTAB (Ad 372). The postulated blocking effects of CTAB might be similar to those suggested by Walsh and Lee (Ad 408), who noticed a conduction-decreasing action of CTAB on the propagation of impulses along the giant fibers of the squid axon. On the other hand the action of the anionic detergents SDS and SDB may depend upon the increased conduction of impulses from the cell surface to the motile components of the cilia. The action of ionic detergents can be easily explained by the new theory of Eckert by evoking blocking effects of CTAB and activating effects of SDS and SDB on the calcium conductance of the cell membrane of *Paramecium*.

The effect of chloropromazine (CPZ) on the motile behavior of *P. aurelia* (51) was studied recently by Dryl and Masnyk (Ad 374). It was found that this anesthetic induces ciliary reversal in quite a different way as was known from observations on the effects of cations and other chemical agents producing ciliary reversal. In strongly acting, sublethal concentrations of CPZ (1–1.2×10^{-3} % w/v) there was no visible change of behavior during the first 30–40 seconds of exposure. Afterwards, typical PCR responses appeared and these lasted for the next 20–30 seconds. They were followed by CCR, which persisted during the next 60–80 seconds on the average. Then the animals showed again PCR until they recovered a forward movement at a reduced rate within 20–30 minutes. This characteristic sequence of motor responses induced by CPZ was not much influenced by the presence of Ca^{2+} in the external medium, suggesting a different action of this drug on the cell membrane and/or locomotor organelles from that known from studies with other substances. In more recent studies on *P. caudatum* (Ad 369) similar effects of CPZ were observed. It is evident that the CPZ-induced "delayed" CCR response appeared almost at the same time after the exposure of paramecia adapted to a medium containing a high ratio of K/Ca, or CTAB (2×10^{-6} g/ml) as in those in the control medium devoid of above mentioned agents. This suggests that the CCR-inducing action of CPZ is not affected by the depolarization of the membrane due to the action of K/Ca or by conduction blocking effects of CTAB. The unique mode of action of CPZ on *Paramecium* is also evident from the fact that animals in the stage of CPZ-induced "delayed" CCR continue this response during a relatively long time (30–50 seconds or more), even after having been washed in a medium without CPZ. It is difficult at present to explain the mechanism of CPZ action on *Paramecium*, but it is believed that a careful analysis of the membrane potential in the various stages of CPZ treatment will provide some important information, which will help to elucidate the problem whether we deal with some peculiar changes of membrane excitability or whether the "delayed" CCR is caused by a release of calcium ions from its storage sites within the cortex. It should be pointed out that some authors recently postulated that under some conditions the release of Ca^{2+} may take place from the cortical reticulum of ciliates (Ad 361a, 361b).

External stimuli may affect not only the physiological properties of the cell membrane and locomotor organelles but also the intracellular movement of cytoplasm. Recently, Kuźnicki and Sikora (Ad 393, Ad 394) made an attempt to follow the pattern and rate of cyclosis in *P. aurelia*, 51, immobilized by homologous antiserum. In contradiction to the view of some authors (Ad 386–388) they found that cyclosis does not occur in the whole volume of the endoplasm, but is limited to some determined regions of the paramecium body. The cytoplasmic streaming shows various velocities depending on localization, the width of the cytoplasmic stream and the intensity of the stimulation. The occurrence of spontaneous cessation of cyclotic movement and its sudden recovery was observed by means of microcinematographic analysis (Ad 396). On the basis of their own experimental results and in agreement with some data reported by Yamada (Ad 410) Kuźnicki and Sikora (Ad 394) proposed a hypothesis of an inverse relation between the rate of cyclosis and locomotion activity of *Paramecium*. They noticed a high speed of cyclosis in thigmotactic or immobilized animals while no visible signs of cyclosis could be observed in the cytoplasm of free swimming paramecia. With regard to the effect of temperature on the course of cyclosis, Kuźnicki and Sikora indicated (Ad 395) that the Arrhenius plot of temperature–velocity relation for cytoplasmic streaming in *P. aurelia* resembles plots for the Mg^{2+} activated ATPase activity of actomyosin from rabbit muscle (Ad 383, Ad 384). This may be an indirect proof for the existence of an actomyosin-like system involved in the process of cyclosis in *Paramecium*.

REFERENCES
(Abstracts are indicated by *)

361a Allen, R. D. 1971. Fine structure of membraneous and microfibrillar systems in the cortex of *Paramecium caudatum*. *J. Cell Biol.* **49**, 1–20.

361b *— & Eckert, R. 1969. A morphological system in ciliates comparable to the sarcoplasmic reticulum transverse system in striated muscle. *J. Cell Biol.* **43**, 4a.

361c Andrivon, C. 1970. Preuves de l'existence d'un transport actif de l'ion Ni^{++} à travers la membrane cellulaire de *Paramecium caudatum*. *Protistologica* **6**, 445–455.

362 — 1972. The stopping of ciliary movements by nickel salts in *Paramecium caudatum;* the antagonism of K^+ and Ca^{++} ions. *Acta Protozool.* **11**, 373–386.

363 Bancroft, F. W. 1906. On the influence of the relative concentration of calcium ions on the reversal of the polar effects of the galvanic current in *Paramecium*. *J. Physiol.* **34**, 444–463.

364 Bianchi, C. P. 1961. Calcium movements in striated muscle during contraction and contracture. In Shanes, A. M., *Biophysics of Physiological and Pharmacological Actions*, American Association for the Advancement of Science, Washington, D.C., 281–292.

365 Bowler, K. & Duncan, C. J. 1967. Evidence implicating a membrane ATPase in the control of passive permeability of excitable cells. *J. Cell Physiol.* **70**, 121–126.

366 *Bujwid-Ćwik, K. & Dryl, S. 1971. The effects of detergents on motile behavior of protozoa. *J. Protozool.* **18** (Suppl.), 92.

367 — & — Unpublished data.

368 Cooper, J. 1965. A fast-swimming "mutant" in Stock 51 of *Paramecium aurelia, variety 4. J. Protozool.* **12**, 381–384.

369 Dryl, S. Unpublished data.

370 — & Bujwid-Ćwik, K. 1972. Effects of detergents on excitability and motor response in protozoa. *Acta Protozool.* **11**, 367–372.

371 — & — 1972. Effects of detergent cetyl trimethyl ammonium bromide on motor reactions of *Paramecium* to potassium/calcium factor in external medium. *Bull. Acad. Polon. Sci., Ser. Sci. Biol. (Classe II).* **20**, 551–555.

372 — & Kaliński, J. Unpublished data.

373 — & Łukowicz, M. Unpublished data.

374 *— & Masnyk, S. 1971. Observations on the effects of chlorpromazine on the motility of *Paramecium aurelia, stock 51. J. Protozool.* **18** (Suppl.), 91.

375 Eckert, R. 1972. Bioelectric control of ciliary activity. *Science* **176**, 473–481.

376 — & Murakami, A. 1971. Control of ciliary activity. In Podolsky, R., *Contractility of Muscle Cells and Related Processes*, Prentice-Hall, Englewood, N.J.

377 *— & Naitoh, Y. 1969. Graded calcium spikes in *Paramecium. 3rd Int. Biophys. Congr. Int. Union Pure and Applied Biophys., Cambridge, Mass.*, 257.

378 — & — 1970. Passive electrical properties of *Paramecium* and problems of ciliary coordination. *J. Gen. Physiol.* **55**, 467–483.

379 — & — 1972. Bioelectric control of locomotion in ciliates. *J. Protozool.* **19**, 237–243.

380 Gibbons, I. R. 1966. Studies on the adenosine triphosphatase activity of 14S and 30S dynein from cilia of *Tetrahymena. J. Biol. Chem.* **241**, 5590–5596.

381 — & Rowe, A. J. 1965. A protein with ATP-ase activity from *Tetrahymena pyriformis. Science* **149**, 424–425.

382 *Gibbons, B. H., Fronk, E. & Gibbons, I. R. 1970. Adenosine triphosphatase activity of Sea urchin sperm. *J. Cell Biol.* **47**, 71a.

383 Hasselbach, W. 1952. Die Umwandlung von Aktomyosin-ATPase in L-Myosin-ATPase durch Aktivatoren und die resultierenden Aktivierungseffekte. *Z. Naturforsch.* **7b**, 163–174.

384 Hartshorne, D. J., Barns, E. M., Parker, L. & Fuchs, F. 1972. The effect of temperature on actomyosin. *Biochim. Biophys. Acta* **267**, 190–202.

385 Hildebrand, E. 1972. Avoiding reaction and receptor mechanisms in protozoa. *Acta Protozool.* **11**, 361–366.

386 Jahn, T. L. & Bovee, E. C. 1969. Protoplasmic movements within cells. *Physiol. Rev.* **49**, 793–862.

387 Koenuma, A. 1954. Study on cyclosis of *Paramecium. J. Shinshu Univ.* **4**, 49–57.

388 — 1963. The velocity distribution of the cyclosis in *Paramecium. Annot. Zool. Japon.* **36**, 66–71.

389 Kung, C. 1971. Genic mutants with altered system of excitation in *Paramecium aurelia*, Part 1 (Phenotypes of the behavioral mutants). *Z. Vergleich. Physiol.* **71**, 142–162.

390 — 1971. Genic mutants with altered system of excitation in *Paramecium aurelia*. II. Mutagenesis, screening, and genetic analysis of the mutants. *Genetics*, **69**, 29–45.

391 — & Eckert, R. 1972. Genetic modification of electric properties in an excitable membrane. *Proc. Nat. Acad. Sci. U.S.A.* **69**, 93–97.

392 Kuźnicki, L. 1966. Role of Ca^{++} ions in the excitability of protozoan cell. Calcium factor in the ciliary reversal induced by inorganic cations in *Paramecium caudatum. Acta Protozool. (Warsaw)* **4**, 241–256.

393 — & Sikora, J. 1971. Cytoplasmic streaming within *Paramecium aurelia*. I. Movements of crystals after immobilization by antiserum. *Acta Protozool. (Warsaw)* **8**, 439–446.

394 — & — 1972. Cytoplasmic streaming within *Paramecium aurelia*. III. The effect of temperature on flow velocity. *Acta Protozool. (Warsaw)*, **12**, 143–150.

395 — & — 1972. The hypothesis of inverse relation between ciliary activity and cyclosis in *Paramecium. Acta Protozool. (Warsaw)* **11**, 243–250.

396 —, — & Fabczak, S. 1972. Cytoplasmic streaming within *Paramecium aurelia*. II. Cinematographic analysis of the course and reversible cessation of cyclosis. *Acta Protozool. (Warsaw)* **11**, 237–242.

397   Machemer, H. 1970. Primäre induzierte Bewegungs-stadien bei Osmiumsäurefixierung vor-
      wärtsschwimmender Paramecien. *Acta Protozool. (Warsaw)* **7**, 531–535.
398   — 1972. Properties of polarized ciliary beat in *Paramecium*. *Acta Protozool. (Warsaw)* **11**,
      295–300.
399   Naitoh, Y. 1969. Ciliary orientation: controlled by cell membrane or by intracellular fibrils?
      Science **166**, 1633–1635.
400   — & Eckert, R. 1971. Electrophysiology of the ciliate protozoa. In Kerkut, G. A., *Experiments
      in Physiology and Biochemistry, Vol. 5*, Academic Press, London.
401   — & Kaneko, H. 1972. Reactivated Triton-extracted models of *Paramecium*. Modification of
      ciliary movements by calcium ions. *Science* **176**, 523–524.
402   —, Eckert, R. & Friedman, K. 1972. Regenerative calcium responses in *Paramecium*. *J. Exp.
      Biol.*, **56**, 667–681.
403   Pado, R. 1972. Spectral activity of light and phototaxis in *Paramecium bursaria*. *Acta Protozool.
      (Warsaw)* **11**, In press.
404   Satir, P. 1965. Studies on cilia, Part 2 (Examination of the distal region of the ciliary shaft and
      the role of the filaments in motility). *J. Cell. Biol.* **26**, 805–834.
405   — 1972. Sliding microtubules of cilia. *Acta Protozool. (Warsaw)* **11**, 279–286.
406   Shanes, A. M. 1961. Correlation of calcium uptake and contractility in frog rectus abdominis
      muscle. In Shanes, A. M., *Biophysics of Physiological and Pharmacological Actions*, American
      Association for the Advancement of Science, Washington, D.C. 309–316.
407   Tawada, K. & Oasawa, F. 1972. Responses of *Paramecium* to temperature. *J. Protozool.* **19**,
      53–57.
408   Walsh, R. R. & Lee, J. P. R. 1962. Action of surface active agents on axonal conduction.
      Amer. J. Physiol. **202**, 1241–1247.
409   Winegrad, S. 1960. The relationship of calcium uptake to contraction in guinea pig atria.
      Physiologist **3**, 179.
410   Yamada, K. 1969. A comparative study on the cyclosis in *Paramecium*. *J. Sci. Hiroshima Univ.*,
      Ser. B, Div. 1 **22**, 127–153.

Growth Patterns and Morphogenetic Events in the Cell Cycle of *Paramecium aurelia*

MASAO KANEDA* AND EARL D. HANSON

Shanklin Laboratory, Wesleyan University, Middletown, Conn. 06457 (U.S.A.)

I. INTRODUCTION

There are no exceptions to Virchow's fundamental axiom of biological continuity that all cells come from pre-existing cells. And nowhere is this axiom more beautifully illustrated than in the strikingly differentiated cellular systems of the ciliated protozoa. Such differentiations offer special opportunities to biologists concerned with the fundamental processes of cellular reproduction. First, the distinctive differentiations of ciliated cells can be exploited as topographical markers. Such markers, which are absent in other cell systems, allow us to follow in unprecedented detail the behavior of the cell and its parts as the cell cycle unfolds from one fission to the next. Second, these exceptional differentiations offer extraordinary opportunities not only to observe but also to analyze intracellular development. These differentiations arise from and are liable to much the same regulatory and control mechanisms as other organelles. The characteristic measure of visible diversity and complexity found in the ciliates affords unique opportunities to study fundamental mechanisms of cellular differentiation and morphogenesis (36, 83, 87, 88).

The primary goal in this paper is to present a coherent review of the developmental events of the cell cycle of *Paramecium aurelia* as seen in the cortical structures, including the feeding and osmoregulatory organelles, and the macronucleus. The further problem of analyzing the control of these events will be discussed only as it arises from the data presented here.

The present status of our knowledge of paramecium is of course compounded from a variety of studies. Starting with the monumental works of European protozoologists, exemplified by Hertwig (39) and Maupas (60), interest in paramecia has led to an enormous variety of studies which are more recently summarized in Wichterman's (89) monograph on the genus *Paramecium*. This range of research is both the delight and the dismay of anyone trying to obtain an overview of our present understanding

This work was supported by a grant from the National Science Foundation (GB-3937).
* Present address: Biological Institute, Department of General Education, Hiroshima University, Hiroshima (Japan).

References p. 258–261

of these forms. That is to say, the amount and variety of information available is welcome, but because it is generated from so many special points of view there are the inevitable differences of terminology and of research perspective. These points are nowhere more evident than in the history of the ciliate cortex.

The great advance in the study of the surface morphology of ciliates in general and paramecium in particular came with development of the silver impregnation techniques and the publication of Klein's authoritative papers (54, 55) and of von Gelei's detailed study of *Paramecium* (31). This was followed by very extensive acceptance of this approach to cell morphology and morphogenesis (5, 59). The larger implications of these studies are reviewed in Corliss' (9) monograph on *The Ciliated Protozoa* which contains an invaluable bibliographical compilation. As regards cortical structures in paramecium, the next major turning point was the work of Ehret and Powers (22) on *P. bursaria*, wherein the confusing diversity of fiber systems and surface units revealed by the silver staining studies was resolved with the aid of critical electron microscopy into the basic concept of the ciliary corpuscle. This structural unit has been further conceptualized into ciliary units or, simply, cortical units (12, 68, 83) and finally, most elegantly restated by Pitelka (69, 70) as kinetosomal territories.

The term "kinetosomal territories" derives operationally from electron microscopy, and the term "cortical unit" is commonly associated with silver impregnation studies. In that the work reported in this paper draws heavily on the use of silver impregnation for visualizing the cortical structures, it is arguably accurate and consistent to retain the term "cortical unit" and treat it as interchangeable with "kinetosomal territory". Both terms refer to the basic unit of organization of the ciliate cortex.

Although morphological and morphogenetic studies of the oral or feeding structures in *Paramecium* have been reported by many authors (1, 17, 18, 20, 21, 31, 33, 37, 41, 57, 67, 71, 72, 73, 78, 79, 83, 86, 89, 94), developmental patterns of the other cortical structures which have important relations to stomatogenesis have been described by relatively few authors (68, 73, 83, 89). Many important features of the differentiation and morphogenesis of the cortical structures in *P. aurelia* still need further clarification, particularly in reference to the time sequence and spatial relations involved. However, in providing clearer descriptive details it is not enough simply to add to older points of view. What is needed are statements and concepts that utilize the insights derived from fine structure, genetic and developmental studies and anticipate the day when we can account for organelle formation as an assembly problem utilizing gene products to form functionally integrated cells.

Using carefully timed cells, sequential changes in the cortex and related structures of living and silver-impregnated *P. aurelia* have been followed from the beginning of the interfission cycle to final separation of the daughter cells. Observations include changes in the number and location of already existing and of newly developing entities and, also, quantitative changes in selected structures and dimensions of the cell which reveal some unexpected activities of the cell surface. In addition, sequential changes in the macronucleus, in terms of varying form and location in the cytoplasm, are seen to have important correlations with the differentiation and morphogenesis of the cortical structures. These data permit accurate formulation of the relations between proliferation of the cortical territories and associated organelles and the size changes of the cell. In turn, this leads to reformulation of possible organizational constraints acting in the cell and to a reappraisal of some of the postulated factors controlling morphogenetic changes.

II. MATERIALS AND METHODS

A. Cultures

The organisms used in this study are *P. aurelia* stock 51, sensitive, syngen 4 (80, 82). They were cultured in a Cerophyl infusion inoculated with *Aerobactor cloacae* (35) and grown at room temperature (25 °C) or at 31 °C, as needed.

B. Timing of Interfission Events

In order to time developmental events occurring in the interfission period, the sister cell technique (51) was used (32, 38). Autogamous animals were obtained by standard techniques (81) and one or two were isolated into culture depressions containing fresh culture medium, usually in the afternoon preceding the day they were needed for study. The next morning, one of these paramecia, having completed its 3rd or 4th fission after autogamy, was again isolated into fresh culture medium in the central depression of a 3-spot depression slide. Immediately after the next fission, the sister paramecia were separated, one each to fresh medium in the side depressions on the same slide, and the time of the fission was recorded. Whenever possible, the proter was transferred to the left depression and the opisthe to the right one. At a certain time after fission, one fission product was observed or fixed and the other kept and the duration, T, of its interfission interval ascertained. The unknown duration of the interfission interval of the fixed or observed paramecium was estimated as $\hat{T}$ from the following formula (51):

$$\hat{T} = rT + (1-r)\,\overline{T}$$

$\overline{T}$ is the average of T for the sample studied. The intraclass correlation coefficient, r, measures the degree of similarity between the interfission periods of sister cells. The time at which the paramecium was observed or fixed, t, was divided by $\hat{T}$ and this value was used as the interfission age of the experimental cell. For example, cells examined at 3.0 and 3.5 hours after the prior fission will both be close to inter-fission age 0.5 if their estimated interfission periods are 6 and 7 hours, respectively. This provides a basis for grouping cells for purposes of comparison when the absolute interfission periods show considerable variation in length.

C. Silver Impregnation

Individual paramecia were impregnated with silver using a technique based on the

Chatton–Lwoff wet silver method, but incorporating subsequent modifications by Corliss (7) and Gillies (33). The preparations are often remarkably transparent with both ventral and dorsal silver line systems clearly visible in the same paramecium. In addition, such preparations permit clear observation of the macronucleus by means of phase contrast microscopy.

There is some problem in distinguishing the spot of reduced silver which marks the location of the trichocyst from the cluster of spots marking the ciliary base or bases and the associated parasomal sac. This is especially true for low power microscopy. There are three criteria, however, which are applicable in identifying these silver impregnated structures. The trichocyst spot or dot is usually smaller and fainter than the cluster of ciliary base and parasomal sac material. Further, the trichocyst dot lies in a different focal plane for this spot appears at the junction of two territories, whereas the ciliary materials lie in the depressed center of these units. Finally, the trichocyst is a unitary structure and the other material can appear as a doublet (one ciliary base, one parasomal sac) or a triplet (two bases, one sac) and this compound nature is commonly resolvable under higher powers of the light microscope. Using these criteria it is almost always possible to distinguish clearly between trichocysts and ciliary sites (see also 10, 83). These bases and the adjacent parasomal sacs lie at the right of the center and are the most obvious surface manifestation (in silver prep-arations) of the kinetosomal territory. We have never observed more than one sac per unit except at times of unit proliferation. Hufnagel's electronmicroscopical studies (41), give clear evidence of two sacs in units that do not appear to be proliferating. The longitudinal alignment of the cortical units forms the ciliary rows or kineties (59) while more horizontal alignments have been termed paratenes (19). These ciliary rows form highly characteristic patterns on the cell surface.

D. Measurements

In particular, counts of unit territories have been used to describe events on the dorsal surface of paramecia by following the number of these units on the contractile vesicle meridian, at various times throughout the cell cycle. The contention (90) that contractile vacuoles do not generate their own contraction is still open to doubt (74), but the term *vesicle* does seem preferable to *vacuole*, because of the latter's connotation of emptiness. Hence, we will refer to contractile vesicles throughout this paper, rather than to contractile vacuoles. In order to measure the growth, extension and contraction of the cortex of paramecium, the contractile vesicle meridian is divided into three regions prior to the appearance of new vesicles (Fig. 1A): anterior dorsal region or (ad) extends from the anterior end of the cell body to the level of the pore of the anterior contractile vesicle; the middle region (md) lies between the levels of the pores of the anterior and the posterior vesicles; and the posterior region (pd) extends from the level of the posterior vesicle pore to the posterior end of the cell body (Fig. 1A). After new contractile vesicles have appeared, the vesicle meridian is divided into five or six regions (Fig. 1B): adp = the new anterior dorsal region of the proter extending from the anterior end of the cell to the level of the new anterior

contrasted to germinal parts of an organism and not to oral parts.] Because oral or feeding structures are the basis for the terminology we are here proposing, and, more importantly, are the basis for the anatomical orientation of the cell—ventral and dorsal, right and left are all defined in relation to the oral apparatus— we will start with a review of the buccal organelle and then describe the cortical zones and their component fields.

B. Buccal Structures

The buccal organelle is a curved, funnel-like structure opening in the middle of one side of a paramecium. Details of its structure have been published elsewhere (31, 37, 41, 54, 58, 68, 71, 72, 76, 78, 83, 89, 94). These reports are all in essential agreement except for the terminology used (Table 2). We will here follow Corliss' widely accepted proposals (8, 9). The term gullet which is popular only among workers with paramecia, and which includes the cytostome, buccal cavity and buccal opening, will be abandoned in favor of the phrase *buccal structures* or buccal organelle.

TABLE 2

Terminology used for feeding structure. Horizontal rows contain terms applied to the same structural part of the feeding organelle, e.g. the bottom row includes the terms applied to the oral groove. In Kahl's terminology, Schlund and Trichter are also lumped as Schlundtrichter (throat, gullet or pharynx). Von Gelei includes all items down through Vestibulum as Cytopharynx. Items in rows down through mouth are termed gullet by Wichterman and others working with paramecia. Corliss collectively refers to those same parts as oral structures.

Kahl (49)	von Gelei (31)	Wichterman (89)	Corliss (8, 9)
Nahrungs-vakuole	Ösophagus mit der	food vacuole forming region	cytopharynx
Schlund	Empfangs-vakuole		cytostome
Pharynx	Pharynx	esophagus pharynx or cytopharynx	buccal cavity
Mund	Mund	mouth	buccal overture
Mundtrichter	Vestibulum	vestibulum	vestibule
Grube	Mulde	oral groove	oral groove

The major features of this organelle are given in Fig. 2 and only a brief summary is needed here. There are 12 kineties, collected into three groups of 4 kineties each, which lie on the wall of the buccal cavity. The ventral and dorsal peniculi lie largely on the left wall, with the posterior end of the dorsal peniculus wrapping around the posterior ventral wall of the cavity. The quadrulus lies mostly on the dorsal wall of

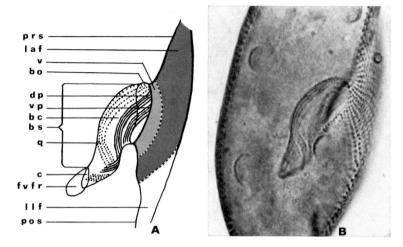

Fig. 2. Feeding structures. *A:* diagram illustrating terminology used for feeding structures. The right side of the ventral surface of the cell is removed to reveal the buccal organelle and part of the left ventral surface. bc = buccal cavity; bo = buccal opening; bs = buccal structures; fvfr = food vacuole forming region; laf = left adoral field (which also includes v); llf = left lateral field; pos = postoral suture; prs = preoral suture; v = vestibule. The dotted lines mark the boundaries between the vestibule (light gray) and the rest of the left adoral field (dark gray, not shown in its entirety) and also mark the boundary between this latter field and the adjacent left lateral field which is part of the anoral zone. On the left wall of the buccal cavity are the ventral and dorsal peniculi (vp, dp) each made up of four closely packed kineties (solid lines). On the dorsal buccal wall is the quadrulus (q), also composed of four kineties (dotted lines). *B:* photograph of silver-stained specimen.

the buccal cavity, with its kineties spread somewhat apart anteriorly. These kineties converge as they, too, wrap around the posterior ventral side of the buccal cavity. Not shown in Fig. 2 is the ribbed wall, first seen in *P. bursaria* by Ehret and Powers (21) and clearly described recently in *P. aurelia* by Hufnagel (41). This structure is not prominent in silver-stained specimens. A last kinety associated with the buccal structure is the endoral kinety, first described by von Gelei (31). This lies along the right border of the buccal opening and is better described as part of the vestibular ciliature.

It is also pertinent to record here some facts regarding the dimensions of the feeding organelle which have not been previously reported in a systematic fashion. The buccal overture or opening is 15μ long by 8μ wide. The buccal cavity extends into the cell about 15μ from the buccal opening and then curves posteriorly for another 15μ. This latter part tapers from a width of about 10μ just behind the buccal opening to the narrow posterior end where the cytostome lies. Food vacuoles are formed more on the dorsal side of this end (Fig. 2) rather than at its posterior tip (58) as is so commonly figured in texts.

C. *Adoral Zone and Sutures*

Turning next to the cell surface adjacent to the buccal opening, that part which is thought to aid in feeding is termed the adoral zone. The evidence for the postulated feeding function is incomplete. Certainly for the oral groove there is clear observational evidence (44) that food particles pass along this groove and into the buccal cavity. In all probability the peculiar pattern of ciliation (see below) associated with the groove and the whole anterior left side of the cell, is what determines this flow of particles. Arguing similarly for the characteristic pattern on the right side of the buccal opening, we infer that this area of ciliation is also part of the adoral zone. This zone can be subdivided into the two fields already referred to and these are designated the *left* and *right adoral fields*.

The two fields are separated by suture lines extending anteriorly and posteriorly from the buccal opening—the *pre-* and *postoral sutures*, respectively. The appearance of these lines has been most clearly described by Sonneborn (83) who emphasized the fact that the preoral suture is devoid of material on which silver will deposit and represents, therefore, a clear linear space on the anterior ventral surface between the anterior termini of certain ventral kineties. There are 20 to 23 of these kineties on the left side of this suture. The postoral suture, on the other hand, shows close abutting of the posterior ends of right and left ventral kineties and the suture is simply seen as this line of abutment.

In the posterior half of this line, silver deposits heavily and marks the position of the cytoproct (also called, but less accurately, cytopyge—meaning "cell hips", or anal pore, though there is no typical pore). This line of silver deposition marks the slit-like site where undigested contents of food vacuoles are eliminated from the cell. The cytoproct is somewhat variable in length and in the few cases where elimination through the cytoproct has been found in fixed and stained cells, it appears that the elimination site is smaller than the length of the dark line marking the cytoproct. Perhaps slit size is adjusted to the amount of material being released and our specimens did not show the largest possible opening.

A new feature of the suture lines is the fact that both extend around the ends of the cell and terminate on the dorsal side of the cells. The preoral suture, as previously described, terminates just below the anterior end of the cell. Indeed, the suture as a narrow line free of silver impregnation does terminate as described, but then a line of abutment of kineties (like the postoral suture) continues anteriorly. It passes over the end of the cell, swings to the left for a short distance on the dorsal side and then stops. In a similar manner the postoral suture extends around the posterior end of the cell, but just to the left of the apex, and then swings briefly to the right and terminates on the dorsal side (Fig. 3). These sutures can be demonstrated most readily by the dry silver method (56). (Fig. 2 of Kozloff's paper illustrates the polar end of the postoral suture on the dorsal side of the cell but is simply designated as "a suture in paramecium".) In wet silver preparations the sutures are hard to see and become apparent when kineties are traced out and the exact positions of their termini are identified. (Compare Figs. 3B and D, for example.)

The *left adoral field* lies wholly to the left of the sutures and the buccal opening. It is best defined by a cortical pattern where paratenes [rows of cortical units in adjacent kineties (19)] extend largely latitudinally. Only at the left posterior margin

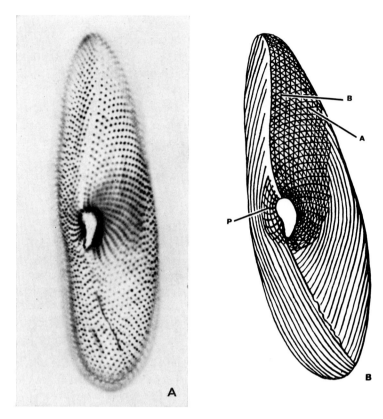

Fig. 3. Selected aspects of paramecia illustrating the surface zones and their constituent fields. *A:* ventral view of a silver impregnated cell. *B:* diagram of a ventral view illustrating the adoral zone with its kineties and paratenes. A = an A paratene, B = a B paratene. Both paratenes lie in the adoral groove field. P = paratene in the right adoral field. *C:* dorsal view of a silver-impregnated cell showing the two contractile vesicle pores. *D:* diagram of the dorsal surface illustrating the largely meridianal kineties of the anoral zone. Note also the sutures at each end of the cell. cvp = contractile vesicle pores. (Figures *A* and *C* courtesy of Dr. R. V. Dippell.)

of the buccal opening do the paratenes bend posteriorly until, at the posterior edge of the opening, the paratenal rows are running longitudinally. All of these will be called the A-paratenes. This field (Fig. 3A) lies for the most part in the oral groove of a paramecium. In this area the kineties arise anteriorly against the left edge of the preoral suture. They swing laterally and posteriorly in a strong curve of parallel, kinetal rows around the buccal opening. Porter (72) reports 33 kineties arising along the preoral suture; our observations agree closely with this figure. Not all of these kineties lie entirely within the left adoral field. Only the ten or so kineties closest to the buccal opening lie entirely within the field. The others have only their anterior end in the field, the rest of the kinety emerging from the field and becoming part of the left lateral field of the anoral zone. In addition to the predominantly latitudinal A-paratenes one can also identify another paratenal pattern, the B-paratenes, which run longitudinally in the area just to the left of the preoral suture. More laterally in the oral groove, these B-paratenes curve their posterior ends to the right into the vestibular

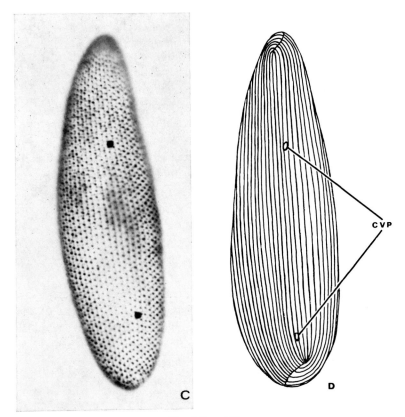

Fig. 3 C and D.

depression which surrounds the buccal opening. It is the conjunction of the kinetal and the two paratenal patterns that most clearly defines the left adoral field (Fig. 3A). At its left border there are a few cortical units in each A-paratene which are hard to assign, either to the left adoral field or to the left lateral field. Such an area of uncertainty also borders the right adoral field, the other field of the adoral zone.

The *right adoral field* lies wholly to the right of sutures and the buccal opening. It is smaller than the adoral groove field, but shows a paratenal pattern similar to that of the A-paratenes in its sister field (Fig. 3A). However, the right adoral field does not show the analog of the B-paratene pattern. The kineties in this field, about 10 in number, arise anteriorly at the right side of the posterior end of the preoral suture. They swing around the buccal opening in a shallow curve and those 5 to 7 closest to the opening terminate against the right side of the anterior end of the postoral suture. The other kineties continue on posteriorly as part of the right lateral field of the anoral zone.

One notable feature shared by both adoral fields is that the cortical units are typically of the triplet type, apparently containing two cilia and one parasomal sac as Hufnagel (41) has also noted. The rest of the surface, the anoral zone, typically shows doublet units. This observation disagrees with Roque's (78) claim that triplets cover the surface of *P. aurelia* and it differs from the situation in *P. trichium* (33) where triplets lie in a girdle around the mid-third of the cell, with doublets occurring as two large polar caps, one anterior and the other posterior.

Before leaving the adoral zone, special mention must be made of details in the vestibule (9) or circumoral depression (83). The vestibule sinks about 5μ below the surrounding cell surface, with the buccal opening at its bottom. The curved left part of the vestibule forms the anterior and left margin of the buccal opening and grades into the oral groove. There are usually three kineties seen on the left anterior surface and five kineties on the left lateral surface. The anterior end of the innermost, first kinety is situated on the middle portion of the left margin of the buccal opening; the anterior end of the 2nd kinety starts on the anterior left corner of the opening; the anterior end of the 3rd kinety starts midway on the anterior margin of the buccal opening; and those of the 4th and 5th kineties appear close to each other on the posterior tip of the preoral suture. The units on the left wall of the vestibule often appear as two ciliary bases without a parasomal sac. However, careful scrutiny of properly oriented, well-stained specimens, under oil immersion, reveals that faintly impregnated parasomal sacs are present.

The right wall of the vestibule is very steep, resulting in its outer edge—away from the buccal opening—being the most conspicuously raised portion of the ventral surface. Five kineties are normally observed on this right wall, the ones closest to the buccal opening being the shortest. The parasomal sacs in these kineties are difficult to see, as is the case on the opposite side of this depression.

Finally, von Gelei's endoral kinety is located on the innermost margin of the right wall. This is a single row of unitary ciliary bases usually 20 to 25 in number. They appear much darker and are therefore more distinct than the other basal granules. [Roque (78) has seen 15 basal granules in the endoral membrane and Yusa (94) reports 22.] The ciliary membranelle formed by this kinety lies on the edge of the buccal opening, overhanging the right side of the buccal cavity, and effectively demarcates the cavity from the adoral depression. The anlage field for a new feeding organelle arises between the endoral kinety and the posterior ends of the kineties to its right, on the right surface of the vestibule.

Before turning next to the anoral zone, the relation of the vestibule and oral groove to the two adoral fields should be made explicit. The vestibule is part of both fields. The right adoral field is coextensive with the right wall of the vestibule but then extends beyond it onto the cell surface proper. The left adoral field covers the left wall of the vestibule and then extends anteriorly and to the left (Fig. 2A), filling the oral groove and extending beyond it to the anterior end of the cell. Also, to the left side of the vestibule, the left adoral field extends onto the surface of the cell beyond the edge of the vestibular depression for a short distance (see Figs. 3A and B).

D. Anoral Zone

This zone covers the remaining cell surface not included in the adoral zone. As the name implies, this zone is without direct structural or functional relations to the feeding organelle. Its surface is largely characterized by doublet units forming quite regular kinetal rows; this regularity is modified at the ends of the cell where the kineties often curve slightly and are usually terminated along the sutures (Figs. 3C and D). Also notable are the two contractile vesicle pores which lie on the dorsal or aboral surface of the cell (Fig. 3D). These pores lie between kineties, sometimes the same two kineties. When they lie in different interkinety spaces, they are not more than a few kineties apart. The position of the pores is useful in defining a dorsal meridian separating, in a somewhat arbitrary way, the right and left lateral fields of the anoral zone. We find about 70 cortical units in the kinetal line running closest to this contractile vesicle meridian. Porter (72) has reported about 80 such units; and King (53) and Downing (15) about 66.

The caudal tuft of cilia (54, 89) was not identifiable in the stained preparations we used and, therefore, we are unable to report on its exact location and behavior. Protargol preparations (46) were unable to reveal it.

The *left lateral field*, on its more ventral part, carries the kineties which extend posteriorly from the left adoral field. The more ventral kineties terminate on the suture line, not far posteriorly from the vestibular depression. The more lateral kineties run mostly to the posterior tip of the cell, abutting on the postoral suture as it wraps over the end of the cell. Not all kineties, however, reach the suture; some terminate abruptly on the flank of the cell and do not reach the zone of close-packed units that characterize the tip (6).

On the aboral surface of the cell, it can be seen that the ends of the kineties terminate at the ends of both suture lines (Fig. 3D) and do not reach the apices of the cell.

The *right lateral field* extends from the aboral, contractile vesicle pore meridian around to the right side of the pre- and postoral sutures and to the right edge of the right adoral field. Even more so than with the left lateral field, this right field is characterized by largely meridional kineties. The most conspicuous exceptions are the kineties that lie close to the sutures; these curve around the right adoral field, with their termini on the two suture lines (Fig. 3A).

Paratenes can be found in the anoral zone and are seen as diagonals, about 55 in number in the dorsal, mid-region of the cell. This mid-region pattern is less regular and less dense than that at the poles of the cell. Furthermore, the regularity of the diagonal paratenes is very often interrupted—one out of five or so stops suddenly. This can be seen by attempting to impose consecutive paratenal lines on the units shown in Fig. 3D. Also, it should be noted that convincing latitudinal paratenes, as compared to the diagonal ones, do not exist. This observation is in sharp disaccord with Ehret's (19) description of latitudinal paratenes on the dorsal surface of *P. aurelia*.

IV. GROWTH PATTERN OF THE CELL

Changes in cell length and cell breadth relative to the interfission ages are shown in Fig. 4. Based on the general shape of these curves, the interfission interval of *P. aurelia* is divided into three parts: the post-fission, morphostatic, and morphogenetic phases.

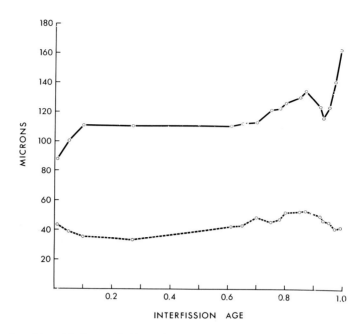

Fig. 4. Length and breadth changes in *P. aurelia*. In this and subsequent graphs, measurements were made on cells fixed and stained by the silver impregnation method at different times in the interfission period. Upper curve denotes length changes; lower curve shows changes in breadth.

A. Post-fission Phase

This phase continues to about 0.1 of the interfission interval and is characterized by rapid increase in length and decrease in breadth of the cell-body, which tends to change its form from an ovoid to an ellipsoid. This is accompanied by anterior–posterior elongation of the ciliary units and by a change in some of the doublet ciliary units at the edge of the right adoral field to the triplet type.

Although, in the new fission products, the new preoral and postoral sutures are much shorter than the old ones, shortly after division rapid elongation of the new sutures results in their attaining lengths very similar to the old ones—about 46μ—by the end of the post-fission phase. A more detailed study of the elongations of the preoral and postoral sutures (Fig. 5) shows a difference in their growth patterns. Elongation of the preoral suture is rapid and is performed in the earlier part of the post-fission phase, but that of the postoral suture takes place rather slowly and in the later part of this phase. Furthermore, measurements on the dorsal surface of paramecia in this phase show that the anterior and middle regions of the dorsal surface elongate extensively, but the posterior region remains almost the same length, about 20μ (Fig. 6). Therefore, rapid increase of cell length in this phase is attributable to the elongations of the anterior and the middle regions of the cell body.

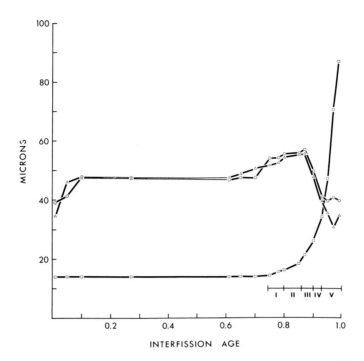

Fig. 5. Changes in the length of the preoral suture (triangles), the postoral suture (circles) and in the length of the buccal opening (squares). In the latter, after age 0.84 the length indicated includes the whole distance from the anterior edge of the old opening to the posterior edge of the new one. The line in the lower right corner of the figure shows the duration of stomatogenesis and its constituent stages, indicated as I-V.

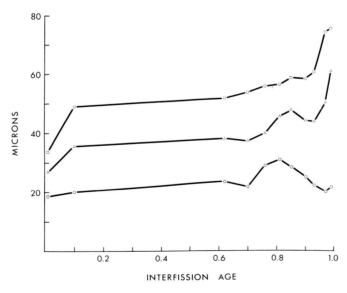

Fig. 6. Length of three dorsal regions. Anterior (triangles); mid-region (squares); posterior (circles).

References p. 258–261

B. Morphostatic Phase

The morphostatic phase, extending from about age 0.1 to age 0.7, is characterized by a slender cell outline, little, if any, growth in length, and a very wide buccal opening. The breadth of the central part of the cell body decreases to less than 35μ part way through this phase. Then, the breadth gradually increases to around 40μ at about age 0.6, changing the cell from somewhat cigar-shaped to more spindle-shaped. The length of the buccal opening is almost constant, about 15μ throughout the post-fission and morphostatic phases, but the width of the buccal opening changes significantly. Paramecia in the morphostatic phase have the widest buccal openings, about 8μ in width.

The kinetosomal territories vary somewhat in size depending on where they lie on the cell surface. Examining cells close to interfission age 0.7 and looking at the territories lying just anterior to the posterior contractile vesicle pore, the average size is 1.8μ long and 1.5μ wide. This agrees closely with Hufnagel's (41) dimensions of 2.0μ by 1.7μ, taken from pellicular fragments. Both of these are smaller than Klein's (54) values of 3.0μ by 2.5μ. Hufnagel states that the longitudinal or longest axis can vary by a factor of about 2, but this does not explain the great breadth reported by Klein.

C. Morphogenetic Phase

This phase extends from close to age 0.7 to the separation of the fission products. Its beginning is seen as a significant increase in length (Figs. 4 and 6) accompanied by

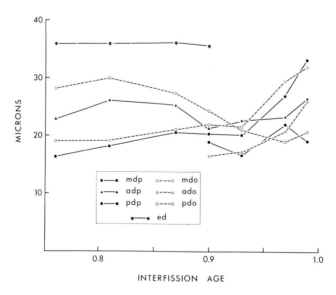

Fig. 7. Curves showing changes in the lengths of seven regions on the dorsal cell surface. Abbreviations are those given earlier (see Fig. 1B), e.g. ed = equatorial region on dorsal side, and pdo = posterior, dorsal side of opisthe, etc.

the appearance of new contractile vesicles and their pores. These changes presage the subsequent rapid lengthening of the cell which anticipates the bulk needed to form two daughter cells. The new organelles are the gross manifestation of internal events (32, 38) which have prepared the cell for the many, complex steps necessary to form the various organelles needed to fully equip the division products. We will here complete the description of changes in size and cell form before proceeding to the details of organelle formation. The gross changes in length and breadth are seen in Fig. 4 and these are broken down into changes on the ventral and dorsal surfaces in Figs. 5 and 6, respectively. Fig. 7 shows details of the dorsal surface that reflect the presence of the new contractile vacuoles and also, at age 0.9, the appearance of the fission plane.

The rapid increase in cell length between ages 0.7 and about 0.86 is due largely to extension at the ends of the cell. The posterior end grows quite rapidly (Fig. 6) and then starts to shorten slightly, whereas the anterior end continues growth somewhat longer and then starts to contract. The mid-region shows little increase until around 0.86 when the ends start to shorten and the mid-region starts rapid extension. The same picture is confirmed from measurements on the ventral surface (Fig. 5). Here, both sutures show an increase in length while the buccal opening remains constant in length to about age 0.8.

Close to age 0.86, when the sutures show a dramatic shortening, the buccal opening is at the start of rapid expansion. This is the result of formation of the new oral area and the concomitant enlargement of the old buccal opening. The two openings are connected though moving apart. As the fission line appears, the two openings separate. The measurements of the size of the buccal opening specify, at this time, the distance from the anterior edge of the old structure in the developing proter, to the posterior edge of the new opening in the new opisthe, plus the surface space between the two openings. By the time the cell divides, this composite distance averages about 90μ (Fig. 5), whereas the lengths of the most anterior and posterior sutures is about 36 to 40μ, i.e. the expected post-fission length.

If we look yet closer at the changes in cell length (Fig. 7), we can more precisely localize the source of these changes. In the developing anterior end, the area between the new and old anterior contractile vesicles (mdp) shows a steady increase from age 0.75 to 0.85, where the most anterior part, in that same period, shows an initial increase and then stops lengthening. A similar pattern is seen in the developing opisthe. The equatorial region shows no real change.

After age 0.87, the extremities of the cell, which is now close to developing a constriction, shorten further (opisthe end) or stay much the same (proter end) except for a final, new elongation (0.97–1.0). The significant changes after the start of constriction are seen in elongation of both mid-regions, especially from age 0.94, and in the new proter posterior end (pdp) and the new opisthe anterior end (ado).

In brief, in the morphogenetic period, there is first (age 0.7–0.87) a lengthening of the anterior and posterior portions of the cell, largely localized in the mid-regions of the developing proter and opisthe. Then, as the extreme ends of the dividing cell continue a contraction begun around age 0.8, and when the constriction furrow appears (age 0.9), the mid-regions go into very rapid elongation as do also the new, developing ends of the presumptive daughter cells. The older data of Jennings (43) for *P. caudatum* also show that this period, when the constriction has just begun, is a period of very rapid lengthening of the cell.

During the morphogenetic period, the length of the cell increases from about 110μ to about 135μ (age 0.87) and then decreases to 115μ before elongating to 165μ as the combined length of the fission products, just before fission. Changes in width are less dramatic, but it is worth noting that the maximum width (age 0.87) occurs close to the time of maximum length prior to visible evidence of constriction (age 0.9). Then, while the cell is lengthening rapidly as constriction proceeds, the breadth tends to decrease, also quite rapidly.

V. MORPHOGENETIC EVENTS

In this section we are especially concerned with defining the time when new organelles appear. In addition, there are descriptive details on organelle formation where these have previously been lacking or meager or where contradictory reports have been published. The appearance of new contractile vesicles, feeding structures, kinetosomal territories, and new suture lines will be presented.

A. Contractile Vesicles

New contractile vesicles appear from about age 0.71 to age 0.81, but mostly at around age 0.75. The new anterior and posterior vesicles always emerge, respectively,

TABLE 3

Distribution of new contractile vesicle pores relative to interkinetal meridian of pre-existing ones. Data are recorded separately for anterior (proter) and posterior (opisthe) pores. The one undetermined case (?) refers to a new pore lying anterior to two old pores, the new pore being to the right of one old pore but in the same interkinetal space as the other old one. New double pores were classified as lying to the right if one new one was to the right of the old one. Similarly, double pores were classified as lying to the left if one new pore was to the left of the old one. Data supplied through the courtesy of Dr. Mikio Suhama.

Position	Anterior		Posterior	
	number	%	number	%
Right	37[a]	58	19[b]	30
Same	23	36	28	44
Left	3	5	17[c]	27
?	1	2		
Total	64		64	

[a] Includes one double pore.
[b] Includes one double pore.
[c] Include six double pores.

anterior to the old anterior and posterior vesicles (53), and their pores usually lie in the same interkinety space where the old pores are situated. However, the new pores also appear, not uncommonly, in the space between the two kineties running to the right or left side of the interkinety space in which the old pores are situated (Table 3). More will be said subsequently on this important point.

With the emergence of the new contractile vesicles, the old anterior vesicle becomes the posterior contractile vesicle of the proter, and the old posterior vesicle is the posterior contractile vesicle of the opisthe (53).

When the new vesicles have appeared in the anterior and middle regions, 3 or 4 ciliary units on the contractile vesicle meridian have disappeared in each region as determined by reduction in the number of groups of doublets. The ciliary units in the posterior region do not change in number or structure except for a slight elongation of the ciliary corpuscles (Fig. 16, p. 243).

The new and old contractile vesicle pores have never been seen closer together than 8μ. This distance is always greater in the opisthe. These observations contradict those reports which refer to a division of the pores of the contractile vesicles (71). Also we have never observed the neuronemes reported by King (53).

It is worth pointing out that though the distance from the posterior pore to the end of the cell changes, increasing then decreasing (Fig. 6), the number of units in this meridian varies very little (Fig. 16). At the other end of the cell, the distance from pore to cell apex increases overall and so does the unit number (Fig. 6 and 16).

B. Stomatogenesis

The formation of the new feeding organelle can be divided into five stages, the first three of which are comparable to those found in other hymenostome ciliates (24, 29, 33), while the last two differ somewhat from other accounts (71–73, 78, 94). Further comparative comments are postponed until the discussion.

Stage I. Visible differentiation of new buccal parts starts, at about age 0.75, at the inner right posterior edge of the vestibule as an active proliferation of kinetosomes as evidenced by an increase of silver stainable dots (presumed of course to be the surface evidence of underlying kinetosomes) to the right of the endoral kinety. This agrees with Yusa's (94) account in terms of pre-existing kineties, but Yusa incorrectly refers to this location as "the right wall of the buccal cavity". Also, Porter's (71) account is slightly misleading for he refers to this initial proliferation as occurring "along the right edge of the buccal overture between the endoral membranelle and the vestibulum". These events occur *in* the vestibulum or circumoral depression. Further proliferation of kinetosomes spreads anteriorly. The first silver stainable structures to appear in this anlage field are irregular in size, arrangement, and intensity of silver impregnation (Fig. 8) as reported by Yusa (94). Hufnagel (41), using pellicular fragments and phase contrast and electron microscopy, reports evidence that kinetosomes are present in the anarchic field area throughout the interfission period. This is also found by Dippell (personal communication) on the basis of electron microscopical observations. The

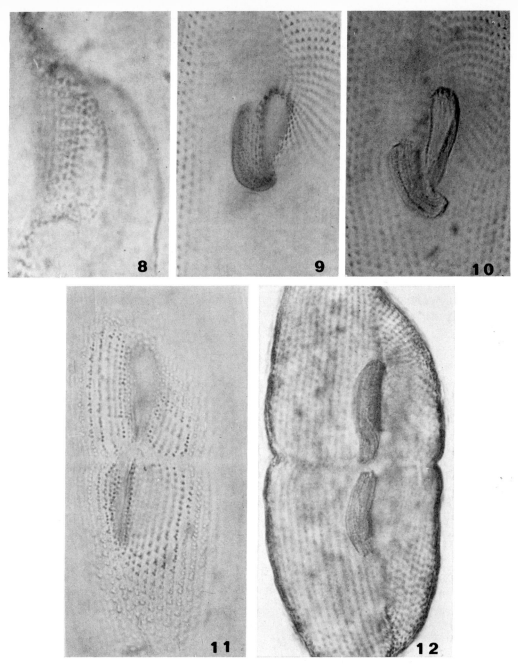

Figs. 8–12. Stomatogenesis in *P. aurelia*.

Fig. 8. Proliferation of kinetosomes in the anlage at age 0.78. Stage I. About 2,500 ×.

Fig. 9. Longitudinal rows of kinetosomes in the enlarged anlage field close to age 0.85. Stage III. About 1,300 ×.

Fig. 10. Further differentiation of the new buccal structures at about age 0.9. Stage IV. About 1,300 ×.

Fig. 11. The buccal openings of the proter (upper) and the opisthe, showing that the left wall of the opisthe's buccal opening is aligned on the meridian which runs past the right wall of the proter's buccal opening. Note evidence of quadruplication of units on either side of the posterior end of the proter's buccal opening. About age 0.92 and 1,200 ×.

Fig. 12. The proter's buccal parts show slight contraction at their posterior end. The opisthe's buccal organelle has started to elongate the posterior ends of the quadrulus and dorsal peniculus. About age 0.93 and 1,200 ×.

difference between these latter reports and the observations reported here could depend on the techniques used to visualize the presence of kinetosomes.

Stage II. This stage starts close to age 0.80 when the full anlage area appears as a rather disorganized field—anarchic field (7, 30) or "champ ciliaire" (77, 78)—of silver dots along the whole inner edge of the right wall of the vestibular depression. Invagination occurs towards the end of this stage. At this time the sites of deposited silver become uniform in size (about 0.5μ in diameter) and closely packed in the anlage field which has now extended to the inner, anterior right border of the vestibule. During this period, the kinetosomes on the posterior ends of the inner kineties of the right depression wall and on the endoral kinety also show proliferation, while at the same time the vestibule itself is elongating (Fig. 5). There is no convincing evidence of the dedifferentiation of vestibular kineties and part of the endoral membranelle as reported by Yusa (94); nor is there any evidence of incorporation of the old endoral kinety into the new anlage and subsequent differentiation of a new kinety for the old oral organelle (71). We agree with Roque (78) that the old endoral kinety stays with the proter. After the anlage field has extended to the anterior end of the vestibule, it begins to enlarge at its posterior end, becoming somewhat spoon-shaped. This enlargement of the anlage field develops further as it invaginates into the cytoplasm which underlies the right depression wall, resulting in a C-shaped structure when seen in a frontal view around age 0.82. This stage was clearly described by Hertwig in 1889 (39).

Stage III. At about age 0.85, the kinetosomes which have been distributed somewhat randomly in the whole anlage field begin to be arranged in longitudinal rows starting from the anterior portion of the field. This is the beginning of the third stage. In addition, the developing buccal parts pass around the posterior end of the old buccal opening (Fig. 9). At about age 0.87, the buccal opening narrows, becoming close to 20μ in length and 3μ in width at its anterior part. The cell is now at its greatest length. The anterior end of the new buccal cavity is located beneath the right wall of the vestibular depression, being separated by about 8μ from the anterior margin of the buccal opening. At this time, the wall of the new buccal cavity distinctly shows six rows of closely packed kinetosomes on its inner surface and one row of intensively stained kinetosomes to the right of the six rows with a curved membrane between (Fig. 10). The former six rows represent the primordial quadrulus and peniculi, the latter row the endoral kinety, and the membrane between them becomes the ribbed wall of the opisthe gullet. Yusa (94) proposes that, of these six kineties, the first two to the left will become the ventral and dorsal peniculus and the remaining four become the quadrulus. We are unconvinced of this and will tend to the view that two kineties go to each peniculus and the quadrulus. Further work is needed here as other accounts are also unclear on this important point.

Stage IV. This stage is initiated close to age 0.92 by the appearance of two distinct buccal openings. They share a common vestibule at first but this soon ceases to be the case as they move apart. Around age 0.90, the length of the buccal opening is 28μ to 30μ and the new buccal parts are located at the posterior right side of this elongated opening. The appearance of two distinct buccal openings is definite when the fission line, which is apparent at age 0.92, starts to show a constriction. As this constriction of the cortex progresses, the original, single, elongated buccal opening is separated into anterior and posterior parts, which are, respectively, 15μ and 13 to 15μ in length.

When the invagination of the fission furrow advances further, between ages 0.92 and 0.95, the posterior section of the old buccal parts, which has been lying across the fission line at the left side of the new feeding organelle, seems to be pushed forward. The reason for saying this is because there is a few microns difference between the anterior margin of the proter's buccal opening and the slightly more anterior margin of the underlying, old buccal cavity which seems shoved up under the preoral suture. This apparent anterior movement also separates the posterior end of the old buccal structures from the anterior end of the new one. Furthermore, the new buccal parts, lying at the right side of the elongated buccal opening, move posterior to the old structure while the fission furrow constricts further. And as a consequence of these events, the left wall of the opisthe's buccal cavity is aligned with the right wall of the buccal cavity of the proter (Fig. 11) (83).

Stage V. The two oral structures become aligned in tandem fashion by age 0.95 and the constriction furrow is now very conspicuous. Subsequently, in the new buccal parts of the opisthe, the posterior ends of the quadrulus and the dorsal peniculus begin to extend spirally from the left side to the right side over the posterior inner ventral surface of the buccal cavity, increasing the distance between the constituent kinetosomes (Fig. 12). The ribbed wall, however, develops concurrently with the extension of the quadrulus. The silver-impregnated dots which are often distributed on the surface of the wall, gradually become less stainable. Thus, with development of the new buccal cavity in the opisthe, stomatogenesis in the dividing paramecium completes its whole course just before separation of the proter and the opisthe. This agrees with Porter's (71, 73) account and not with Yusa's (94) wherein final molding of the new oral structures reportedly continues well into the next interfission period.

C. *Proliferation of Kinetosomal Territories*

The first visible evidence of the on-coming proliferation of kinetosomal territories, apart from the oral anlage, appears at about age 0.8 on the posterior part of the vestibule as new loci within the pre-existing kinetosomal territories, stainable with silver. These new loci of silver deposition are interpreted as follows: New loci lying directly in the axis of a kinety and immediately anterior to a similar pre-existing locus indicate the presence of a new underlying kinetosome. New loci lying to the right and anterior of an old kinetosomal locus are parasomal sacs and new sacs appear just anterior to the old ones and to the right anterior of the new kinetosome. Though the silver impregnation technique is very useful in identifying these new parts of the territories, it does not make clear when these parts are partitioned to result in complete new units. In what follows we are able to describe the proliferation of ciliary bases and the associated parasomal sacs. When these bases and sacs become conspicuously separated, we infer that partitions are forming and that completion of new units is occurring. As will be seen, kinetosome and parasomal sac proliferation occur significantly earlier than the final formation of the units. This has also been seen in *P. trichium* (33).

Proliferation of these ciliary bases in the vestibular depression progresses anteriorly, concurrently with elongation of the buccal opening by extension of both right and left depression walls. Further proliferation of the ciliary bases spreads out more or less along latitudinal ciliary lines or paratenes, about 10 in number, running from both the left and right margins of the buccal opening. Previously, this early proliferation onto the right adoral field had not been reported.

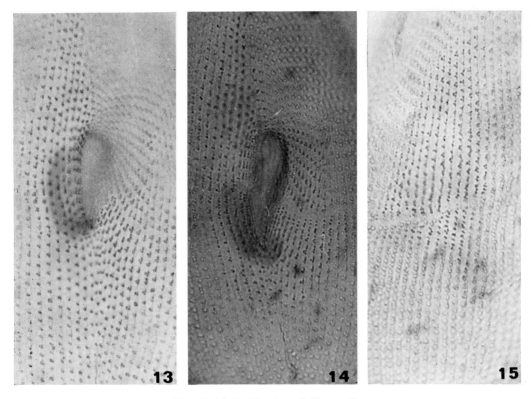

Figs. 13–15. Proliferation of ciliary units.

Fig. 13. Proliferation of units spreads from the vestibule to the left and right ventral surfaces at about age 0.88. About 1,300 ×.

Fig. 14. The fission line appears on the left ventral surface and then a little later, on the right ventral surface as a clear space between proliferating units. About age 0.92 and 1,300 ×.

Fig. 15. Transverse separation of new units starts in the units just posterior to the fission line at about age 0.93. Approximately 1,300 ×.

Further proliferation of the ciliary bases onto the left adoral field now occurs close to the left posterior corner of the adoral depression at about age 0.85. It then continues in these paratenes (the A-paratenes) running at the level of the posterior end of the buccal opening, to the dorsal or aboral side. Proliferation of the ciliary bases on the right adoral surface starts around age 0.88 on paratenes at the middle of the vestibule and progresses to the dorsal side (Fig. 13). As proliferation on the left and right surfaces extends to the dorsal side, it also spreads anteriorly and posteriorly along that portion of the kineties lying in the mid-third of the cell body. Proliferation progresses more rapidly on the left surface than on the right one, and posteriorly directed proliferation is more rapid than that which is directed anteriorly.

Dippell (13) reports that maximum proliferation of kinetosomes occurs in a period of 20 minutes duration starting 50 minutes before final separation of daughter cells. This would be between ages 0.86 and 0.92 which agrees well with the observations reported here.

At age 0.92, proliferation of the ciliary bases on the ventral surface extends anteriorly to the level of the anterior end of the buccal opening and posteriorly to the level of the anterior end of the cytoproct progressing somewhat slower on the right surface. On the lateral and dorsal surfaces, the posteriorly directed proliferation ceases at the level of the pore of the old posterior contractile vesicle, but anteriorly directed pro-liferation extends beyond the level of the pore of the old anterior contractile vesicle. This is around age 0.95.

Close to age 0.92 the fission line appears first on the left ventral surface, and a little later on the right ventral surface as a clear space between the proliferating ciliary bases and sacs (Fig. 14). This is accompanied by a rapid extension of the mid-third of the cell body. Finally the fission line extends to the dorsal side. This account differs from Sonneborn's (83) where the fission line is described as starting on the left of the vestibule and continuing right around the cell to finally terminate on the right side of the buccal opening.

Following the appearance of the fission line, partitioning of the proliferating structures apparently starts in the territories lying just posterior to the fission line and spreads posteriorly and anteriorly, resulting in new territories of the doublet type (Fig. 15). The apparent transverse partition results in an increase of the numbers of unit territories and an extension of the ciliary rows or kineties. It is to be noted here that increase in territories does not proceed simply by duplication, since more than one new unit appears anterior to the old, pre-existing ones (especially Figs. 11 and 14). The situation seems similar to that analyzed in detail in *P. trichium* (33) where tri-plication and quadruplication of the ciliary units or territories occurs. Though the details are not complete for *P. aurelia*, it appears that quadruplication of the ciliary units can occur close to the fission line (see Fig. 1) (83). Further from the fission line triplication of these units takes place. The duplication reported by others (71, 78) may occur only at the limit of the proliferation area farthest away from the fission line.

This period is now the period of the most rapid increase of complete territories (as distinct from an increase in number of ciliary bases) (Fig. 16). At this time there is a rapid elongation of the kineties since the cell is elongating as a whole [Figs. 4–7, (73)].

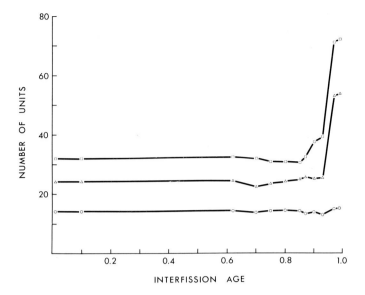

Fig. 16. Changes in number of ciliary units in three regions along the contractile vesicle meridian: anterior (triangles); mid-region (squares); and posterior (circles).

In *P. aurelia* proliferation of the territories in the anterior region of the opisthe is more active than that in the posterior region of the proter. Continuing from about age 0.93, the middle and anterior regions of the proter and the opisthe show a heavy increase in unit numbers accompanied by an active extension of the ciliary rows or kineties in each region (Figs. 16 and 17).

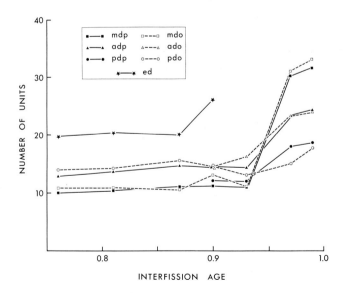

Fig. 17. Curves showing changes in unit number along the contractile vesicle meridian. Abbreviations as in Fig. 1B.

Although initially, at the beginning of the morphogenetic phase, the numbers of the kinetosomal territories in the middle regions of the proter and the opisthe are smaller than their number in other regions, they finally show the largest number of territories at the end of this phase. On the other hand, the posterior region of the opisthe shows the largest number of territories at the middle of the morphogenetic phase, but this region finally ends up with the smallest number of these units (Fig. 17). It is to be noted that comparable regions in the proter and the opisthe show almost the same number of territories at the end of the morphogenetic phase notwithstanding that these numbers are quite different earlier. For instance, although the anterior region of the opisthe starts to differentiate near the equator at about age 0.92, and begins proliferating territories at an earlier stage than those in the anterior region of the proter, the numbers of the territories in the anterior region of both the proter and the opisthe ultimately attain almost identical values (Fig. 17).

D. New Preoral and Postoral Sutures

When the fission plane begins to constrict, there appears a clear space between the posterior end of the buccal opening of the proter and the fission furrow (83). This space is then promptly filled by the extension of kineties actively elongating posteriorly from both sides of the buccal opening. This is the beginning of the formation of a new postoral suture at age 0.92. Then, from approximately age 0.93 to close to age 0.95, the cortex of the posterior half of the proter actively grows posteriorly. During this period, the posterior ends of the kineties in the left adoral field extend obliquely to the right accompanied by vigorous proliferation of the ciliary units, and abut on the left margin of the right adoral surface. Here as in all cases of typical unit proliferation, new units appear anterior to pre-existing ones (12, 13, 19, 20, 22). The kineties in the right adoral field extend almost directly posteriorly, accompanied by separation of the new kinetosomal territories. Later, from about age 0.95 to age 0.97, a new cytoproct appears as a clear area at the posterior end of the new postoral suture. The posterior ends of the 7th to 9th kineties, extending posteriorly on the ventral surface of the left lateral field of the proter, abut slantingly on the left margin of the new cytoproct, and the other more lateral kineties extend posteriorly parallel to the cytoproct.

The new preoral suture first appears as a space devoid of ciliary units in front of the buccal opening of the opisthe close to age 0.93 (73, 83). At this stage, the new postoral suture in the proter has already developed to about 5μ in length. The empty space in front of the opisthe's buccal opening continues anteriorly to the left fission furrow (Fig. 19). As time elapses, between ages 0.93 and 0.95, the anterior ends of the kineties on the developing left adoral field of the opisthe gradually turn to the right side, forming parallel curves along the anterior and left margin of the buccal opening of the opisthe, and meet the anterior ends of the kineties from the right ventral surface, which extend anteriorly, showing longitudinally stretched ciliary territories and forming large spaces between adjacent kineties (Fig. 11). This has been described in great detail by Sonneborn (83), who concluded that there are differential rates of elongation of the kineties on either side of the sutures. It turns out that the anterior ends of about 20 kineties on the left ventral surface meet the ends of about 11 kineties on the right ventral surface with a clear area between them which is the new preoral suture. From close to age 0.95, oral groove formation commences by depression of the anterior part of the left ventral surface of the opisthe. This is accompanied by further differentiation of the territories from the doublet to the triplet type. The final development of these territories progresses from the left, anterior margin of the buccal opening of the opisthe to the left anterior cell surface. The postoral suture and the cytoproct of the parental cell are retained intact in the opisthe.

From age 0.97 on, continuing constriction of the fission furrow brings together the ends of the opisthe and proter kineties which now lie in both lateral fields. These kineties meet without a clear area between them, resulting in formation of the anterior pole of the opisthe and the posterior pole of the proter and their respective sutures.

When the new postoral suture in the proter and the new preoral suture in the opisthe are about 35μ long, the constriction on the fission plane separates the dividing cell into two daughter cells. Immediately after fission the proter has the longer preoral suture and the cell outline has a long triangular form. On the other hand, the opisthe has the longer postoral suture and that cell shows an ovoidal outline.

VI. MACRONUCLEAR EVENTS

The macronucleus in the morphostatic phase is situated just internally to the feeding structures, in the most compact form observable throughout the entire interfission period.

Changes in the external form and location of the macronucleus during the morphogenetic phase are first seen around age 0.7, as a movement away from the dorsal side of the buccal cavity to the central part of the cell and as a progressive increase in volume. From age 0.75 to age 0.85, the swollen macronucleus retains a spherical shape and maintains a central location in the cytoplasm (Fig. 18). During this period, the growth of the cortex spreads gradually from the left and right ventral sides to the dorsal side, in the mid-third of the cell body. Concomitantly there is an elongation of the ciliary units and, somewhat later, a proliferation of the kinetosomes in the anlage field for the new gullet. At this time, according to Roque, the micronuclei begin swelling in the anticipation of mitosis (78).

Following age 0.85, the macronucleus starts to elongate, showing a slightly spiral form, and moves towards a position beneath the dorsal cortex. At about age 0.90, just after the paramecium has attained its maximum size in length and breadth, the macronucleus becomes rod-shaped (about 70μ in length) and lies parallel to the longitudinal axis of the cell beneath the dorsal cortex (Fig. 19).

While constriction of the cortex is occurring at the fission plane (age 0.92), the macronucleus starts to contract at its ends. With progressive contraction, the macronucleus gradually becomes shorter and fatter and gives some evidence of being twisted

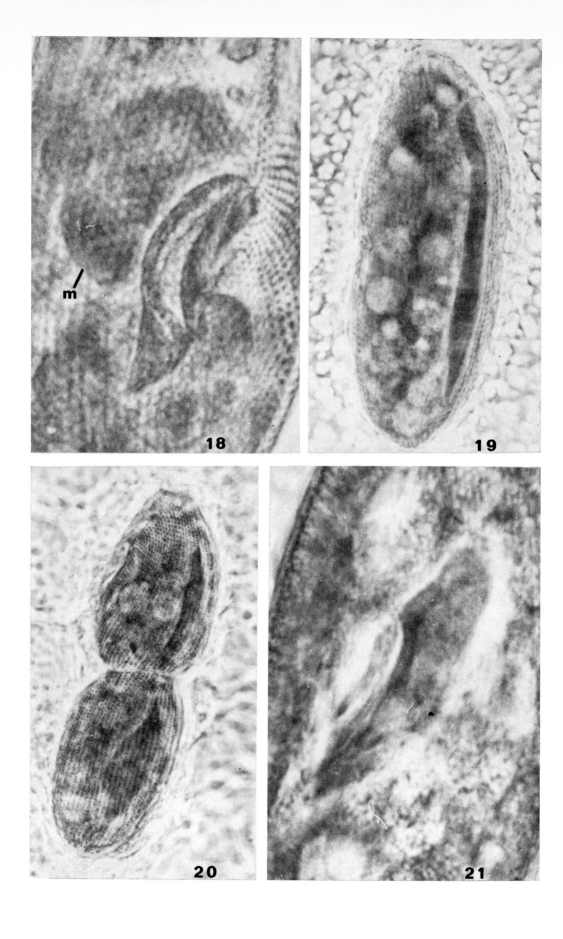

18

19

20

21

m

at each end. The equatorial portion of the elongated macronucleus becomes gradually more slender as a result of constriction (Fig. 20).

As soon as the macronucleus has divided, the paramecium separates into two daughter paramecia, and each daughter macronucleus rapidly loses its twisted sausage form and contracts into a short ovoid form. Immediately post-fission, the daughter macronucleus leaves its central position in the daughter cell, and returns to its typical position at the dorsal side of the buccal parts (Fig. 21).

VII. DISCUSSION

The foregoing report raises three sets of problems. Regarding the macronucleus, only brief remarks are needed in relation to changes in shape, size, and location. Stomato-genesis, however, must be looked at in the light of similar processes in other well-studied hymenostomes. The aim is to define basic similarities and to expose and understand pertinent differences. Finally, the surface or cortical organization of paramecium needs to be viewed from three points of view. First, the enigmatic distribution of new vesicle pores demands comment; second, there is a need to see precisely how kinetosomal territories are used to cover the cell; and third, the relation of territory proliferation to changes in cell size and shape will have to be examined. This latter study will force us to conclude that territories cannot be the agents which determine growth and change in cellular form, but rather that the behavior of these units is correlated with some underlying organizational feature of the cell which is the determinative factor with regard to the cellular form throughout the life of a cell.

A. Macronucleus

Thus far no one knows the functional significance of the interesting behavior of the macronucleus apart from the generally obvious fact that bipartition is necessary in a dividing cell. Certainly, nothing in our knowledge of its fine structure, both in its intact form (11, 14, 47, 48, 68, 75) and after careful disruption (91) is at present helpful in elucidating the changes in position and form that occur between ages 0.8 and 1.0. Also, it is important to know how the nucleus is functioning at this time. It is certainly carrying out DNA and RNA syntheses (3, 4, 52, 93). It has completed transcription

---

Fig. 18. Macronucleus (m) at age 0.75. Note swollen size and central location in the cell, just off the dorsal side of the buccal organelle.

Fig. 19. Macronucleus at age 0.92, when it is rod-shaped and lying beneath the dorsal cortex.

Fig. 20. At age 0.97, the macronucleus becomes shorter and fatter. The tendency to become ovoid is more pronounced in the opisthe.

Fig. 21. The cell and its macronucleus have recently divided (age 0.05). The nucleus has returned to its position along the dorsal side of the buccal structures.

of RNA needed for the polypeptides concerned with the oncoming fission (32) and is starting on the transcription needed for the following fission (38). But none of this tells us what is controlling the movement away from the feeding structures, the swelling, the twisting, elongation and subsequent constriction, division, and final rounding up with resumption of the original fixed position near the dorsal side of the buccal region.

These further comments may help define certain problems that are emerging here. The movement of the macronucleus and the cessation of food-vacuole formation, which also occurs towards the very end of the interfission period (34, 35), indicate changes in the organization of the cytoplasm of the cell. These may well be factors entirely outside of the macronucleus and it may be these which determine the changes in macronuclear location towards the end of the cell cycle. On the other hand, the changes in size and shape of the macronucleus might well result from processes going on within this organelle. Not only are DNA and RNA being synthesized but proteins, at least in the form of microtubules, are appearing and some of the latter are almost certainly concerned with the distribution of materials within the macronucleus and with its bipartition (69, 75). Sonneborn (personal communication) has observed that macronuclei displaced from their normal location in dividing cells go ahead and divide at their normal mid-region as well as dividing at the region lying in the fission plane, indicating an intramacronuclear determination of the division plane. That the macronucleus is behaving in a special way at this time, above and beyond its synthetic activities, and changes in form and location, is also seen from its response to actinomycin D at different times in the cell cyle (32, 38). And certainly other ciliate macronuclei show striking internal changes during the cell cycle and especially just prior to division (50, 75). Someday, it should be possible to understand macronuclear behavior as the outcome of both internal synthetic activities and organizational changes and of external forces from the surrounding cytoplasm, all aimed at duplicating and distributing the genetic material stored in this organelle.

B. Stomatogenesis

The five stages of stomatogenesis described here for *P. aurelia* can be compared to those first defined in *Glaucoma chattoni* (24, and personal communication, J. Frankel) and then, also in *Tetrahymena pyriformis* (26, 27) and *P. trichium* (33). Carefully timed data are available only on *P. aurelia* and the last two species (Table 4) but certain comparisons can be made for all four forms.

Essentially, homologous relations seem to hold in the matter of kinety and membrane formation. The sequence of, first, anlage field, then initial emergence of double kineties, and then three triple- or quadruple-kinety membranes in the mature oral area, is common to all four forms. However, it should be noted that Yusa's (94) interpretation of the fate of the six earlier kineties may not be correct, as mentioned earlier.

Some differences in development of the buccal structures are seen in the relative locations of old and new structures and in the timing of invagination of the new buccal

TABLE 4

A comparison of stomatogenic stages and their duration in certain hymenostome ciliates. Decimal figures refer to interfission age, with the upper row referring to the age when a given stage is initiated for a designated species and the lower row referring to the duration of the stage in that species. The figures for P. trichium *are from data wherein, due to irregular growth, the interfission period initially appeared to be 1.17. The data are recalculated on the basis of the interfission period as being 1.0. Stage 0 refers to no visible stomatogenesis. Stage I is the time of an anarchic or stomatogenic field. Stage III marks the time of appearance of kineties. The remaining stages differ in the features which characterize them. In Tetrahymena, Stages IV, V, and VI are identified, respectively, by differentiation of buccal (= AZM) membranelles, appearance of oral cilia, and invagination. In* P. trichium, *Stages IV and V are characterized by invagination and separation of two buccal openings. In* P. aurelia *IV and V are identified by separation of two buccal openings and tandem alignment of the openings. Invagination occurs in the stages marked by* *.

			Stages			
	0	I	II–III	IV	V	VI
Tetrahymena pyriformis	0	0.68	0.77	0.80	0.85	0.87
(Frankel, 26)	0.68	0.08	0.03	0.05	0.02	0.13*
Paramecium trichium	0	0.59	0.78	0.87	0.92	0.97	—
(Gillies and Hanson, 33)	0.59	0.19	0.09	0.05*	0.05	0.03	—
Paramecium aurelia	0	0.75	0.80	0.85	0.90	0.92	—
	0.75	0.05*	0.05	0.05	0.02	0.08	—

cavity. During pre-fission morphogenesis in *Tetrahymena* and *Glaucoma*, the new primordium first appears well posterior to the old oral area but it does typically lie between kineties which are associated at their anterior termini with the right side of the buccal opening. This is similar to the case in the two species of paramecia, with *P. trichium* being somewhat intermediate to *Tetrahymena* and *P. aurelia* (33). [It is of interest to note that during oral replacement in *Tetrahymena* (28) and *Glaucoma* (24) the new oral area develops close to the old area in ways very similar to oral development in *P. trichium*.]

Invagination, in these different forms, occurs in a sequence which repeats that just described for the location of new buccal structures. Again *Tetrahymena* is at one extreme with late invagination and *P. aurelia* is at the other with relatively early invagination. In between are *Glaucoma* and *P. trichium*, with the former closer to *Tetrahymena* and the latter closer to *P. aurelia* (Table 4, see asterisks). The reason for this sequence may be related to problems of elongation of the primordium within the confines of the adoral depression, as suggested by Gillies (33). In *Tetrahymena* the primordium is free to extend itself along the flat, mid-ventral surface of the cell. In *P. trichium* a shallow adoral depression somewhat confines the primordium and to continue its growth it must invaginate to remain within the available space. The deep adoral depression of *P. aurelia* imposes severe limitations to the extension of the primordium. After it runs from one end to the other of the right wall of the depression (stage II) it can only elongate further by curving or by invaginating, which it finally does. Thus, invagination is interpretable as the largely mechanical consequence of growth within restricted space. This is not the whole story, for it does not account for the invagination in *Tetrahymena* where space is apparently unrestricted. It may, however, account for the difference in timing of invagination.

In conclusion, it appears that the formation of structures like the cilia and their alignment into kineties runs a common course in hymenostomes. The mechanics of formation of the buccal cavity appears to be different but this may largely be, according to Gillies' important suggestion, a simple consequence of local differences in the cell surface at the site of stomatogenesis. The real differences are reflected in those, as yet not understood, aspects of stomatogenesis that finally cause the visible differences which distinguish a tetrahymenal buccal area from that in paramecium, and buccal parts in *Tetrahymena* from those in *Glaucoma* and those in *P. aurelia* from those in *P. trichium*. Apart from such terminal differences, stomatogenesis may be remarkably similar in all four species.

C. Cortical Organization: Contractile Vesicle Pores

The most striking feature of the placement of new pores (Table 3) is that regarding anterior ones; in more than half of the cases the new pores are to the right of the interkinetal meridian where the old anterior pore is located. If this phenomenon occurred without any compensatory movement of other structures, clearly one would expect the anterior contractile vesicles to migrate around the right side of the cell, and even on to the left side. In a random sample of such cells we would, then, predict finding the anterior vesicle at any position on the cell circumference in the latitude appropriate to them. Clearly this is not the case. The alternative must be that the cortex shifts to the left and compensates for the rightward shift of the anterior vesicles. This certainly seems to be the case as seen from studies on the distribution of cortical markers, i.e. inverted or reversed kineties (2). In this study Beisson and Sonneborn found that in subsequent asexual generations the abnormal kinety could be found in any location on the cell circumference and the sequence of these changes demonstrated a migration of the cortex from the right, around the dorsal side of the cell, and on to the left side. (This is referred to here as a leftwards migration although Beisson and Sonneborn refer to the reversed kinety as being "located progressively further to the right, eventually reaching the left side of the cell".)

Cortical migration is of course known from other studies in the ciliates. Nanney's studies on cytogeometry and cortical slippage in *Tetrahymena* (62, 63) represent a detailed description of cortical movement towards the left with the oral area and contractile vacuole pores retaining a relatively fixed relation to each other. Other similar cases are those of Frankel (25) and Tartar (87) in *Glaucoma* and *Stentor*, respectively. And returning to *Paramecium*, King (53) reported a strain in which he regularly found that the new anterior pore was located 0.13 meridians to the right of the posterior pore.

All of this raises important questions as to how new kineties are presumably generated at the right side of the suture lines and resorbed at the left side. These presumptions are based simply on the fact that kineties are not observed to pile up on the left side of the sutures which is the logical end-point of the migration, and new kineties may be generated on the right side in paramecium as part of the anarchic field. Clearly, more careful experimentation is needed to resolve these questions.

Additionally, the question arises as to why the new posterior vesicle is distributed differently from the anterior one. The new posterior pore lies equally often, in our material, to either the left or the right of the old pore. This suggests that cortical migration is not a fact in the opisthe. How can this be? At present there are not enough observations to suggest a definite answer; only suggestions are possible. One suggestion is to note that the new oral area generates no new kineties as it develops; these are all provided by the old oral area. Hence the opisthe may generate no new kineties and therefore never shows cortical migration. Such a suggestion reinforces the notion that perhaps the anarchic field supplies kinetosomes for more than just the oral anlage and indeed may generate a new surface kinety once every two fissions or so. If this is correct, there is a specially important problem relating to the generation of cortical kineties, with their characteristic genetic and developmental potentials, from the loosely organized anarchic field which also generates the highly differentiated kineties of the buccal cavity. The morphogenetic potential of the anarchic field may have been grossly underestimated.

D. *Cortical Organization: Surface Zones and Unit Territories*

Next, there is the specific pattern which we have described as adoral and anoral zones with their constituent fields. What is the relation of the surface units or territories, both structurally and functionally, to this pattern?

As was first emphasized by Ehret and Powers (22), the cortical unit or kinetosomal territory is, at the microscopic level of organization, the basic unit of the cell surface. In *P. aurelia* these units are fairly constant in size, with the variation occurring mostly in the longitudinal axis. There is also some variation in constituent parts, i.e. one or two cilia and one or two parasomal sacs per unit. These units notably cover surfaces of ciliates by being organized into the linear arrays called kineties. Kineties are units of organization as seen in their constancy of pattern, genetic continuity, and developmental potencies (2, 59, 63, 64, 83, 85, 87, 88 and others referred to in these articles). Kinetal rows can be laid side by side to form a flat, two-dimensional coverlet, which would wrap around a cylinder readily, leaving the ends open, as Ehret (19) has pointed out. However, paramecia are obviously not cylinders. They are irregular ovoids and this necessitates special tailoring of the coverlet at its ends and along one side in some sort of a seam.

It is readily apparent that the suture lines of a paramecium can be interpreted as such a seam. Further, in that the oral structures and cytoproct must be present as breaks in the surface, the location of these on the suture line is an efficient solution to placing them where there will be minimal disruption of the kinetal rows of cortical units. However, certain cilia of some of the nearby kineties are needed to generate feeding currents for the buccal opening. Thus there are two constraints on certain cortical units relative to the feeding organelle: (1) to cover the surface around a permanent opening and (2) to aid in feeding. The cytoproct only needs space to open intermittently and the units around it show little relation to the cytoproct as an organelle; rather, they seem to show only the packing problems related to the suture line. At the poles of the cell, the packing problem along the suture is very severe and we see its expression as regular diagonal rows of small units and the dropping out of kineties before they reach the end of the cell.

On the other hand, the contractile vesicles exert little influence on their nearby surface units, for their pores are small enough to open between the kineties and the vesicles, like the cytoproct, do not need ciliary action to help them in their function. Hence we see no real evidence of cortical pattern responding to the presence of the contractile vesicles.

Finally, in realizing that the units not only provide surface covering to paramecia, but also are the location of locomotory organelles, we can understand the strong tendency to longitudinal orientation of the kineties. This orientation results, function-ally, in a swimming pattern that is certainly adequately efficient for the needs of the cell, but probably has, also, some arbitrary elements resulting from wrapping an irregular ovoid in relatively regular rows of locomotory organelles. Not all of these organelles can fit into neat longitudinal rows and the disruption of this highly or-ganized pattern will cause spiralling and other special aspects of swimming (45). Since paramecia use spiralling as a device in their food searching (44), this postulated by-product of cortical organization might be viewed as a beneficial pre-adaptation.

The foregoing can be summarized as seen from the viewpoint of the adoral and anoral zones. In the adoral zone, packing patterns of territories and kineties reflect their interaction with the buccal organelle to achieve feeding as well as to contribute to swimming, and to cover the special surface configurations of vestibule and oral groove. The patterns of the anoral zone reflect the requirements of covering the rest of the ovoid cellular form, and the achievement of swimming, with little disruption from other cortical organelles, namely, the cytoproct and the contractile vesicles. Throughout, the sutures represent the line of closure necessary to an irregular ovoid when covered with regular linear arrays of kinetosomal territories.

This view of the surface morphology of *P. aurelia* better integrates the observed features of territories, kineties and other cortical organelles with their functional interaction than do other classifications of these structures (83). It can also be extended to other species of paramecia and probably also to related forms such as *Neobursari-dium gigas* which, as Nilsson (65, 66) has carefully redescribed it, shows striking morphological affinities to the peniculine hymenostomes (16) rather than to the heterotrichs with which it was originally classified. [See especially Fig. 15 (66).]

E. Cortical Organization: Cellular Form and Unit Territories

The problem can be brought into focus by the following passage from Tartar's pioneering paper on intracellular patterns on ciliates (85): "Is there some as yet invisible groundwork which determines the shape and pattern of the cell following which the fiber system simply spreads over this contour and adapts itself accordingly, or is the fiber system itself the active immediate agent in determining pattern and form?"

Tartar's review goes on to favor the "fiber system" whose modern counterpart is the ciliate cortex composed of kinetosomal territories. However, in what follows we will be forced to the other view; namely, the cortex reflects the contours of a groundwork which we suspect as being an endoplasmic gel.

The analysis begins with the question whether change in numbers of cortical units can be the cause of change in cellular size and form. If the answer is affirmative, then we can expect to find positive correlations between the two types of change, i.e. units increase when the cell is growing. As we shall see, some correlations do exist, in a general sense, but on closer examination these do not offer convincing evidence in support of the hypothesis that unit increase is the basis of cellular growth. We will then go on to discuss briefly alternate views.

It is clear from the results reported here (Figs. 4, 6, and 7) that paramecia lengthen during both the post-fission and morphogenetic periods and that units increase significantly in number only during the latter part of the morphogenetic period. These conclusions immediately indicate that if there is a correlation, it is not a precise one. However, let us look in greater detail at the events in the cell by examining specific parts of the cell at definite interfission periods. Such changes are summarized for the morphogenetic period in Fig. 22, where change in length and in unit number are both plotted against interfission age.

Let us look first at the posterior end of the cell. It can be seen that the number of units is essentially constant and there is no evidence of unit proliferation here. If one

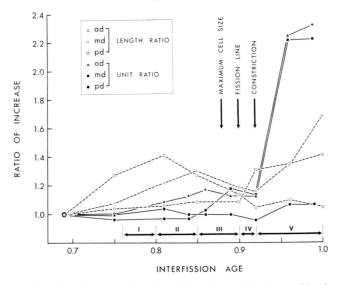

Fig. 22. Changes in length and unit number of specific regions of the dorsal side of cell expressed as ratio of increase. That is, the lengths or unit numbers of the given regions (see abbreviations on graph) are expressed in relation to their length or unit number at age 0.69 taken as 1.0. Duration of stomatogenic stages is also given along with time of maximum cell size, appearance of the fission line, and start of constriction.

References p. 258–261

looks next at the length of the posterior end, there is clear evidence of increase with a peak at age 0.81. The curves for the length of the posterior end and for the unit number are clearly uncorrelated and we must conclude that something other than unit proliferation is causing the size changes.

Looking next at the anterior end, there is also lack of correlation but of a different sort. Initially, from age 0.69 to age 0.92, length and unit number are behaving similarly but with length increasing and then decreasing faster than the changes in unit number. After age 0.92 there is an extraordinary burst of proliferation wherein the number more than doubles. This increase is much more rapid than the increase in length which only completes its course in the post-fission period. In this case, then, territory proliferation proceeds so much faster than linear extension of the cell that a direct causal relation between the two is unlikely. The tight packing of territories that occurs around age 0.96 is alleviated finally by post-fission growth of the proter.

It might be argued that rapid increase in number of territories is due to a posteriorly directed movement of the old anterior contractile vesicle pore, which does occur (53), thus allowing more units to be anterior to the pore. This is not a sufficient explanation since there is, at this same time, an increase in units in the mid-region, just posterior to the anterior region. (The same is true regarding the posterior region, too.) Fig. 17 shows that most of the increase in units of the anterior region comes from proliferation in the mid-region of the proter, as would be expected from the earlier description of the pattern of territorial unit proliferation.

In the mid-region, between the two old contractile vesicle pores, a situation exists similar to that found in the anterior region. First (age 0.7 to 0.87), there is slight lengthening but no unit proliferation. However, starting at age 0.8 there is evidence of an increase in ciliary bases and therefore the units are probably elongating. Here, then, a case could be made for increase in length being dependent on unit activity—that is, unit elongation prior to unit proliferation. Actual proliferation starts at about age 0.84 and continues at a constant rate until age 0.92 when it shows a burst of activity, more than doubling the number of units. This increase is much more rapid than the increase in length. After age 0.94 there is little increase in territories, though lengthening of the mid-region continues on into the post-fission period, and the tight packing of new territories is thereby relieved.

It is clear that no consistent, direct relationship, suggesting a causal situation, holds between the proliferation of territories and the change in cell length. Apparently, flexibility in size of the individual territories allows for tight packing and for considerable elongation. When these limits are exceeded, territories must presumably be resorbed to allow for any further reduction in space, or, conversely, new territories must be formed to allow increase in space. To that extent there is an obvious connection between territory number and changes in cell length. But expressing the problem in that way reverses the postulated causal connection we were examining: it now suggests that changes in cellular form might elicit changes in the number of territories. Before turning to that possibility we have to finish examining our original postulate, since it can be restated in terms of changes in territory size, rather than territory number.

The next question is: Do territories determine cell size simply on the basis of changes in their own size? Here, of course, size of territories will correlate with cell size changes because of the fact that units form a kind of elastic skin on the cell. A more meaningful question then has to be the causal one: Can territories generate size changes? For this to be so, there must be evidence that (a) territories form a stiff skin, sufficiently rigid to determine the form of the cell, and (b) that this stiff skin is capable of both localized and general size changes.

At this point in the discussion by the loose term "skin" we here mean the surface layer of ciliary units and associated trichocysts. It seems that electron microscopists use cortex in that way too (19). Developmental biologists, however, refer to the cortex as a superficial layer of the cell which carries developmental information (36, 87, 88); such a definition is obviously imprecise in structural terms. For present purposes we will define the cortex following the electron microscopists, including also the sutures, cytoproct, and contractile vesicle pores as special cortical components and organelles. The feeding organelle can also be considered a cortical organelle, because of its intimate relation—developmental, functional, and morphological—with the cortex proper, but it is not, strictly speaking, part of the cortex in the sense that sutures, cytoproct, and vesicle pores are part of the cortex. Pellicle seems an outmoded term and will not be used here. Similar comments can be made with regard to ectoplasm and endoplasm. Ectoplasm would include the cortex and some underlying material; it would not be synonymous with the cortex even though it has been so defined (83). The transparent or clear nature of the ectoplasm, so often cited in reports using the light microscope, seems nowhere apparent in electron micrographs. Ectoplasm is, then, a term of limited precision. Endoplasm, the rest of the cytoplasm inside the ectoplasm, is distinguishable by light microscopy and in some cases also by electron microscopy. However, because of possible ambiguity we will not refer to ecto- or endoplasm further.

Returning now to the problem of territory elongation as a basis for cell size changes, we can say that kinetosomal territories may provide a cortex of some rigidity but there is no experimental test demonstrating it. Nor is there, at present, any positive information on how changes in the size of individual units might be brought about to cause changes in the form of the cell. Turning next to negative evidence, most protozoologists have often observed blistering of the superficial layers of paramecia while the underlying cytoplasm retains its shape. This suggests that the superficial layer of the cell is delicate and flexible and that the shape of the cell is being preserved by something below the cortex.

The fact that ciliates are not simple spheres means that work is being done to hold them in their unique forms. It is generally assumed that there exists some kind of gel layer in these cells, though there appears to be no critical analysis of its extent in the cell. Porter (72) has evidence that a gel layer solates when oral morphogenesis and constriction occur. It is, therefore, plausible to argue that there is a gel-like aspect of the cytoplasm of paramecia which provides these cells with the rigidity necessary to keep their shape. That this rigidity does not, in any case, reside in the "pellicle", was long ago suggested by Nadler's (61) observation that *Blepharisma* exposed to strychnine, apparently, literally climbed out of their pink pellicle via the buccal opening, and emerged as colorless cells *still retaining their characteristic shape*. Scattered throughout the literature one can find various other observations consistent with the concept of a peripheral gel layer which determines the shape of the cell. Hirshfield *et al.* (40) speak to the point that the cytoplasm subjacent to the cell surface "displays an organized cortical gel". "Synchronous rounding" in *Tetrahymena* (Frankel, personal communication) occurs in a manner which seems to exclude changes in territory number and strongly suggests some change beneath the cortex. In this same regard Tamura (84) reports that the rounding is reversible and associated with changes in –SH groups of proteins.

Perhaps some of the most interesting observations come from the early work of Jennings (42). His reports of the transmission of cytoplasmic projections on the surface of *P. caudatum* show that such distortions of shape are passed passively from one cell generation to another by simply following the patterns of distribution which are contingent on cellular location. For example, when the projection or "spine" is in the mid-region of a cell, it passes to the posterior end of the proter if it is in front of the fission line, and to the anterior end of the opisthe if it is back of the fission line. The fact that the projections do not change shape when lying in the mid-region where we know there is quadruplication of territories clearly tells us that the protrusion is not controlled by the cortex. Furthermore the protrusion shows no evidence of migrating around the cell when it lies in anterior fission products; Jennings' diagrams show a constant relation of projection to oral area. These careful observations clearly indicate that the cortex is a skin overlying some framework which determines the general form of the cell, including abnormal projections.

As yet there is no clear evidence from electron microscopy as to whether the postulated gel is visible or not. Various workers (41, 69, 92) report filaments in the superficial layers of ciliates. Pitelka (69) says, "Filaments often make up a complete or perforate layer separating ectoplasm from endoplasm in ciliates" (p. 328). Whether or not this could be the physical basis of the gel-layer postulated here remains to be seen.

Some notion of the possible distribution of the postulated gel can be guessed at from watching the motion of the inner cytoplasm of paramecia and, in particular, watching the food vacuoles. Their motion suggests that the gel material would be heaviest around the periphery of the cell and that the center is in a more fluid sol-state. If such a peripheral gel layer could add to itself, it is possible to conceive of it determining various changes in cell shape and size, including even the generation of the constriction at fission by means of local contraction of the gel. Such additions might occur by local deposition of proteins, analogous to formation of the gel in amebae.

We conclude then, on present evidence, that neither cortical unit proliferation *per se*, nor change in size of units can plausibly account for changes in cell size. In fact, we would argue that cellular form itself is not determined by the cortex (as defined here). We urge the view that a paramecium has its cellular form determined by the "as yet invisible groundwork" of a semi-solid shell surrounding the more fluid, sol-like interior where nuclei and food-vacuoles lie. (The constant location of the macronucleus through the post-fission and morphostatic phases of the interfission period probably results from its attachment to the gel layer or cortex on the dorsal side of the buccal area). External to the gel is the outermost, cortical wrapping of kinetosomal territories. Whether the postulated gel-layer might play any determinative or regulative role in the assembly of new cortical territories is at present purely speculative. It could certainly be part of the proposed "gullet maintenance area" necessary for stomatogenesis (35, 37). And it might also be part of the factors and gradients concerned in "cortical morphogenesis" (87, 88), "cytotaxis" (83) or "generative control" (36).

The point of formulating the foregoing questions, problems and concepts has been, of course, to clarify further various issues which are central to understanding the cell cycle in *P. aurelia*. Clearly much rewarding work remains to be done in exploiting the challenging complexity of the ciliates in the study of cellular reproduction.

VIII. SUMMARY

(1) The surface of paramecium is a system of kinetosomal territories or cortical units associated in kineties. Overall, the kinetal pattern is divided into two zones. The adoral zone is concerned with feeding and contains the external opening of the feeding organelle. The anoral zone covers the rest of the cell. Each zone has two fields. The former contains the left and right adoral fields which exhibit distinctive paratenal patterns. The latter zone contains the right and left lateral fields. There are distinctive sutures at the ends of the cell, newly described here.

(2) The interfission period can be divided into three phases—post-fission, morphostatic, and morphogenetic—depending on the nature of size changes in paramecia. Changes in length are significant in the first and last stages.

(3) Changes in number of cortical units are largely seen as rapid increase towards the end of the morphogenetic phase. However this proliferation is anticipated by the appearance of new basal granules and parasomal sacs earlier in the morphogenetic phase.

(4) The time of first appearance of new contractile vesicle pores is described. The timing of stomatogenesis is reported and stages are described conforming to stages seen in other hymenostome ciliates. The behavior of the macronucleus is outlined.

(5) Major conclusions are the following: (a) In stomatogenesis, the formation of new ciliary structures and membranelles seems homologous in the several hymenostomes reviewed. The process of invagination may also be similar with apparent differences ascribable to location of the primordium on a flat surface or confined in a vestibular depression. Detailed modelling to give species differences is, however, thought to be due to different processes. (b) The kinetal pattern is interpreted as showing patterns jointly determined by the functional needs of the cell and the nature of the unit territories as building blocks. The size changes and shape of the cell cannot be explained by patterns of increase in the cortical units, nor by changes in the size of these units. (c) It is useful to postulate an underlying framework, possibly a gel, as determining the form and overall growth patterns of paramecia. On this view, the

cortex, composed of kinetosomal territories and other special structures, is a somewhat elastic, specialized wrapping which surrounds the postulated framework.

IX. ACKNOWLEDGEMENTS

The authors have received generous help from colleagues in the preparation of this
review. Especially to be noted are Drs. Ruth V. Dippell, Joseph Frankel, Tracy M.
Sonneborn, and Mikio Suhama, each of whom provided previously unpublished
information. Drs. Dippell, Frankel and Sonneborn also read through the entire
manuscript and supplied many useful criticisms and comments. To all of them we are
especially grateful. Our sincere thanks go also to Mrs. Concettina Gillies and Miss
Jane Sibley for their special technical assistance.

X. REFERENCES
(Abstracts are indicated by *)

1   Beams, H. W. & Anderson, E. 1961. Fine structure of Protozoa. *Ann. Rev. Microbiol.* **15**, 47–68.
2   Beisson, J. & Sonneborn, T. M. 1965. Cytoplasmic inheritance of the organization of the cell
    cortex in *Paramecium aurelia. Proc. Nat. Acad. Sci. U.S.* **53** (2), 275–282.
3   Berger, J. D. 1969. *Nuclear Differentiation and Nucleic Acid Synthesis*, Ph.D. Thesis, Indiana
    University, Bloomington, Ind.
4   — & Kimball, R. F. 1964. Specific incorporation of precursors into DNA by feeding labelled
    bacteria to *Paramecium aurelia. J. Protozool.* **11**, 534–537.
5   Chatton, E. & Lwoff, A. 1935a. Les ciliés apostomes. I. Aperçu historique et général: étude
    monographique des genres et des espèces. *Arch. Zool. Exp. Gen.* **77**, 1–453.
6   Chen-Shan, L. 1969. Cortical morphogenesis in *Paramecium aurelia* following amputation of the
    posterior region. *J. Exp. Zool.* **170**, 205–228.
7   Corliss, J. O. 1953. Comparative studies on holotrichous ciliates in the Colpidium-Glaucoma-
    Leucophrys-Tetrahymena group. II. Morphology, life cycles and systematic status of strains
    in pure culture. *Parasitology* **43**, 49–87.
8   *— 1955. Proposed uniformity in naming "mouth parts" in ciliates. *J. Protozool.* **2** (Suppl.), 12.
9   — 1961. *The Ciliated Protozoa: Characterization, Classification, and Guide to the Literature*,
    Pergamon Press, New York, N.Y.
10 *Dippell, R. V. 1962. The site of silver impregnation in *Paramecium aurelia. J. Protozool.* **9**
    (Suppl.), 24.
11 *— 1963. Nucleic acid distribution in the macronucleus of *Paramecium aurelia. J. Cell Biol.* **19**,
    20A.
12 *— 1965. Reproduction of surface structure in *Paramecium aurelia. In Progress in Protozoology*
    (Proc. 2nd Int. Conf. Protozool., London), 65.
13 — 1968. The development of basal bodies in *Paramecium. Proc. Nat. Acad. Sci. U.S.* **61**, 461–468.
14 *— & Sinton, S. E. 1963. Localization of macronuclear DNA and RNA in *Paramecium aurelia.*
    J. Protozool. **10** (Suppl.), 22.
15 *Downing, W. L. 1951. Structure and morphogenesis of the cilia and the feeding apparatus in
    Paramecium aurelia. J. Protozool. **2** (Suppl.), 14.
16 Dragesco, J. & Tuffrau, M. 1967. *Neobursaridium gigas* Balech, 1941, cilié holotriche hymeno-
    stome pan-tropical. *Protistologica* **3**, 138–148.
17 *Ehret, C. F. 1958. Replication of ciliated organelles during gullet development in *Paramecium.*
    J. Protozool. **5** (Suppl.), 11.
18 — 1960. Organelle systems and biological organization. *Science* **132**, 115–123.
19 — 1967. Paratene theory of the shapes of cells. *J. Theor. Biol.* **15**, 263–272.

20   — & De Haller, G. 1963. Origin, development and maturation of organelle systems of the cell surface in *Paramecium. J. Ultrastruct. Res.* **6** (Suppl.), 1–42.

21   — & Powers, E. L. 1957. The organization of gullet organelles in *Paramecium bursaria. J. Protozool.* **4**, 55–59.

22   — & — 1959. The cell surface of *Paramecium. Intern. Rev. Cytol.* **8**, 97–133.

23   Frankel, J. 1960. Effects of localized damage on morphogenesis and cell division in a ciliate, *Glaucoma chattoni. J. Exp. Zool.* **143**, 175–194.

24   — 1960. Morphogenesis in *Glaucoma chattoni. J. Protozool.* **7**, 362–376.

25   — 1961. Spontaneous astomy: loss of oral areas in *Glaucoma chattoni. J. Protozool.* **8**, 250–256.

26   — 1962. The effects of heat, cold, and *p*-fluorophenylalanine on morphogenesis in synchronized *Tetrahymena pyriformis* GL. *Compt. Rend. Trav. Lab. Carlsberg.* **33**, 1–52.

27   — 1967. Studies on the maintenance of oral development in *Tetrahymena pyriformis* GL-C. *J. Cell Biol.* **34**, 841–858.

28   — 1969. Participation of the undulating membrane in the formation of oral replacement primordia in *Tetrahymena pyriformis. J. Protozool.* **16**, 26–35.

29   — 1970. The synchronization of oral replacement without cell division in *Tetrahymena pyriformis* GL-C. *J. Exp. Zool.* In press.

30   Furgason, W. H. 1940. The significant cytostomal pattern of the *"Glaucoma-Colpidium Group"* and a proposed new genus and species. *Arch. Protistenk.* **94**, 224–226.

31   Gelei, J. von. 1934. Der feinere Bau des Cytopharynx von *Paramecium* und seine systematische Bedeutung. *Arch. Protistenk.* **82**, 331–362.

32   Gill, K. & Hanson, E. D. 1968. Analysis of prefission morphogenesis in *Paramecium aurelia. J. Exp. Zool.* **167**, 219–236.

33   Gillies, C. & Hanson, E. D. 1968. Morphogenesis of *Paramecium trichium. Acta Protozool. (Warsaw)* **6**, 13–31.

34   Hanson, E. D. 1955. Inheritance and regeneration of cytoplasmic damage in *Paramecium aurelia. Proc. Nat. Acad. Sci. U.S.* **41**, 783–786.

35   — 1962. Morphogenesis and regeneration of oral structures in *Paramecium aurelia:* An analysis of intracellular development. *J. Exp. Zool.* **150**, 45–65.

36   — 1967. Protozoan development. In Florkin, M. & Scheer, B., *Chemical Zoology, Vol. 1,* Academic Press, New York, N.Y., 395–539.

37   —, Gillies, C. & Kaneda, M. 1969. Studies on oral structure development and nuclear behavior during conjugation in *Paramecium aurelia. J. Protozool.* **16**, 197–204.

38   — & Kaneda, M. 1968. Evidence for sequential gene action within the cell cycle of *Paramecium. Genetics* **60**, 793–805.

39   Hertwig, R. 1889. Über die Konjugation der Infusorien. *Abh. Bayer. Akad. Wiss.* **17**, 150–233.

40   Hirshfield, H. I., Zimmerman, A. M., Landau, J. V. & Marsland, D. 1957. Sensitivity of UV-irradiated *Blepharisma undulans* to high pressure lysis. *J. Cell. Comp. Physiol.* **49**, 287–294.

41   Hufnagel, L. A. 1969. Cortical ultrastructure of *Paramecium aurelia. J. Cell Biol.* **40**, 779–801.

42   Jennings, H. S. 1908. Heredity, variation, and evolution in Protozoa. I. The fate of new structural characters in *Paramecium* in connection with the problem of inheritance of acquired characters in unicellular organisms. *J. Exp. Zool.* **5**, 577–632.

43   — 1908. Heredity, variation and evolution in Protozoa. II. Heredity and variation in size and form in *Paramecium* with studies of growth, environmental action and selection. *Proc. Amer. Phil. Soc.* **47**, 393–546.

44   — 1923. *Behavior of the Lower Organisms,* Columbia University Press, New York.

45   Jensen, D. D. 1959. A theory of the behavior of *Paramecium aurelia* and behavioral effects of feeding, fission, and ultra-violet microbeam irradiation. *Behaviour* **15**, 82–122.

46   Jerka-Dziadosz, M. & Frankel, J. 1970. The control of DNA synthesis in macronuclei and micronuclei of a hypotrich ciliate: A comparison of normal and regenerating cells. *J. Exp. Zool.* In press.

47   Jurand, A., Beale, G. H. & Young, M. T. 1962. Studies on the macronucleus of *Paramecium aurelia.* I. With a note on ultra-violet micrography. *J. Protozool.* **9**, 122–131.

48   —, — & — 1964. Studies on the macronucleus of *Paramecium aurelia.* II. Development of macronuclear anlagen. *J. Protozool.* **11**, 491–497.

49   Kahl, A. 1931. Wimpertiere oder Ciliata (Infusoria) (2) Holotrichia. In Dahl, *Die Tierwelt Deutschlands und der angrenzenden Meeresteile,* Gustav Fischer, Jena.

50   Kaneda, M. 1961. Fine structure of the macronucleus of the gymnostome ciliate, *Chlamydodon pedarius*. *Japan. J. Genetics* **36**, 223–234.

51   Kimball, R. F., Caspersson, T. O., Svenson, G. & Carlson, L. 1959. Quantitative cytochemical studies on *Paramecium aurelia*. I. Growth in total dry weight measured by scanning interference microscope and X-ray absorption methods. *Exp. Cell Res.* **17**, 160–172.

52   — & Perdue, S. W. 1962. Quantitative cytochemical studies on *Paramecium*. V. Autoradiographic studies of nucleic acid synthesis. *Exp. Cell Res.* **27**, 405–415.

53   King, R. L. 1954. Origin and morphogenetic movements of the pores of the contractile vacuoles in *Paramecium aurelia*. *J. Protozool.* **1**, 121–130.

54   Klein, B. M. 1932. Das Ciliensystem in seiner Bedeutung für Lokomotion, Koordination und Formbildung mit besonderer Berücksichtigung der Ciliaten. *Ergeb. Biol.* **8**, 75–179.

55   — 1933. Silberliniensystem und Infraciliatur. *Arch. Protistenk.* **79**, 146–169.

56   Kozloff, E. N. 1964. A simple and rapid method for demonstrating the arrangement of kinetosomes in ciliates. *Carolina Tips* (Carolina Biological Supply Co., Burlington, N.C.) **27**, 9–12.

57   Lund, E. E. 1933. A correlation of the silverline and neuromotor systems of *Paramecium*. *Univ. Calif. (Berkeley) Publ. Zool.* **39**, 35–76.

58   — 1941. The feeding mechanisms of various ciliated protozoa. *J. Morph.* **69**, 563–571.

59   Lwoff, A. 1950. *Problems of Morphogenesis in Ciliates*, Wiley & Sons, New York, N.Y.

60   Maupas, E. 1889. La rejeunissement karyogamique chez les ciliés. *Arch. Zool. Exp. Gen. (Sér. 2)* **7**, 149–517.

61   Nadler, E. J. 1929. Loss and regeneration of pellicle in *Blepharisma undulans*. *Biol. Bull.* **56**, 327–330.

62   Nanney, D. L. 1966. Cortical integration in *Tetrahymena:* An exercise in cytogeometry. *J. Exp. Zool.* **161**, 307–318.

63   — 1967. Cortical slippage in *Tetrahymena*. *J. Exp. Zool.* **166**, 163–170.

64   — 1968. Cortical patterns in cellular morphogenesis. *Science* **160**, 496–502.

65   Nilsson, J. R. 1962. Observations on *Neobursaridium gigas* Balech, 1941 (Ciliata Heterotrichida). *J. Protozool.* **9**, 273–276.

66   — 1969. The fine structure of *Neobursaridium gigas*. *Compt. Rend. Trav. Lab. Carlsberg* **37**, 49–76.

67   Pitelka, D. R. 1961. Fine structure of the silverline and fibrillar systems of three tetrahymenid ciliates. *J. Protozool.* **8**, 75–89.

68   — 1963. *Electronmicroscopic Structure of Protozoa*, Pergamon Press, Oxford.

69   — 1969. Fibrillar systems in Protozoa. In Chen, T. T., *Research in Protozoology*, Vol. 3, Pergamon Press, New York, N.Y., 282–388.

70   — 1969. Fibrillar structures in the ciliate cortex: The organization of kinetosomal territories. In *Progress in Protozoology* (Proc. 3rd Int. Cong. Protozool., Leningrad), 44–46.

71   Porter, E. D. 1960. The buccal organelles in *Paramecium aurelia* during fission and conjugation with special reference to the kinetosomes. *J. Protozool.* **7**, 211–217.

72   *— 1961. Observations on the effect of mechanical pressure on *Paramecium caudatum*. *Bull. Georgia Acad. Sci.* **19**, 5.

73   — 1962. Morphogenetic migration of the buccal cavity of *Paramecium aurelia*. *J. Protozool.* **7**, 211–217.

74   Prusch, R. D. & Dunham, P. B. 1967. Electrical and contractile properties of the isolated contractile vacuole of *Amoeba proteus*. *J. Gen. Physiol.* **50**, 1083.

75   Raikov, I. B. 1968. Macronucleus of ciliates. In Chen, T. T., *Research in Protozoology*, Vol. 3, Pergamon Press, New York, N.Y., 1–128.

76   Roque, M. 1956. L'évolution de la ciliature buccale pendant l'autogamie et la conjugaison chez *Paramecium aurelia*. *Compt. Rend.* **242**, 2592–2595.

77   — 1956. La stomatogénèse pendant l'autogamie, la conjugaison et la division chez *Paramecium aurelia*. *Compt. Rend.* **243**, 1564–1565.

78   — 1961. Recherches sur les infusoires ciliés: les hyménostomes péniculiens. *Bull. Biol. France Belg.* **95**, 431–516.

79   Sedar, A. W. & Porter, K. R. 1955. The fine structure of cortical components of *Paramecium multimicronucleatum*. *J. Biophys. Biochem. Cytol.* **1**, 583–604.

80   Sonneborn, T. M. 1947. Recent advances in the genetics of *Paramecium* and *Euplotes*. *Advanc. Genetics* **1**, 263–358.

81   — 1950. Methods in the general biology and genetics of *Paramecium*. *J. Exp. Zool.* **113**, 87–147.

82 — 1957. Breeding systems, reproductive methods, and species problems in Protozoa. In Mayr, E., *The Species Problem*, American Association for the Advancement of Science, Washington, D.C., 155–324.

83 — 1963. Does preformed cell structure play an essential role in cell heredity? In Allen, J. M., *Biological Diversity*, McGraw-Hill, New York, N.Y., 165–221.

84 Tamura, S., Toyoshima, Y. & Watanabe, Y. 1966. Mechanism of temperature-induced synchrony in *Tetrahymena pyriformis*. *Japan J. Med. Sci. Biol.* **19**, 85–96.

85 Tartar, V. 1941. Intracellular patterns: facts and principles concerning patterns exhibited in the morphogenesis and regeneration of ciliate protozoa. *Growth*, **5** (Suppl.), 21–40.

86 — 1954. Anomalies of regeneration in *Paramecium*. *J. Protozool.* **1**, 11–17.

87 — 1961. *The Biology of Stentor*, Pergamon Press, New York, N.Y.

88 — 1968. Morphogenesis in Protozoa. In Chen, T. T., *Research in Protozoology, Vol. 2*, Pergamon Press, New York, N.Y., 1–116.

89 Wichterman, R. 1953. *The Biology of Paramecium*, Blakiston, New York, N.Y.

90 Wigg, D., Bovee, E. C. & Jahn, T. C. 1967. The evacuation mechanism of the water expulsion vesicle ("contractile vacuole") of *Amoeba proteus*. *J. Protozool.* **14**, 104–108.

91 Wolfe, J. 1967. Structural aspects of amitosis: A light and electron microscope study of the isolated macronuclei of *Paramecium aurelia* and *Tetrahymena pyriformis*. *Chromosoma* **23**, 59–79.

92 — 1970. Structural analysis of basal bodies of the isolated oral apparatus of *Tetrahymena pyriformis*. *J. Cell Sci.* **6**, 1–22.

93 Woodward, J., Gelber, B. & Swift, H. 1961. Nucleoprotein changes during the mitotic cycle in *Paramecium aurelia*. *Exp. Cell Res.* **23**, 258–264.

94 Yusa, A. 1957. The morphology and morphogenesis of the buccal organelles in *Paramecium* with reference to their systematic significance. *J. Protozool.* **4**, 128–142.

XI. ADDENDUM

(Additional references are indicated by the prefix Ad)

For culturing and experimentally manipulating paramecia the reader should consult two methods papers (Ad 100, Ad 104). Unit territories are discussed in further detail and as a concept generally applicable to ciliated protozoa by Pitelka (Ad 99). Sonneborn (Ad 101) has examined in detail the distribution of doublet (one cilium, one parasomal sac) and triplet (two cilia, one sac) territories on the surface of *paramecium* and the account here should be corrected in the light of these new data. (See his Figs. 27 and 28.) Triplet units are found throughout the adoral zone, as stated in this paper, but additionally they are scattered throughout the anterior two thirds of the anoral zone, which is a new observation. The posterior third of the anoral zone is made up exclusively of doublets. Furthermore, Sonneborn (Ad 102) reviews in detail the formation of new territories and their genetic continuity, including further comments on the possibility of kinetosomal DNA. In this latter regard see also Hufnagel (Ad 98). Further work on the restoration of damaged cortical patterns is given by Chen-Shan (Ad 95). The control of contractile vesicle formation has been studied by exposing cells to cycloheximide, an inhibitor of protein synthesis (Ad 103). It was found that the new anterior vesicle is affected earlier in the interfission cycle than the new posterior one. There is an apparent sequential development of these new organelles. Further work with the UV microbeam (Ad 97) shows that the ability to form new buccal parts can be damaged anytime during the interfission period by irradiating the site of primordium formation (right posterior wall of the vestibule). This is experimental

References p. 262

confirmation of Dippell's (personal communication) and Hufnagel's (41) claim that the anarchic field persists throughout the cell cycle. Further work on the possible occurrence of a cytoplasmic gel has been encouraging (Ad 96). We have been able to dissect off the superficial layers of paramecia by various physical, chemical, and biological agents. When the cortex with associated kinety patterns is separated from the rest of the cell, we can find that the cellular form remains intact. This is clearly what would be expected from the presence of a gel. Preliminary electron microscopy indicates the presence of subcortical microfilaments which could play the role of a gel, or, better, of a cytoskeleton.

REFERENCES

95 Chen-Shan, L. 1970. Cortical morphogenesis in *Paramecium aurelia* following amputation of the anterior region. *J. Exp. Zool.* **174**, 463–478.

96 Hanson, E. D., Kaneda, M. & Sibley, J. 1972. A subcortical cytoskeleton in *Paramecium aurelia? J. Protozool.* **19** (Suppl.), 23.

97 — & Ungerleider, R. 1972. The formation of the feeding organelle in *Paramecium aurelia. J. Exp. Zool.* **185**, 175–187.

98 Hufnagel, L. A. 1969. Properties of DNA associated with raffinose-isolated pellicles of *Paramecium aurelia. J. Cell Sci.* **5**, 561–573.

99 Pitelka, D. R. 1970. Ciliate ultrastructure: Some problems in cell biology. *J. Protozool.* **17**, 1–10.

100 Sonneborn, T. M. 1970. Methods in *Paramecium* research. In Prescott, D. W., *Methods in Cell Physiology*, *Vol. 4*, Academic Press, New York, N.Y., 241–339.

101 — 1970. Gene action in development. *Proc. Roy. Soc. London, B.* **1761**, 347–366.

102 — 1970. Determination, development, and inheritance of the structure of the cell cortex. *Symp. Int. Soc. Cell Biol.* **9**, 1–13.

103 Suhama, M. & Hanson, E. D. 1971. The role of protein synthesis in prefission morphogenesis of *Paramecium aurelia. J. Exp. Zool.* **177**, 463–478.

104 Van Wagtendonk, W. J. & Soldo, A. T. Methods used in the axenic cultivation of *Paramecium aurelia.* In Prescott, D. W., *Methods in Cell Physiology*, *Vol. 4*, Academic Press, New York, N.Y., 117–130.

The Structure of *Paramecium* as Viewed from its Constituent Levels of Organization*

CHARLES F. EHRET AND EUGENE W. MCARDLE**

Division of Biological and Medical Research, Argonne National Laboratory, Argonne, Ill. 60439 (U.S.A.)

Prospectus: The course towards structural understanding chosen here proceeds from a consideration of the parts, to a consideration of the whole. Although this reverses the usual approaches to anatomical interpretation, it yields rewarding perspectives about eukaryotic complexity, especially if the reader makes frequent reference to the course-plan charted in the table of contents (q.v.).

I. MACROMOLECULAR AND SUBORGANELLAR LEVELS

A. Structural Proteins

1. Trichynin

The ultrastructure of the extruded form of the freshly isolated trichocyst is shown in Fig. 1 (97); when examined directly by negative staining, it appears to be made up of zones of alternating density (period ∼ 550 Å) as described earlier (48). Steers, Beisson and Marchesi have shown that the trichocyst is composed of a protein which contains no detectable carbohydrate or nucleic acid moieties. Polyacrylamide disc gel electrophoresis reveals the presence of two forms of the protein, estimated to have molecular weights of 17,000 and 36,000 respectively; furthermore, reduction with 2-mercaptoethanol results in a single unit of 17,000 molecular weight, suggesting a polymeric complex dependent upon disulfide bonding.

Because of the thickness of individual trichocysts, the structural elements responsible for banding cannot be resolved in freshly isolated trichocysts (Fig. 1); dialysis of trichocysts against distilled water causes their gradual disruption, resulting in the beautifully resolved image in Fig. 2.

Molecular and ultrastructural data taken together suggest that the trichocyst (at least its residual elements) may be constructed of as few as two polypeptide chains, each with a molecular weight in the 17,000–20,000 range. The resulting dimer, the structural protein of the trichocyst, is called Trichynin by Steers and his associates.

2. Dynein

Two forms of an axonemal ATPase protein, named 14S dynein and 30S dynein

---

This work was supported by the United States Atomic Energy Commission.
* The illustrations are placed at the end of this paper to facilitate easy reading of the text (pp. 291–338).
** Present address: Northeastern Illinois University, Chicago, Ill., U.S.A.

References p. 286–290

(on the basis of their sedimentation constants) have been separated from the cilia of *Tetrahymena pyriformis* by zonal centrifugation through sucrose density gradients (39). Gibbons' electron micrographs of 14S dynein show globular particles, ellipsoids with axes 85 Å, 90 Å and 140 Å; 30S dynein appears more rod-like, with dimensions of $\sim$ 80 Å by $\sim$ 1700 Å, and with a repeating globular structure (period $\sim$ 140 Å). Approximate molecular weights for 14S and 30S dynein are 600,000 and 5,400,000 respectively. A further general similarity in enzymatic relationships (ATPase activity) is consistent with a monomer-polymer relationship between the 14S and 30S molecules.

Dynein appears to be localized in the axoneme because (a) it is solubilized during dialysis, and therefore not in the outer fibers, and (b) purified 30S dynein recombines with outer fibers *in vitro* to reconstitute the arms on subfibers (subfiber A) of axonemal outer fibers.

3. Axonemal fiber protein

Gibbons also characterizes the molecular properties of isolated outer fibers of ciliary axonemes in *Tetrahymena*. Analytical centrifugation of protein extracts in 5M guanine hydrochloride yields a single peak with a sedimentation constant of $\sim$ 2.1S, and a molecular weight of $\approx$ 55,000. Disc electrophoresis in polyacrylamide gels gave a number of minor bands, with about 90% of the protein in a single major band. Reconstitution experiments have been somewhat successful, yielding 40 Å protofilaments in the resulting fibrous precipitates.

Although Gibbons' work unfortunately has not yet been extended to *Paramecium*, we mention it here not only for heuristic reasons, but also to point out the general similarities in amino acid composition between ciliary outer-fiber protein, and *Trichynin* (Table 1).

TABLE 1

Amino acid composition of structural proteins of Paramecium *and of* Tetrahymena.

Amino acid	Trichynin (97) (residues/10^5 mol. wt.)	Axonemal outer fiber protein (39) (residues/10^5 g protein)
Lysine	48	51
Histidine	17	22
Arginine	37	41
Aspartic acid	108	94
Threonine	45	46
Serine	50	54
Glutamic acid	176	117
Proline	14	39
Glycine	37	80
Alanine	92	56
Cysteine	9	13
Valine	50	53
Methionine	5	26
Isoleucine	50	49
Leucine	87	66
Tyrosine	20	29
Phenylalanine	33	39

B. Nucleic Acids

1. DNA

Fine structure identification of DNA directly in thin sections of *Paramecium* has (surprisingly) not yet been accomplished with any degree of certainty, although cyto-chemical techniques *have* been successful at a grosser level to be discussed later. Molecular studies by a number of workers (44, 92) have characterized the nuclear DNA of *P. aurelia* by buoyant densities in CsCl of from 1.687 to 1.691 $\times$ 10^3 kg m^{-3}, with some question regarding precision of measures *vis à vis* possible intraspecific stock differences (45). Fig. 3 is a low power electron micrograph of the high molecular weight DNA isolated from *P. multimicronucleatum*, applied to electron microscope grids by the monolayer technique of Freifelder and Kleinschmidt (35), and stained with uranyl acetate. Continuous stretches well in excess of 85 μm in length (non-circular) have been measured.

C. Karyoplasmic Elements

1. Chromatin bodies

The small densely granular bodies or chromatin bodies (cb) in Fig. 4 are numerous throughout the macronucleus, range in size from $\sim$ 0.1 to 0.2 μm and contain DNA (16, 63, 77). Thicker sections contain patterns that suggest either a catenary linkage between or a snake-like continuum of small bodies (Fig. 80). Thinner sections at high resolution reveal 100 Å thick fibrils; considered together, the observations suggest that the chromatin bodies are sections of macronuclear chromosomes (15, 16, 51).

2. Nucleoli

The larger less numerous granular bodies of Figs. 4 and 5 are the cup-like basophilic nucleoli of the macronucleus (24). These bodies are Feulgen-negative and RNA-containing (16, 77), are extremely variable in size, form and degree of aggregation, but in *P. bursaria* show patterned consistencies within the life cycle (23, 24, 49, 85, 103, 104); granules resembling cytoplasmic ribosomes (150–200 Å) are sometimes present (54, 77); the matrix of the cup resembles the karyoplasmic matrix (Fig. 6) sometimes containing the tightly coiled microfibrils discussed below (Fig. 5).

3. Karyoplasmic matrix

Tightly coiled microfibrils ($\sim$ 100 Å diameter) are seen in high resolution micrographs of the otherwise Feulgen-negative (77) karyolymph. These microfibrils have a helical periodicity of about 400 Å and show an ultrastructural continuity with the chromatin bodies (Figs. 6, 7) much like that described by Frenster (36) as characteriz-ing the zones of transition between condensed masses of genetically repressed heterochromatin (chromatin bodies?) and extended microfibrils of genetically active euchromatin. Also seen in the matrix (Figs. 87–90) are fibrillar and microtubular inclusions (98) probably associated with spindle fiber production and nuclear division.

D. Microtubules, Tubules, Fine Filaments and Fibrils, Fibers and Lattices and Striated Root Fibrils

1. Microtubules

Hollow cylinders with diameters of 15–30 nm are present throughout the cytoplasm. These appear most frequently as kinetosomal derivatives, either singly (Fig. 8, arrow), or in pairs or bands termed *ribbons*, or in groups and bundles termed nemadesma. Ribbons composed of two or three microtubules in intimate contact appear in the ciliary axoneme (Fig. 11, c) and kinetosome (Fig. 11, ks) respectively (68); near the epiplasmic layer, surrounding the trichocyst tip, over 30 small circles are seen which we interpret to be a short cylinder of microtubules (Fig. 11, t). Fig. 8 shows ribbons of five and of seven microtubules near the naked ribbed wall sawtooth band of the gullet (53, 68). In raffinose-isolated kinetosomes, the composition of collapsed kinetosomal microtubules has been resolved by Hufnagel (46); each microtubule is made up of a number of longitudinal fine filaments [resembling the protofilaments of Gibbons discussed earlier, (39)] about 45 Å wide and having a beaded appearance (Fig. 9).

2. Tubules

Much larger membranous tubules are seen in the orthotubular system (Fig. 10, ots) (23, 53) and in the membranous tubules of the mitochondria (Fig. 11, m) (73, 74, 107, 109). Jurand and Selman describe the dimensions of ots tubules as 50 nm × 1.3 μm; outside each tubule, knob-like projections of about 10 nm diameter appear to be joined to the membrane by a short neck about 5 nm long, suggesting subunits similar to those seen in mitochondrial membranes (34, 53).

3. Fine filaments and fibrils

The fine filaments (45 Å) that compose microtubules have already been discussed in Fig. 9; other filaments (85 Å × ~450 Å) have been identified in ethanol-isolated pellicles by Hufnagel (46). These 85 Å fine filaments are found in fibrils composed of three or four fine filaments, fused at their centers to form a bow-tie or cat's whisker-like cluster (Fig. 13).

4. Fibers and lattices

The cat's whiskers compose the larger bundles and networks that underlie the pellicular epiplasm and constitute an extensive network termed the infraciliary lattice [Fig. 12, (46)]. Hufnagel has also identified a globular element in ethanol-isolated infraciliary lattice material (Figs. 12, 13, 105). The complex networks of fine filaments, fibrils and fibers are difficult to visualize by means other than those ingeniously resorted to by Hufnagel, although paradoxically they had been grossly mapped years earlier with light microscopy (38). In a thin section, the infraciliary lattice (il) may appear as a deceptively simple bundle, as seen in Fig. 11; we shall return to the lattice aspect in a later section on organelle complexes of the cell envelope.

5. Striated root fibrils

In suitable thin sections (68) or in raffinose or ethanol-isolated preparations, striated root fibrils (46) appear to be made up of longitudinal fine filaments with a center to center period of $\sim$ 50 Å [horizontal patterns, left edge Fig. 14, (46)]. According to Hufnagel's studies of isolated striated root fibrils, each is about 410 nm wide near the kinetosome (left edge, Fig. 14 and lower left, Fig. 11); about 700 nm away from its 3 attachment points to the kinetosome it tapers precipitously, and then more gradually. A repetitive substructure appears across their lengths (66), whose periodicity ranges from 350 to 400 Å in raffinose-isolated fibrils, and from 290 to 340 Å in ethanol-isolated fibrils (46). The major period contains two minor bands: (a) a pale band $\sim$ 180 Å wide, contains five fine bright lines, the central line appears slightly narrower than the other four; and (b) a darker band $\sim$ 135 Å wide, which contains a slightly off-center fuzzy bright band (Fig. 15).

As we shall see later, the striated root fibril of *Paramecium* (kinetodesmal fibril, Figs. 11, 14, 15), itself a complex of subunits, acts as a unit component in a bundle of overlapping fibrils, the kinetodesmos (Figs. 28, 38, 39, 40, 107, 110).

E. Ribosomes and Glycogen Granules

1. Ribosomes and endoplasmic reticulum

At low magnification, ribosomes appear as somewhat regular, and at high resolution, as slightly bilobed homogeneous dense structures $\sim$ 200 Å in diameter. The introduction of glutaraldehyde in the early 1960's as an effective prefixative marks a transition between poorly fixed and beautifully fixed material in the *Paramecium* literature (compare the poorly preserved ribosomes and ribosome-associated membranes in reference 23 with Figs. 16 and 17). In some cells, and in nearly all cells near the nuclear envelope and cell envelope, well-ordered patterns of distribution of ribosomes are evident along and around membranous components, and sometimes appear as small spirals and rosettes (Figs. 16, 17 *et seq.*). Although the endoplasmic reticulum is not developed dramatically into the extensive sheets seen in some metazoan secretory cells, it is well represented in *Paramecium* in numerous smaller territories or cytoplasmic domains, not as yet adequately mapped as a function of region or stage or mode of growth (30) in the cell cycle.

2. Glycogen granules

Food storage granules, which appear to be glycogen (53), are found throughout the cytoplasm; when compared with ribosomes, glycogen granules are more irregular in shape, less homogeneous in texture (appear "peppery" at high resolution, Fig. 17), more randomly distributed and larger; their diameters are in the neighbourhood of 40 nm.

F. Unit Membranes and Other Membranes

1. Three-ply unit membranes

A three-ply unit membrane (79, 80) is common to many (but not all) membrane-bounded structures in *Paramecium*. Each unit membrane consists of 2 dense layers, each ~ 25 Å thick with a less dense layer sandwiched between, about 30 Å thick, thus totalling about 80 Å of three-ply unit membrane. In Fig. 16, reading left to right from the last layer of cytoplasmic ribosomes toward the clear bay of fluid outside the cell, one first observes a finely granular line (the epiplasmic layer) then, in turn, three unit membranes. The outermost unit membrane is the outer layer of the entire cell envelope; though claimed to be continuous (67) mappings of continuities and interruptions remain to be done. The inner 2 unit-membranes surround a small space (~ 0.5 × 1 μm, arrow) and, in so doing, compose a considerable fraction of a kinetosomal territory (discussed below). The continuities and interruptions of these alveolar spaces have also been discussed by Allen (2).

2. Other membranes

Other membranes, not obviously composed of 3 layers (in well-fixed high resolution micrographs), surround many organelles. The microbody-like organelle (Fig. 17) (23) is surrounded by only a single membrane, composed apparently of only a single opaque layer; the conventional mitochondrion is surrounded by 2 outer membranes, each one like the one described above. In Jurand and Selman's electron micrographs (53) these membranes sometimes have a beaded appearance, produced by corpuscular units about 8 nm in diameter.

II. ORGANELLES

A. Chondriome, Large Inclusions and Symbionts

1. Mitochondria

Paramecium mitochondria are typically 2 to several microns long, and nearly a micron in diameter (Figs. 16–23). They contain no cristae (ridges) but instead have numerous cylindrical tubules about 50 nm in diameter (53, 73, 74, 105, 107) that branch (Fig. 20) throughout the matrix. Between the tubules are ribosomes remarkably similar in size and electron-density to those of the cytoplasm (n.b. Figs. 18, 20). Tubules and membranous protrusions are common (Figs. 18, 19 at the top of each, and Fig. 20, at the top of the small mitochondrion in the lower right field), especially in actively dividing cells.

2. Microbody-like organelles

These organelles are large, granular vesicles (Figs. 17, 20, 22) somewhat smaller than mitochondria, but bounded by only a single dense membrane, devoid of ribosomes, and containing only sparse concentrations of mitochondrion-like tubules seen near the cell envelope in rapidly dividing cells. These peroxisome-like organelles (44) had been called "small-tubuled mitochondria" by Ehret and De Haller (23) who found them to be especially evident in pre-fission cells, and are undoubtedly the same bodies discussed by Jurand and Selman (53) as "undifferentiated membrane bound vesicles". Following Wohlfarth–Bottermann's early speculations (53, 107, 108) they presented a case for these organelles as progenitors of mature mitochondria, leaving open the questions of their *de novo* origin, as well as of other modes of chondriomal mass-increase.

3. Large inclusions: crystals and fingers

A wide variety of other objects in the cytoplasm, are discussed here only because they are *about the size of* the unit organelles of the cytoplasm (mitochondrion, cilium and trichocyst). Figs. 21 and 22 show the remarkable membrane-bound fingers of cytoplasm located in positions formerly occupied by trichocysts. De Haller and Heggler (43) have shown such fingers as regular occurrences in UV-irradiated clones having "unstable trichocysts". In predividing cells (Fig. 22), small circular vesicles about 100–200 nm in diameter appear at about the level of kinetosomes; these vesicles may be bodies previously identified as "kinetosomal precursors" (23), and are also seen underlying new kinetosomes (arrow, lower left, Fig. 109). On the other hand they may represent membranes derived from pinocytotic activity of parasomal sacs (1). In Fig. 23, the membrane-bound inclusions of crystals, long well-known by light microscopists, are seen.

4. Symbionts

Also organelle-like in size, and even somewhat so in integral relationship to the genome (see the review by Beale, Jurand and Preer, 5) a wide variety of symbionts, including algae, bacteria and bacteria-like symbionts inhabit the cytoplasm (5, 102). With but few exceptions only one kind of symbiont is found at a time in a given cell or clone; the best known of these, kappa of *P. aurelia*, syngen 4, stock 51, is pictured here (Figs. 23–26). Specifically kappa particles frequently contain highly refractile (light microscopy) R-bodies associated with the ability to kill other *Paramecia* that lack R-type kappa; however, certain mutants of kappa are known to possess R-bodies that are not killers. More generally, a substantial proportion of wild *Paramecia* contain symbionts in both the cytoplasm and the macronucleus, thereby confronting the geneticist, whatever his calling, with the classical problem of Koch's postulates. Most observers have now accepted the view that the symbionts are bacteria; the remarkably flagellated lambda particles (52) are reported to have been cultivated in a *Paramecium*-free medium (101).

The first symbiont discovered and named by Sonneborn (93, 94) is *kappa*. A kappa particle is an endosymbiont that contains, or can produce R-bodies or is derived from such an endosymbiont. An R-body is a hollow membranous structure that consists of a long ribbon wound into a tight spiral of about 10 turns (Fig. 25). There typically is only one R-body per kappa, but occasionally 2 may be seen [Fig. 24, (10, 75)].

The summary paper by Beale *et al.* (5) relates the myriad manifestations of "kappas" and of "R-bodies", and the other classes of generally gram-negative bacteria that one may encounter on a journey through the plasms of *Paramecium*.

In Fig. 26, a kappa particle containing an R-body is seen near the cytoproct, a possible exit site.

B. Kinetosomes, Kinetosomal Territories and Ciliary Corpuscles

1. Archetypal view

The single kinetosome and the small territory organized around it (consisting of the microtubules, fine filaments and fibrils, fibers and lattices and striated root fibers; sometimes including alveolar membranes and axonemal elements of the suborganellar levels) represents the unit organelle *par excellence*, not only for the theoretician inclined to regard the biosphere from a *levels of organization* viewpoint (19, 58), but also for the protozoologist-anatomist who regards it as the unit corpuscle of the cell envelope of *Paramecium* (26, 53, 68) and the comparative anatomist who recognizes it as the homologue of the morphogenetically dynamic and ubiquitous centriole (23, 40, 58, 69, 70). The kinetosomal unit may appear singly (as if a macromonomer) or in groups of 2 or 4 (as if macrodimers and tetramers). The units (whether monomeric or polymeric) form well-ordered complexes of organelles, and these, in turn, form systems composed of organelle-complexes to be described later.

Returning to a "type-view" of the unit, consider first how its already familiar parts (Figs. 11–16) are assembled (Figs. 27–37).

a. Kinetosomes.

a. Kinetosomes. A cross-sectioned kinetosome, when viewed outward, as if *from within* the cell, appears as shown at the lower right of Fig. 27. We employ the numbering convention of Grain but make cross reference to the earlier one of Pitelka (68) to whom we are indebted for her pioneering mappings and insights into the comparative anatomy of kinetosomal microtubules (66, 67). At this lowest cross-sectional level, microtubules occur in ribbons of 3 with contiguous microtubules sharing a common wall (79). Nine of these triplets are assembled in "cartwheel" formation around a central space to form the cylindrical base of the kinetosome—a short tube $\sim 0.25\ \mu m$ in diameter. The striated kinetodesmal fibril attaches to triplets 1 to 3 (numbered clockwise) and most definitively to 2 and 3 (Figs. 11, 35, 37) and then proceeds surfaceward to the right and anteriorly (Figs. 38–40). A postciliary ribbon of microtubules *originates at triplet 5*, and at this level its three to five microtubules appear to be coextensive with those of triplet 5, this is also shown clearly in Fig. 7 of Dippell (14). *At a slightly higher level of sectioning, the postciliary ribbon then appears in the cytoplasm adjacent to triplets 4 and 5* (Fig. 27, center, second cross-section from the bottom, and Figs. 35, 36, 40, 43, 106); this is also the relationship described by Pitelka in her Fig. 18 (68). It is worthwhile now to note that the nine cartwheel triplets are skewed inward *clockwise* when viewed looking outward from within the cell, and they are skewed inward *counterclockwise*, when viewed inward towards the cell; even topologically indoctrinated microscopists carry maps like these (Fig. 27, lower right) with them when reading cartwheels on their micrographs, because handedness is better pictured than described!

A transverse ribbon (arrows, Figs. 28, 34) appears to originate near triplets 8 and 9, and like the postciliary, rises to attach itself to the epiplasm at the lower alveolar surface [Figs. 29, 30, 31, (45)]. When kinetosomes appear in pairs (Figs. 11, 36, 37, 40, 106, 108, 111), the kinetodesmal fibril usually arises from the posterior kinetosome, but filamentous interkinetosomal connectives also appear (Figs. 106, 108). It is important to note that the numbering convention employed by Grain (40) not only differs from that of Pitelka (Pitelka Convention, P.C.), but as practiced sometimes lacks internal consistency. Triplet 6, in the Grain Convention (G.C.), corresponds to triplet 1 (P.C.), if we can agree that the kinetodesmal fibril attaches to triplets 1, 2, 3 (P.C.) or to triplets 6, 7, 8 [G.C. (40), p. 66]; and further, that the transverse ribbon *(fibres transverses)* appears next to triplets 8 and 9 (P.C.) or next to triplets 4 and 5 (G.C.). The postciliary ribbon *(fibres post-ciliaires)*, which originates at triplet 5 (P.C.) and at a higher level of sectioning appears near triplets 5 and 4 (P.C.), is claimed by Grain to *originate* at triplet 4 (P.C.) ≡ triplet 9 (G.C.). This procedure maps the postciliary ribbons inconsistently. The situation becomes more confusing when the kinetodesmal fibril is defined [as in Grain (40), p. 67] as a fiber that originates at triplets *5, 6, 7 and 8;* claiming then to adopt the Grain Convention, Noirot–Timothée (64) goes on to place the kinetodesmal fibrils at triplets 5, 6, and 7 (≡ 1, 2, 3 P.C.), the transverses *(fibres transverses)* next to 4 and 5 (≡ 9, 1 P.C.), and the postciliaries *(fibres post-ciliaires)* emerging from triplet 9 (≡ 5 P.C.). This procedure maps the transverse ribbons inconsistently. Finally, because the theoretical grounds for mapping in the Grain Convention (p. 57) are doubtful [i.e., the assertion that the axial fibers of a cilium near its base are perpendicular to the direction of the kinety, (40)] as also recently shown by Fauré-Fremiet (33) we prefer the more practical approach of Pitelka, which in effect assigns numbers to triplets by virtue of the fibrils and microtubular ribbons one finds there, and at least results to date in a self-consistent system of mapping. However in the interest of uniform and popular usage (e.g. 4) we have adopted that of Grain. Pitelka suggests that the *postciliary ribbon* at its point of origin is so *distinctive and consistent* that it *ought to be accepted as the reference point* for whatever enumeration system is used (71): it is *triplet 5* (P.C.) or triplet 9 (G.C.), in agreement with some otherwise premature recommendations of an international commission (76).

b. Kinetosomal territories and ciliary corpuscles. The most conspicuous proliferation of a kinetosome is the ciliary shaft or axoneme, which rises for a distance of about 10 μm above the transverse plate in *Paramecium*. The nine ribbons of microtubules are now represented by peripheral doublets; in a central location above the axosome and apparently connected to it by only one microtubule, a central doublet rises to continue (even beyond the peripheral ring of nine) to the end of the shaft. At about the levels of the transverse plate, which resembles a disc of cytoplasmic epiplasm penetrated by nine perforations (46) and an axosome, 2 membrane-bound alveoli entirely surround the centrally located kinetosome for a radial distance of $\sim$ 0.5 μm. The closely opposed inner membranes of the paired alveoli form a double septum (ds, Figs. 11, 27, 40) along the B–B axis (Fig. 94), the locus of the so-called "primary meridian" of interciliary fibrils of the light microscopist (65). This *line of contiguity* between alveoli has been identified in ethanol-isolated pellicles and in raffinose-isolated pellicles which have experienced partial degradation of fibrous structures, and is then composed of two closely spaced membranes (width: 110 Å) separated by a width of 100 Å (46); the latter may represent the same structure identified in silver-impregnated cells (11), and isolated from silver-impregnated cells after treatment with NaOH and termed "interciliary fibrils" (65). Within the right alveolus, to the right of the kinetosome and anteriorly, a parasomal sac appears (26), as a dimpled invagination of the cell envelope about 80 nm in diameter. The parasomal sac is connected to the kinetosome by a thin sheet of membrane-bound cytoplasm (46).

According to Allen (2) the aveolar boundaries do not follow the crests of the polygonal ridges in a regular fashion. Because of this he asserts that their staining with silver cannot account for the external lattice so well known via light microscopy; the granulo-fibrillar material present within the peaks of the ridges, (if it is indeed argentophilic) could account for it. Perforations occur between adjacent alveoli which makes the whole mosaic of alveolar spaces within the cell's cortex a continuum. Taken all together and contrary to the critique of Párducz (65), an intimate association *is* seen to exist between the kinetosome and its surrounding alveoli; this domain of the kinetosome [a "territory" (68, 70) of microtubules, striated root fibrils, unit membranes and their derivatives with a kinetosome at its center] constitutes one of the unit organelles that composes the cell envelope; the kinetosomal territory of the pellicle system of *Paramecium* is sometimes called, more specifically, the *ciliary corpuscle* (19, 26). If the domains are ideally visualized as nearly spherical (as isolated corpuscles are at low magnification), they never appear this way *in vivo;* instead, we see the hexagonal and rhomboidal patterns classically associated with close-packing or with skewed rectilinear packing, patterns well described in terms of unit-dynamics and geometrically defined limits in such widely divergent fields as crystallography, virology, ecology and architecture.

2. Multiplication of kinetosomes

The work of Ehret and De Haller (23) clearly established that new kinetosomes in *P. bursaria* do not arise from old ones by bipartition; the anisotropic character of kinetosomal territories also has permitted discovery of the rule of antecorpy: that new kinetosomes arise anterior to old ones (Figs. 41, 42). Unfortunately, in that study poor resolution of such essential structures as microtubules limited further conclusions; although the striking quantitatively measured changes in time experienced by vacuolated and membranous elements surrounding the kinetosomes in precisely-staged cells should not be ignored as irrelevant artifacts in future work.

Dippell's study (13, 14) of kinetosome (or basal body) development in *P. aurelia* has provided the higher resolution needed to recognize important earlier stages (Figs. 43–49). It also signals once again the dramatic service of *Paramecium* to the cause of cell biology: as Ehret and de Haller, as well as Dippell realized, in *Paramecium* thousands of new kinetosomes appear almost synchronously in precisely mappable positions on the surface of a staged cell. Thus, the investigator is given many opportunities to witness the capture of a fleeting synthetic event. In Stage 1 of the development of a new kinetosome (nks) according to Dippell, a flat disc of fibrous material appears near the left side of the kinetodesmal fibril, and anterior to the old kinetosome and

its transverse ribbon of microtubules (Fig. 43). At Stage 2, a ring of singlet microtubules, the *A-set* of microtubules, is formed, proceeding at first from the upper right
quadrant of the new kinetosome adjacent to the kinetodesmal fibril (Fig. 44) con-
secutively around the disc (Fig. 45) until a circle of nine is completed. At Stage 3,
when six or seven tubules are present, a second round of tubule development (the
B-set) begins, again in the upper right quadrant (Fig. 46). At Stage 4, triplet formation
occurs (Fig. 47) with the addition of the *C-set* of microtubules; at this stage cytoplasmic
microtubules appear to originate from a "pad" (arrow, Fig. 47) alongside of the new
kinetosome. At this stage, Dippell goes on to note that the cartwheel microtubules of
the new kinetosome have a pitch that is steeper than those of the mature kinetosome;
that it lacks the filamentous A–C linkers around the nine triplets; and that it lacks
the organized central mass, vesicular and fibrous components and accessory outer
appendages.

The old and new kinetosomes then move apart as the new kinetosome begins to
tilt upward, and the microtubules continue to elongate (Fig. 48); a "cap" is elaborated
at this time, and the new axosome forms near this cap before the new kinetosome
surfaces. In Fig. 49, newly-formed kinetosomes are seen alongside a mature one
(with cilium attached), and, one may say, in a single kinetosomal territory, namely
that of the *elder*! The newly-formed kinetosomes lack the helically disposed large
dense RN'ase-sensitive bodies found inside the mature kinetosome (14, 15). Further
relationships of multiplying kinetosomes to expanding territories will be discussed
later, in the section on organelle complexes and systems of the cell envelope.

C. Trichocysts

1. Archetypal view

The trichocyst consists of a body shaped like a carrot, a tip that looks like a golf-tee,
and a cap that snugly covers the tip (Fig. 53). The cell contains thousands of these
organelles, in number roughly equal to the number of ciliary corpuscles (Fig. 50).
Trichocysts are readily discharged by means of geranium tannic acid; a technique
that leaves behind a live and healthy cell capable of growing a new crop in less than
an hour [Fig. 51, (84)]. When viewed with the scanning microscope (89), the discharged shafts appear not to be perfectly cylindrical (Fig. 52); a major period of
$\approx$ 200 nm (88) is also discernible, corresponding to the major period seen in light
microscopy and in some electron microscope sections (Figs. 54, 55). The other more
characteristic periods ($\approx$ 55 nm) seen in Fig. 55 correspond to those already discussed
in the work of Steers *et al.* (97). Cells fixed with slowly penetrating fixatives or by
slow freezing down to $-10\,^{\circ}$C (61) frequently show extensive trichocyst discharge
internally as well as externally; well-fixed cells rarely discharge trichocysts, and one
very rare trichocyst caught in the act of early emergence is shown on the left of Fig. 53.
In cross-sectioned material, the caps and tips look like bull's eyes, while the body
looks like a pale circular field $\approx$ 1.5 μm in diameter. In Fig. 56, a number of cross-
sections of tips from apexes to bases are evident—as if one had sliced across the long-
sectioned view of Fig. 57; glancing sections of the cap reveal longitudinal periodicities

of ~ 25 nm (Fig. 57); the tip, itself, is strongly osmiophilic, with a crystalline structure (periodicity ≈ 8 nm, Figs. 58, 59); while the body of the mature trichocyst appears nearly structureless (53, 86, 110), but of a density reminiscent of the non-tubular regions of microbody-like organelles discussed earlier. Extremely short cytoplasmic microtubules, about 32 in number, surround the cap near its locus of insertion at the cell surface (arrows, Figs. 59, 60). As one descends the tip in consecutive sections (Figs. 61, 62) the cross-sectional appearance becomes squarish or clover-lobed (Fig. 62), an appearance that extends even into the region of transition from tip to body (arrow, upper left, Fig. 56). In longitudinal and cross-sections, the cap is composed of a dense granular material ~ 50 nm thick; this latter material appears to be compressible, even by so delicate an object-type as a microbody (Fig. 59, lower right). A plausible reconstruction of the cap and tip appears in Fig. 63A.

2. Multiplication of trichocysts

During the course of the cell cycle of precisely staged single cells of *P. bursaria*, Ehret and de Haller (22, 23) have observed high concentrations of subspherical organelles, about the size of sparsely-tubuled mitochondria, which contain, within the amorphous, faintly filamentous and finely granular matrix, a dense central core of crystalline material about 0.2 μm in diameter. These bodies have been identified as *pretrichocysts* because of (a) their large numbers in pre-fission cells immediately prior to the appearance of many new mature trichocysts in the new cell-envelopes of young daughter cells, (b) the orderly sequence of relatively complex structural states represented, and (c) the absence of other evidence for trichocyst-like bodies during the cell cycle. From pretrichocyst to maturity (Figs. 63B–74), seven developmental stages are recognized: (a) the subspherical pretrichocyst has a central crystalloid with spacings of 7–15 nm in different sections; (b) the crystalloid elongates, and the fine-tubuled matrix diminishes in width; (c) the crystalloid acquires the approximate shape of a trichocyst, while the surrounding matrix diminishes; (d) the outer shell of tip and body forms around a central core, while a cap begins to cover the tip; (e) only traces of the early pretrichocyst matrix material remain, the cap is nearly complete; (f) dense granules with spacings of ≈ 22 nm appear to form a two dimensional lattice about the tip (not observed in *P. aurelia*); these dense granules of ~ 13 nm diameter condense around the body (53) and, finally, (g) the body loses its dense osmiophilic character, while the tip remains crystalline. The terminal stage, correctly recognized by Yusa (110) in *P. caudatum*, and by Jurand and Selman (53, 86) in *P. aurelia*, also appears (sporadically?) in *P. bursaria*, where it may represent a later state of development maturation highly responsive to fixation differences.

3. Time constants in the birth and death of trichocysts

Kinetic parameters of the morphological studies of development (23) and measures of the time required to incorporate tritiated leucine and proline into mature trichocysts (27, 29) show that the developmental process of pretrichocyst to trichocyst, as shown in Fig. 63B, can take place in less than 30 minutes. According to Selman and Jurand (86), nearly all of the trichocysts in *P. aurelia* are newly synthesized at each cell division, with only a few old trichocysts remaining at the extreme ends of the animal, remote from the fission furrow. That some trichocysts are conserved through a number of fissions is shown by another isotope experiment with *P. bursaria* (27). Cells grown in unlabeled medium, but derived from labeled cells (^{3}H-leucine) show the bulk of the residual label to be associated with mature trichocysts and also with other stable components of structural organelles of the cell envelope. One can designate the fissionts derived from an initial cell (termed f_0) as f_1 (two cells), f_2 (four cells), f_3 (eight cells), etc. An f_0 cell, derived from a clone grown in tritiated leucine for many generations and then transferred to unlabeled medium, shows virtually all of its trichocysts heavily labeled. At f_1 trichocysts as heavily labeled as ever persist; but newly synthesized trichocysts (sparsely labeled because of residual pools of tritiated precursors) begin to appear (Fig. 75). The phenomenon continues at least into f_3 (Fig. 78), at which time labeled trichocysts are rare, but *as radioactive as they were at f_0*. Sections of cells reveal comparable specific activities and distributions (e.g., f_2 cells in Figs. 76, 77 show sparsely distributed but intensely radioactive trichocysts). Strongly labeled pretrichocysts are never seen in such sections, but do appear in f_0; labeled pretrichocysts are also seen after a short pulse of a tritiated precursor amino acid (27, 29). Other experiments, employing electron microscopic autoradiography of cells labeled with ^{3}H given in thymidine (28, 29), too complex to report in detail here, have *suggested* the presence of small amounts of DNA and RNA, but only on cytochemical grounds, especially through the application of specific nucleases to preparations like those seen in Fig. 75. Attempts to identify trichocyst DNA in CsCl gradients have, to date, been unsuccessful.

III. ORGANELLE COMPLEXES AND SYSTEMS

A. The Nucleome (Fig. 79)

1. Macronucleus

Having already described the suborganellar properties of nucleoli, chromatin bodies and karyoplasmic matrix, it seems strange so little is left to say. Is this sac (Fig. 79) homogeneously filled? Where are the component organelles (chromosomes)? Dippell's thick section (Fig. 80) and autoradiographs (Figs. 81, 82) may be helpful at this level of resolution. Chromatin bodies appear to be members of catenaries, or coiled snake-like strands (Fig. 80). After the feeding of tritiated thymidine, silver grains are found predominantly in the chromatin body regions and matrix region of the karyoplasm, and not over the nucleoli (Fig. 81). After the feeding of tritiated uridine, silver grains are distributed over cytoplasmic structures, and on macronuclear nucleoli (Fig. 82). As shown in Fig. 83, the macronucleus dominates the central region of the cytoplasm, flanked only by the gullet (see in closed cross-section) commanding nearly equal territory. The relatively rigid fixity of the macronucleus alongside of the gullet appears paradoxical to many workers, in the presence of cytoplasmic cyclosis. According to De Haller and Heggler (42), the two systems are joined together by fine karyophoric fibers (Fig. 84) extending from the macronuclear envelope to interdigitate with bundled groupings of thicker cytopharyngeal fibers extending from the gullet. Fig. 84 shows well-resolved pores of the macronuclear envelope, seen also in cross-section in Fig. 85.

2. Micronucleus

a. Vegetative and mitotic. *P. aurelia* has two small micronuclei. Surrounding the osmiophilic crescent of very dense and compact granules (Fig. 86) is a region of low density-containing networks of finely dispersed fibers. During very early stages of division (Figs. 87–89), and well evidenced in early anaphase (Fig. 90), a longitudinally oriented network of karyoplasmic microtubules appears, and the nuclear envelope remains intact (54, 99). Like the macronuclear envelope, the envelope surrounding the micronucleus appears to have two separate simple (not three-ply) membranes, penetrated by nuclear pores at regular intervals (Figs. 86, 87 top right, 88).

b. Meiosis, pronuclear exchange and nuclear anlagen. Though chromosomes continue to be our "missing organelles", they *do* appear conspicuously during conjugation, and show up especially well in metaphase of the first pregamic division in *P. bursaria* (Fig. 91). After several (species-dependent) prezygotic divisions, zygote formation occurs following the reciprocal exchange of migratory pronuclei, long recognized genetically, but beautifully captured in conjugating *P. multimicronucleatum* with an electron micrograph by Inaba *et al.* [Fig. 92, (47)]. After pronuclear fission, and in *P. bursaria*, following 2 post zygotic divisions, 2 micronuclear and 2 macronuclear anlagen are present in each exconjugant. It is at just this stage that in certain other ciliates (*Stylonichia* and *Euplotes*) giant chromosomes become evident in the macronuclear anlagen (3, 78); but alas, thus far, not identified in *Paramecium* (Fig. 93)!

B. The Cell Envelope or Cortex

1. The pellicle system of organelles

With the exception of the large invagination of the cell envelope by the food-intake system or *gullet* (described below) the cell is entirely bounded by its elaborately patterned pellicle system. Its chief organellar components have already been described as kinetosomal territories of the pellicle (or ciliary corpuscles) and as trichocysts. These pack together in the regular arrays represented diagrammatically in Fig. 94.

a. Mosaics, kineties and paratenes. The entire pellicle may be regarded as a mosaic of kinetosomal territories (26, 70), each with one or two kinetosomes as its morphogenetic center. Under the light microscope, the morphogenetic centers are especially conspicuous (Figs. 95–97) and in some light microscope preparations and thick electron microscope sections the territorial boundaries as well (Figs. 97–99). Single kinetosomes and hexagonal packing are common on the dorsal surface (Figs. 96 left, 99); double kinetosomes and territories shaped like lozenges or parallelograms are common on the ventral surface, especially in the cell's left anterior field, and on the inward sloping central fields—the vestibular region of the pellicle—that descend into the gullet opening (Fig. 96 right figure, cell's left = LFT; Fig. 97, right side of figure, cell's left = LFT; and Fig. 100, in which the cell is viewed *from within*, so that the lozenges are mirror-images of those seen in Fig. 97LFT). Underlying the mosaic (and *viewed from within* Figs. 101, 102, 104), the kinetodesmal fibril from a given territory extends to the right laterally a short distance and then anteriorly for a distance of several territories, usually overlapping but sometimes spiralling with the anteriorly neighboring kinetodesmal fibrils, to form a longitudinally oriented bundle of fibrils called a *kinetodesmos*. The many bundles of *kinetodesma* (3 are shown in a cross-section of 3 ridges, Fig. 107) around a *Paramecium* form longitudinally oriented patterns of linked territories known as *kineties*. In earlier usage (58) the term *kinety* focussed upon the fibrous kinetodesma; currently it is extended to mean a column of territories. Depending upon the species, a *Paramecium* may have from roughly 60 to over 100 kineties in its pellicle (cf. Figs. 95, 96). In the cell's left anterior field, the kineties assume a sharp arc of curvature, and strike the right field (to form the stripe-contrast zone or anterior suture) at an angle of about 45°. Four kineties from this region are seen in Fig. 100 (viewed from within). Another pattern running across the kineties, but at variable angles (approaching 60° in the hexagons of the dorsal surface) as one proceeds around the cell, is definable as the sets of adjacent kinetosomal territories located roughly parallel to the path of the future fission furrow: these sets are called *paratenes* (20), 2 of which cross Fig. 100 from lower left to right. Precise mappings (corticotyping) (62) of the pellicle's mosaics, kineties and paratenes are now within practical reach through scanning electron microscopy (Figs. 114, 115) but remain to be done.

b. Infraciliary lattice. The bewildering patterns of the infraciliary lattice seen by light microscopists to underlie and apparently to cross the territorial boundaries [Fig. 103, (38)], are also intimately associated with the territorial units. In ethanol-isolated pellicles, the fibrillar elements of the infraciliary lattice discussed earlier (46) appear attached to the epiplasm underlying the alveolus of a ciliary corpuscle (Fig. 105). A rare surface slice through several territories at just this level is seen in Fig. 106A; in a slightly more superficial section, the lattice-like quality of the pattern is lost almost entirely, and the fibrillar elements appear (deceptively) bundle-like (Fig. 108). The presence of cartwheels and of postciliary ribbons, emerging at the number 5 triplet, is sometimes seen in sections that include the lattice (Figs. 106B, 109).

Allen (2) using the kinetosome as a sort of depth indicator, draws the conclusion that there are two distinct and separate systems of microfibrils below the granulo-fibrillar meshwork within the crests of the polygonal ridges: the "deeper" of these (i.e., in a plane passing through the proximal half of the kinetosome) is the infraciliary lattice and it appears as a branching system of fibers (Fig. 106A) distinct from the more "superficial" system of striated bands (Figs. 106B). However, the kinetosome is a valid depth indicator only in pellicular regions in which the alveoli surrounding a kinetosome assume the more conventional dihedral (V-shaped) angle to it (as seen in cross-section); under those circumstances, planar sections will always show alveolar-associated fibrils (46) at levels "superficial" to the "deeply rooted" kinetosome. If, however, the alveolar spaces surrounding a kinetosome assume the rarer cathedral (inverted V) configuration, then in planar sections the peripheral borders of alveoli will appear "deeper" than their centrally located kinetosomes; contiguous edges of these alveoli would thus form hexagonal patterns of the "infraciliary lattice" rarely seen in electron microscope sections (2, 106). If we further assume that alveoli in the less conventional cathedral configuration have pleated bottoms (rather like the pleated folds of a cup-cake container) then Hufnagel's complex network of fine filaments, fibrils, and fibers (46) distributed somewhat evenly over their lower surfaces would appear as striated bands in planar sections. Allen's Figure 15 (2) had been presented as cross-sectional evidence to support his conclusion for separate levels and distinct systems of microfibrils. Analysis of the section reveals it to be diagonally longitudinal (as if through view E arrow, our Fig. 94). This results in sections through a posterior kinetosome and a parasomal sac of the right posterior ciliary corpuscle, and *across* alveolar epiplasm; and through the line of contiguity between alveoli of the left anterior ciliary corpuscle so close to the ascending edge of the alveolar epiplasm that the section of it is more "planar" than "cross"; in effect, somewhat like the left portion of our section D–D, Fig. 94.

It is beyond the scope of this paper to present yet another historical review of the indispensable contributions of light microscopy and early electron microscopy to our present perspective. For this, the reader should consult the paper of Ehret and Powers (26), of Pitelka (67), of Párducz (65), and especially of Allen (2).

     c. Elongation and multiplication of territories. Continuing Dippell's view (12, 13, 15) of new kinetosome development to the organelle complex level (in the mosaic of kinetosomal territories), each territory increases in size by elongation while the new structures form within; membranes form transversely across the elongated units, and each old territory "divides" into one or more units. Such a process of partitioning is beautifully shown in Fig. 109; this is in an area of the pellicle originally having 2 cilia per unit territory. In the non-partitioned region if we number the kineties 1 to 5 (left to right), then 3 cilia are now seen in each unit of kinety 5; one may infer from the deeper region of this tangential section at the lower left of the figure that each of the kinetosomes of these 3 has associated with it a newer kinetosome which has not yet grown its cilium (see especially kineties 1–3). Thus, where the original mosaic earlier had a single 2-kinetosomed unit, we now see three 2-kinetosomed units; in other words, the kineties have elongated 3-fold *in this region* and the paratenes have multiplied in number 3-fold.

     d. Suprakinetodesmal microtubules. During fission, and shortly thereafter ($\approx$ 15 minutes) new ribbons of microtubules appear at the apexes of the longitudinal ridges through which the bundles of kinetodesmal fibrils pass (21, 53): these have been termed *suprakinetodesmal microtubules* [Figs. 110–113, (21)]. In ridges where the supra-kinetodesmal microtubular ribbons are flattened, the ribbons contain 5–7 microtubules (Figs. 110, 112); in other ridges where the ribbons assume a jelly-roll configuration in cross-section from 20 to 28 microtubules can be seen (Fig. 113). Suprakinetodesmal microtubules are only seen in ridges containing few kinetodesmal fibrils in a bundle (0–3), whereas the interphase cell has 4 to 7 fibrils in a bundle. From the above relations one may speculate, that these are the best candidates to appear to date as possible progenitor elements for new kinetodesmal fibrils.

e. Vestibular region of the pellicle system: an introduction to the gullet. The funnel-shaped vestibular region is especially well illustrated in the scanning electron micro-graphs of Small (87). The segment of nearly circular arcs assumed by kineties from the left anterior field, and their relations to the region of stripe contrast (anterior suture) are especially clear (Fig. 114). At a higher magnification (Fig. 115), the paths of kineties are even more evident (ky→→ky) aided by the presence of paired cilia present along each kinety axis; crossing the kineties, *paratenes* are separated by conspicuously ray-like interparatenal ridges (par→→par). Extending along the suture line, bundles of kinetodesmal fibrils terminate regularly, with fibrillar elements of right and left fields playing criss-cross [Fig. 116, (67, 68)]. The scanning electron micrographs reveal no elements of the gullet system *within* the opaque cell but suggest several dozen cilia of the thousand that reside within its endoral and penicular complexes. On the other hand, in silver stained light microscope preparations, the relation of the finger-like gullet invagination to the ray-like paratenes is transparently evident (Figs. 95, 117).

2. The gullet system of organelles

a. An invagination of the cell envelope. The pellicle system provides a nearly complete wrapper for the cell, only to terminate abruptly in the oval shaped terminus of the vestibulum. At the perimeter of this gaping hole in the surface (the buccal overture of the gullet system) a new system of organelle complexes begins. The gullet provides a food-intake pathway through a long tube 10 μm in diameter lined by nearly a thousand cilia, and composed of kinetosomal, microtubular and fibrillar elements, generally, more densely packed than comparable structures in the pellicle. Finally, unlike the pellicle, the walls of the gullet are not reinforced by trichocysts (26).

b. Complexes of the gullet system in the region of the buccal overture (p1, p2, p3, ribbed wall and endoral ridge). A cross-sectional view of the whole cell in the region of the gullet's opening to the outside (the buccal overture) illustrates clearly its relationship to the pellicle system, and its role as part of the cell envelope (Fig. 118). The cell is viewed from its inside anteriorly: the upper part is *its* right, the lower part *its* left-hand side. Reading clockwise from 6 o'clock to 12 o'clock along the shoreline of the vestibular bay, the gullet is seen to consist of 3 complexes of ciliated organelles, p1, p2 and p3, followed by a non-ciliated ribbed wall, and an endoral ridge containing 2 columns of kinetosomes, only 1 of which is ciliated. Each ciliated complex (some-times called *peniculus*, latin for "brush") has 4 cilia abreast, in columns that run in a dextral spiral nearly the length of the gullet tube ($\sim$ 30–40 μm). The cilia of p1 and p2 are closely packed; those of p3 are more widely spaced, showing a 1-1-2 pattern in

References p. 286–290

the buccal overture region. Posteriorly (and embryonically) p3 (formerly called the *quadrulus*) appears exactly like p1 and p2. There are about 90 p3 complexes in *P. bursaria*, somewhat fewer p2's and fewest p1's; the total number of penicular kinetosomes that compose the gullet wall is thus nearly 1000. Posteriorly a naked ribbed wall (devoid of alveoli, but saw-toothed in appearance) appears between p3 and the ribbed wall.

Figs. 119 and 120 are optical sections *of the same cell* seen in Fig. 117. The gullet's relation to the arcs and rays of vestibular kineties and paratenes is again evident; the S-shape (Figs. 117, 119) is seen to have its substance in the dextral spiral of p3 (Fig. 120). Kinetosomes of p1 and p2 are more difficult to resolve, but the foreshortening of p1 can also be seen in Fig. 120; p1, p2 and p3 are all resolved in Fig. 122, reading from right to left. Now, imagine the gullet of Fig. 122 to be shaped like the left half of a pea-pod opened along its "hinge-line" so that we may peer into its left inner wall, and that Fig. 121 represents the *right* side of the pod, a view of the inner right wall (the white line separating Figs. 121, 122 representing the "hinge line" in our "dissection"). This, then, is the ribbed wall of the gullet seen under phase microscopy in a freshly compressed, unfixed section. Note well its periodicities, and the absence of any fibers extending away from it (although penicular cilia are easily resolvable around the buccal overture). The 8 sets of cilia of p1 and p2 are also visible, using ultraviolet microscopy in unstained sections (Figs. 123, 124) prior to viewing neighboring sections to these under the electron microscope (Fig. 125). Reading from top to bottom, Fig. 126 shows ciliated vestibular kinetosomes of the pellicle system on the cell's right side (top), a few aciliated kinetosomes of the prepenicular zone (ppz) (23), the endoral kinetosomes and ridge, the ribbed wall, p3, p2, p1 and the ciliated vestibular kinetosomes of the cell's left side. At higher resolution, one sees the microfibrillar reticulum (68, 82, 83) underlying the ribbed wall, the ribbed wall alveoli, and the parasomal sacs surrounding p1 and p2 and on the right of p3 (Figs. 127, 134, 135). Only the outer kinetosomes (more distal from the ridge) of the 2 columns of kinetosomes of the endoral ridge are ciliated (Fig. 128); nodes and filaments are seen to criss-cross in the microfibrillar reticulum "which embraces the entire buccal cavity, except for a dorsal zone comprising the cytostome and its lips" (Figs. 129, 132) (68) in a periodic pattern closely coupled to the ectoplasmic ribs of the ribbed wall. Raffinose-isolated preparations of the ribbed wall show it to consist of 0.5 μm wide ribs, which reveal a cross-striated appearance and contain fibrous knots of material (46). Underlying the ribs, bundles of hexagonally packed microtubules, nemadesma (68) are seen (Figs. 129, 130, 131, 139); these structures, taken altogether (ribs, reticulum and bundles), account for the gross integrity of the isolated ribbed wall and its inherent periodicities as seen with the phase contrast microscope (Fig. 121) (25, 26, 46).

c. Complexes of the gullet system in the region of the buccal cavity. In sections taken more posteriorly than those of Figs. 126–131, several changes become evident. Most strikingly, the C-shaped buccal overture becomes an O-shaped buccal cavity. At first, most anteriorly (Fig. 83), the closure is accomplished on the ventral surface of the cylindrical buccal cavity *by pellicular material* of the posterior vestibular slope (V, Fig. 83; Fig. 147), including some barren kinetosomes of the prepenicular zone (ppz), as well as kinetosomal territories of the most anterior region of the left posterior field. More posteriorly (Fig. 143), the closure is accomplished by material of the gullet itself, in the form of an extension of the microfibrillar reticulum and an over-lying alveolar surface. This *alveolar ventral wall* (avw, Figs. 143, 147) is nearly as broad anteriorly as the width of the buccal overture, but rapidly diminishes in expanse posteriorly as the distance narrows between the right rib of the ribbed wall and the parasomal sac of the left edge of p1 (centrally) or of p2 (posteriorly). Anteriorly in the region of the buccal overture, kinetosomes of p3 are broadly separated from one another by distances of 1 micron or more, giving the 1-1-2 pattern already mentioned; posteriorly in the buccal cavity, they are as close to one another ($\sim$ 0.75 μm center to center) as the kinetosomes of p1 and p2 (Figs. 133, 134); the gap between p2 and p3 continues to close more posteriorly, until p2 is separated from p3 only by the space of a parasomal sac (Figs. 132, 135). At this level, p1 is absent entirely, but its extent and foreshortening have already been referred to in the light and phase contrast micrographs of whole gullets in Figs. 120 and 122. The nearly hexagonal patterns of close-packing of *mature* penicular kinetosomes are also evident in Fig. 135, in striking contrast to the rectangular packing that will be seen in *prepeniculi* of the developing gullet. In raffinose-isolated preparations, Hufnagel (46) has shown posteriorly directed bundles of fibers, consisting of fibers originating from the penicular kinetosomes as well as from the left vestibular kineties; the appearance of these in sectioned material is shown in Figs. 136 and 137. Additional fibrous links between penicular kinetosomes, extending in several directions and at different levels, are also seen (Figs. 135, 136, 137) (40, 64).

d. Microtubular and vesicular elements: the naked dorsal wall. Also posterior to the buccal overture and composing the dorsal wall of the cavity (Fig. 118) the naked dorsal wall is seen. It consists of a strikingly distinct type of surface organization, devoid of alveoli and epiplasm, and bounded by only a single unit membrane (53, 68). It lies as a triangular surface between p3 and the ribbed wall (Figs. 118, 133, 134, 138, 139), is rather narrow at its anterior apex (on the dorsal buccal wall, Fig. 118), and broadens posteriorly, separating p3 from the ribbed wall by a broad strip. The naked dorsal wall (ndw, Fig. 118) consists of three components: (1) a right dorsal groove subtended by narrow ribbons of microtubules, (2) a left sawtooth band (or naked ribbed wall) characterized by ridges, with broad and narrow ribbons of microtubules and (3) the cytostomal area, an irregular ameboid surface between (1) and (2) and within which food vacuoles form (68); for this reason the right groove and left sawtooth band are also called the right lip and left lip of the cytostome (68). Sections of the cytostome are seen in Figs. 139 and 150. The right lip (or right dorsal groove) is relatively long and narrow, and marks the dorsal edge of the ribbed wall (r1, Figs. 126, 127, 139, 150). The left lip is between p3 and the right lip and forms the bulk of the naked dorsal wall (Figs. 133, 134, 138, 139, 150, 151). It consists of as many as 24 sawtooth ridges (68), each ridge containing parallel broad and narrow ribbons of microtubules. The broad ribbons consists of 10 or 15 microtubules, the narrow ribbons of 2 or 3 (Figs. 133, 134, 138), the longest of these extend into the cytoplasm for 3–6 μm (Fig. 139). Scattered between the ribbons of microtubules are numerous disc-shaped bodies 0.3 μm in diameter and about 60 nm thick (Figs. 138, 139) (53). According to Pitelka (68), the microtubules of the naked ribbed wall appear to arise as radial ribbons from kinetosomes of the nearest row of p3.

e. Development of the new gullet. The new gullet system is formed by means of a developmental process that takes place in the prepenicular zone [ppz, Figs. 118, 147 (23)] the "anarchic field" of Hufnagel [Fig. 140 (46)]. The non-ciliated kinetosomes of this zone appear to organize into prepeniculi (pp1, pp2 and pp3) entirely outside of the endoral ridge of the old gullet (Fig. 141). Figs. 142 and 143 are sections located posteriorly in the same cell, showing that invagination commences anteriorly (Fig. 141), and that the prepenicular ribbons (pp1, pp2, pp3) are entirely outside of the old gullet in posterior sections (Fig. 143); in this condition, the cell's swimming is noticeably influenced by the presence of the ciliature of the new gullet at the pellicular surface. At a slightly later stage, each peniculus is equipped with 4 kinetosomes, but only 2 bear cilia (Fig. 144), the younger nonciliated kinetosomes are always located on the left of the older ciliated ones (i.e., in the direction of the ribbed wall). The precise staging of cells in this study of *P. bursaria* (23) allows estimates of the growth of penicular cilia to be about 10 minutes, or at a rate roughly equivalent to that for wool-keratin in sheep of about 170 Å per second. At this stage of development, and slightly later (Fig. 145), the 3 prepeniculi are similar, and the kinetosomes of each are rectangularly packed; at maturity p3 will show the 1-1-2 pattern, and hexagonal packing will dominate the kinetosomes of p1 and p2. In *P. bursaria*, macronuclear and micronuclear events are closely coupled to stages in gullet development (Fig. 147). Just before proliferation of ppz kinetosomes, in State I of the macronucleus, nucleoli are in grape-like clusters, and the surrounding matrix appears coarse-textured. About 20 minutes before cell division, at macronuclear State II, the nucleoli appear unbunched (single) and the karyoplasmic matrix appears fine-textured (owing to the diminished average size of the chromatin bodies); immature gullets of the ribbon stage (Figs. 144–145) are always observed in cells whose macronuclei are State II. In State III, about 10 minutes later, the nucleoli are single or in small groups, and the karyoplasmic matrix is again like that of State I. At this time, the peniculi have already begun to separate (p3, from p1 and p2) and a fission-furrow is evident. The micronucleus has virtually completed its division, is in late anaphase, but is still connected to its fissiont micronucleus by a spindle bundle (arrow, Fig. 146); it then moves to its new normal position in the opisthe (posterior fissiont cell) along with the new gullet. The old gullet remains with the proter, or anterior fissiont. A more extensive view of gullet development, including subtle differences between *P. bursaria* and *P. aurelia* and *P. trichium* is given elsewhere in this volume (55) and in the papers of Hanson and coworkers cited therein.

In a tracer study designed to appraise the molecular stability of the organelles of the old and new gullets, cells of known age (10 minutes) were grown in medium containing tritiated leucine for a single generation interval (12 hours). When these cells divided, their proters and opisthes were identified and transferred into medium containing unlabeled leucine as chaser, where they remained 2 hours. Each cell was then embedded and sectioned by the single-cell technique (41). Figs. 148 and 149 are autoradiographs of the gullets of proter and opisthe, respectively. It is clear that in these f_1 cells the gullet cilia of each fissiont are labeled. The axonemes and the kineto-somes of the opisthe, however, are about twice as heavily labeled as those of the proter. A statistical study (21) of nearly 10,000 such axonemal cross-sections on 100 cell sections shows, for example, 0.19 ± 0.008 grains per axoneme on proter cilia (over 14 hours old) and 0.40 ± 0.031 grains per axoneme on opisthe cilia (less than 2 hours old). Thus, one may conclude that about half of the newly synthesized protein of the ciliature of the opisthe's (new) gullet, and all labeled protein of the proter's (old) gullet is transient, nonstructural and relatively short-lived. Kinetic studies of this sort are badly needed to establish the decay constants of transient proteins, and to appraise the degree to which "structural" proteins may be conserved.

f. Food vacuole formation, digestion and expulsion. At the posterior end of the gullet on its naked dorsal wall, food vacuoles form a circular opening into the cytoplasm $\sim 4\ \mu$m in diameter, bounded on the right by elements of the ribbed wall and the right lip of the cytostome, and on the left by microtubular and disc-like elements of the left lip (Figs. 150, 151). In older food vacuoles, the vacuole membrane becomes extremely irregular in shape, and numerous pinocytotic vesicles appear (Fig. 152) (50). Surrounding the food vacuoles (Figs. 150–152) are small "enzyme granules" (83) that appear to be in some way involved with digestion. Short-lived enzymes within older food vacuoles are also shown by a tracer experiment (Figs. 153, 154). In this case, a single cell of *P. bursaria* was exposed to tritiated leucine in *bacteria-free* medium for 1 hour, was then washed free of the isotope, and was transferred to unlabeled bacterized medium for 5 hours. Figs. 153 and 154 are representative regions *of the same cell* (27). Nuclei and cytoplasm appear randomly labeled; the new food vacuole (smoother surface, Fig. 153) contains unlabeled *intact* bacteria; the old food vacuole (Fig. 154) shows bacteria in different stages of digestion, their cell membranes expanded and lifted away from the bacterial cytoplasm. The many silver grains covering these bacteria represent tritium-labeled *Paramecium* enzyme attached to the bacteria of the *older* vacuoles only, and newly synthesized within the past 6 hours. In another ultrastructural study (31) the relations of acid phosphatase activity to the digestive process in *P. caudatum* were investigated cyto-chemically using the Gomori lead-phosphate method. Positive reactions occurred in older food vacuoles only; the enzyme may originate in sacculated endoplasmic reticulum (interpreted to be Golgi material).

Older vacuoles contain fine granular material and "ghosts" of bacterial cell walls (50); these come to the vicinity of the cytoproct, where they are extruded from the cell (Figs. 155–158). The cytoproct (actually a special differentiation of the *pellicle system*) is found to border the terminal kinetosomal territories of about 15 paratenes along 1 kinety of the right posterior field a distance of about 14 μm along the postoral suture (Fig. 158). Although the precise location and length of the cytoproct may be variable, observations on its capacity to regenerate in cut cells (8, 9) are consistent with Sonneborn's conclusion that it is induced by an interaction between right and left posterior ventral fields (95). Its wavy ridge-like edge is slightly reminiscent of the endoral ridge of the gullet; in electron microscopic sections, this *endoproctal ridge* is also seen to contain 2 non-ciliated kinetosomes (Figs. 156, 157). The 14 μm long canyon into the cell between the endoproctal ridge and the corresponding territories of the *left* posterior field is frequently lined by a spongy reticulum, similar to that seen surrounding food vacuoles at advanced stages of pinocytosis (cf. Figs. 152, 157). Other representations of cytoproct structure and function (31), however, may present a structure somewhat like that of the contractile vacuole.

3. The contractile vacuole system

Each of the 2 contractile vacuoles (Figs. 159–164) occupies a position on the mid-dorsal surface of the pellicle system somewhat relative to that of the axes of an ellipse (Fig. 165). The 2 are generally found to open near or upon the same kinety through a discharge channel about the size of a single kinetosomal territory (Figs. 159, 162, 164, 165). Ribbons of microtubules, running in 2 directions, surround the walls of the discharge channel (Figs. 159, 162) (53, 81). The main body of the contractile vacuole (Figs. 162, 163) is fed by 6 to 10 nephridial canals that surround it radially (Fig. 147); nephridial canals are pentagonal in cross-section (Figs. 160, 161). These canals are surrounded by narrow ribbons of 3 or 4 microtubules on each side (Fig. 161) and by an extensive spongelike network of tubules, the nephridial tubules; the latter is frequently surrounded (especially conspicuous in pre-dividers) by a more regular network of tubules, the orthotubular system (Figs. 160, 161) (23). The nephridial canals show numerous connections with the inner lumens of the spongelike nephridial tubules (Figs. 160, 161). Each canal then enters a wider ampulla (Figs. 159, 162, 163, 164) before entering the main vacuole (Fig. 147).

Concluding our hierarchical approach to the cell, the contractile vacuole presents a small problem in categories: it may be thought of as *an organelle system* composed of 2 relatively simple complexes of organelles (the anterior and posterior vacuoles); each of these, at its discharge channel, occupies a single kinetosomal territory, and, *in toto*, may represent at most one to several organelles conjoined. Apart from this flimsy bit of comparative anatomy, it may on the other hand be "more correct" to think of each separate contractile vacuole as *an organelle per se*, built up of sub-organellar tubules, microtubules and membranes, and as distinctive in this sense as "mitochondrion", "kinetosome" or "trichocyst". It is perhaps rather late, but still too early to discuss the criteria for resolving such ultimately useful and subtle distinctions (18, 19).

IV. THE CELL LEVEL

A. Recapitulation

One can arrive at another understanding of the structure of *Paramecium*, our subject of long-time interest and study, by pretending to build it up, level by level, from its constituent building blocks (as if the whole *is* the sum of its parts... if one only could know all of the connectives!). Perhaps this is a futile exercise when one reflects upon the synthetic approaches to a "more likely" subject, such as bacteriophage T_4 (17, 56). Optimism about utility can at once be restored for those investigators whose intellectual curiosity requires them to employ subjects that, even if they do not *epitomize*, at least *include* the bizarre diversity of the eukaryotes, with their large multirepliconic genomes, nuclear and cytoplasmic envelopes, kinetosomes and cilia, and their related organelle complexes and derivatives. Such phenomena can only be studied at or above their lowest level of manifestation, and frequently one finds *Paramecium* a candidate with credentials that make it choice.

Considering the size of its genome, and the number of proteins it can code for, the number listed here, as identified to date, is scandalously small (1: Trichynin)! Still, the coupling of the discovery of trichocyst mutants (72) at the same time to the methods of molecular genetics and of ultrastructural cytology offers great promise. Continuing the orientation on the consequences of molecular genetics and a large multirepliconic genome, it is important to note that our considerations of microtubules, fibers, mitochondria, kinetosomal territories and trichocysts, and the complexes and systems of these elements, have been guided by *gross* structural criteria. An evident challenge is to appraise and map the degree to which each "set" of elements at each level of organization is genetically homogeneous or heterogeneous.

This approach to structure gives a fresh view to "function" [forseen by D'Arcy Thompson (100)]; or rather than say "function" (which in classical cytology has usually appeared as a form of teleological crystal-ball gazing), one might ask instead, "what are the *consequences* of this structure to the cell?" Viewed in this way, irrespective of their "function", *trichocysts*, for example, have the unmistakable consequence of fitting into the ordered pattern of kinetosomal territories in such a way as to enormously thicken and strengthen the pellicle system of the cell envelope, without imposing excessive rigidity upon the structure, much in the manner of the trusswork of a geodesic dome (19, 37, 53).

B. Paramecium as a Free-Living Eukaryotic Cell

The initial reactions to electron micrographs of *Paramecium* by physicists and other laymen is commonly something like "how unexpectedly complex!", where "complex" in such context usually means something like "mysteriously incomprehensible". The cell *is* complex, but attractively and somewhat comprehensibly so: built up of defineable subunits. In the words of Pitelka (70), "only *Paramecium*, the all-purpose ciliate, has provided sufficient details of ... fibrillar structures to yield a reasonably

References p. 286–290

clear picture of their architecture ...". She then goes on to enumerate the critical pieces of evidence still missing.

And so here in our last plate (Fig. 166) viewed by the promising new technique of scanning electron microscopy (87, 90), we see our hero subject *Paramecium au naturel*, in surprisingly furry ciliated glory. Will we really be able to understand him so well as to be able to build him up out of bits and pieces someday? In a small but very real way, we have already begun (6, 95, 96).

V. ACKNOWLEDGEMENTS

We are grateful to our many colleagues present and past who have viewed this subject *Paramecium* like blind men seeing the elephant so ingeniously and from so many technologically difficult but scientifically useful vantage points. In *this* paper, with the help of our friends, we have tried to see our subject from every vantage point employed to date. Especially we thank Richard Allen, Gregory Antipa, L. H. Bannister, Audrey Barnett, Tze-Tuan Chen, Ruth V. Dippell, Gérard De Haller, Linda Hufnagel, Fumie Inaba, Dorothy Pitelka, E. Lawrence Powers, Frederick L. Schuster, Eugene B. Small, Tracy M. Sonneborn, Edward Steers, George Svihla, Sidney L. Tamm, Harry Wessenberg and Arlene Zadylak Dobra for providing us so generously with the "best of *Paramecium*", and for granting permission to use material, in many cases unpublished. For their scholarly volume on *Paramecium* anatomy which has served us as an indispensable guide, we are indebted to Artur Jurand and G. G. Selman. New drawings were done by John Mitchell of the Argonne Graphic Arts Department, and plates and prints were prepared by Rita Januszyk and Arlene Zadylak Dobra.

VI. REFERENCES

(Abstracts are indicated by *)

1   Allen, R. D. 1967. Fine structure, reconstruction and possible functions of components of the cortex of *Tetrahymena pyriformis*. *J. Protozool*. **14**, 553–565.
2   — 1971. Fine structure of membranous and microfibrillar systems in the cortex of *Paramecium caudatum*. *J. Cell Biol*. **49**, 1–20.
3   Ammermann, D. 1965. Cytologische und genetische Untersuchungen an dem Ciliaten *Stylonychia mytilus* Ehrenberg. *Arch. Protistenk*. **108**, 109–152.
4   Antipa, G. 1971. Structural differentiation in the somatic cortex of a ciliated protozoan, *Conchophthirus curtus* Engelmann 1862. *Protistologica* **7**, 471–501.
5   Beale, G. H., Jurand, A. & Preer, J. R. 1969. The classes of endosymbionts of *Paramecium aurelia*. *J. Cell Sci*. **5**, 65–91.
6   Beisson, J. & Sonneborn, T. M. 1965. Cytoplasmic inheritance of the organization of the cell cortex in *Paramecium aurelia*. *Proc. Nat. Acad. Sci. U.S.* **53**, 275–282.
7   Chen, T. T. 1957. Unpublished.
8   Chen-Shan, L. 1969. Cortical morphogenesis in *Paramecium aurelia* following amputation of the posterior region. *J. Exp. Zool*. **170**, 205–228.
9   — 1970. Cortical morphogenesis in *Paramecium aurelia* following amputation of the anterior region. *J. Exp. Zool*. **174**, 463–478.
10 Dippell, R. V. 1958. The fine structure of kappa in killer stock 51 of *Paramecium aurelia*. Preliminary observations. *J. Biophys. Biochem. Cytol*. **4**, 125–128.

11 *— 1962. The site of silver impregnation in *Paramecium aurelia*. *J. Protozool.* **9** (Suppl.), 24.

12 *— 1964. Perpetuation of cortical structure and pattern in *P. aurelia*. *Proc. 11th Int. Congr. Cell. Biol.* **77**, 16–17.

13 *— 1965. Reproduction of surface structure in *Paramecium*. In *Progress in Protozoology* (Proc. 2nd Int. Conf. Protozool., London), 65.

14 — 1968. The development of basal bodies in *Paramecium*. *Proc. Nat. Acad. Sci. U.S.* **61**, 461–468.

15 — 1970. Personal communication.

16 *— & Sinton, S. E. 1963. Localization of macronuclear DNA and RNA in *Paramecium aurelia*. *J. Protozool.* **10** (Suppl.), 22–23.

17 Edgar, R. S. 1969. The genome of bacteriophage T$_4$. *The Harvey Lectures* **63**, 263–281.

18 Ehret, C. F. 1958. Information content and biotopology of the cell in terms of cell organelles. In Yockey, H. P., Platzman, R. L. & Quastler, H., *Symposium on Information Theory in Biology*, Pergamon Press, New York, N.Y., 218–229.

19 — 1960. Organelle systems and biological organization. Structural and developmental evidence leads to a new look at our concepts of biological organization. *Science* **132**, 115–123.

20 — 1967. Paratene theory of the shapes of cells. *J. Theor. Biol.* **15**, 263–272.

21 —, Alblinger, J. & Savage, N. 1964. Developmental and ultrastructural studies of cell organelles. *Argonne Nat. Lab. Biol. Med. Res. Div. Ann. Rep.* 6971, 62–70.

22 *— & Haller, G. de. 1961. Formation des organelles et des systèmes d'organelles cytoplasmique au cours de la division chez *Paramecium*. In *Progress in Protozoology* (Proc. 1st Int. Conf. Protozool., Prague), 419–420.

23 — & — 1963. Origin, development and maturation of organelles and organelle systems of the cell surface in *Paramecium*. *J. Ultrastruct. Res.* **6** (Suppl.), 3–42.

24 — & Powers, E. L. 1955. Macronuclear and nucleolar development in *Paramecium bursaria*. *Exp. Cell Res.* **9**, 241–257.

25 — & — 1957. The organization of gullet organelles in *Paramecium bursaria*. *J. Protozool.* **4**, 55–59.

26 — & — 1959. The cell surface of *Paramecium*. *Intern. Rev. Cytol.* **8**, 97–133.

27 —, Savage, N. & Alblinger, J. 1964. Patterns of segregation of structural elements during cell division. *Z. Zellforsch. Mikrosk. Anat.* **64**, 129–139.

28 —, — & Schuster, F. L. 1965. The incorporation into cytoplasmic organelles of tritium given in thymidine. *Argonne Nat. Lab. Biol. Med. Res. Div. Ann. Rep.* 7136, 205–210.

29 —, — & — 1966. Incorporation into cell organelles of tritium given in leucine and thymidine. *Amer. Zoologist* **6**, 3.

30 — & Wille, J. 1970. The photobiology of circadian rhythms in protozoa and other eukaryotic microorganisms. In Halldal, P., *Photobiology of Microorganisms*, Wiley, London, Ch. 13, 369–416.

31 Estève, J. C. 1969. Observation ultrastructurale du cytopyge de *Paramecium caudatum*. *Compt. Rend.* **268**, 1508–1510.

32 — 1970. Distribution of acid phosphatase in *Paramecium caudatum*. Its relations with the process of digestion. *J. Protozool.* **17**, 24–35.

33 Fauré-Fremiet, E. 1970. Microtubules et mécanisms morphopoiétiques. *Année Biol. (France)* **9**, 1–61.

34 Fernández-Morán, H., Oda, T., Blair, P. V. & Green, D. E. 1964. A macromolecular repeating unit of mitochondrial structure and function. Correlated electron microscopic and biochemical studies of isolated mitochondria and submitochondrial particles of beef heart muscle. *J. Cell Biol.* **22**, 63–100.

35 Freifelder, D. & Kleinschmidt, A. K. 1965. Single-strand breaks in duplex DNA of coliphage T$_7$ as demonstrated by electron microscopy. *J. Mol. Biol.* **14**, 271–278.

36 Frenster, J. H. 1969. Biochemistry and molecular biophysics of heterochromatin and euchromatin. In Lima-de-Faria, A., *Handbook of Molecular Cytology*, North-Holland Publishing Company, Amsterdam, Ch. 12, 251–276.

37 Fuller, R. B. 1959. Dome structures: A philosophy of space and shape. *Consulting Eng.* **13**, 90–94.

38 Gelei, G. von. 1937. Ein neues Fibrillensystem im Ectoplasma von *Paramecium*. *Arch. Protistenk.* **89**, 133–162.

39 Gibbons, I. R. 1968. The structure and composition of cilia. *Symp. Int. Soc. Cell Biol.* **6**, 99–113.

40 Grain, J. 1969. Le cinétosome et ses dérives chez les ciliés. *Année Biol. (France)* **8**, 53–97.

41 Haller, G. de, Ehret, C. F. & Naef, R. 1961. Technique d'inclusion et d'ultramicrotomie destinée à l'étude du développement des organelles dans une cellule isolée. *Experientia* **17**, 524–526.

42 — & Heggler, B. ten. 1968. Détails de l'ultrastructure des ciliés. I. Fibres karyophores chez *Paramecium bursaria. Compt. Rend. Soc. Phys. Hist. Natur. Genève* **3**, 31–32.

43 — & — 1969. Morphogénèse expérimentale chez les ciliés. III. Effet d'une irradiation U.V. sur la génèse des trichocystes chez *Paramecium aurelia. Protistologica* **5**, 115–120.

44 Hruban, Z. & Rechcigl, Jr., M. 1969. Microbodies and related particles. Morphology, biochemistry and physiology. *Intern. Rev. Cytol.*, Suppl. 1, Academic Press, New York, N.Y.

45 Hufnagel, L. 1969. Properties of DNA associated with raffinose-isolated pellicles of *Paramecium aurelia. J. Cell Sci.* **5**, 561–573.

46 — 1969. Cortical ultrastructure of *Paramecium aurelia. J. Cell Biol.* **40**, 779–801.

47 Inaba, F., Imamoto, K. & Suganuma, Y. 1966. Electron microscope observations on nuclear exchange during conjugation in *Paramecium multimicronucleatum. Proc. Japan Acad.* **42**, 394–398.

48 Jakus, M. A. 1945. The structure and properties of the trichocysts of *Paramecium. J. Exp. Zool.* **100**, 457–485.

49 Jankowski, A. V. 1966. The conjugation processes in *Paramecium putrinum* Clap et Lachm. IX. On "necrochromatin" and the functional significance of macronuclear fragmentation (in Russian with English summary). *TSitologiya* **8**, 725–735.

50 Jurand, A. 1961. An electron microscope study of food vacuoles in *Paramecium aurelia. J. Protozool.* **8**, 125–130.

51 — & Jacob, J. 1969. Studies on the macronucleus of *Paramecium aurelia.* III. Localization of tritiated thymidine by electron microscope autoradiography. *Chromosoma* **26**, 355–364.

52 — & Preer, L. B. 1968. Ultrastructure of flagellated lambda symbionts in *Paramecium aurelia. J. Gen. Microbiol.* **54**, 359–364.

53 — & Selman, G. G. 1969. *The Anatomy of Paramecium aurelia*, Macmillan, London and St. Martin's Press, New York, N.Y.

54 — & — 1970. Ultrastructure of the nuclei and intranuclear microtubules of *Paramecium aurelia. J. Gen. Microbiol.* **60**, 357–364.

55 Kaneda, M. & Hanson, E. D. 1971. Growth patterns and morphogenetic events in the cell cycle of *Paramecium aurelia.* In Van Wagtendonk, W. J. (Ed.), *Paramecium—A Current Survey*, Elsevier, Amsterdam, Ch. 5, 219–262.

56 Kellenberger, E. 1968. Studies on the morphopoiesis of the head of phage T_4. Ch. 7. Polymorphic assemblies of the same major virus subunit. In: *Nobel Symposium. Symmetry and Function of Biological Systems at the Macromolecular Level*. Wiley, New York, N.Y., 349–366.

57 Kimball, R. F. & Gaither, N. 1951. The influence of light upon the action of ultraviolet on *Paramecium aurelia. J. Cell. Comp. Physiol.* **37**, 211–233.

58 Lwoff, A. 1950. *Problems of Morphogenesis in Ciliates*, Wiley, New York, N.Y.

59 McArdle, E. & Svihla, G. 1970. Unpublished.

60 — & Zadylak, A. 1970. Unpublished.

61 Matsusaka, T. 1969. Influence of extracellular freezing on the ultrastructure of *Paramecium. Low Temp. Sci. Ser. B.* **27**, 67–72.

62 Nanney, D. L. 1968. Cortical patterns in cellular morphogenesis. *Science* **160**, 496–502.

63 Nilsson, J. R. 1970. Macronuclear changes in *Tetrahymena pyriformis* GL. during the cell cycle and in response to alterations in environmental conditions. *Compt. Rend. Trav. Lab. Carlsberg* **37**, 285–300.

64 Noirot-Timothée, C. 1969. Discussion sur le système de fibres tangentielles. *Année Biol. (France)* **8**, 92–95.

65 Párducz, B. 1962. On a new concept of cortical organization in *Paramecium. Acta Biol. Hung.* **13**, 299–322.

66 Pitelka, D. R. 1961. Fine structure of the silverline and fibrillar systems of three tetrahymenid ciliates. *J. Protozool.* **8**, 75–89.

67 — 1965. New observations on cortical ultrastructure in *Paramecium. J. Microscopie* **4**, 373–394.

68 — 1969. Fibrillar systems in protozoa. In Chen, T. T., *Research in Protozoology*, Pergamon Press, Oxford, New York, Ch. 3, 279–388.

69 — 1969. Centriole replication. In Lima-de-Faria, A., *Handbook of Molecular Cytology*, North-Holland Publishing Company, Amsterdam, Ch. 15, 1199–1218.

70   *— 1969. Fibrillar structures of the ciliate cortex: the organization of kinetosomal territories. In *Progress in Protozoology* (Proc. 3rd Int. Congr. Protozool., Leningrad), 44–46.

71   — 1970. Personal communication.

72   *Pollack, S. 1969. Trichocyst mutants in *Paramecium aurelia*. *Amer. Zool.* **9**, 1136–1137.

73   Powers, E. L., Ehret, C. F. & Roth, L. E. 1955. Mitochondrial structure in *Paramecium* as revealed by electron microscopy. *Biol. Bull.* **108**, 182–195.

74   —, —, — & Minick, O. T. 1956. The internal organization of mitochondria. *J. Biochem. Cytol.* **2** (Suppl.), 341–346.

75   Preer, J. R. & Stark, P. 1953. Cytological observations on the cytoplasmic factor "kappa" in *Paramecium aurelia*. *Exp. Cell Res.* **5**, 478–491.

76   Puytorac, P. de. 1970. Definitions of ciliate descriptive terms. *J. Protozool.* **17**, 358.

77   Raikov, I. B. 1968. Macronucleus of ciliates. In Chen, T. T., *Research in Protozoology*, Pergamon Press, Oxford, New York, N.Y., Ch. 3, 1-128.

78   Rao, M. V. N. & Ammermann, D. 1970. Polytene chromosomes and nucleic acid metabolism during macronuclear development in *Euplotes*. *Chromosoma* **29**, 246–254.

79   Ringo, D. 1967. The arrangement of subunits in flagellar fibers. *J. Ultrastruct. Res.* **17**, 266–277.

80   Robertson, J. D. 1959. The ultrastructure of cell membranes and their derivatives. *Biochem. Soc. Symp.* **16**, 3-43.

81   Schneider, L. 1960. Elektronenmikroskopische Untersuchungen über das Nephridialsystem von *Paramecium*. *J. Protozool.* **7**, 75–90.

82   — 1964. Elektronenmikroskopische Untersuchungen an den Ernährungsorganellen von *Paramecium*. I. Der Cytopharynx. *Z. Zellforsch.* **62**, 198–224.

83   — 1964. Elektronenmikroskopische Untersuchungen an den Ernährungsorganellen von *Paramecium*. II. Die Nährungsvakuolen und die Cytopyge. *Z. Zellforsch. Mikrosk. Anat.* **62**, 225–245.

84   Schuster, F. L., Prazak, B. & Ehret, C. F. 1967. Induction of trichocyst discharge in *Paramecium bursaria*. *J. Protozool.* **14**, 483–485.

85   Schwartz, V. 1956. Nukleolenformwechsel und Zyklen der Ribosenucleinsäure in der vegetativen Entwicklung von *Paramecium bursaria*. *Biol. Zentralbl.* **75**, 1–16.

86   Selman, G. G. & Jurand, A. 1970. Trichocyst development during the fission cycle of *Paramecium*. *J. Gen. Microbiol.* **60**, 365–372.

87   Small, E. B. 1970. *Scanning Electron Microscopy of the Protista*, Academic Press, New York, N.Y., In preparation.

88   — 1970. Personal communication.

89   — & Antipa, G. 1970. Unpublished.

90   — & Marszalek, D. 1969. Scanning electron microscopy of fixed, frozen, and dried protozoa. *Science* **163**, 1064–1065.

91   — & — 1970. Unpublished.

92   Smith-Sonneborn, J., Green, L. & Marmur, J. 1963. DNA base composition of kappa and *Paramecium aurelia*, stock 51. *Nature* **197**, 385–387.

93   Sonneborn, T. M. 1945. The dependence of the physiological action of a gene on a primer and the relation of primer to gene. *Amer. Naturalist* **49**, 318–339.

94   — 1961. Kappa particles and their bearing on host-parasite relations. In Pollard, M., *Perspectives in Virology*, Burgess Press, Minneapolis, Minn., Ch. 2, 5–12.

95   — 1963. Does preformed cell structure play an essential role in cell heredity? In Allen, J. M., *The Nature of Biological Diversity*, McGraw-Hill, New York, N.Y., 165–221.

96   — 1970. Gene action in development. *Proc. Roy. Soc. B.* **176**, 347–366.

97   Steers, E., Beisson, J. & Marchesi, V. 1969. A structural protein extracted from the trichocyst of *Paramecium aurelia*. *Exp. Cell Res.* **57**, 392–396.

98   Stevenson, I. & Lloyd, F. P. 1971. Ultrastructure of nuclear division in *Paramecium aurelia*. II. Amitosis of the macronucleus. *Austr. J. Biol. Sci.* **24**, 977–987.

99   — & — 1971. Ultrastructure of nuclear division in *Paramecium aurelia*. I. Mitosis in the micronucleus. *Austr. J. Biol. Sci.* **24**, 963–975.

100 Thompson, D. W. 1942. *On Growth and Form*, Cambridge University Press, New York, N.Y.

101 Van Wagtendonk, W. J., Clark, J. A. D. & Godoy, G. A. 1963. The biological status of lambda and related particles in *Paramecium aurelia*. *Proc. Nat. Acad. Sci. U.S.* **50**, 835–838.

102 *— & Soldo, A. T. 1965. Endosymbiotes of ciliated protozoa. In *Progress in Protozoology* (Proc. 2nd Int. Congr. Protozool., London), 244–245.

103   Vivier, E. 1960. Cycle nucléolaire en rapport avec l'alimentation chez *Paramecium caudatum*. *Compt. Rend.* **250**, 205–207.

104   — 1963. Étude au microscope électronique des nucléoles dans le macronucléus de *Paramecium caudatum*. In *Progress in Protozoology* (Proc. 1st Int. Congr. Protozool., Prague), 421.

105   — 1966. Variations ultrastructurales du chondriome en relation avec le mode de vie chez des protozoaires. *Proc. 6th Int. Congr. Electron Microscop. Kyoto* 247–248.

106   Wessenberg, H. 1970. Unpublished.

107   Wohlfarth-Bottermann, K. E. 1956. Protistenstudien. VII. Die Feinstruktur der Mitochondrien von *Paramecium caudatum*. *Z. Naturforsch.* **11b**, 578–581.

108   — 1957. Cytologische Studien. IV. Die Entstehung, Vermehrung und Sekretabgabe der Mitochondrien von *Paramecium*. *Z. Naturforsch.* **12b**, 164–167.

109   — 1958. Cytologische Studien, IV. Die Feinstruktur des Cytoplasmas von *Paramecium*. *Protoplasma* **49**, 231–247.

110   Yusa, A. 1963. An electron microscope study on regeneration of trichocysts in *Paramecium caudatum*. *J. Protozool.* **10**, 253–262.

Note Added in Proof

Glycogen can now be identified in *Paramecium* and other ciliates with some degree of certainty under the electron microscope by application of a periodic acid/thiosemi-carbazide/silver protein method to thin sections (Sutherland *et al.*, 112). Fig. 63A and 166A are from the work of Bannister (111) and Tamm (113) respectively.

111   Bannister, L. H. 1972. The structure of Trichocysts in *Paramecium caudatum*. *J. Cell Sci.* **11**, 899–929.

112   Sutherland, A., Antipa, G. A. and Ehret, C. F. 1973. Ultracytochemical observations of glycogen synthesis and storage during infradian growth of *Tetrahymena pyriformis*. *J. Cell Biol.* **58**, 240–244.

113   Tamm, S. 1972. Ciliary motion in *Paramecium*. *J. Cell Biol.* **55**, 250–255.

FIGURE LEGENDS

General information

Figs. 1–166. In all figures, the manner of pretreatment of the cells is symbolized as follows: PTA = phosphotungstic acid; UrAc = uranyl acetate; G = glutaraldehyde; Os = osmium tetroxide; Epon, Vestopal = embedding medium; Pb = the sections were stained with lead, usually lead citrate; Il L-4 = a nuclear track emulsion from Ilford was used to cover the specimen for autoradiography. Please note that we have relabeled the donated photos, and that the new labels do not necessarily agree with those of the donors. We have tried to be objective, but flaws in interpretation may be present owing to compromises with operationalism, in deference to tradition, consensus, and innocent bias.

p. 298   Figs. 1, 2. (PTA negatively stained) *P. aurelia*, 51 VIII S; isolated trichocysts. (Steers *et al.* 97).
Fig. 1. Freshly isolated; note 550 Å periodicity.
Fig. 2. After dialysis in water; note subunits.

p. 299   Fig. 3. (UrAc stained) *P. multimicronucleatum*, syngen 2; DNA. (Barnett and Zadylak, personal communication).

p. 299   Figs. 4–7. (G/Os/Epon/Pb) *P. aurelia*, 51 VIII K; macronucleus. Large cup-shaped nucleoli, smaller chromatin bodies (cb), and coiled ∼ 100 Å microfibrils with ∼ 400 Å helical periodicity within the karyoplasmic matrix (arrows, Figs. 4, 7) and within a nucleolar cup (arrow, Fig. 5). (McArdle and Zadylak, 60).

p. 300   Figs. 8, 10, 11. (G/Os/Epon/Pb) *P. aurelia*, 51 VIII K. (McArdle and Zadylak, 60).
p. 300   Figs. 9, 12–15. (Os/PTA) *P. aurelia*, CD syngen 1. (Hufnagel, 46).
Fig. 8. Microtubule sectioned lengthwise (arrow), and cross-sectioned ribbons of 5 and 7 microtubules. Note that each of the microtubules contains subfibrils.
Fig. 9. Raffinose-isolated microtubules, showing 45 Å subfibrils.
Fig. 10. Large tubules (∼ 50 nm diameter) of orthotubular system (ots).
Fig. 11. Large tubules of mitochondrion (m), microtubules of ciliary axoneme (c) and kinetosome (ks), and surrounding trichocyst cap (t); fiber-bundle of fibrils of infraciliary lattice (il); striated kinetodesmal fibril (kd).
Fig. 12. Ethanol-isolated infraciliary lattice (il) material pulled away from pellicle.
Fig. 13. Ethanol-isolated 85 Å fine filaments grouped to form "cat's whiskers" fibrils of infraciliary lattice fibers.
Figs. 14, 15. Ethanol-isolated striated kinetodesmal fibrils. At the left, note the tripartite "root" end, the bend in the fibril, and the reduction in diameter. Major periodicities range from 290–340 Å (right end).

p. 301–304   Figs. 16–28. (G/Os/Epon/Pb) *P. aurelia*, 51 VIII K. (McArdle and Zadylak, 60).
Fig. 16. Three 3-ply unit membranes and epiplasmic layer of cell envelope. The outer two unit membranes are separated from the inner one by an alveolar space (arrow).
Fig. 17. Glycogen granules (arrows) and endoplasmic reticulum with ribosomes; microbody (sm).
Figs. 18, 19. Mitochondria with tubular protrusions.
Fig. 20. Mitochondrial ribosomes, anastomosing mitochondrial tubules; microbody; ribosomes and endoplasmic reticulum.
Figs. 21, 22. Cytoplasmic fingers (f) in usual trichocyst loci; alveoli (a), kinetosome (ks), microbodies and normal mitochondria. Note the small circular vesicle consisting of a dense spherical body (∼ 100 nm diameter) surrounded by concentric layers and within irregular vesicle (arrow).
Fig. 23. Several crystalline inclusions in the cytoplasm between trichocyst bodies and mitochondria.
Figs. 24–26. Kappa particles containing the cylindrically coiled bright bodies. At the left, two brights are seen, each with 7 coils; 10 coils are present in the center figure, and 12 at the right; the latter appears about to be extruded through the cytoproct (cp). The crystalloid at the upper right of Fig. 24 is the core of a pretrichocyst.
Fig. 27. The kinetosome and its territory. Cross-sectional diagrams are shown at several levels.

Axonemal microtubules are doublets; ad = axial doublets; pd = peripheral doublets; kinetosomal microtubules are triplets, each is lettered A, B, C, and Linkers = L; hub and spokes of the cartwheel = hs and foot = F. At the right, the view is outward from within the cell, the triplets skew inward clockwise and number 1–9 clockwise. The left cross sections are viewed downward from outside the cell, the triplets skew inward counterclockwise and number counterclockwise. The postciliary ribbon (pr) and the transverse ribbon (tr) rise to the alveolar (a) surface as shown. Epiplasm (e) and transverse plate (tp) appear as a common stratum. Kinetodesmal fibril (kd) attaches to triplet ribbons 5–7; axosome (ax) attaches to axonemal axial doublet (ad) by means of a single microtubular connective. A transverse fibril (tf) appears next to triplet 3. Parasomal sac penetrates the left alveolus. Peripheral doublets (pd) have dynein arms (dy) attached to the A-microtubule in the distal regions of the cilium.

Fig. 28. Kinetosomal territory: one ciliary corpuscle. Alveoli are flanked by ectoplasmic ridges; the ridge on the right contains 3 overlapping kinetodesmal fibrils. The kinetosome is sliced longitudinally, so that its peripheral microtubules are seen. Other cytoplasmic microtubules include the transverse ribbon (tr) underlying the epiplasm of the right alveolus (left in this figure, but see diagram, Fig. 27). The two electron dense blebs at the tip of the cilium are probably artefacts.

p. 304   Figs. 29–31. (Ethanol/Os/PTA) *P. aurelia*, CD syngen 1. Postciliary (pr) and transverse (tr) ribbons of kinetosomal microtubules appear to adhere closely to the pellicular membranes (arrows). (Hufnagel, 46).

p. 304–305   Figs. 32–39. (G/Os/Epon/Pb) *P. aurelia*, 51 VIII K. (McArdle and Zadylak, 60).
Figs. 32, 33. Ultrastructural details diagrammed in Fig. 27 are shown well here; only one of the two central axonemal tubules touches the axosome; the upper half of the central kinetosomal core contains a dense spiral of granules.
Figs. 34–37. Postciliary ribbons (pr) of microtubules originate at triplet 9; transverse ribbons (tr) near triplets 3 and 4; kinetodesmal fibrils (kd) attach to triplets 5, 6 and 7.
Figs. 38, 39. In cross-sections, bundles of overlapping striated kinetodesmal fibrils (kd) are seen in cytoplasmic ridges bounding the kinetosomal territories laterally.

p. 305   Fig. 40. (G/Os/Epon/UrAc/Pb) *P. aurelia*, 51 VIII S. Cross-sectional views of ciliary corpuscles showing two cilia or two kinetosomes per territory. The view is from outside looking in, anterior is upper left. At the kinetosome level, lower right, note the kinetodesmal fibril attachment at triplets 5, 6, 7 and the postciliary ribbon (pr) near triplet 9; at a higher level (above the infraciliary lattice, il) the postciliary (pr) passes to the right alveolar epiplasm attached by one edge; on the same alveolus, a parasomal sac is seen anteriorly. On the same corpuscle, and its nearest neighbor, transverse ribbons (tr) are also seen, but parallel to the left alveolus of each (right in the figure). A portion of the double septum (ds) between alveoli is also seen. (Dippell, 15).

p. 306   Figs. 41, 42. (Os/UrAc/Vestopal/Pb) *P. bursaria*, Jamestown syngen 1, mt D. Sections tangential to the pellicle of a pre-division cell near the fission furrow. New cilia and new kinetosomes (arrows) form in positions directly posterior to old trichocysts, and anterior to old kinetosomes (rule of antecorpy). (Ehret and de Haller, 23).

p. 306   Figs. 43–49. (G/Os/Epon/UrAc/Pb) *P. aurelia*, 51 VIII S. The development of kinetosomes. (Dippell, 14).
Fig. 43. Longitudinal sections through new kinetosomes (nks) about 1/4 grown, and associated with an original territory of 2 old kinetosomes, the latter cross-sectioned.
Fig. 44. Section at right angles to plane of pellicle. Formation of the first microtubules of the new kinetosome.
Fig. 45. An almost complete ring of A-microtubules.
Fig. 46. Doublets of microtubules begin to form in the existing ring of A-microtubules.
Fig. 47. Completed triplet ring, but lacking the A–C linkers, and showing a steeper skew than the adult kinetosome. Arrow indicates "pad" whence originate cytoplasmic microtubules.
Figs. 48, 49. New kinetosomes begin to tilt surfaceward. The new kinetosomes lack the organized central mass of fibrous granules present in the old kinetosomes.

p. 307   Fig. 50. (G/Os/Epon/UV Micrograph) *P. aurelia*, 51 VIII K. Thick section shows the dense occupancy of the cell envelope by trichocysts in their natural state. (McArdle and Svihla, 59).

p. 307 Fig. 51. (Live/Geranium Tannic Acid/Phase Contrast) *P. bursaria*, Argonne syngen 1, mt C. Discharged trichocysts impair the mobility of the cell, though it continues to feed. Each extends 15 to 20 μm from the cell envelope. (Schuster *et al.*, 84).

p. 307 Fig. 52. (Scanning Electron Micrograph) *P. caudatum*. The discharged trichocyst shafts do not appear perfectly cylindrical, and show a major period of about 200 nm. (Small and Antipa, 89).

p. 307–308 Figs. 53–58, 60–62. (G/Os/Epon/Pb) *P. aurelia*, 51 VIII K; long and cross-sections of trichocysts. (McArdle and Zadylak, 60).

Fig. 53. Long section of 3 trichocysts, left one in process of extrusion.

Figs. 54, 55. Extruded trichocysts, showing helically spiralled major periods (marked; $\approx$ 220 nm) and minor periods (unmarked, $\sim$ 55 nm); minor periods contain 4 bands, ultrastructurally resolved in Fig. 2.

Fig. 56. Trichocysts (t) cut transversely at progressively deeper levels from upper right to lower left. Distal sections of trichocyst tips resemble "bull's eyes". Arrow indicates rectangular zone of junction between a tip and the larger, less dense trichocyst body (upper left).

Fig. 57. Longitudinal section through a trichocyst cap exhibiting a longitudinal periodicity of $\sim$ 25 nm.

Fig. 58. Longitudinal section showing the crystalline tip and the granular cap of a trichocyst.

Fig. 59. (G/Os/Epon/UrAc/Pb) *P. aurelia*, 51 VIII S. Crystalline tip and microtubules surrounding the cap (arrow). m = microbody. (Dippell, 15).

Figs. 60–62. Distal to proximal cross-sections through trichocyst tips. The most distal section shows 32 circles that we interpret to be extremely short microtubules in the cytoplasm immediately surrounding the tip (arrow).

---

p. 309 Fig. 63A. Schematic reconstruction of the cap and tip of a trichocyst by L. H. Bannister, 1972. *J. Cell. Sci.* **11**, 899–929.

p. 309 Fig. 63B. Diagrams of stages in the development of the mature trichocyst from the pretrichocyst (see text). (After Ehret and de Haller, 22, 23, and Selman and Jurand, 86).

---

p. 310 Figs. 64–73. Trichocyst development.

Figs. 64, 66–71, 73. (Os/UrAc/Vestopal/Pb) *P. bursaria*, Jamestown, syngen 1, mt D. (Ehret and de Haller, 23).

Figs. 64, 65. Stage 1: the subspherical pretrichocyst with a central crystalloid.

Fig. 66. Stage 2: elongation of the crystalloid.

Figs. 67, 68. Stage 3: crystalloid assumes the shape of a trichocyst.

Fig. 69. Stage 4: trichocyst cap formation.

Figs. 70–72. Stage 5: cap nears completion while only traces of the pretrichocyst matrix remains.

Fig. 73. Stage 6: dense granules form a two-dimensional lattice around the tip.

p. 310 Figs. 65, 72, 74. (G/Os/Epon/Pb) *P. aurelia*, 51 VIII K. Note the remarkably different effect on the appearance of pretrichocyst and trichocyst crystalloids when glutaraldehyde treatment preceeds osmium fixation. (McArdle and Zadylak, 60).

Fig. 74. Stage 7: loss of dense osmiophilic character from mature trichocyst body.

---

p. 311–312 Figs. 75–79. (EM Autoradiograph/Il L-4) *P. bursaria*, Cal 3 (R. W. Siegel), mt D. Vermiform tracks in emulsion overlying trichocysts that had been labeled with tritiated leucine some cell generations earlier, at "f_0". (Ehret *et al.*, 27).

Fig. 75. (Whole Mount/Dried/Formaldehyde) Discharged trichocysts from a cell at f_1. A heavily labeled trichocyst (made during f_0) and two lightly labeled ones (made since f_0) are seen.

Figs. 76, 77. (Os/UrAc/Vestopal/Il L-4/Pb) Sections of f_2 cells descended from a cell labeled at f_0 with tritiated leucine. The non-random distribution of label is evident; some of the trichocysts are very heavily labeled (those made at f_0).

Fig. 78. Whole mount specimen at f_3, showing persistent heavily labeled trichocyst (made at f_0) accompanied by other very lightly or unlabeled younger organelles: tt = trichocyst tip; ts = trichocyst shaft.

Fig. 79. (G/Os/Epon/Pb) *P. aurelia*, 51 VIII K. The *Paramecium* nucleome: micronucleus and macronucleus. (McArdle and Zadylak, 60).

---

p. 313 Figs. 80–82. (Os/Epon/UrAc/Pb) *P. aurelia*, 51 VIII S; macronuclei. (Dippell, 15).

Fig. 80. Thick section shows continuity between chromatin bodies.

Figs. 81, 82. (EM Autoradiograph/Il L-4) Nucleolar fusion induced by ultraviolet pretreatment. (Kimball's method, 57).

Fig. 81. Chromatin bodies and matrix labeled with tritiated thymidine.

Fig. 82. Nucleoli labeled with tritiated uridine.

p. 314–315   Figs. 83–89. (G/Os/Epon/Pb) *P. aurelia*, 51 VIII K. (McArdle and Zadylak, 60).

Fig. 83. Topological relationship of the macronucleus to the gullet (G). A portion of the vestibular region of the pellicle (V) is at the right; er = endoral ridge; ppz = prepenicular zone.

Fig. 84. Macronuclear envelope and pores (in planar sections left) and kinetosome of peniculus (right); between these, karyophoric and/or cytopharyngeal microtubules; ma = macronucleus.

Fig. 85. Macronuclear pores (arrows) in cross-section.

Fig. 86. Interphase micronucleus.

Figs. 87–89. Early prophase micronuclei.

Figs. 87, 88. Pores in the micronuclear envelope.

Fig. 89. Microtubules and fine filaments of the micronuclear karyoplasm.

p. 316   Fig. 90. (G/Os/Epon/Pb) *P. aurelia*, 51 VIII S. Micronucleus in early anaphase. (Dippell, 15).

p. 316   Fig. 91. (Fe Hematoxylin/Light Micrograph) *P. bursaria*, syngen 1, mt C X mt D; conjugating cells. *Paramecium* chromosomes:metaphase (left) and late prophase (right) of the first pregamic division of the micronuclei. (Chen, 7)

p. 316   Fig. 92. (Os/Epon/UrAc/Pb) *P. multimicronucleatum*, CH 323 (mt III) X CH 326 (mt IV). Transverse section showing nuclear exchange between 2 conjugating cells; mp = migratory pronucleus; sp = stationary pronucleus; kinetosomes of the conjoined kinetosomal territories are evident. (Inaba *et al.*, 47).

p. 316   Fig. 93. (Unfixed/Phase Contrast) *P. bursaria*, exconjugant. Twenty-four hours after conjugation the new macronuclei begin to form (center, top and bottom). A new micronucleus (mi) and the old macronucleus (ma) are also evident. (Ehret and Powers, 24).

p. 317   Fig. 94. The pellicle system of organelles. Interpretive diagrams of kinetosomal territories as they relate to hexagonal packing patterns, to kineties (longitudinal line B–B) and to paratenes (lateral-diagonal line C–C). The diagram is an idealized composite, showing dextrally spiralling kinetodesma (simple overlapping patterns are the more common arrangements) and two kinetosomes per territory (on the dorsal surface one per hexagon is more common). Kinetosomal microtubules that may regulate the size of a territory (Pitelka, 70) are not shown here; but see Fig. 27. The epiplasmic layer is represented by a broken line; in sections through the epiplasmic borders of adjacent alveoli (e.g., D–D, left portion) infraciliary fibers (Hufnagel, 46) may be seen. (After Ehret and Powers, 26).

p. 318   Fig. 95. (Air-Dried/Nigrosine) *Paramecium* sp. Kineties of the right anterior and left anterior fields converge at the preoral suture. Note the large number of kinetosomal territories when compared with those of the organism in Fig. 96. (McArdle and Svihla, 59).

p. 318   Fig. 96. (Ag Impregnation) *P. aurelia*, 51 VIII S. Dorsal (left) and ventral (right) surfaces. Note hexagonal packing patterns on dorsal surface, as well as the preoral and postoral sutures on the ventral surface. (Dippell, 15).

p. 318   Fig. 97. (Light Micrograph/Ag Impregnation) Hexagonal patterns of kinetosomal territories in the right anterior field; in the left anterior field kinetosomal territories are shaped like lozenges or parallelograms. (von Gelei, 38).

p. 318   Figs. 98–99. (Os/Methacrylate/Thick Section) *P. bursaria*. Close packing patterns formed by contiguous ciliary corpuscles in the pellicle system, the territory is roughly hexagonal, each contains one or two cilia (tt = trichocyst tip; a = alveolus; c = cilium).

p. 318   Fig. 100. (G/Os/Epon/Pb) *P. multimicronucleatum*. The left anterior field, viewed from inside the cell. Cell's right is right on the figure. Anterior is toward top of the page (or toward the preoral suture for the kineties). Each territory is lozenge-shaped, and contains two cilia, paired alveoli, one postciliary (pr) and one transverse (tr) microtubular ribbon and a parasomal sac (ps). Kinetodesmal fibrils (kd) run through the pellicular ridges, linking the longitudinally contiguous territories into kineties. Laterally contiguous territories (lower left to upper right) are called paratenes. The figure shows parts of four kineties (longitudinal) crossing parts of two paratenes (lateral-diagonal). (Pitelka, 68).

p. 319   Figs. 101, 102, 104, 105. (Isolated pellicles) *P. aurelia*, CD syngen 1. (Hufnagel, 46).

Fig. 101. (Raffinose/Phase Contrast) A complex of kinetosomal territories.

Fig. 102. (Ethanol/Phase Contrast) Kinetosomes (ks), kinetodesmal fibrils overlapping to form kinetodesma and trichocyst attachment sites (t) are seen. In this view from inside the cell, right on the figure is the cell's right. The rule of desmodexy, that the kinetodesma lie to the right of the kinetosomes is shown.

Fig. 103. (Light Micrograph/Ag Impregnation) A classical view of the relationship of the so-called infraciliary lattice system to the so-called outer lattice (heavy lines overdrawn by G. von Gelei). (von Gelei, 38).

Fig. 104. (Raffinose/Os/PTA) Relationship of kinetodesmal fibrils (kd) to one another in contiguous territories is seen; pr = postciliary ribbon of microtubules.

Fig. 105. (Ethanol/Os/PTA) Fibrous and granular portions of infraciliary lattice (il) adherent to epiplasm of a single kinetosomal territory.

p. 319   Fig. 106. (G/Os/Epon Araldite/UrAc/Pb) *P. caudatum* (Allen, 2).

A. Planar view of the infraciliary lattice (il) composed of bundles of microfibrils. These bundles undergo branching at loci that are characterized by having an electron-opaque post (arrow). The kinetosomes are sectioned through the level at which kinetodesmal fibrils (kd) are attached to triplets 5, 6, and 7. Postciliary microtubular ribbons (pr) are seen to emerge from triplet number 9. A transverse fibril (tf) (Antipa, 4) appears next to triplet 3.

B. Two microfibrillar systems, the striated bands (sb) and the infraciliary lattice (il) can be distinguished. Electron-opaque posts can be seen at the center of each branching locus of the infraciliary lattice. A parasomal sac (ps), kinetodesmal fibrils (kd) and transverse microtubules (tr) are indicated. (Allen, 1).

---

p. 320   Figs. 107, 108. (G/Os/Epon/Pb) *P. aurelia*, 51 VIII K. Kinetodesma reside in the ridges between territories (kd = kinetodesmal fibril); in Fig. 108, fibrous elements of the infraciliary lattice (il) follow nearly the same course as they underlie and flank the corpuscular alveoli. (McArdle and Zadylak, 60).

---

p. 321   Fig. 109. (G/Os/Epon/UrAc/Pb) *P. aurelia*, 51 VIII S. Partitioning of alveolar and membranous elements of kinetosomal territories following kinetosomal multiplication. (Dippell, 15; Sonneborn, 96).

---

p. 322   Figs. 110–113. (Os/UrAc/Vestopal/Pb) *P. bursaria*, Argonne, syngen 1, mt C. Suprakinetodesmal microtubules (arrows) appear in the ridges above the kinetodesma of a recently divided cell. The ribbons are sometimes coiled lengthwise into jellyroll configurations (Fig. 113). (Ehret *et al.*, 21).

---

p. 323   Figs. 114, 115 (Scanning Electron Micrographs) *P. caudatum*. Vestibular region of the pellicle system, showing conspicuous ridges between paratenes in the left anterior field, and the strong curvature of the kineties. (Small, 87).

p. 323   Fig. 116. Diagram of the kinetosomes and kineties of the preoral suture, to illustrate the angle at which kinetodesmal fibrils meet and cross over one another in this region. (Pitelka, 68).

p. 323–325   Figs. 117, 119, 120. (Ag Impregnation/Light Micrograph) *Paramecium* sp. Optical sections showing relationship of gullet system (G) to pellicle system of cell organelles; arrows indicate path of a vestibular paratene. (McArdle and Svihla, 59).

---

p. 324   Fig. 118. Diagram of cross-section of *Paramecium* through the region of the buccal overture, showing the relationship of pellicular to gullet organelles. An observer within the cell is looking anteriorly. The vestibular region of the pellicle system is on the cell's ventral surface; the gullet system includes p1, p2, p3 the gullet peniculi; ndw = naked dorsal wall; rw = ribbed wall; er = endoral ridge. ppz = prepenicular zone of the vestibular region; kt = kinetosomal territory; t = trichocyst; pt = pretrichocyst; *k* = kappa; fv = food vacuole; ma = macronucleus; mi = micronucleus; ots = orthotubular system.

---

p. 325   Fig. 121. (Unfixed/Phase Contrast) *P. bursaria*, HS 1, mt B. View from within of the right ribbed wall of the gullet. (Ehret and Powers, 25).

p. 325   Fig. 122. (Ag Impregnation/Light Micrograph) *P. bursaria*, HS 1, mt B. View from within of the gullet's left wall. Eight columns of p1 and p2 kinetosomes are closely spaced and two of the four columns of p3 kinetosomes are widely separated. (Ehret and Powers, 25).

---

nuclear State II (II, single nucleoli, finely textured karyoplasm). (After Ehret and De Haller, 23; Jurand and Selman, 53; Small, 87).

---

p. 333 Figs. 148, 149. (Os/UrAc/Vestopal/Pb) *P. bursaria*, Cal-3 (R. W. Siegel), mt D. Autoradiographs of proter and opisthe of a ³H-leucine-labeled parent cell. (Ehret *et al.*, 21).

Fig. 148. Proter cell (anterior fissiont). Silver grains overlying old axonemes and kinetosomes of proter's gullet originate from tritium incorporated during the previous fission generation *after* the old gullet's (OG) structural proteins were synthesized.

Fig. 149. Opisthe cell (posterior fissiont). Tritium-induced silver grains overlie structural, as well as transient protein present in the newly formed axonemes and kinetosomes of a newly formed (2 hours old) gullet (NG).

---

p. 334 Figs. 150–152. (G/Os/Epon/Pb) *P. aurelia*, 51 VIII K. Food vacuole (fv) formation (Figs. 150–151) and pinocytosis (pc) in mature food vacuole. Right and left lips (rl, ll) of the cytostome (cs) are shown; b = bacterium. (McArdle and Zadylak, 60).

---

p. 335 Figs. 153, 154. (Os/UrAc/Vestopal/Pb/Il L-4) *P. bursaria*, Cal-3 (R. W. Siegel), mt D. Single-cell pulse-labeled with tritiated leucine at age 50 minutes for one hour, and fixed 5 hours after end of pulse. Nucleome and cytoplasm are somewhat uniformly labeled; but new food vacuoles (Nfv) are unlabeled (Fig. 153) whereas bacteria (b) in old food vacuoles (Ofv) are labeled, owing to presence of labeled *Paramecium* enzymes. (Ehret *et al.*, 27).

p. 335 Figs. 155–157. (G/Os/Epon/Pb) *P. aurelia*, 51 VIII K. Transverse sections through a cytoproct (cp). Note the spongy-network of cytoplasm that interrupts the pellicle and is bounded by non-ciliated kinetosomes (ks) in the endoproctal ridge (ep). (McArdle and Zadylak, 60).

p. 335 Fig. 158. (Scanning Electron Micrograph) *P. caudatum*. The sinuous endoproctal ridge of the cytoproct in the postoral suture between left posterior and right posterior fields. (Small, 87).

---

p. 336–337 Figs. 159–164. (G/Os/Epon/Pb) *P. aurelia*, 51 VIII K. Contractile vacuole (CV) system. The discharge channel (dc) is surrounded by microtubules that run in two directions. The nephridial canals (nc, Figs. 160, 161) connect directly through openings at their pentagonal apexes with surrounding nephridial tubules. Walls of nephridial canals are lined by microtubular ribbons. The irregular mass of nephridial tubules is in turn surrounded by the very regular orthotubular system. Although the system extends into the depths of the cytoplasm, it's openings through the pellicle each occupy the space of a single kinetosomal territory. (McArdle & Zadylak, 60).

p. 337 Fig. 165. (Scanning Electronmicrograph) *P. caudatum*. Contractile vacuole pores or discharge channels open on the mid-dorsal pellicular surface, upon or near the same kinety. (Small and Marszalek, 90, 91).

---

p. 338 Fig. 166. (Scanning Electronmicrograph). Paramecium *LIVES!* A: *P. multimicronucleatum* fixed during forward swimming; metachronal ciliary waves travel from posterior (P) to anterior (A); note mouth opening ventrally (V) (Tamm. 1972. *J. Cell Biol.* **55**, 250–255). B: *P. caudatum* in a more relaxed mood (Small and Marszalek, 90).

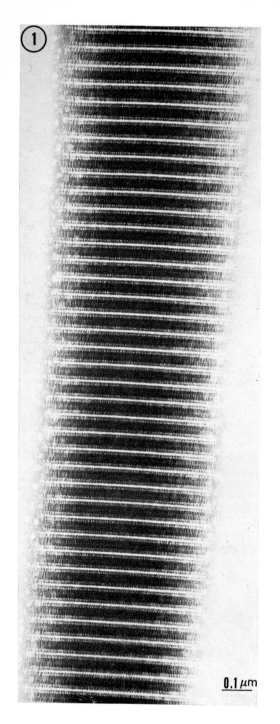

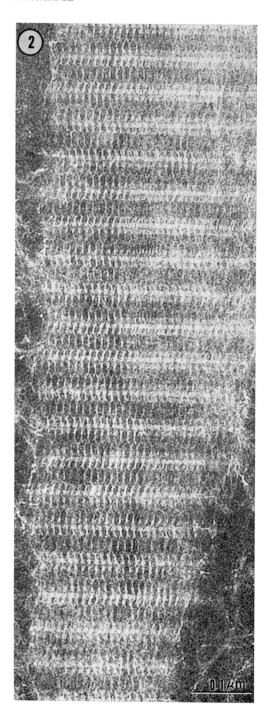

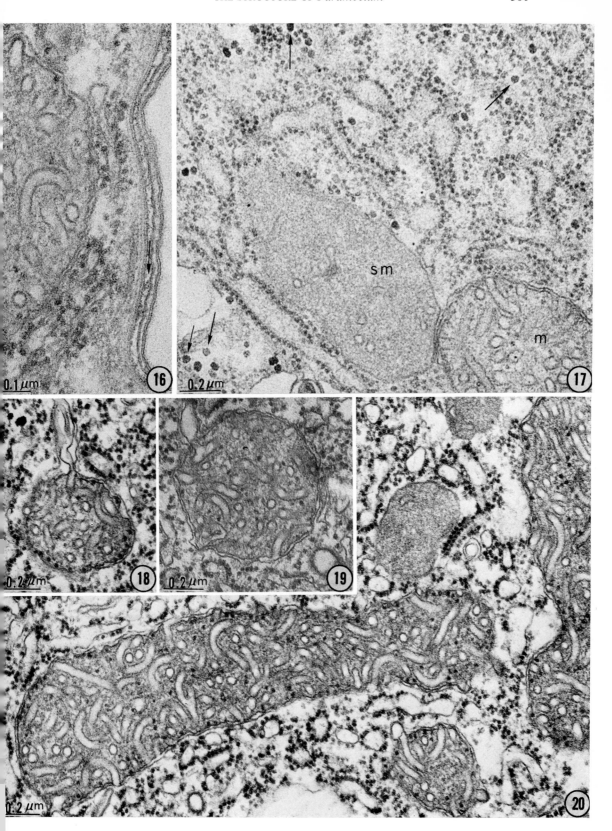

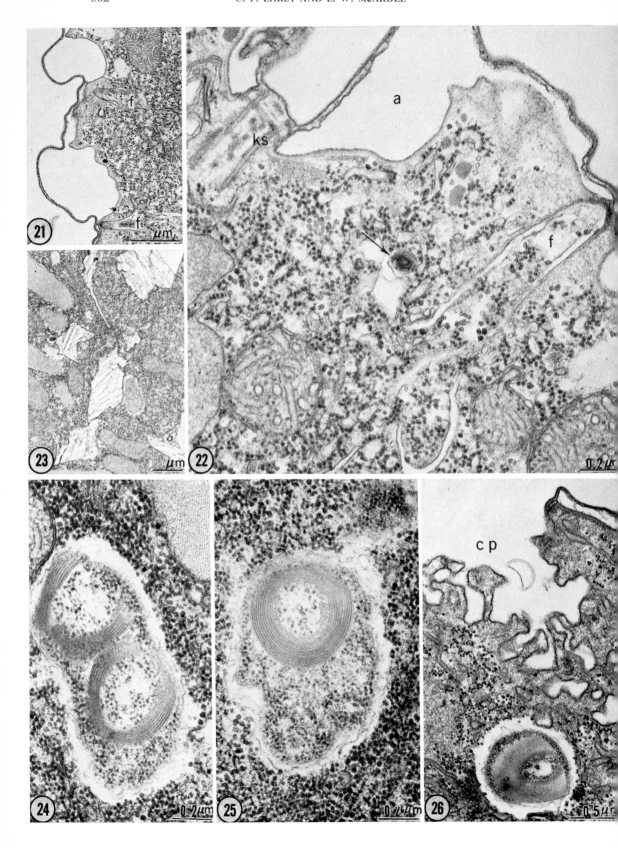

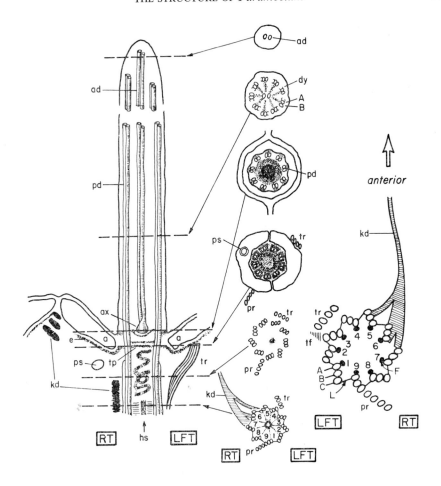

27

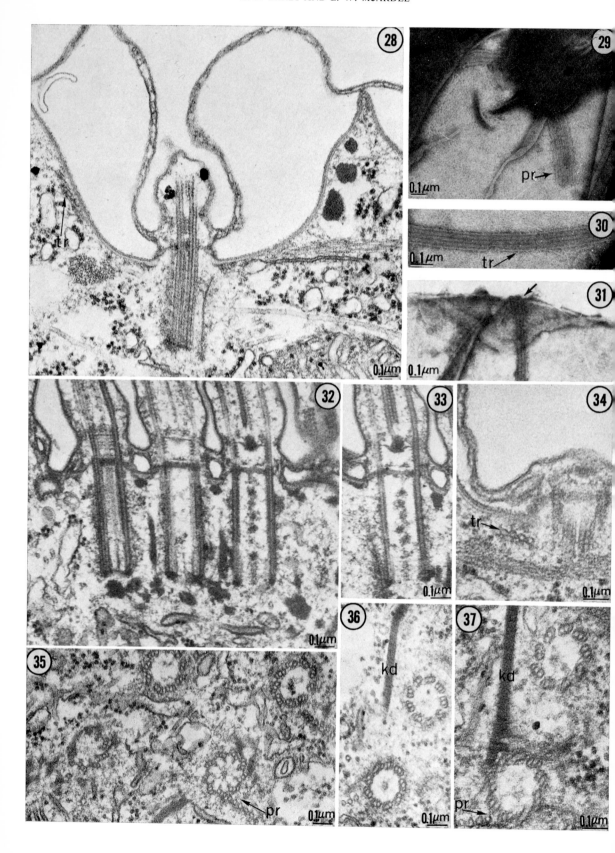

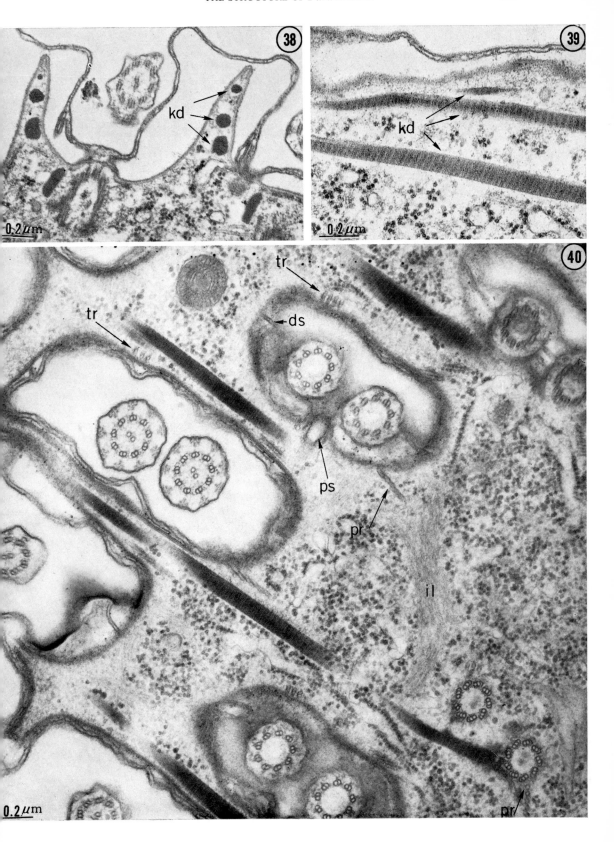

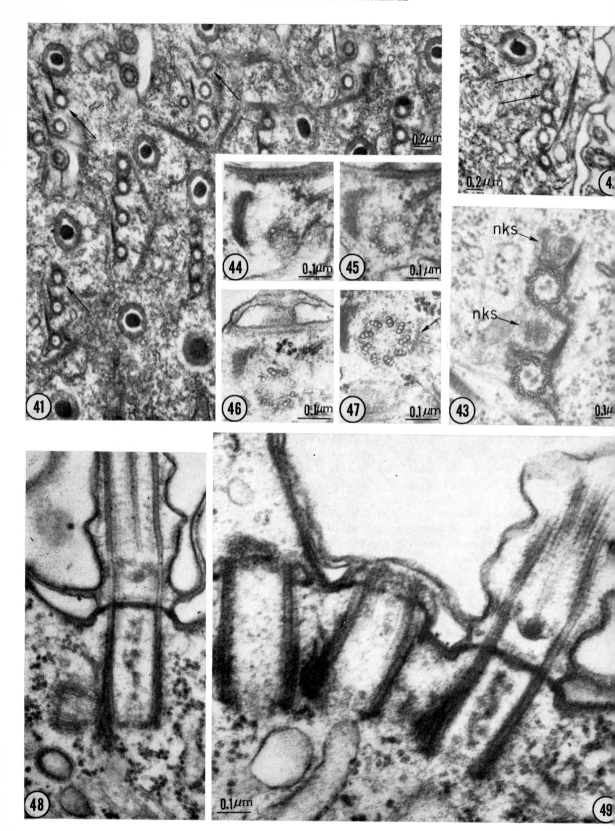

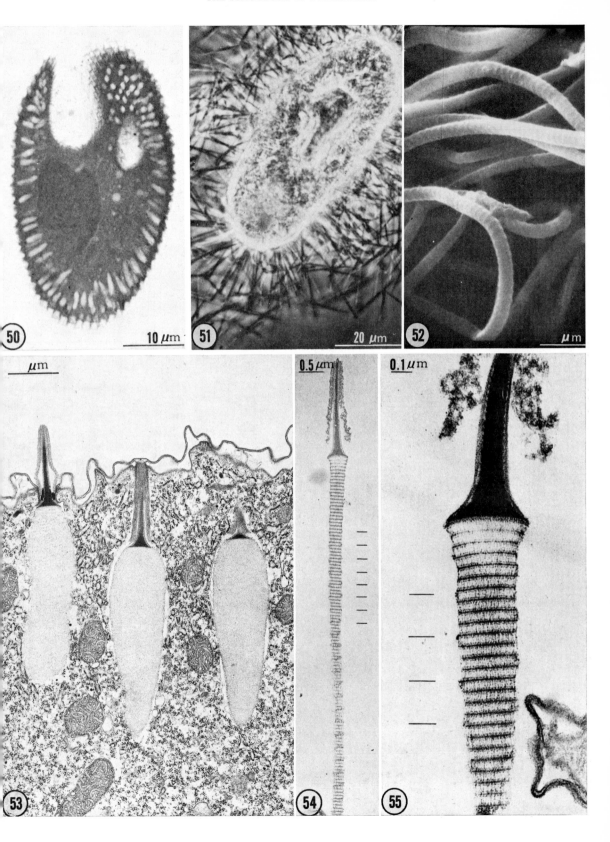

50   10 μm

51   20 μm

52   μm

53

54   0.5 μm

55   0.1 μm

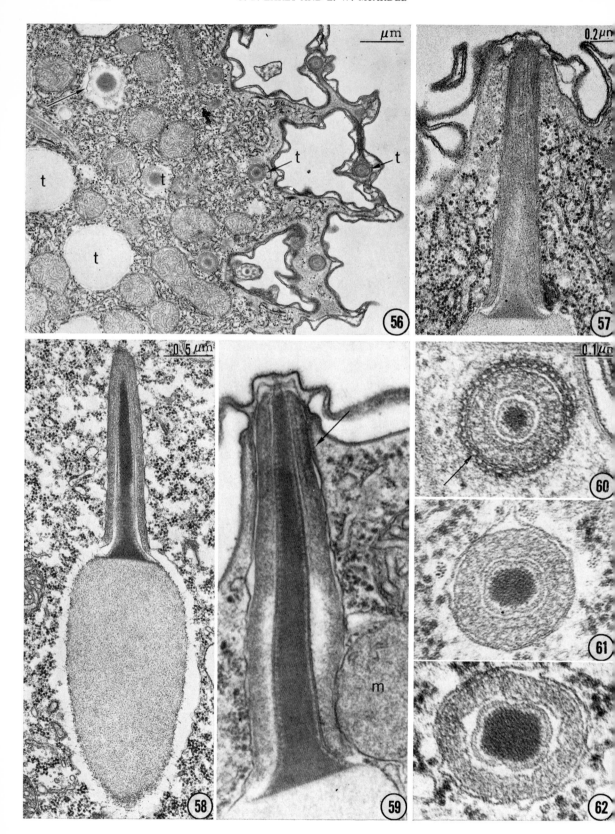

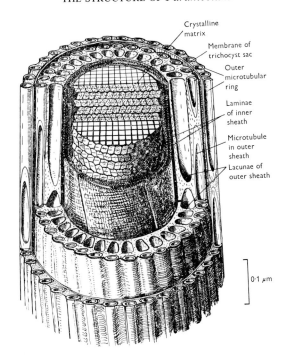

Crystalline
matrix

Membrane of
trichocyst sac

Outer
microtubular
ring

Laminae
of inner
sheath

Microtubule
in outer
sheath

Lacunae of
outer sheath

0·1 μm

63A

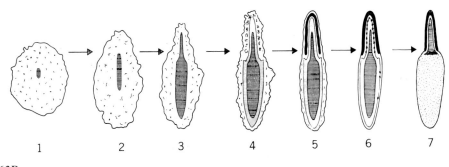

1   2   3   4   5   6   7

63B

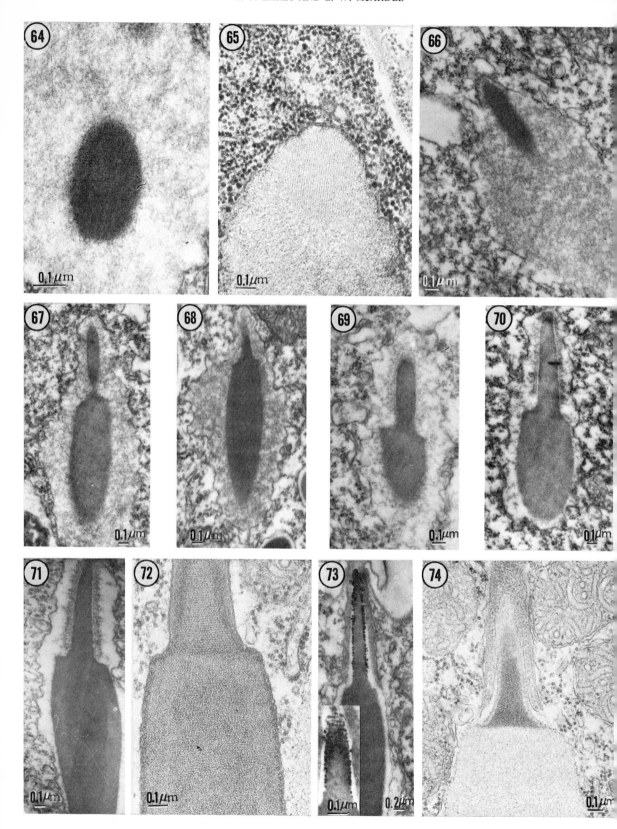

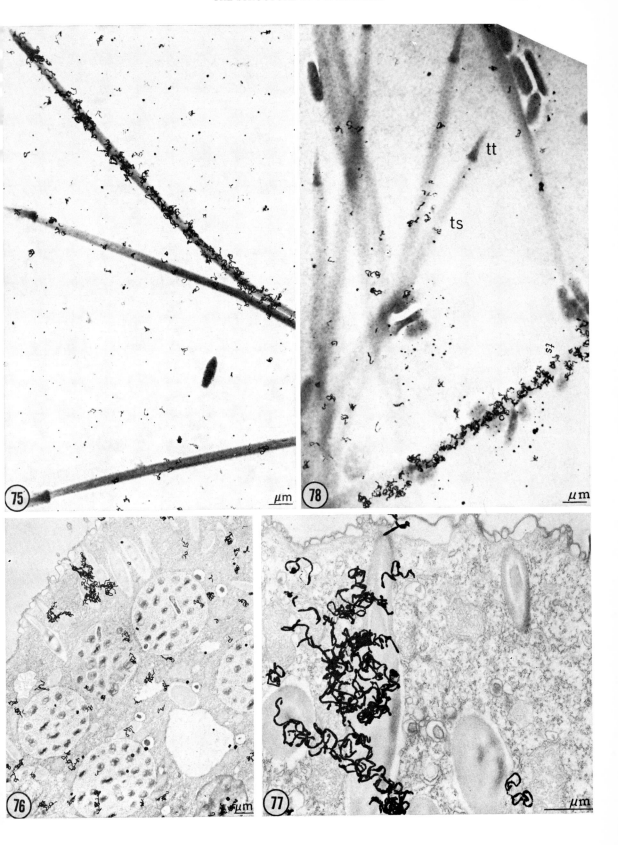

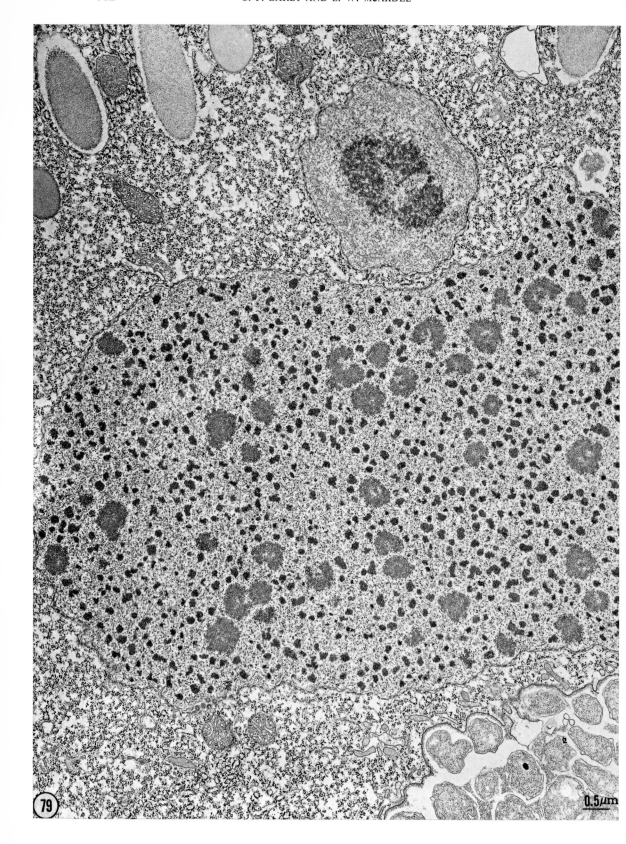

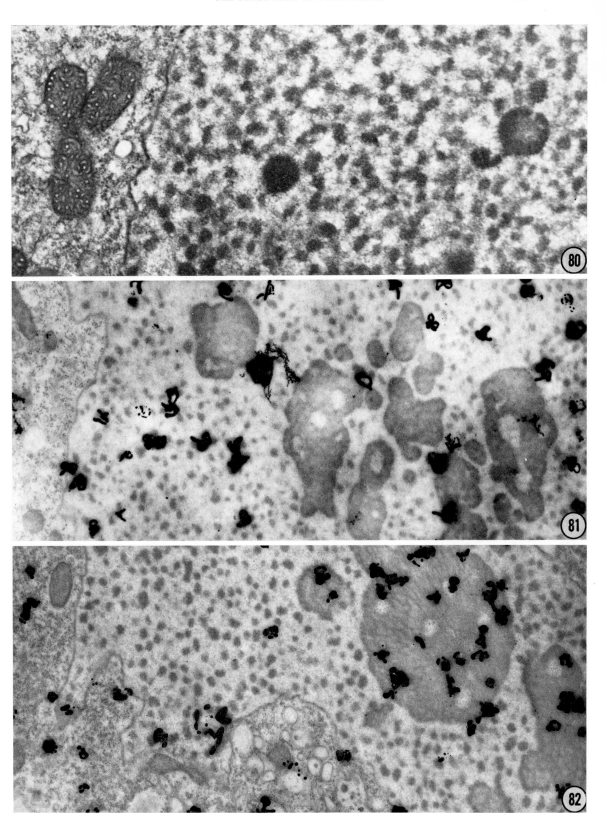

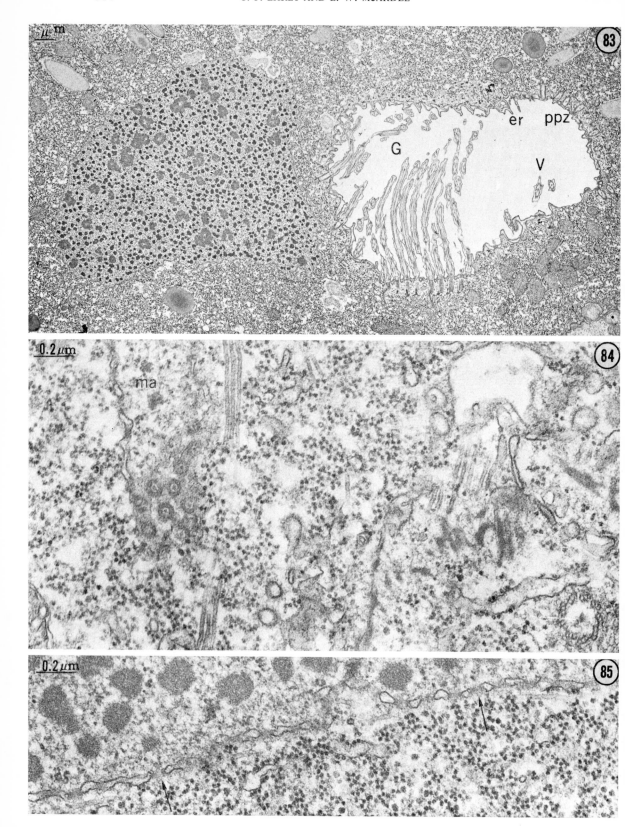

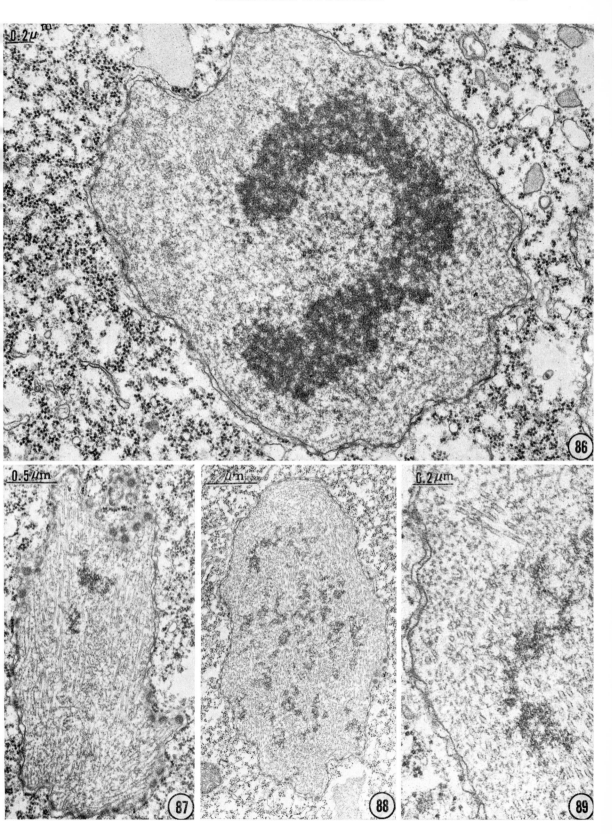

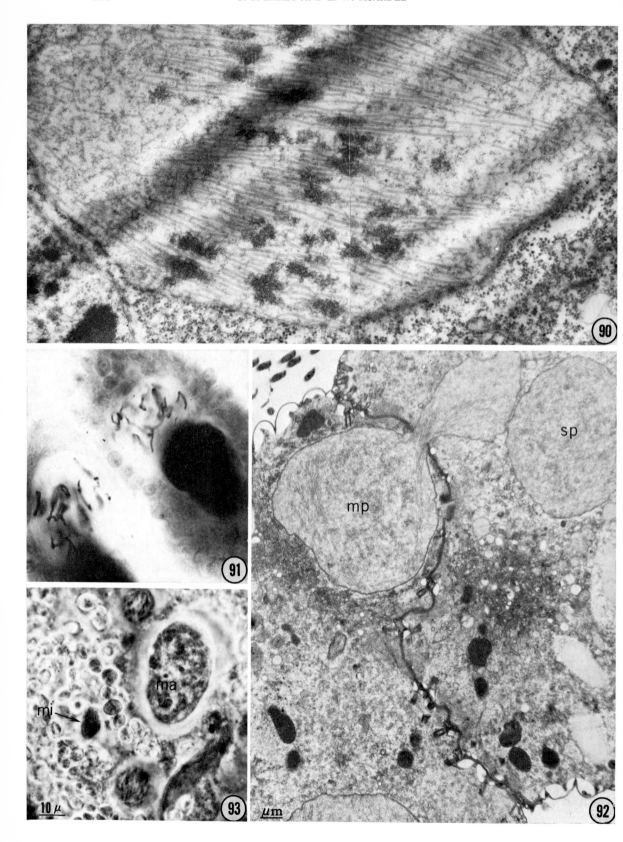

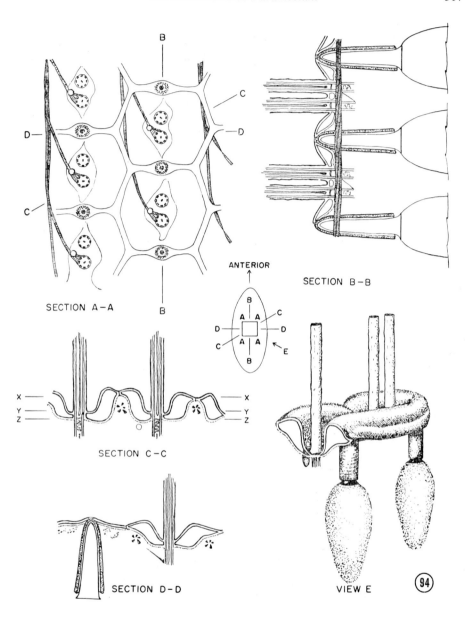

SECTION A–A

SECTION B–B

SECTION C–C

SECTION D–D

VIEW E

ANTERIOR

94

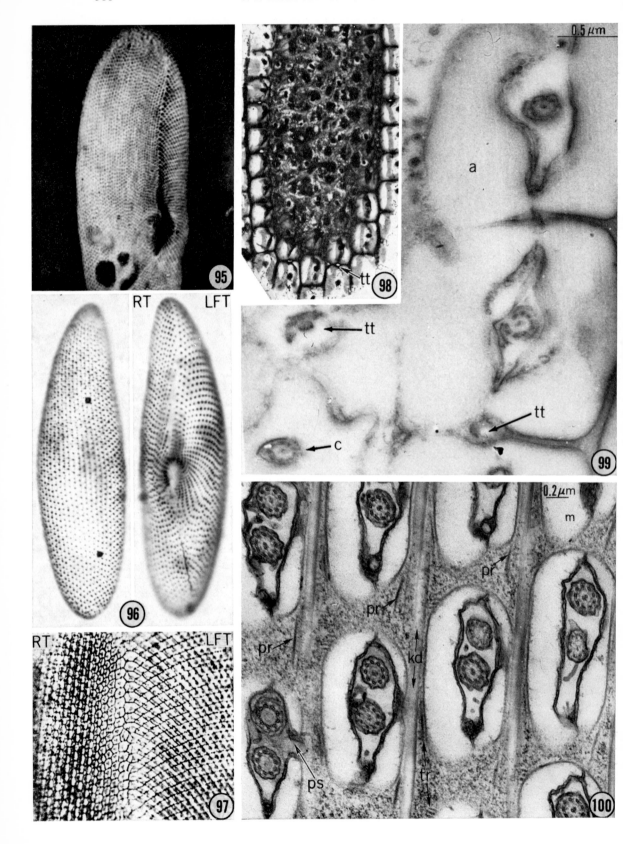

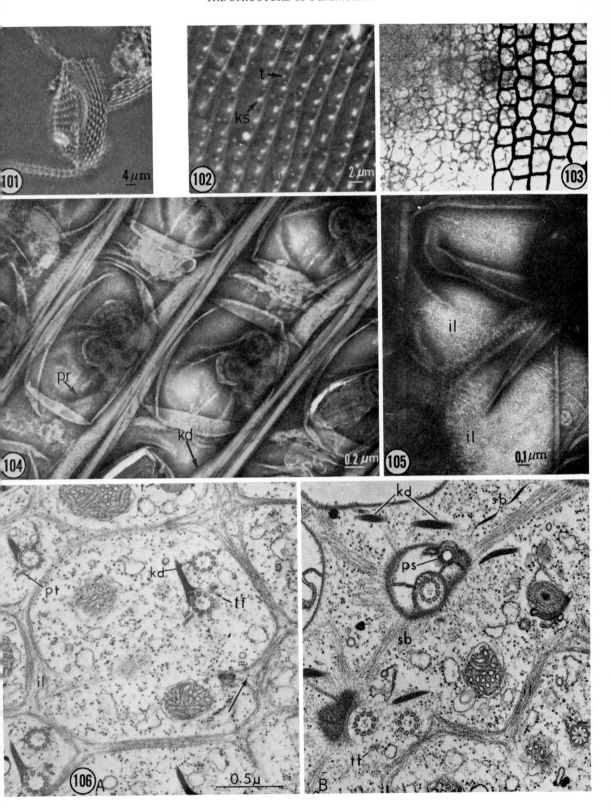

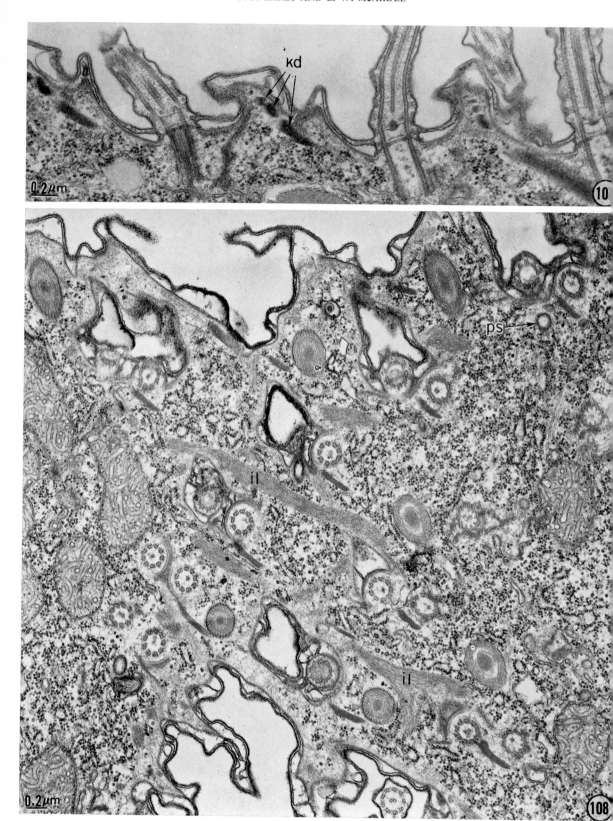

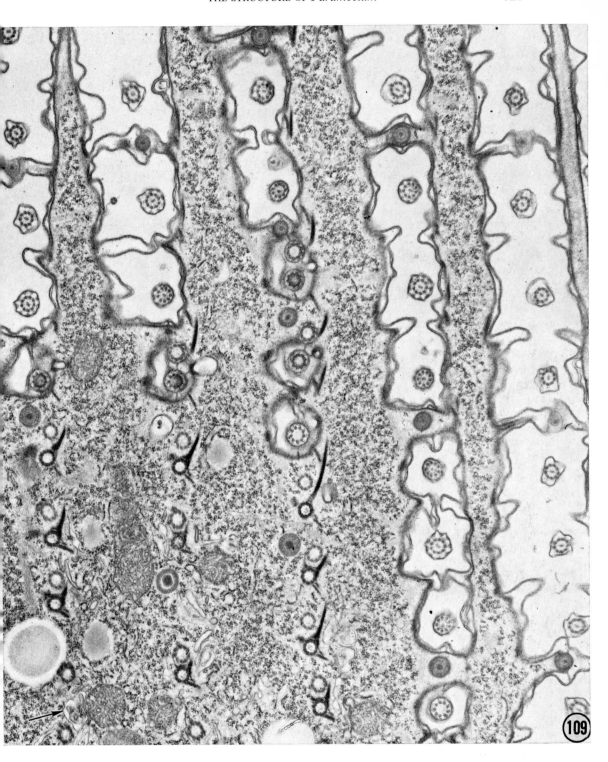

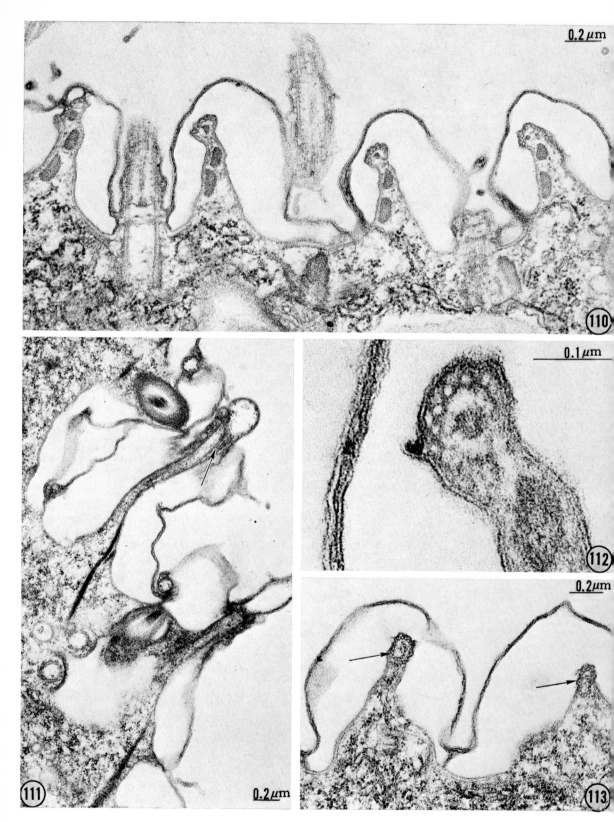

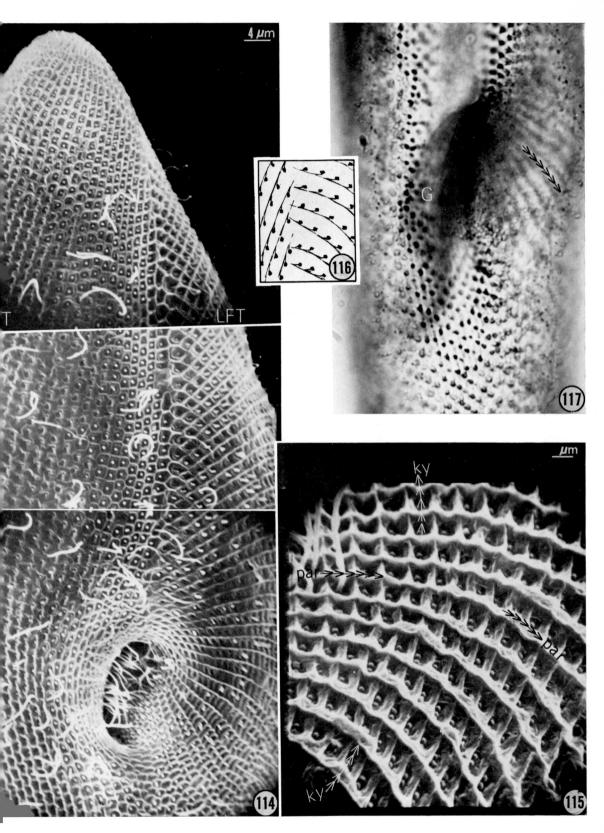

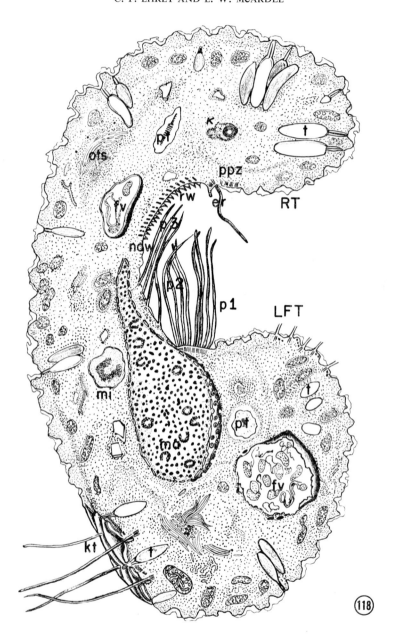

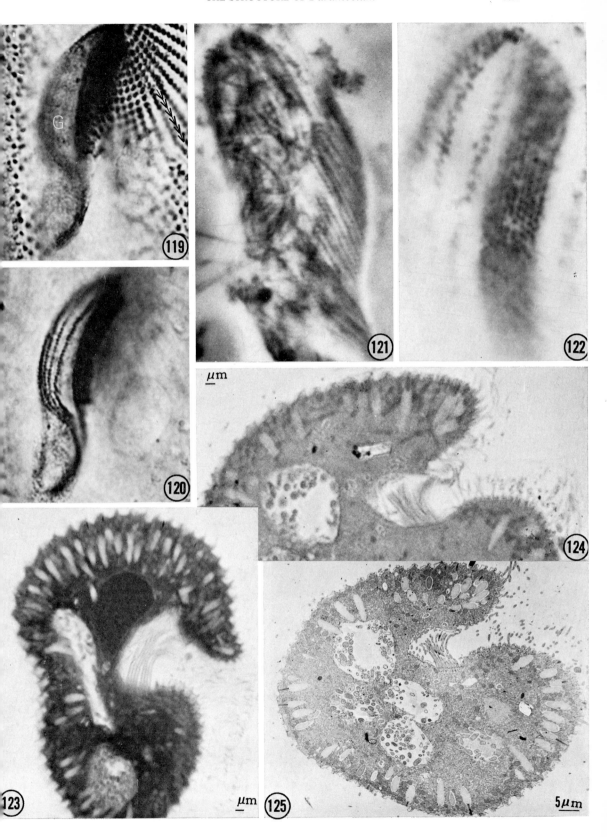

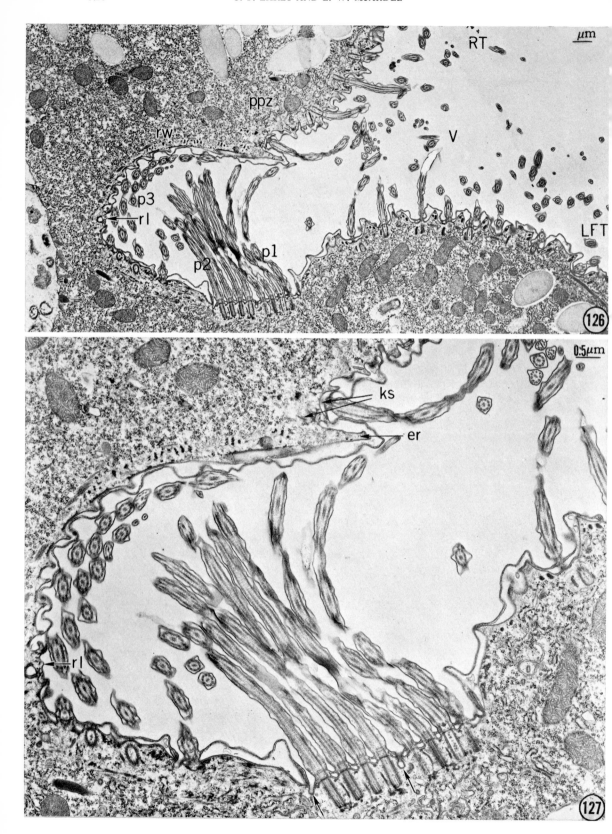

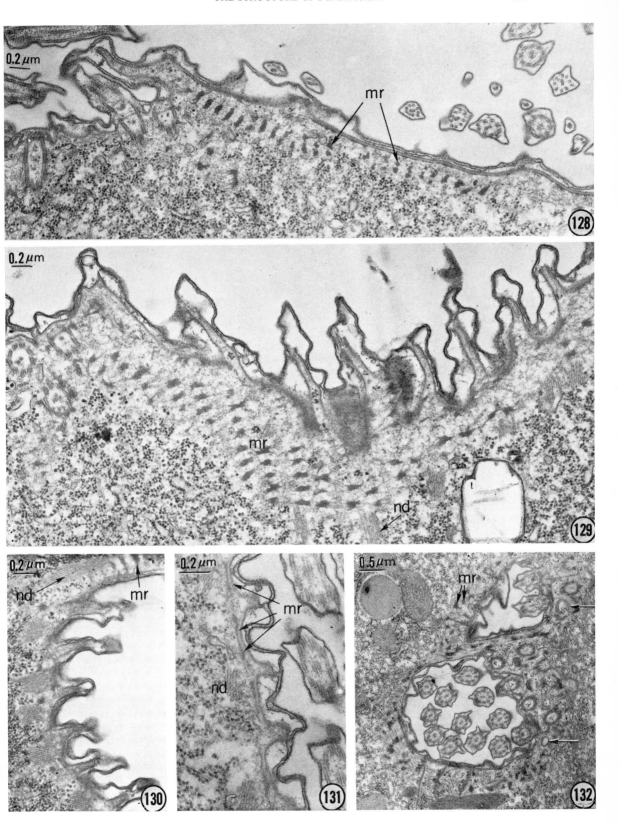

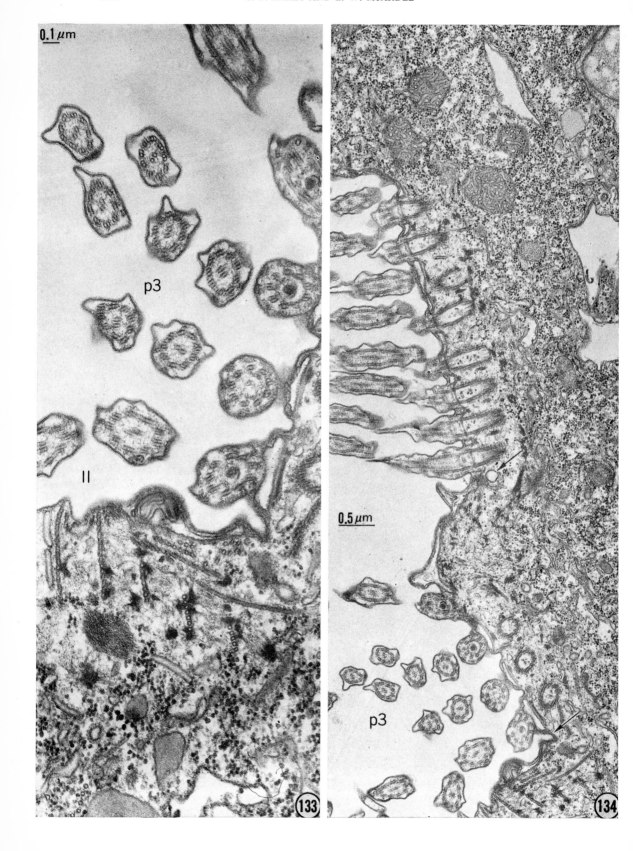

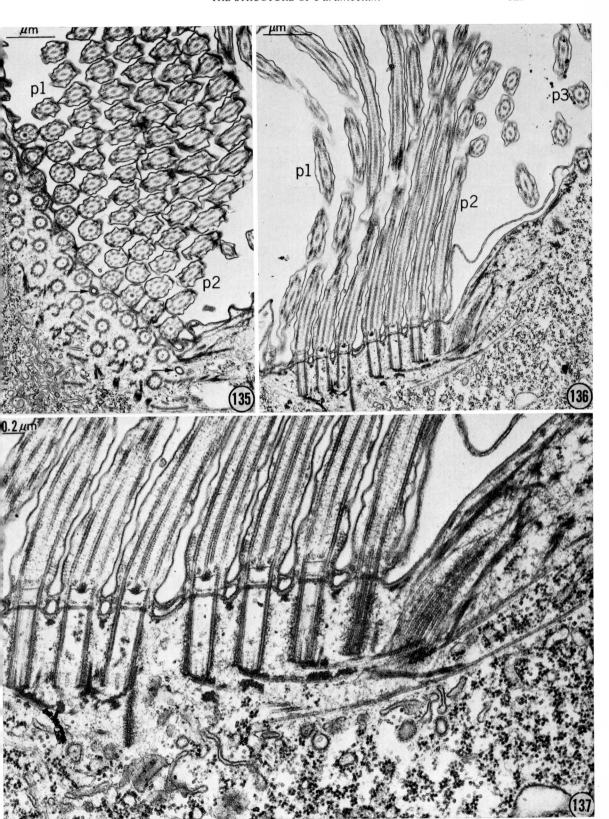

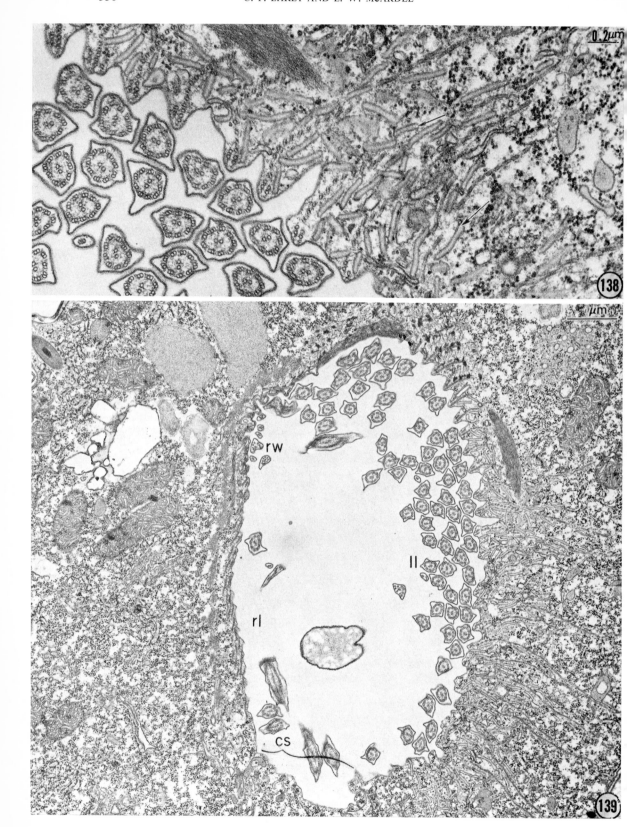

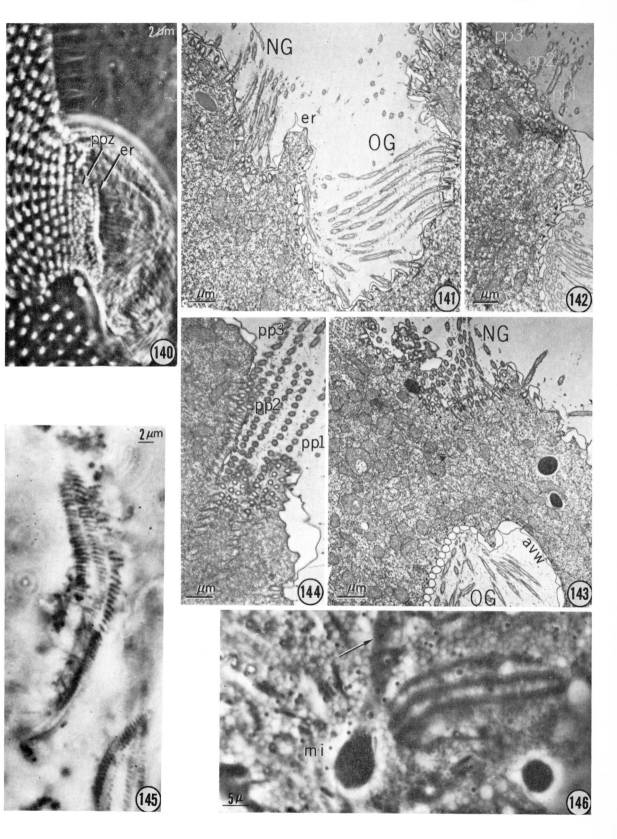

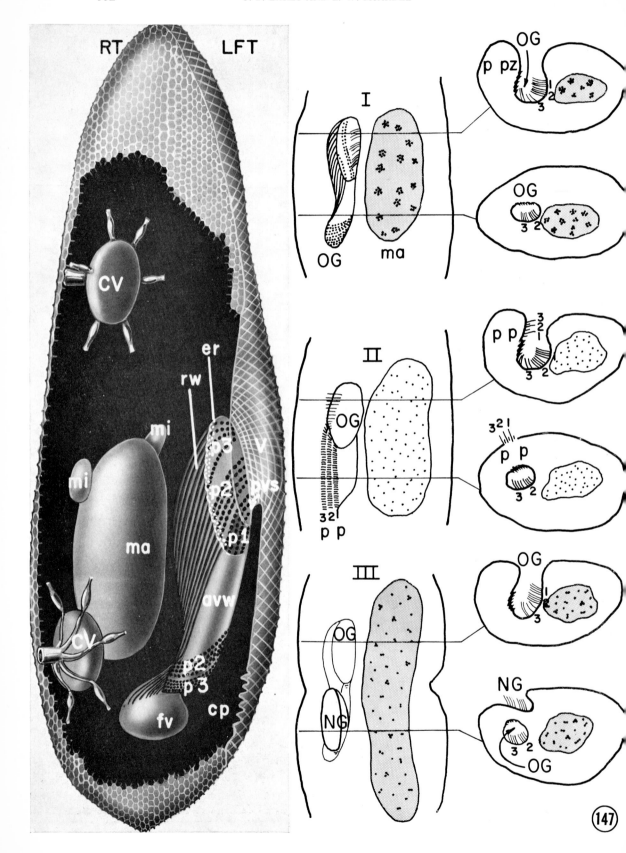

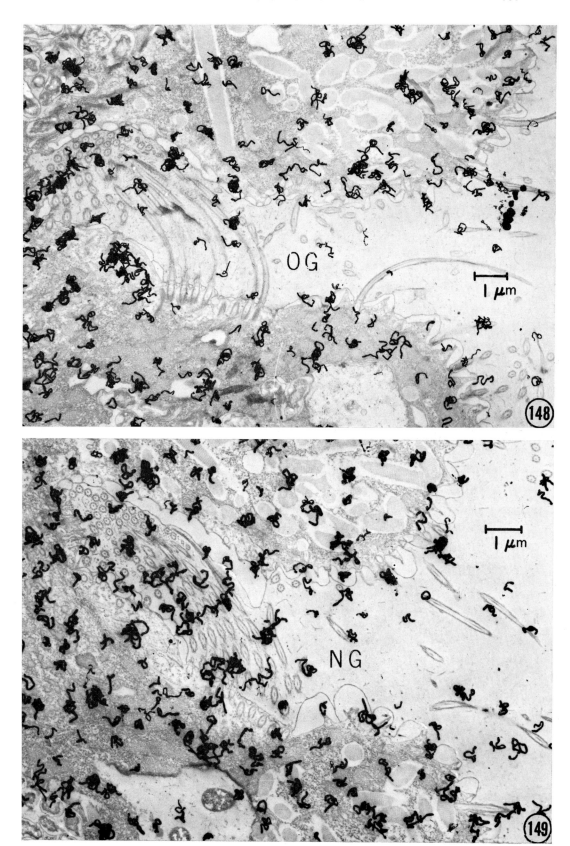

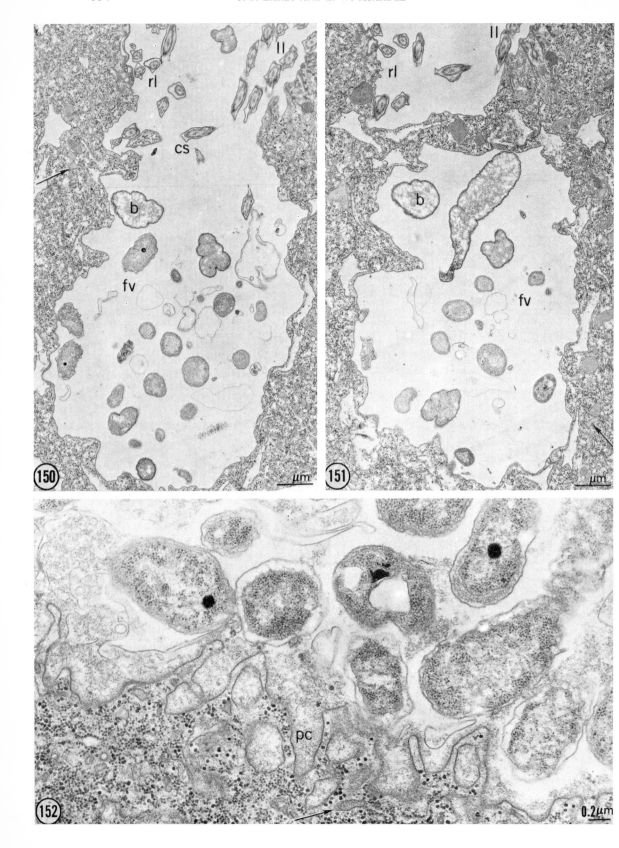

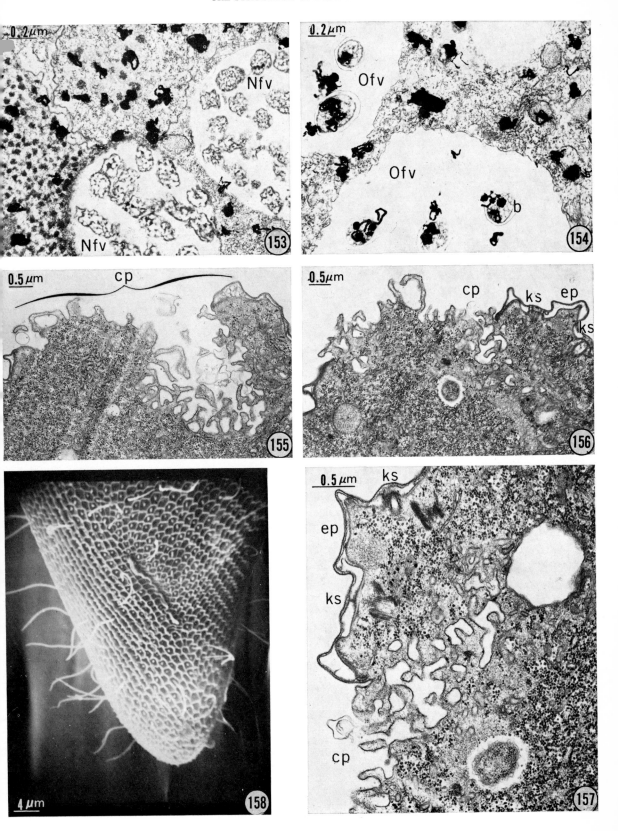

0.2 μm

Nfv

Nfv

(153)

0.2 μm

Ofv

Ofv

b

(154)

0.5 μm   cp

(155)

0.5 μm   cp   ks   ep

ks

(156)

0.5 μm   ks

ep

ks

cp

(157)

4 μm

(158)

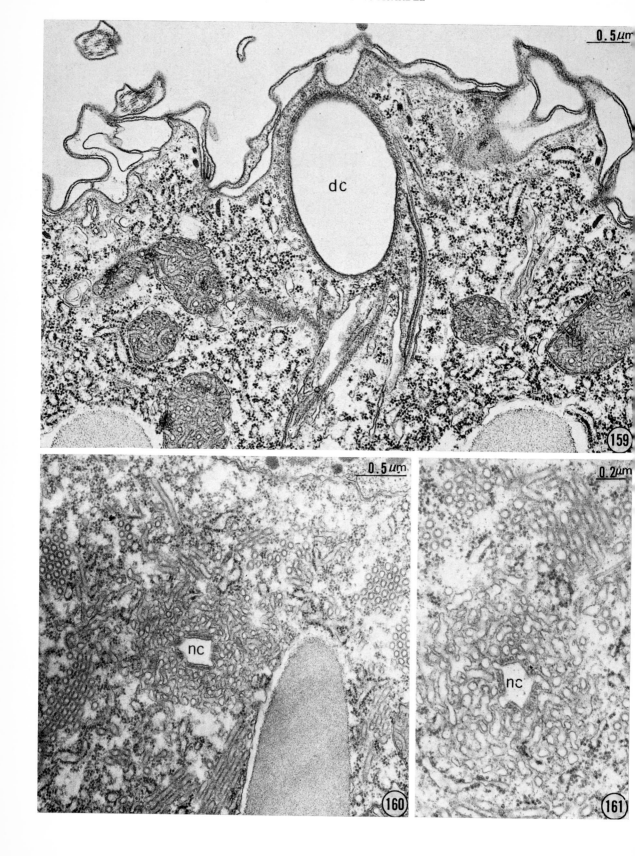

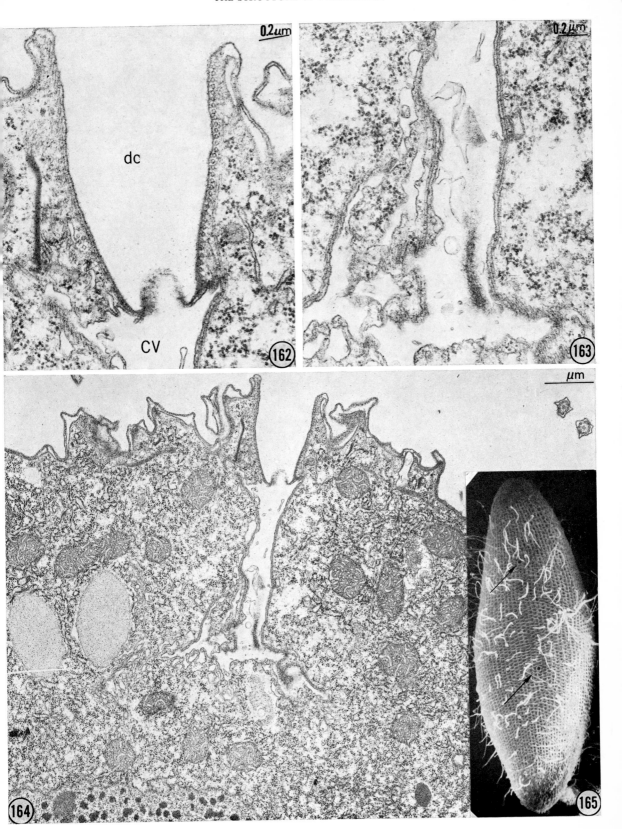

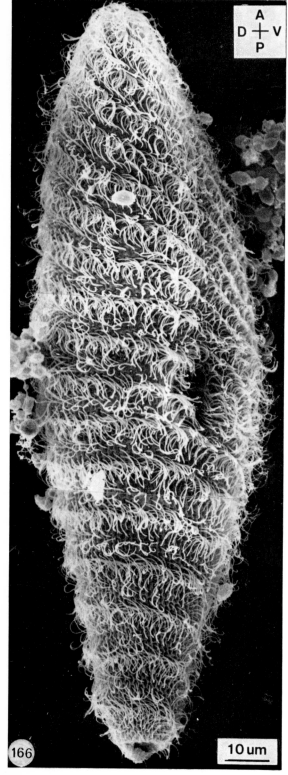

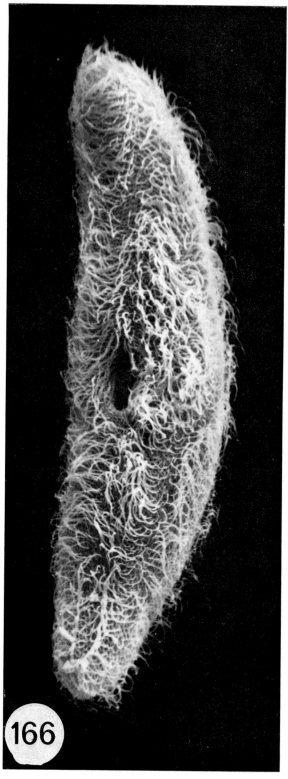

A

B

Nutrition of *Paramecium*

W. J. VAN WAGTENDONK*

Research Division of the Veterans Administration Hospital, Miami, Fla. 33125 (U.S.A.)

"... ende desespereert niet"

Jan Pieterszoon Coen, (1587–1629) in a letter to the
United East Indian Company, Amsterdam, Netherlands

*This chapter is dedicated to Professor Ir. E. J. G. Schermerhorn who introduced me to
the wonders of chemistry and to the memory of Professor Dr. Fritz Kögl who directed my
interest to the nutrition of microorganisms.*

I. INTRODUCTION

For many investigations the culture of *Paramecium aurelia* and related strains in
bacterized media [baked lettuce and Cerophyl infusions, inoculated with *Klebsiella
aerogenes (Aerobacter aerogenes)*] creates no problems. The classic studies by Sonneborn and his associates on genetics, mating types, inheritance of antigenic characters,
and cortical inheritance were carried out in bacterized cultures. A requisite for nutritional and biochemical studies on *Paramecium*, however, is the availability of axenically
grown organisms. Only then can an investigator be sure that the measured parameters
are those of *Paramecium alone*, and not those of the entire microcosmos. The first
barrier to overcome, after sterilization of the ciliate (73) is the construction of an
axenic medium, however crude it may be, in which the ciliate will multiply. After such
a medium has been established it will be possible to refine it and thus eventually to
arrive at a defined medium. The criteria for an adequate medium are: (1) successful
transfer and subsequent growth of mass cultures, and (2) the ability of enabling a single
transfer animal to grow up into a mass culture. It will become evident that the development of a defined medium from the original crude medium will necessitate the use of
various undefined media, in order to evaluate the contributions of certain components

The work of the author has been supported at various times by grants from the Jane Coffin Childs
Memorial Fund for Medical Research, Eli Lilly and Co., the National Science Foundation, the Office
of Naval Research and the National Institutes of Health (most recently by grant AI 03644 from the
National Institutes of Health).
* Present address: Division of Mathematics and Natural Sciences, Talladega College, Talladega,
Ala. 35160 (U.S.A.).

References p. 373–376

of the medium. Only by a step-wise definition of the contribution of the various components of the medium can one arrive at a defined medium. This sometimes necessitates back-tracking on previous findings, such as resubstituting an amino acid mixture by proteose peptone or trypticase. Once the contribution of component X has been established, it is possible to incorporate this compound in the more defined medium previously used. The road to success is long and often tortuous. For example, in 1949, Van Wagtendonk and Hackett (72) established *Paramecium aurelia* in a crude axenic medium. Twenty years later the often frustrating efforts culminated in the formulation of a defined medium for one strain of *P. aurelia* (64).

The nutritional requirements of three members of the genus *Paramecium; P. aurelia, P. multimicronucleatum*, and *P. caudatum* are in general identical. This is best demonstrated by the intertwining of the research efforts of three groups of investigators. At the time that Van Wagtendonk and Hackett (72) established *P. aurelia* in a medium which contained only heat-stabile factors, Johnson was able to grow *P. multimicronucleatum* in a medium containing a heat-labile factor (Johnson and Tatum, 37). In 1952 Johnson (32) reported the replacement of the heat-labile factor of yeast press juice by hydrolyzed nucleic acid or a mixture of guanylic acid and cytidylic acid in a medium containing, besides these components, an autoclaved yeast press juice supernatant, and proteose peptone. In 1953, Conner *et al.* (14) isolated a steroid factor from plant sources which was essential for the growth of *P. aurelia*. Conner and Van Wagtendonk (12) established in 1955 that stigmasterol was the most active steroid. This was followed in 1956 by a report by Johnson and Miller (34) establishing an absolute requirement for stigmasterol by *P. multimicronucleatum*.

A year later Miller and Johnson (47) reported that after removal of nucleic acids by enzymatic digestion and dialysis from an unidentified component of the medium (NDF) *P. multimicronucleatum* had specific purine and pyrimidine requirements. These studies were extended to *P. aurelia* by Tarantola and Van Wagtendonk in a paper published in 1959 (70), in which they reported similar requirements for purine and pyrimidines for *P. aurelia*. Miller and Johnson reported in 1960 (48) that *P. multimicronucleatum* required fatty acids for growth. The lipid interrelationships in the growth of *P. aurelia* were subsequently worked out by Soldo and Van Wagtendonk (62, 63). These efforts finally led to the formulation of a defined medium (64).

Lilly and Klosek were able to grow *P. caudatum* in axenic medium and established in 1961 that stigmasterol was an essential metabolite for this strain (44).

No defined media have been constructed for either *P. multimicronucleatum* or *P. caudatum*. Comparable studies have not been made with *P. bursaria, P. calkinsi, P. polycaryum, P. trichium, P. woodruffi*, and *P. jenningsi*.

II. THE EARLIER WORK

The early studies on the nutritional requirements of *Paramecium aurelia* were almost entirely conducted with Strain 51 (sensitive) and Strain 47 of syngen 4. Other syngens were investigated later, after more defined media had been devised.

The first successful axenic medium was formulated by Van Wagtendonk and Hackett (72). This medium consisted of equal volumes of 0.5% yeast autolyzate (Basamin-Busch) sterilized by autoclaving for 20 minutes at 121 °C and an autoclaved lettuce infusion which had been inoculated with *Klebsiella aerogenes*, 24 hours previously. The average fission rate in this medium was 1.7 fissions per day. The medium met the criteria for a successful environment for the axenic culture of *P. aurelia*.

Two approaches to the problem of the nutrition of *P. aurelia* were now possible: (1) an analysis of the plant extract for essential nutrilites, and (2) an analysis of the yeast extract. Both avenues were explored. An axenic medium, consisting of a salt solution, dialyzed yeast extract, a mixture of water-soluble vitamins, proteose peptone, and guanylic acid and cytidylic acid was developed by Van Wagtendonk *et al.* (Table 1). *P. aurelia* was unable to grow in this medium unless plant extracts were added. Among the various plant sources tested for growth promoting activity (Table 2),

TABLE 1

Composition of the axenic stock medium (76).

	Final concentration per ml			Final concentration per ml	
NaCl	150	mg	Thiamine	5	μg
KCl	25	mg	Riboflavin	2.5	μg
CaCl$_2$·H$_2$O	25	mg	Niacin	18	μg
MgSO$_4$·7 H$_2$O	50	mg	Ca-pantothenate	5	μg
K$_2$HPO$_4$	100	mg	Pyridoxine	5	μg
Fe(NH$_4$)$_2$ (SO$_4$)$_2$·6 H$_2$O	5	mg	Pyridoxal	2.5	μg
Water	1	liter	Pyridoxamine	2.5	μg
Yeast extract	0.75	mg/ml	Folic acid	2.5	μg
Proteose peptone	8	mg/ml	Biotin	1.25	μg
Guanylic acid	50	μg/ml	Choline	5	μg
Cytidylic acid	50	μg/ml			

TABLE 2

The growth promoting activity of various plant sources (13).

	Relative activity
Cerophyl extract[a]	1
Water extract of desiccated lettuce (Difco)	3
Water extract of baked lettuce	4
Orange juice (fresh)	10
Lemon juice (fresh)	15
Lemon concentrate[b]	60
Lemon peel infusion	0

[a] Obtained from the Cerophyl Laboratories through the courtesy of Dr. G. A. Köhler.
[b] Obtained from the California Fruit Growers Exchange through the courtesy of Dr. W. E. Baier.

References p. 373–376

Chemical characteristics of the factor from lemon juice (14).

Melting point: 136.5–137.5° (uncorrected)	Carbon and hydrogen analysis	
	C	H
Lemon factor:	83.60	12.32
Calculated for: $C_{27}H_{46}O$	83.80	11.91
$C_{29}H_{48}O$	84.46	11.65
$C_{29}H_{50}O$	83.92	12.07
Color tests: Liebermann-Burchard +	Salkowski + Rosenheim −	

lemon concentrate was found to be the most active source and was selected for the extraction and purification of the growth factor. The extraction procedure included the following steps: (1) precipitation of the active fraction with acetone; (2) extraction of the precipitate with hot ethanol; (3) saponification of the extract with methanolic KOH; (4) phase distribution; (5) chromatography (for details see 14). A pure compound, representing 0.025% of the original solids and 20% of the original activity, was obtained. The chemical characteristics of this compound are given in Table 3. The infra-red adsorption curves of the isolated compound were very similar to those of β-sitosterol.

The growth-promoting activity of the factor from lemon juice was compared with

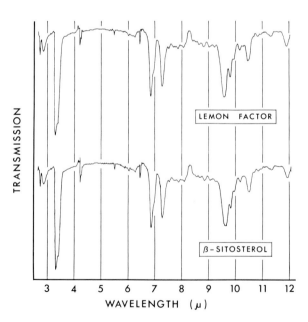

Fig. 1. Infrared adsorption curves of the "lemon juice" factor and of β-sitosterol.

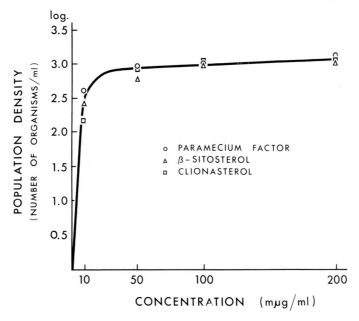

Fig. 2. Growth response of *P. aurelia*, var. 4, stock 51.7 (s) in axenic culture to increasing concentrations of the steroid, isolated from lemon juice, β-sitosterol and clionasterol (14).

that of β-sitosterol and clionasterol (inferred to be the 24α-epimer of β-sitosterol (2, 23). It can be seen from Fig. 2, that the three compounds have very similar growth-promoting activities.

Because of (1) the close similarity of the infra-red spectra of the factor from lemon juice and β-sitosterol, and (2) the fact that the factor from lemon juice, β-sitosterol and clionasterol had similar growth-promoting activities, it was tentatively concluded that the growth factor for *P. aurelia*, var. 4, strain 51.7 (s) was closely related to, or identical with, either β- or γ-sitosterol. The possibility was not excluded, however, that the fraction isolated from lemon juice was a mixture of these two isomers.

Subsequent investigations by Conner and Van Wagtendonk (12) revealed that the requirement for a steroid could be satisfied by the following compounds: β- and γ-sitosterol, fucosterol, brassicasterol, stigmasterol, poriferasterol, or $\Delta^{4,22}$-stigmastadienone (Fig. 3). The same population density was reached with the two sitosterols and fucosterol. When stigmasterol, poriferasterol or $\Delta^{4,22}$-stigmastadienone were substituted for the sitosterols, 10 times higher population densities were obtained. The response of *P. aurelia* to the addition of brassicasterol was intermediate to that of the sitosterol and the stigmasterol group.

The response of *P. aurelia* to increasing concentrations of stigmasterol and stigmasteryl acetate, and brassicasterol and its acetate are given in Fig. 4. The acetates, at low concentrations, were less efficient in promoting the growth of *P. aurelia*. At higher concentrations the growth-promoting activity of stigmasteryl acetate and brassicasteryl acetate approached that of the corresponding free steroids.

The medium developed by Van Wagtendonk *et al.* (76) would support the growth of kappa-less animals of strain 51, but not that of kappa-bearing animals. The kappa-bearing paramecia would go through one to four fissions and die. During this period of vegetative fission, the kappa population would increase from 400–1000 particles per cell to well over 3000 particles per cell.

It was possible to demonstrate that folic acid, riboflavin, and thiamine were essential nutrilites for *P. aurelia*. The yeast extract used undoubtedly contributed other essential vitamins to the medium. Its presence and the use of proteose peptone made an investigation of the amino acid requirements unfeasible. Miller and Van Wagtendonk (49) replaced the proteose peptone by a mixture of amino acids, based upon the amino acid composition of proteose peptone, as determined by acid hydrolysis of the proteose peptone and subsequent two dimensional paper chromatography of the hydrolysate. The composition of the mixture is given in Table 4. By decreasing the concentration of the non-dialyzable fraction (NDF) of the yeast extract from 2–4 mg/ml to 1 mg/ml it was possible to determine with the single omission technique the essentiality of the following amino acids: tryptophan, methionine, threonine, isoleucine, leucine, lysine, histidine, arginine, serine, phenylalanine and tyrosine. The lowering of the NDF concentration also unmasked the requirement for three additional vitamins: pantothenic acid, nicotinamide and pyridoxal. The complexity of the

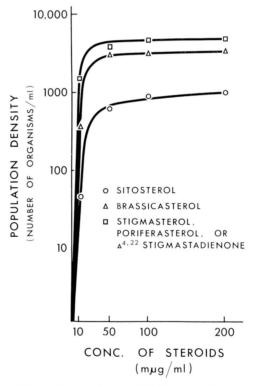

Fig. 3. Growth response of *P. aurelia*, var. 4, strain 51.7 (s) in axenic culture to increasing concentrations of the sitosterol group or $\Delta^{4,\ 22}$-stigmastadienone (12).

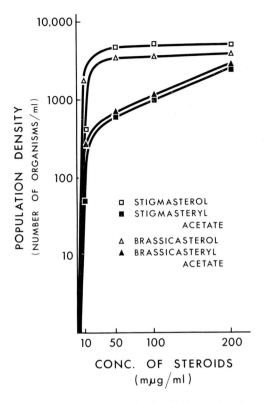

Fig. 4. Growth response of *P. aurelia*, var. 4, strain 51.7 (s) in axenic culture to increasing concentrations of: stigmasterol, stigmasteryl acetate, brassicasterol and brassicasteryl acetate (12).

TABLE 4

Mixture of amino acids used to replace proteose–peptone in the basal medium for the growth of Paramecium aurelia, *var. 4, stock 47.8 (49).*

	Final concentration (μg/ml)		Final concentration (μg/ml)
L-Tryptophan	50	DL-Serine	200
DL-Methionine	150	L-Phenylalanine	75
DL-Threonine	150	L(−)-Tyrosine	50
DL-Isoleucine	150	L-Proline	50
L-Leucine	150	L-Alanine	25
L-Lysine	125	L-Aspartic acid	50
L-Histidine	50	L(+)-Glutamic acid	75
L-Arginine	100	Glycine	25
DL-Valine	75	L(−)-Hydroxyproline	100
		L-Cystine	50

NDF fraction, which contained polysaccharides, some protein, and nucleic acid derivatives, prevented at this time further investigation of the amino acid and vitamin requirements for *P. aurelia*.

By acid and alkali extraction of the NDF fraction, Tarantola and Van Wagtendonk (70) isolated an essential fraction for growth, containing purine and pyrimidine derivatives. This fraction could be replaced by the following combinations (in decreasing order of activity): guanosine + cytidine, guanosine + uridine, guanylic acid + uridylic acid. Each combination was maximally effective when the molar purine: pyrimidine ratio was ∼0.4. On a molar basis, the minimal riboside combinations were ∼1.3 × more active than the ribotides. The free bases were inactive.

Further progress was made by Soldo (60, 61) who was able to grow lambda-bearing *P. aurelia* (strain 299, syngen 8) in a medium consisting of proteose peptone, a non-dialyzable component of a hot water extract of Baker's yeast, Edamine S (an enzymatic digest of lactalbumin), vitamins, stigmasterol and salts (Table 5). Kappa-bearing animals could be grown tenuously in the medium while maintaining their kappa population by frequently (every 2 or 3 days) subculturing the animals. Large populations of kappa-bearing *P. aurelia* could not be obtained in this manner.

All the media used for the axenic growth of *P. aurelia* contained the undefined, but indispensible component derived from Bakers' yeast. Soldo *et al.* (65) were able to replace the "yeast fraction" with a mixture of trypticase, yeast nucleic acids and TEM-4T (diacetyl tartaric esters of tallow monoglycerides). This medium which supported the growth of several particle-free stocks and particle-bearing stocks (while maintaining the full particle population) of *P. aurelia* is given in Table 6. The growth responses of various stocks to this medium are listed in Table 7.

TABLE 5

Composition of axenic medium[a] *(µg/ml unless otherwise specified)* (60).

	Final concentration		Final concentration
Ca-pantothe-nate	2.5	K$_2$HPO$_4$	250.0
			250.0
Nicotinamide	2.5	Na$_2$-ethylenediamine-tetra-acetate	5.0
Pyridoxal·HCl	2.5		
Riboflavin	2.5	MgSO$_4$·7 H$_2$O	25.0
Folic acid	1.25	Fe(NH$_4$)$_2$ (SO$_4$)$_2$·6 H$_2$O	6.25
Thiamine·HCl	7.5	Proteose peptone	5.0 mg
Biotin	0.625 mg	Edamine S	10.0 mg
Stigmasterol	0.2	NDF (yeast fraction)	5.0 mg
MnCl$_2$·4 H$_2$O	0.125		
ZnCl$_2$	0.015		
CaCl$_2$·2 H$_2$O	12.5		
CuCl$_2$·2 H$_2$O	1.25		
FeCl$_3$·6 H$_2$O	0.31		

[a] Proteose-peptone, Edamine S, salts and stigmasterol may be autoclaved together. NDF and vitamins are added asceptically. pH is 6.9.

TABLE 6

Components of the medium for the axenic cultivation of P. aurelia, *stock 299* (65).

Component	Concentration	
	(mg/ml)	(μg/ml)
Proteose peptone	10	
Trypticase	5	
Yeast nucleic acid	1	
MgSO$_4$·7 H$_4$O		500
TEM-4T[a]		100
Stigmasterol[a]		5
Ca-pantothenate		5
Nicotinamide		5
Pyridoxal·HCl		5
Pyridoxamine·HCl		2.5
Riboflavin		5
Folic acid		2.5
Thiamine·HCl		15
Biotin		0.00125
DL-Thioctic acid		0.05

[a] TEM-4T and stigmasterol were dissolved in absolute ethanol as a 100 × concentrated solution and added to the medium before autoclaving. Vitamins were prepared as 100 × concentrated aqueous solutions. Both were stored at −20 °C until ready for use. After the addition of all components, the pH of the medium was adjusted to 7.0 with 1N NaOH and sterilized at 121 °C for 20 minutes.

TABLE 7

Growth and maintenance of particles of several stocks of P. aurelia (65).

Syngen	Stock	Particle designation	Growth of animals[a]		Maintenance of particles
			particle-bearer	particle-free	
2	114	sigma	−	+	lost after 4–6 fissions
4	51	kappa	−	+ + + +	lost after 4–6 fissions
	139	pi	+ +	+ +	maintained indefinitely
6	101	none	−	+ +	
	225	kappa	−	+	lost after 4–6 fissions
8	138	mu	+ + +	+ + +	maintained indefinitely
	151	none		+ + +	
	299	lambda	+ + + +	+ + + +	maintained indefinitely
1	540	mu	−	−	
3	92	none		+	
	152	none			
5	87	nu	−	−	
	135	none		−	
7	38	none		−	
	227	none		−	
9	204	none		−	

[a] − = no growth; + = ∼ 2000/ml; + + = ∼ 4000/ml; + + + = ∼ 6000/ml; + + + + = ∼ 8000/ml (all in tube culture at 27 °C).

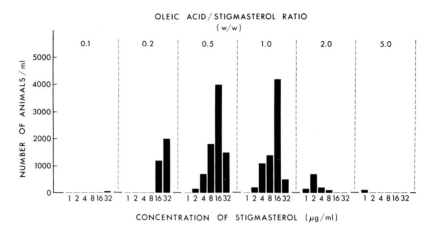

Fig. 5. Growth response of *P. aurelia* to various ratios of oleic acid and stigmasterol. Data pooled from 3rd, 7-day serial transfer of 3 experiments.

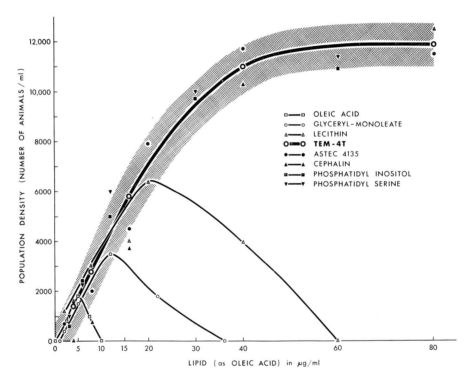

Fig. 6. Growth response of *P. aurelia* to lipids as a function of their oleic acid content. Data pooled from 3rd, 7-day serial transfer of 2 experiments.

Utilizing the medium given in Table 6, Soldo and Van Wagtendonk (62, 63) undertook a detailed investigation of the role of the lipids in the nutrition of *P. aurelia*. A definite balance of fatty acids and stigmasterol was necessary for optimal growth. Highest ciliate populations were obtained at oleic acid/steroid ratios of 0.5 and 1.0,

TABLE 8

Growth response of Paramecium aurelia *stock 299 to TEM-4T and stigmasterol* (63).

	Number of weekly subcultures		
	1	2	3
Component of medium	population density		
Basal[a]	0	—	—
Basal + stigmasterol	1	0	—
Basal + TEM-4T	7	0	—
Basal + stigmasterol + TEM-4T (autoclaved separately)	102	98	99
Basal + stigmasterol + TEM-4T (autoclaved together)	100	100	100

[a] The basal medium consists of the components given in Table 6 minus TEM-4T and stigmasterol. Population density after 7 days expressed as % of control.

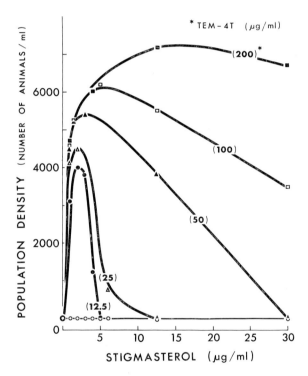

Fig. 7. Growth response of *P. aurelia* stock 299 to TEM-4T and stigmasterol.

especially at relatively high concentrations of stigmasterol (Fig. 5). *P. aurelia* has a specific need for oleic acid. Only oleic acid and oleic acid-containing lipids such as TEM-4T, lecithin, Astec 4135 (a crude preparation of soybean phosphatides), crude animal cephalin and purified phosphatides (i.e., phosphatidyl serine and phosphatidyl inositol) effectively support its growth (Fig. 6). The regulation of the growth of

P. aurelia can be achieved by varying the relative amounts of TEM-4T and stigmasterol (63). In the absence of stigmasterol or TEM-4T growth is reduced to near zero in the first weekly subculture and fails completely in the second (Table 8). Growth is dependent upon the relative concentration of stigmasterol and TEM-4T in the culture medium (Fig. 7). Small quantities of TEM-4T support growth over a narrow range of stigmasterol concentrations. Increasing the concentration of TEM-4T in the culture medium permits growth of the ciliates over wider ranges of stigmasterol concentrations. At a given amount of TEM-4T in the medium the growth response is optimal at only certain concentrations of stigmasterol; at others growth is either suboptimal or fails completely. Optimal growth is observed when the proportion of TEM-4T:stigmasterol is 10:1. Ratios of TEM-4T:stigmasterol of 2:1 or less are inhibitory. The rate of growth decreases and the final population attained is lower when the ratio of TEM-4T:stigmasterol decreases (Fig. 8). At a ratio of TEM-4T:stigmasterol of 2:1, an actual decrease in population occurs. The length of the period of growth increases as the ratio of TEM-4T:stigmasterol decreases.

Inhibition of division brought about by placing the organisms in a medium containing a ratio of TEM-4T:stigmasterol of 2:1 can be annulled without a lag period of growth even after the ciliates have remained under these inhibitory conditions for as long as 6 days, by transferring them to a fresh medium containing TEM-4T and stigmasterol in a ratio of 10:1 (Fig. 9).

Neither TEM-4T nor stigmasterol added singly to the medium inhibits division.

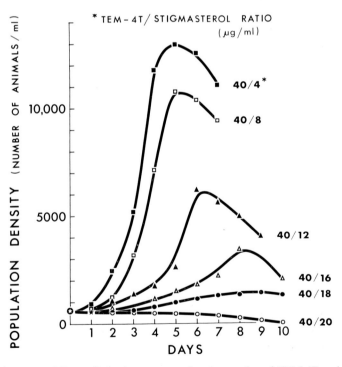

Fig. 8. Growth curves of *P. aurelia* in the presence of various ratios of TEM-4T and stigmasterol.

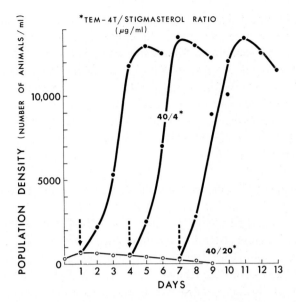

Fig. 9. Annulment of growth inhibition of *P. aurelia* after incubation with an "inhibitory" ratio (2:1) of TEM-4T:stigmasterol. Arrows indicate points at which organisms were transferred to fresh medium containing an "optimal" (w:l) ratio of TEM-4T:stigmasterol.

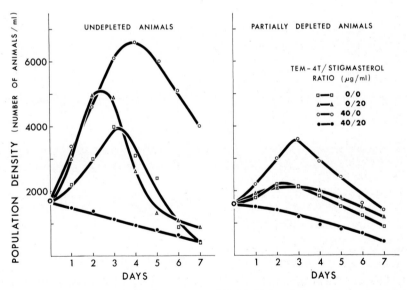

Fig. 10. Growth curves of *P. aurelia* in presence of TEM-4T or stigmasterol alone. Depleted organisms were maintained in salt solution for 3 days before use as inoculum. The composition of the salt solution was as follows (mg/l): NaCl, 30; $CaCl_2 \cdot 2 H_2O$, 300; $MgCl_2 \cdot 6 H_2O$, 200; KH_2PO_4 30; $K_2HPO_4 \cdot 3 H_2O$, 45; adjusted to pH 7.0 with NaOH.

The population increases significantly in the presence of either compound, even at concentrations two or three times higher than those indicated in Fig. 10. In all cases the initial increase in population is followed by a rapid decrease.

"Depleting" the organisms of most of their lipid reserves by maintaining them in a salt solution of 2 to 3 days before use as an inoculum has the effect of diminishing the rate at which they divide and the extent to which they increase in number. These organisms when transferred to media containing a ratio of TEM-4T:stigmasterol of 40:20 (µg/ml) do not divide and the population decreases steadily. Under these conditions the response of "lipid-depleted" and "lipid-undepleted" organisms was the same.

When "lipid-depleted" organisms are incubated in the presence of stigmasterol alone in a concentration of 4 or 20 µg/ml the population decreases steadily over a period of several days. Addition of an amount of TEM-4T on the third day to bring the ratio of TEM-4T:stigmasterol to the optimum for growth (40:4; µg/ml) results in a rapid increase in numbers, and the maximum population density is reached in a few days (Fig. 11). Adjusting the relative proportions of TEM-4T:stigmasterol to the inhibitory amounts (40:20; µg/ml) prevents a significant increase in the population density.

Organisms when pre-incubated with TEM-4T only undergo one or two divisions, after which death occurs. The addition of stigmasterol (on the third day) in an amount to bring about an inhibitory ratio of 40:20 (µg/ml) does not inhibit growth. The organisms multiply at a rate and to an extent comparable to that of the optimal ratio of TEM-4T:stigmasterol of 40:4 (µg/ml).

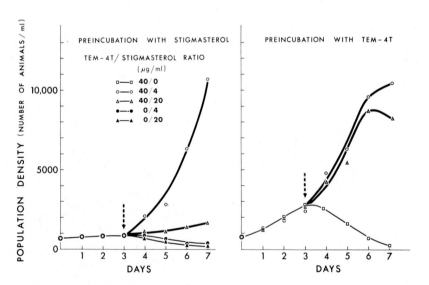

Fig. 11. Growth of *P. aurelia*. Organisms (depleted of lipid by maintaining them in salt solution for 3 days) were incubated in the presence of stigmasterol (4 or 20 µg/ml) or TEM-4T (40 µg/ml) as indicated. After 3 days the ratio of TEM-4T:stigmasterol was adjusted (arrows) to either 40:4 µg/ml (optimum) or 40:20 µg/ml (inhibitory) by addition of appropriate amounts of each lipid directly to the flasks.

III. THE DEFINED MEDIUM

Soldo and Van Wagtendonk (64) were able to substitute each of the crude components of their medium with chemically defined substances. A fatty acid mixture based on the analysis of TEM-4T (62) could replace the latter compound (Fig. 12). Yeast nucleic acid was replaced by a mixture of purine and pyrimidine derivatives, e.g. adenosine, the Na-salt of guanosine-2', 3'-phosphate, cytidine, thymidine and uridine. A mixture composed of sodium acetate, calcium chloride and the trace elements Fe, Zn, Mn and Cu substituted for trypticase. The concentrations of each constituent were adjusted in order to develop a medium, which would permit optimal growth. In this medium proteose peptone was the only undefined component. The medium supported transplantable, but sub-optimal growth, unless supplemented with a phospholipid. For this reason cephalin was included in the various media employed to test the mixtures of amino acids for their ability to replace proteose peptone. Among the amino acids tested were those successfully employed in media for the cultivation of *Tetrahymena pyriformis* (40), *Glaucoma chattoni A* (31), *P. aurelia* (49), and *P. caudatum* (44). None of these mixtures could replace proteose peptone. Consequently a mixture was devised composed of amino acids in approximately the same proportions as those found in the proteins of bacteria, the natural food source for *P. aurelia* [Fig. 13 (77)]. This mixture (Table 9) would support the growth of *Paramecium* in the absence of proteose peptone. On the basis of these results, it was possible to construct a defined medium for the culture of *P. aurelia*, syngen 8, strain 299. The composition of this medium is given in Figure 14. The growth response of *P. aurelia* to several concentrations of the amino acid mixture is shown in Figure 15.

It was now possible to assess the essentiality of the individual components of the medium.

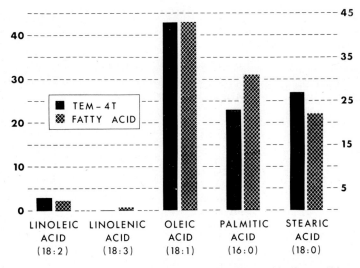

Fig. 12. The fatty acid composition of TEM-4T and of the fatty acid mixture (% total fatty acid present) (62).

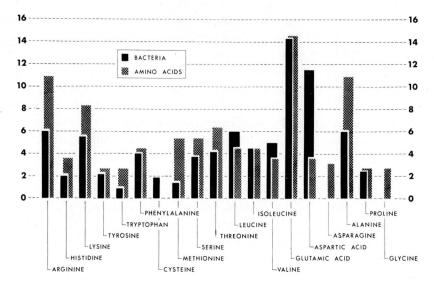

Fig. 13. The amino acid composition of proteins of bacteria and of the amino acid mixture I.

TABLE 9

Amino acid mixture for the axenic cultivation of Paramecium aurelia, *stock 299.*

Amino acid	μg/ml
L-Arginine	600
L-Cysteine	200
Glycine	400
L-Histidine	200
L-Isoleucine	400
L-Leucine	600
L-Lysine HCl	500
L-Methionine	250
L-Phenylalanine	400
L-Proline	200
DL-Serine	800
DL-Threonine	800
L-Tryptophan	150
L-Tyrosine	200
DL-Valine	800

AMINO ACID MIXTURE	µg/ml
AMINO ACID MIXTURE I	**6430**

PURINE & PYRIMIDINE MIXTURE	
ADENOSINE	**225**
GUANOSINE-2′,3′-PHOSPHATE, Na	**375**
CYTIDINE	**225**
THYMIDINE	**200**
URIDINE	**225**

VITAMIN MIXTURE	
BIOTIN	**.001**
FOLIC ACID	**5.0**
THIAMINE·HCl	**15.0**
DL-6-THIOCTIC ACID	**0.1**
NICOTINAMIDE	**5.0**
Ca-d-PANTOTHENATE	**10.0**
PYRIDOXAL·HCl	**5.0**
RIBOFLAVIN	**5.0**

LIPID MIXTURE	µg/ml
LINOLEIC ACID	**1.0**
LINOLENIC ACID	**0.25**
OLEIC ACID	**20**
PALMITIC ACID	**15**
STEARIC ACID	**10**
PHOSPHOLIPID	**250**
STIGMASTEROL	**2**

SALT MIXTURE	
Na — ACETATE	**500**
$CaCl_2 \cdot 2H_2O$	**180**
$MgCl_2 \cdot 6H_2O$	**169**
$(NH_4)_2SO_4$	**20**
K_3PO_4	**112**
Fe	**2**
Zn	**1**
Mn	**0.4**
Cu	**0.08**

Fig. 14. The composition of the defined medium for the axenic cultivation of *P. aurelia*, stock 299.

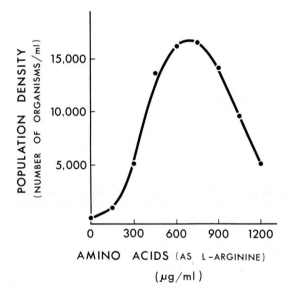

Fig. 15. Growth response of *P. aurelia*, strain 299 to amino acids (64).

IV. THE ESSENTIAL NUTRILITES

The requirements of *P. aurelia* for the individual components of the defined medium were determined by the single omission technique. In most instances the organisms cultured in the stock medium (Table 6) were depleted of the bulk of their endogenous food reserves by incubation in buffered saline solution (64) for 3–4 days at 27 °C in the dark.

A. Amino Acids

No growth is obtained in the first transfer when any one of the following amino acids are deleted from the medium: L-arginine, L-histidine, L-isoleucine, L-leucine, L-lysine, L-methionine, L-phenylalanine, L-proline, DL-threonine, L-tryptophan, L-tyrosine and DL-valine. Growth of *P. aurelia* is sub-optimal when either glycine or DL-serine is deleted from the complete medium and fails in the absence of both amino acids. The growth response to these amino acids individually is dependent upon the presence of thymidine in the medium (Table 10). Although glycine alone supports growth both in the presence and absence of thymidine, higher levels of this amino acid are necessary for optimal growth in the absence of thymidine. DL-Serine, on the other hand, replaces glycine for growth only in the presence of thymidine. Small quantities of serine known to be present in cephalin preparations, the source of phospholipid in these experiments, do not interfere in these assays. Experiments conducted both in the presence of 200 μg/ml of serine-free phosphatidyl ethanolamine, as a replacement for cephalin in the medium, and in the absence of this phospholipid yielded comparable results.

B. Vitamins

P. aurelia requires the following vitamins for growth: folic acid, nicotinamide, Ca-*d*-pantothenate, pyridoxal, riboflavin, thiamine, and DL-6-thioctic acid. When organisms depleted of endogenous reserves are used in the inoculum, growth fails in the first transfer when either folic acid, Ca-*d*-pantothenate, riboflavin, thiamine, or

TABLE 10

Growth response of Paramecium aurelia, *stock 299 to glycine and* DL-*serine in the complete medium and in the medium minus thymidine. Population expressed at % of a control consisting of the components described in Table 1.*

Complete medium[a] plus cephalin (250 μg/ml)				Medium minus thymidine		
glycine (μg/ml)	DL-serine (μg/ml)					
	0	600	1200	0	600	1200
0	0	33	102	0	0	0
100	60	72	89	0	92	88
200	80	104	104	0	102	93
400	102	102	96	65	106	102
600	102	93	98	85	98	104
800	92	100	96	106	102	96

[a] Complete medium contains thymidine (200 μg/ml).

DL-6-thioctic acid are omitted from the medium. In the absence of either nicotinamide or pyridoxal growth fails in the second transfer.

The need for DL-6-thioctic acid by the organisms can be demonstrated in the presence of sodium acetate in the medium. Although the level of growth is reduced to some extent in the absence of sodium acetate, this compound does not replace DL-6-thioctic acid for growth.

Biotin is included in the vitamin mixture although no requirement for this substance could be detected. The addition of avidin to a biotin-free medium does not inhibit growth.

In a medium containing a mixture of nucleosides and nucleotides which does not include thymidine, exceptionally high levels of folic acid are required for optimal growth (Table 11). These high levels can be reduced by the addition of thymidine to the medium. Populations are optimal in media containing folic acid at a concentration of 2 μg/ml in the presence of thymidine at a concentration of 100 μg/ml. A concentration of folic acid of 16 μg/ml is needed for optimal growth when thymidine is omitted from the medium.

C. Purines and Pyrimidines

In the absence of a source of purines and pyrimidines, growth of *P. aurelia* fails in the first transfer. Either guanosine or guanylic acid, but not adenosine, adenylic acid, inosine, inosinic acid, xanthosine or xanthylic acid satisfy the requirement for a purine. Adenosine, however, spares the need for guanosine-2′,3′-phosphate and is therefore included in the medium. The inhibition of the growth of *P. aurelia* by 2,6-diaminopurine, originally established in an incompletely defined medium by Miller and Van Wagtendonk (49) was confirmed using the defined medium. In addition, Soldo and Van Wagtendonk (64) found that 2,6-diaminopurine riboside was also inhibitory. Optimal concentrations of guanylic acid did not reverse the inhibition. *P. aurelia*'s need for pyrimidines is satisfied by either uridine, uridylic acid, cytidine,

TABLE 11

Growth response of Paramecium aurelia, *stock 299 to folic acid in the presence of thymidine. Populations expressed as % of a control consisting of the components described in Table 1 plus cephalin (250 μg/ml).*

Folic acid (μg/ml)	Thymidine (μg/ml)	
	0	100
0	0	0
1	13	56
2	30	102
4	61	112
8	84	105
16	99	102

or cytidylic acid. In accord with the findings of Tarantola and Van Wagtendonk (70) the corresponding free bases of both the purines and pyrimidines do not promote growth in the defined medium. Growth in the presence of the active purine and pyrimidine derivatives (total concentration 1.25 mg/ml), equalled that obtained in the presence of either RNA, DNA or yeast nucleic acid (1 mg/ml).

D. Lipids

Growth of *P. aurelia* is not possible in the absence of lipids. Fatty acids and stigmasterol are essential for growth. The requirements for fatty acids can be satisfied by the following mixture (final concentration in μg/ml): linoleic acid, 1.0; linolenic acid, 0.25; oleic acid, 20; palmitic acid, 15; stearic acid, 10. Transplantable growth is possible with this mixture as the sole fatty acid source, although optimal populations average only about 3000 organisms/ml. Small populations (1000–2000 organisms/ml) can be maintained in serial subculture for an indefinite period in the defined medium containing oleic acid as the sole source of fatty acids (oleic acid:stigmasterol ratios of 2:1 or 1:1 are best). Growth fails in the first or second transfer in the presence of either stearic acid, linolenic acid, or linoleic acid alone.

Higher populations can be obtained when various oleic acid-containing phospholipids are used as a source of fatty acids. In the presence of synthetic 1-oleoyl-2-stearoyl-*dl*-phosphatidyl serine and preparations of either phosphatidyl ethanolamine, phosphatidyl inositol or phosphatidyl serine from animal sources growth in the defined medium is equal to that in the stock medium. Synthetic L-α-dimyristoyl lecithin, synthetic L-α-dimyristoyl cephalin and a preparation of phosphatidyl ethanol-

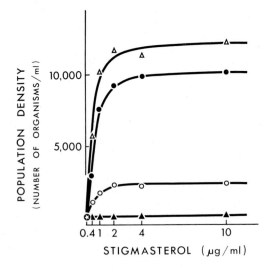

Fig. 16. Growth response of *P. aurelia*, stock 299 to stigmasterol in the presence of fatty acids and cephalin. Medium as in Table 6 minus lipids. No addition (▲——▲); fatty acids (Table 6) (50 μg/ml) (○——○); cephalin (250 μg/ml) (●——●); fatty acids (50 μg/ml) plus cephalin (250 μg/ml) (△——△) (64).

amine from a plant source do not promote growth. Growth response of *P. aurelia* is best in the medium containing the fatty acid mixture supplemented with an active phospholipid.

The need for stigmasterol can be demonstrated in the presence of a mixture of fatty acids or a phospholipid (cephalin) or both (Fig. 16).

E. Inorganic Salts

When either Ca^{2+} or Mg^{2+} ions are omitted from the medium, growth fails in the second transfer. In a medium devoid of inorganic phosphate, the amounts of these elements required for optimal growth are approximately: Ca^{2+}, 50 μg/ml and Mg^{2+}, 10 μg/ml (Fig. 17A). Growth response is unaffected by the addition of EDTA (ethylenediaminetetraacetic acid) to the medium at a concentration of 50 μg/ml (Fig. 17B). On the other hand, addition of 100 μg/ml of inorganic phosphate to the medium

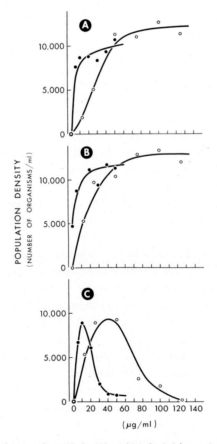

Fig. 17. Growth response of *P. aurelia* to Ca^{2+} (●) and Mg^{2+} (○) ions. *A*: In the absence of inorganic phosphate; *B*: In the presence of EDTA; *C*: In the presence of inorganic phosphate.

markedly affects the growth response of the organisms to Ca^{2+} ions and Mg^{2+} ions (Fig. 17C). No requirement for iron, copper, zinc or manganese could be demonstrated although these trace metals are included in the medium.

V. THE NUTRITION OF *Paramecium multimicronucleatum* AND *Paramecium caudatum*

A. *Paramecium multimicronucleatum*

The strain of *P. multimicronucleatum* used in the nutritional studies was originally isolated by W. H. Johnson near the Stanford University campus in 1941. At that time it had micronuclei, but became amicronucleate under laboratory conditions. This strain has been designated as syngen 2, mating type III of a so-called "*P. aurelia-multimicronucleatum* complex" by Sonneborn. He has also characterized a similar strain of *P. multimicronucleatum* (obtained from Florida) as syngen 2, mating type IV. The two strains will conjugate; however, they differ in the number of micronuclei. The mating type IV strain possesses 4 micronuclei, while Johnson's strain of mating type III is amicronucleate (66).

Paramecium multimicronucleatum was established in 1942 in axenic culture by Johnson and Baker (33). A pressed yeast juice prepared according to Buchner (4) diluted with triple-distilled water did support the growth of *P. multimicronucleatum*, albeit at a rate 50–75% lower than obtained in bacterized cultures. Heated pressed yeast juice would not support the growth.

In 1945 Johnson and Tatum (37) were able to show that two separate fractions of the "pressed yeast juice" were essential for the growth of *P. multimicronucleatum*. One fraction was heat-labile, non-dialyzable, and could be precipitated at three-fourths saturation with ammonium sulfate; the other was heat-stabile and remained water-soluble after autoclaving. Subsequently, Johnson (32) demonstrated that *P. multi-micronucleatum* required guanylic and cytidylic acids when grown in a medium composed of "pressed yeast juice" and proteose peptone. This medium consisted of guanylic and cytidylic acids, autoclaved "pressed yeast juice" supernatant, and proteose peptone. Utilizing the nondialyzable fraction (NDF) of an exhaustively dialyzed yeast extract (49), Johnson and Miller (34) were able to demonstrate that *P. multimicronucleatum* had an absolute requirement for stigmasterol. In the absence of stigmasterol growth declined and eventually failed in successive transfers (Fig. 18). They found that the optimum concentration of stigmasterol for growth was the same as for *P. aurelia*, e.g., 2 μg/ml (Fig. 19). Folic acid, riboflavin, thiamine, and panto-thenic acid were absolute requirements. Some evidence was obtained that nicotin-amide and pyridoxal were also required. The nitrogen requirements could be met with any of the following: proteose peptone, purified casein, crystalline ovalbumin, or a mixture of L-tyrosine, L-phenylalanine, L-tryptophan, DL-methionine, DL-threonine, L-leucine, L-lysine, L-histidine, L-arginine, DL-valine, DL-serine, L-proline, L-alanine, L-aspartic acid, L-glutamic acid, and glycine (35). Detailed studies to determine the

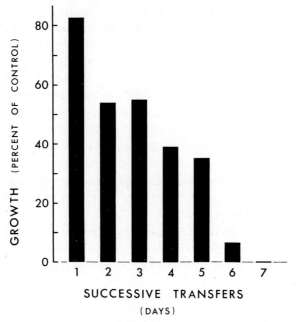

Fig. 18. The decline and failure of growth of *P. multimicronucleatum* in the absence of stigmasterol. Growth is indicated in per cent, taking the growth of the control in the presence of stigmasterol as 100 (34).

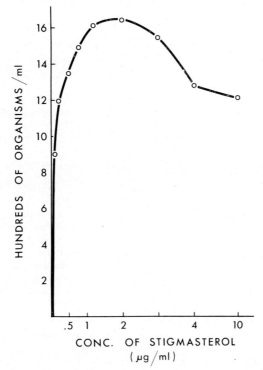

Fig. 19. The response of *P. multimicronucleatum* to varying concentrations of stigmasterol (34).

optimum concentrations of each amino acid were not carried out. However, by testing a number of amino acid mixtures, an optimal mixture was arrived at. In addition, either sodium acetate or sodium pyruvate were required for optimal growth.

After removal of nucleic acids from the NDF by enzymatic digestion and dialysis, Miller and Johnson were able to show specific purine and pyrimidine requirements for this organism (47). The purine requirement was satisfied by guanosine, deoxyguanosine, guanylic acid or deoxyguanylic acid, but not by any of the adenine derivatives. Cytidine, deoxycytidine, cytidylic acid, deoxycytidylic acid, uridine, uridylic acid, or deoxyuridine would serve as the pyrimidine source whereas thymidine and thymine would not. The free bases guanine, cytosine, and uracil did not replace their respective pentose derivatives. Of the active compounds, the pentosenucleosides were more active than the pentosenucleotides. Inosine, xanthine, 2,6-diaminopurine, 5-methylcytosine, and orotic acid did not support growth.

The fractionation of the NDF by Johnson and Miller (36) and Johnson (32) yielded evidence of three additional requirements for the growth of this ciliate. In the basal medium given in Table 12, the fatty acid requirement was readily satisfied by stearic

TABLE 12

Basal medium[a] used by Miller and Johnson (48) to demonstrate the fatty acid requirements of P. multimicronucleatum.

	Final concentration (μg/ml)		Final concentration (μg/ml)
Ethylenediamine tetra-acetic acid (EDTA)	16.0	Folic acid	2
$(MgSO_4) \cdot 7\ H_2O$	40.0	Thiamine·HCl	12
$Fe(NH_4)_2\ (SO_4)_2 \cdot 6\ H_2O$	10.0	L-Tyrosine	50
$MnCl_2 \cdot 4\ H_2O$	0.1	L-Phenylalanine	75
$ZnCl_2$	0.02	L-Tryptophan	50
$CaCl_2 \cdot 2\ H_2O$	20.0	DL-Methionine	150
$CuCl_2 \cdot 2\ H_2O$	2.0	DL-Threonine	150
$FeCl_3 \cdot 6\ H_2O$	0.5	DL-Isoleucine	150
K_2HPO_4	570.0	L-Leucine	150
KH_2PO_4	570.0	L-Lysine·HCl	125
		L-Histidine	50
Sodium acetate·3 H_2O	570.0	L-Arginine·HCl	100
Sodium pyruvate	570.0	DL-Valine	75
		DL-Serine	200
Stigmasterol	2	L-Proline	50
Ca-pantothenate	2	L-Alanine	25
Nicotinamide	4	L-Aspartic acid	50
Pyridoxal·HCl	4	L-Glutamic acid	75
Riboflavin	4	Glycine	25

[a] The EDTA, inorganic salts, acetate, pyruvate and the amino acids or the stigmasterol may be autoclaved together in final concentration. Autoclaving the amino acids and the stigmasterol together or autoclaving NDF together with the amino acids resulted in the inhibition of growth. The B-vitamins were sterilized separately and added aseptically.

acid or oleic acid. Arachidic acid (C_{20}) and palmitic acid (C_{16}) had lower growth promoting activity than did stearic acid. The unsaturated fatty acids, palmitoleic acid ($C_{16}\Delta^9$), linoleic acid ($C_{18}\Delta^{9,\ 12}$), linolenic acid ($C_{18}\Delta^{9,\ 12,\ 15}$) and arachidonic acid ($C_{20}\Delta^{5,\ 6,\ 11,\ 14}$) and the saturated fatty acids myristic acid (C_{14}), lauric acid (C_{12}), capric acid (C_{10}), caprylic acid (C_8) and caproic acid (C_6) were inactive. The methyl esters of several of these compounds gave the same growth response (48). The second fraction of the NDF could be replaced by crystalline ovalbumin. The third "protein" factor remained uncharacterized.

B. *Paramecium caudatum*

Glaser and Coria (26) succeeded in 1933 to establish *P. caudatum* in an axenic medium containing heat-killed yeast cells, liver extract, and sterile slices of rabbit kidney. Burbank (5) showed that *P. caudatum* could grow in the medium developed by Van Wagtendonk and Hackett (72) and thus had similar growth requirements. Sterbenz (68) established *P. caudatum* in an axenic medium based on that of Van

TABLE 13

Growth medium for P. caudatum *(44). Final pH adjusted to 7.0 with 0.1 N NaOH. Undefined portion. Protein factor obtained from peas 50 to 10.*

	μg/ml		μg/ml
L-Alanine	110	Linoleic acid	3.75
L-Arginine	206	Oleic acid	1.25
L-Aspartic acid	122		
Glycine	10	Stigmasterol[a]	2
L-Glutamic acid	233		
L-Histidine	87	Ca-pantothenate[a]	2
DL-Isoleucine	270	Nicotinamide	4
L-Leucine	244	Pyridoxal·HCl	4
L-Lysine	272	Riboflavin	4
DL-Methionine	245	Folic acid	2
L-Phenylalanine	160	Thiamine·HCl	12
L-Proline	250		
DL-Serine	394	Na_2-ethylenediamine tetra-acetate	20
DL-Threonine	238		
DL-Tyrosine	100		
L-Tryptophan	76	$MgSO_4 \cdot 7\ H_2O$	40
DL-Valine	96	$(NH_4)_2 \cdot SO_4 \cdot 6\ H_2O$	10
		$MnCl_2 \cdot 4\ H_2O$	0.1
Guanylic acid	75	$ZnCl_2$	0.02
Adenylic acid	30	$CaCl_2 \cdot 2\ H_2O$	20
Cytidylic acid	75	$CuCl_2 \cdot 2\ H_2O$	2
Uridylic acid	20	$FeCl_3 \cdot 6\ H_2O$	0.5
		K_2HPO_4	570
Sodium acetate	570	KH_2PO_4	570
Sodium pyruvate	570		

[a] The stigmasterol, the mixture of B-vitamins and the protein factor were each sterilized separately and added aseptically to the other components which were mixed and sterilized together.

Wagtendonk *et al.* (76) and found that this organism also had a steroid requirement which could, as in the case of *P. aurelia*, be satisfied by stigmasterol or sitosterol. Lilly *et al.* (45) found that the yeast fraction could be replaced by a non-dialyzable fraction from dried split peas. This factor was purified and concentrated by Lilly and Klosek (44). These investigators developed a medium composed of known chemical components to which 10–50 μg of the pea protein factor had to be added. They were also the first to establish a requirement for the fatty acids linoleic acid and oleic acid for a *Paramecium* species. Their medium is given in Table 13.

The purified pea factor was subjected to an amino acid analysis using both acid and alkaline hydrolysis and two-dimensional paper chromatography. The following amino acids were identified: alanine, arginine, cystine, glycine, histidine, isoleucine, leucine, lysine, methionine, phenylalanine, proline, serine, threonine, tryptophan, tyrosine and valine. Aspartic acid and glutamic acid could not be identified.

The last publication concerned with the nutritional requirements of *P. caudatum* appeared in 1964 (53). Using a medium in which the concentration of serine had been

TABLE 14

Experimental medium for testing the efficiency of solids in the nutrition of P. caudatum. *Final pH adjusted to 7.0 with 0.1 N NaOH. Undefined portion. Protein factor obtained from peas 80.*

Chemically-defined portion		Chemically-defined portion	
L-Alanine	25	Linoleic acid	3.75
L-Arginine	100	Oleic acid	1.25
L-Aspartic acid	50		
Glycine	25		
L-Glutamic acid	75	Stigmasterol	2
L-Histidine	50		
DL-Isoleucine	150		
L-Leucine	150	Ca-pantothenate	2
L-Lysine	125	Nicotinamide	4
DL-Methionine	150	Pyridoxal·HCl	4
L-Phenylalanine	75	Riboflavin	4
L-Proline	50	Folic acid	2
DL-Serine[a]	200	Thiamine·HCl	12
DL-Threonine	150		
DL-Tyrosine	50	Na$_2$-ethylenediamine	20
L-Tryptophan	50	tetra-acetate	
DL-Valine	75		
		MgSO$_4$·7 H$_2$O	40
Guanylic acid	75	(NH$_4$)$_2$SO$_4$	10
Adenylic acid	30	MnCl$_2$·4 H$_2$O	0.1
Cytidylic acid	75	ZnCl$_2$	0.02
Uridylic acid	20	CaCl$_2$·2 H$_2$O	20
		CuCl$_2$·2 H$_2$O	2
		FeCl$_3$·6 H$_2$O	0.5
Sodium acetate (anhyd.)	570	K$_2$HPO$_4$	570
Sodium pyruvate	570	KH$_2$PO$_4$	570

[a] Better growth was later obtained with serine at 400. The undefined portion could be omitted when Celkate was added.

increased twofold (Table 14), Reilly claims to have observed that a synthetic magnesium silicate, Celkate T-21 added in "about 4 mg quantities to 5 ml of culture fluid" resulted in excellent growth in the defined medium. Although Reilly states that the formulation of a chemically defined medium for the ubiquitous and important ciliate should provide a stimulus to research in its nutrition, metabolism and catabolism comparable to that provided by the first establishment of *Tetrahymena* in defined media, no further publications have come forth from the Lilly group.

VI. DISCUSSION

While the original search for a defined medium was carried out with the particle-less strain 51 of syngen 4, the later efforts to design an axenic medium which could support growth of both particle-bearing and particle-free strains of *P. aurelia* used strain 299 of syngen 8 as the test organism. The medium developed by Soldo *et al.* (65) did not support the growth of any strain of the odd-numbered syngens tested, while animals of strains of the even numbered syngens could be subcultured indefinitely. Of the particle-bearing strains, only the pi-, mu- and lambda-bearing strains were able to maintain the particles. Soldo *et al.* (65) hoped that these differences in nutritional requirements of odd- and even-numbered syngens might provide an additional parameter for the taxonomic classification of *P. aurelia*. The results of further investigations by Van Wagtendonk *et al.* (75) could not support this viewpoint (Fig. 20). When the TEM/stigmasterol ratio in medium CT-1 (Table 6) was changed from 100:5 to 40:4 the strains 548 mu, 551 mu, 540 mu of syngen 1, the strain 87 nu of syngen 5, and the strain 204 sigma of syngen 9 could be grown in the CT-1 medium. However,

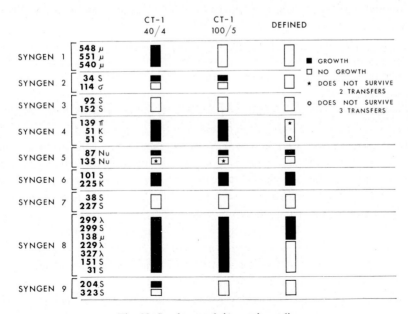

Fig. 20. Stock growth in axenic media.

strains 92 sigma and 152 sigma of syngen 3, the strain 135 nu of syngen 5, strains 38 sigma and 227 sigma of syngen 7, and strain 323 sigma of syngen 9 were not able to grow in the modified CT-1 medium. The kappa-bearing strain 51 of syngen 4 which had resisted all efforts to be grown in axenic medium, could first be grown in the CT-1 medium containing the TEM-4T/stigmasterol ratio of 40:4 and could be adapted to the TEM-4T/stigmasterol ratio of 100:5. Kappa was maintained through all subcultures and single isolation cultures.

The defined medium is even more restricted in its ability to support growth of the various strains. Of the odd-numbered syngens, only the 101 sigma and 225 kappa of syngen 5 can be grown in this medium. The particle-bearing strains of 225 kappa, 138 mu and 87 nu maintain their particles in the defined medium.

It is evident that the various strains of *P. aurelia* vary widely in their responses to the three media. A perusal of Fig. 20 poses many questions which, at present, must go unanswered. The strains of syngen 1 can only be maintained in a medium containing the TEM-4T/stigmasterol ratio of 40:4. Two strains of syngen 3 cannot grow in either of the three media. The particle-bearing strain 87 nu of syngen 5 can grow in all three media, while the particle-bearing strain 135 nu of the same syngen cannot be grown in either of these media. This also holds for strains 38S and 227S of syngen 7. Strain 204S of syngen 9 can be grown in a TEM-4T/stigmasterol ratio of 40:4, while strain 323S of the same syngen cannot be grown at all in any of the media.

Similar variations are evident in the growth response of the members of the even-numbered syngens. Strains 34S of syngen 2 can be cultivated in media containing TEM-4T/stigmasterol ratios of 40:4 and 100:5, but cannot be grown in the defined medium. On the other hand strain 114 of the same syngen cannot be cultured in any of the three media. None of the strains of syngen 4 can grow in the defined medium, while the two strains of syngen 6 will grow in this medium. Only strains 299S and 138 mu of syngen 8 will grow in the defined medium.

Are these phenomena due to imbalances of the medium, or do they reflect genetic differences between the stocks with regard to their synthetic ability? Since Butzel has been able to induce mating in axenic media (Chapter 2, p. 102) the second possibility can now be investigated.

The quantitative need for vitamins for the optimal growth of *P. aurelia* and *P. multimicronucleatum* is substantially greater than that required by the smaller ciliates, *Tetrahymena pyriformis* (40) and *Glaucoma chattoni* (31). This probably reflects differences in the size of the organisms. *Paramecium* is 5 to 10 times larger than either of these ciliates. On the other hand, increased consumption of vitamins by *P. aurelia* may be related to an increased need for these substances for use as coenzymes in the catabolism of amino acids as a source of energy.

Exceptionally high levels of folic acid required for optimal growth of *P. aurelia* can be reduced by addition of thymidine to the medium. Similarly, thymidine spares but does not replace folic acid for the growth of *Tetrahymena* (15). Detailed studies concerned with the mechanism of thymidine-sparing in this ciliate have led to the view that folic acid may serve as a precursor for the synthesis of a cofactor involved in the enzymatic conversion of deoxycytidine to thymidine and possibly as a source of a pteridine participating in the transfer of a one carbon fragment. In *E. coli* the enzyme thymidylate synthetase catalyzes the synthesis of thymidylic acid from deoxyuridylic acid (24). In this reaction:

deoxyuridylic acid $+$ N_5, N_{10}-methylene folic acid $\rightleftarrows$ thymidylic acid $+$ dihydrofolic acid

folic acid serves a dual function in that it acts as both a carrier of the one carbon fragment and as a reducing agent. Presumably, folic acid plays a similar role in the metabolism of *P. aurelia*.

DL-6-Thioctic acid is essential for the growth of *Paramecium* and is probably a requirement of ciliates in general. Previous attempts to demonstrate this requirement for *P. aurelia* and for *P. multimicronucleatum* were unsuccessful because of the need to include a crude component from yeast which may have contributed this factor for growth. A report by Sterbenz (68) suggested that DL-6-thioctic acid was essential for the growth of *P. caudatum*.

The amounts of amino acids required for optimal growth of *P. aurelia* in a defined medium are considerably greater than those originally reported to be necessary by Miller and Van Wagtendonk (49). Undoubtedly, the yeast component of the medium used for the growth of *P. aurelia* contributed substantially to the amino acid needs of the organisms. The proteins of the yeast fraction apparently contained sub-optimal concentrations of arginine, histidine, isoleucine, leucine, lysine, phenylalanine, threonine and tryptophan since requirements for these amino acids could be established only after the organisms were transferred through at least three serial subcultures. Growth failed after the second serial transplant when either methionine or serine was omitted from the medium and was sub-optimal in the absence of either glycine, proline or valine.

The failure to synthesize proline and glycine by both *P. aurelia* and *P. multimicronucleatum* are exceptions to the patterns generally found in animal organisms. However, Weiss and Ball (79) reported that proline and glycine are required by *Trichomonas foetus*, and Fuller (25) found that proline was essential for the growth of *Glaucoma scintillans*. The need for glycine by the growing chick has been known since 1944 (29). Goldberg and De Meillon (27) reported an absolute glycine requirement for *Aedes aegypti*.

In a synthetic medium, first transplant data confirm the requirements for arginine, histidine, isoleucine, leucine, lysine, methionine, phenylalanine, threonine, tryptophan and tyrosine. In addition, proline and valine were found to be essential for the growth of *P. aurelia*.

Growth response to glycine and metabolically related DL-serine is markedly influenced by the thymidine (and presumably folic acid) content of the medium. While glycine alone is adequate for growth, both in the presence or in the absence of thymidine, DL-serine replaces glycine only when thymidine is present in the medium. On the basis of results obtained in a yeast-containing medium, Miller and Van Wagtendonk (49) suggested that DL-serine was essential for the growth of *P. aurelia*, stock 47.8. Failure of the organisms to grow in the absence of this amino acid was probably due to insufficient amounts of glycine (25 μg/ml) and folic acid (1.0 μg/ml) in their medium. Similarly, failure of *Tetrahymena pyriformis* to grow when serine was omitted from a culture medium of known composition (39), was traced to inadequate supplies of both glycine and folic acid in the medium. Several strains were capable of utilizing glycine or threonine for growth provided the concentration of folic acid was high. The ability of DL-serine to replace glycine in the presence of thymidine (and presumably folic acid) indicates that the serine transhydroxymethylase enzyme catalyzing the reaction:

$$\text{DL-serine} + \text{tetrahydrofolic acid} \rightleftarrows \text{glycine} + N_5, N_{10}\text{-methyiene tetrahydrofolic acid}$$

is operative in *P. aurelia*. Since high levels of threonine are always present in the medium it is probable that threonine aldolase, an enzyme capable of catalyzing the formation of glycine from threonine (46) present in some strains of *Tetrahymena* (16) is not functional in *Paramecium*.

The apparent need for serine by *P. multimicronucleatum* as reported by Johnson and Miller (34, 35) was probably also due to the presence of insufficient amounts of glycine (25 μg/ml) and folic acid (2 μg/ml) in their medium.

The inability of *Paramecium aurelia* to oxidize phenylalanine to tyrosine is concluded from the fact that an exogenous source of both amino acids is essential for continued growth. Most animals can meet their tyrosine needs when supplied with a source of phenylalanine, although apparently the reverse does not occur; tyrosine will only spare phenylalanine in such cases. However, Eagle (19, 20) reported that both phenyl-alanine and tyrosine are essential for two mammalian cell lines—a mouse fibroblast and a human uterine carcinoma—in tissue culture. Until recently a one-step oxidation was postulated for the conversion of phenylalanine to tyrosine. Studies of auxotrophic mutants of *E. coli* K12, strain W (57) strongly support the view that the conversion of phenylalanine to tyrosine is indirect, the two amino acids being synthesized from a common precursor by reversible reactions.

At various times reports have appeared in the literature that protozoa need particular food, or large molecules. Tarantola and Van Wagtendonk (70) reported that proteose peptone was active in promoting growth when used in conjunction with the enzymatic-ally digested yeast fraction, and suggested that a specific peptide was involved. Johnson and Miller (34) found that two strains of *P. multimicronucleatum* and two strains of *P. aurelia* (strains 47 and 51) required ovalbumin. Their results suggested that the role of ovalbumin might be that of a carrier of fatty acids and perhaps of steroids to their sites of metabolism in *Paramecium*. *Colpidium campylum* (41) would only grow in the presence of proteins in the molecular weight range of casein, whereas *Glaucoma scintillans* (41) would equally well utilize casein or an enzymic hydrolysate of casein. It was proposed that *Glaucoma scintillans* and *Colpidium campylum* required molecules of a certain minimum size. Reilly claimed that *P. caudatum* required partic-ular material in the form of a synthetic hydrous magnesium silicate. Similarly Ras-mussen and Kludt (52) reported that growth of *Tetrahymena pyriformis* was enhanced by the presence of talcum, quartz, clay and calcium carbonate or water-soluble com-pounds which through hydrolytic decomposition readily produce insoluble particles in aqueous solution, such as ferrous ammonium sulfate, ferrous sulfate and potassium aluminum sulfate. Soldo and Van Wagtendonk (64) observed that their defined medium was slightly turbid and contained small particles throughout the medium. Upon standing, a slight sediment accumulated in the bottom of the tube. Filtrates of the medium, passed through millipore filters (pore size 0.4μ in diameter) did not support the growth of the organisms.

All these observations can probably be explained on the basis of adsorption of essential nutrilites on the large molecules. These compounds would then be released slowly after the large molecules had been ingested. In fact, Reilly found that riboflavin was absorbed on the magnesium silicate. A comparable finding is that the fatty acid requirements of *P. aurelia* are best served in chemically bound forms. It may well be that the statement made by Soldo and Van Wagtendonk (62) "Perhaps as a result of their natural predilection for phagotrophy, ciliates such as *Paramecium* adapted to an environment in which the nutritional needs for fatty acids were satisfied by extracting these substances by enzymatic digestion of food bacteria. The relatively small amounts of these compounds which accumulated under such conditions were assimilated continuously during growth and posed no difficulties for the organisms because inhibitory levels were never reached" can explain the apparent need for particulate matter or large molecules in the nutrition of *Paramecium* and other protists.

The purine and pyrimidine requirements of *P. aurelia* and *P. multimicronucleatum* are essentially the same. Butzel and Van Wagtendonk (6) found that *P. aurelia* was unable to incorporate glycine into the nucleic acids of the macronucleus. The metabolic block in the synthesis of the pyrimidine nucleus is not known, although it can be inferred from the nutritional data (47, 70) that *Paramecium* is unable to convert orotic acid into the pyrimidine nucleus. The inability to synthesize the purine bases is shared by several members of the Phylum Protozoa, *P. aurelia* (64, 70), *P. multimicronucleatum* (32, 35), *Tetrahymena pyriformis* (42) and *Trypanosoma mega* (3). This fact suggested to Butzel and Van Wagtendonk that: "Perhaps it will be found that a biochemical distinction of the Phylum Protozoa will be the inability to form their nucleic acids from simple precursors, resulting in a dependence upon exogenous preformed purines for growth".

P. aurelia, *P. multimicronucleatum* and *P. caudatum* require a steroid for growth. This requirement is satisfied by stigmasterol. Only a few microorganisms and one mammalian species are known to require exogenous steroids for growth. The studies of Cailleau (7–10) indicated a true metabolic requirement for steroids for *Trichomonas columbae*, *T. foetus*, *Entrichomastix colubrorum*, and *T. batrachorum*. Vishniac and Watson (78) studied the steroid requirements of the myxothallophyte *Labyrinthula vitellina* var. *pacifica*. The other microorganisms which require steroids are organisms of the *Pleuropneumonia* group (21), and *Peranema trichophorum* (69). Ross *et al.* (54) and Van Wagtendonk and Wulzen (74) reported that a steroid, later identified as stigmasterol (38), functioned as an essential metabolite for guinea pigs. The specificity requirements for these organisms vary widely, as is evident from Table 15.

On the basis of an exhaustive survey of various steroids, Conner and Van Wagtendonk (12) tried to establish the molecular configuration of the steroid molecule, necessary for its biological activity. Some of the conclusions reached in 1955 may no longer be tenable, because many of the steroids tested were undoubtedly impure. The

TABLE 15

A comparison of the growth-promoting activity of various steroids.

Compound	Trichomonas columbae	Labyrinthula vitellina	Peranema trichophorum	Paramecium aurelia	Trichomonas foetus, T. batra-chorum, Eutricho-mastix colubrorum	Guinea-pig	Pleuro-pneumonia-like organisms
Cholesterol	+	+	+	−	+	−	+
β-Sitosterol	+	+	+	+	.	+	.
γ-Sitosterol	.	−	.	+	.	+	.
Fucosterol	.	+	.	+	.	+	.
Stigmasterol	.	−	+	+	.	+	+
Poriferasterol	.	.	+	+	.	.	.
Brassicasterol	.	−	+	+	.	.	.
Ergosterol	+	−	+	−	.	.	.
5-Dihydro ergosterol	.	−	.	−	.	.	.
Zymosterol	.	−	.	−	.	.	.

+ = active; − = inactive; . = not tested.

present day techniques for purification and identification of impurities were not available at that time. Nevertheless, some of the postulates are still valid today. These are: (1) Steroids having the cholesterol configuration are inactive for *Paramecium;* (2) Addition of either an ethyl, vinyl, or methyl group at C_{24} imparts activity to the molecule, since stigmasterol, the sistosterols, poriferasterol and brassicasterol are active; (3) unsaturation of the side chain at $C_{22, 23}$ enhances the activity; stigmasterol, poriferasterol, and $\Delta^{4,22}$-stigmastadienone are more active than the sistosterols; (4) saturation of all the double bonds destroys activity, as testified by the inactivity of ergostanol and stigmastanyl acetate; (5) esterification of the 3-hydroxyl group lowers the activity; (6) oxidation of the 3-hydroxyl group to a ketone group and the simultaneous rearrangement of the double bond from $C_{5,6}$ to $C_{4,5}$ does not affect the activity since $\Delta^{4,22}$-stigmastadienone is as active as stigmasterol; (7) degradation of the side chain destroys the biological activity; progesterone, testosterone, methyl cholate and desoxycorticosterone are all inactive; (8) changing the side chain of the steroid molecule to that of a saponine or of a cardiac aglycone results in inactivity of the molecule since disogenin and digoxigenin are inactive.

P. aurelia and *P. multimicronucleatum* require in addition to stigmasterol an exogenous source of fatty acids for growth. Unlike *P. multimicronucleatum*, whose nutritional requirements for fatty acids may be satisfied by either oleic or stearic acid, *P. aurelia* has a specific need for oleic acid. Only oleic acid and oleic acid-containing lipids effectively support growth. An increase of the concentration of oleic acid in the medium inhibits the multiplication of *P. aurelia*. These inhibitory effects can be reduced by the addition of increased amounts of stigmasterol. Definite, but non-stoichiometric

TABLE 16

Growth response of P. aurelia *to oleic acid and mixtures of fatty acids based on the composition of TEM-4T and Tween 80.*

Addition	Concentration tested (μg/ml)	Oleic acid concentration (μg/ml)	Number of 7-day serial subcultures		
			1	2	3
None	—	—	0	—	—
Oleic acid	2.5	2.5	1800[a]	700	800
	5	5	2100	1600	1500
	10	10	600	0	—
	20	20	0	—	—
	40	40	0	—	—
TEM-4T	100	43	9900	10,700	10,400
Fatty acid mixture	6	2.6	2100	800	600
based on	12	5.2	6000	2400	3000
TEM-4T	25	10.3	7900	5400	6000
	50	21.5	8000	4800	3900
	100	43	5300	1400	1600
Tween 80	50	44	3600	3900	4000
Fatty acid mixture	6	5.5	2900	0	—
based on	12	11	2400	0	—
Tween 80	25	22	2600	0	—
	50	44	1800	0	—
	100	88	400	0	—

[a] Population density (number of animals/ml).

ratios of steroid: fatty acids would produce good growth. Similarly, a definite balance of fatty acids and stigmasterol is necessary for the optimal growth of *P. multimicronucleatum*.

The inhibition of growth by increased concentrations of oleic acid can be substantially reduced when oleic acid is added in a mixture of fatty acids based on the fatty acid composition of TEM-4T (Table 16). In contrast, the organisms did not

TABLE 17

Growth response of P. aurelia *to oleic acid in presence of Tween 60.*

Oleic acid (μg/ml)	Tween 60 (μg/ml)		
	0	50	100
0	0[a]	200	0
1.0	400	1400	3900
2.0	1100	4600	4500
5.0	1400	7300	5900
10.0	400	5500	7000
20.0	0	4300	3100

[a] Number of animals/ml; data pooled from 3rd, 7-day serial transfer of 3 experiments.

grow when oleic acid was added in a mixture of fatty acids based on the composition of Tween 80 or Tween 85, which contain predominantly oleic acid. The presence of stearic and palmitic acids in a fatty acid mixture apparently reduces the toxicity of oleic acid.

The non-ionic emulsifier, Tween 60, whose major fatty acid component is stearic acid, fails to meet the nutritional requirements for lipids of *P. aurelia*, yet markedly stimulates growth in the presence of very small amounts of oleic acid (Table 17).

Tween 80 which contains predominantly oleic acid, effectively promotes growth of the ciliates, while Span 80, similar in oleic acid content to Tween 80, is less effective in supporting growth. Although it is not yet possible to fully explain these effects, it is clear that oleic acid functions as an essential metabolite for *P. aurelia*. Furthermore, the fatty acid requirements of *P. aurelia* are best served in a chemically-bound form, such as phospholipids, TEM, Tweens or glycerides.

The need for proportioned amounts of lipids for optimum growth is also found in other microorganisms. Shorb and Lund (56) found that the inhibition of growth of *Trichomonas* caused by linoleic acid (an essential nutrilite for this organism) could be overcome by the addition of adequate amounts of cholesterol to the culture medium. Kodicek and Worden (43) observed that inhibition of growth of *Lactobacillus helveticus* by unsaturated fatty acids could be reversed by the addition of cholesterol and related steroids. *Tetrahymena corlissi* TH-X grew better in the presence of certain combinations of oleic acid and cholesterol than in the presence of either lipid alone (30).

The interpretation of the variety of responses to lipids in microorganisms (inhibition of growth–28, 43, 71, 81, 50, 58, 59, 31, 56; stimulation of growth–11, 30, 55; essential nutrilites–1, 22, 48, 51, 56, 80) is complicated because of certain of their physical properties. For example the ability of lipids to lower the surface tension could account for both inhibition and stimulation of growth because of changes in the permeability of the cell wall. However, the recent findings by Conner *et al.* that a derivative of stigmasterol is found to be present in *P. aurelia* predominantly as a fatty acid ester suggests a more direct role for these lipids in the metabolism of the cell.

A supplementary source of exogenous carbon is not essential for *P. aurelia*. Growth which is sub-optimal in the absence of the non-essential amino acids alanine, asparagine, aspartic acid and glutamic acid, can be restored to optimal levels by the addition of sodium acetate or sodium pyruvate. Johnson and Miller (34) reported that *P. multimicronucleatum* required both sodium acetate and sodium pyruvate for growth in their partially defined medium. Kidder and Dewey (40) found that *Tetrahymena pyriformis* has no organic carbon requirement other than the carbon present in the amino acids, although sodium acetate and sodium pyruvate were active in stimulating growth. Apparently, *P. aurelia*, *P. multimicronucleatum* and *Tetrahymena pyriformis* can utilize the carbon skeleton of the amino acids to satisfy their carbon requirements.

VII. CONCLUDING REMARKS

Biochemical investigations with *P. aurelia* have lagged far behind those on *Tetrahymena pyriformis*. This has been due to the fact that the original media for the axenic growth of *Paramecium* were difficult to prepare. Now that easier to prepare media and a synthetic medium are available, an upsurge in biochemical studies on this organism equal to that seen with *Tetrahymena* after Kidder and Dewey developed a synthetic medium, can be expected. It is also hoped that the methods employed in developing a synthetic medium for *Paramecium* will serve as models for the construction of synthetic media for other protozoa. The tasks will never be easy, but the results will be rewarding.

VIII. REFERENCES

(Abstracts are indicated by *)

1 Allen, J. R., Lee, J. J., Hutner, S. H. & Storm, J. 1966. Prolonged culture of the voracious flagellate *Peranema* in antioxidant-containing media. *J. Protozool.* **13**, 103–108.

2 Bergmann, W. & Low, E. M. 1947. Contributions to the study of marine products. XX. Remarks concerning the structure of sterols from marine invertebrates. *J. Org. Chem.* **12**, 67–75.

3 Boné, G. J. & Steinert, M. 1956. Isotopes incorporated in the nucleic acids of *Trypanosoma mega*. *Nature* **178**, 308–309.

4 Buchner, E., Buchner, H. & Hahn, M. 1903. *Die Zymasegärung*, Oldenbourg, Munich, Berlin, 58–66.

5 *Burbank, W. D. 1950. Growth of pedigreed strains of *Paramecium caudatum* and *Paramecium aurelia* on a non-living medium. *Biol. Bull.* **99**, 353–354.

6 Butzel, Jr., H. M. & Van Wagtendonk, W. J. 1963. Autoradiographic studies of the differential incorporation of glycine, and purine and pyrimidine ribosides by *Paramecium aurelia*. *J. Gen. Microbiol.* **30**, 503–507.

7 Cailleau, R. 1937. La nutrition des flagellés tétramitidés. Les stérols, facteurs de croissance pour les trichomonades. *Ann. Inst. Pasteur* **59**, 137–172.

8 — 1938. Le cholestérol et l'acide ascorbique, facteurs de croissance pour le flagellé tétramitidé *Trichomonas foetus* Riedmuller. *Compt. Rend. Soc. Biol.* **127**, 861–863.

9 — 1938. L'acide ascorbique et le cholestérol, facteurs de croissance pour le flagellé *Eutrichomastix colubrorum*. *Compt. Rend. Soc. Biol.* **127**, 1421–1423.

10 — 1939. Le cholestérol, facteur de croissance pour le flagellé *Trichomonas batrachorum*. *Compt. Rend. Soc. Biol.* **130**, 1089–1091.

11 Cohen, S. 1949. Stimulation of the growth of a strain of *Corynebacterium diphteriae* by polyvinyl alcohol. *J. Bacteriol.* **58**, 783–790.

12 Conner, R. L. & Van Wagtendonk, W. J. 1955. Steroid requirements of *Paramecium aurelia*. *J. Gen. Microbiol.* **12**, 31–36.

13 —, — & Miller, C. A. 1953. The isolation from lemon juice of a growth factor of steroid nature required for the growth of a strain of *Paramecium aurelia*. *J. Gen. Microbiol.* **9**, 434–439.

14 Conner, R. L., Landrey, J. R., Kaneshiro, E. S. & Van Wagtendonk, W. J. 1971. The metabolism of stigmasterol and cholesterol by *Paramecium aurelia*. *Biochim. Biophys. Acta* **239**, 312–319.

15 Dewey, V. C. & Kidder, G. W. 1953. Factors affecting the requirement of *Tetrahymena pyriformis* (geleii) for folic acid. *J. Gen. Microbiol.* **9**, 445–453.

16 — & — 1960. The influence of folic acid, threonine and glycine on serine synthesis in *Tetrahymena*. *J. Gen. Microbiol.* **22**, 72–78.

17 — & — 1960. Serine synthesis in *Tetrahymena* from non-amino acid sources; compounds derived from serine. *J. Gen. Microbiol.* **22**, 79–92.

18 *Dippell, R. V. 1954. A preliminary report on the chromosomal constitution of certain variety 4 races of *Paramecium aurelia*. *Proc. 9th Congr. Int. Genet.* 1109.

19 Eagle, H. 1954. The specific amino acid requirements of a mammalian cell (L) in tissue culture. *J. Biol. Chem.* **214**, 839–852.

20 — 1955. The specific amino acid requirements of a human carcinoma cell (strain HeLa) in tissue culture. *J. Exp. Med.* **102**, 37–48.

21 Edward, D. G. & Fitzgerald, W. A. 1951. Cholesterol in the growth of organisms of the *Pleuropneumonia* group. *J. Gen. Microbiol.* **5**, 576–586.

22 Ellinghausen, H. C. & McCullough, W. G. 1965. Nutrition of *Leptospira pomona* and growth of 13 other serotypes. *Amer. J. Vet. Res.* **26**, 39–44.

23 Fieser, L. F. & Fieser, M. 1949. *Natural Products Related to Phenanthrene*, 3rd edition, Reinhold Publishing Corp., New York, N.Y.

24 Friedkin, M. 1963. Enzymatic aspects of folic acid. *Ann. Rev. Biochem.* **32**, 185–214.

25 Fuller, R. C. 1948. *Studies on the Biochemistry of Glaucoma scintillans.* Master's Thesis, Amherst College, Amherst, Mass.

26 Glaser, R. W. & Coria, N. A. 1933. The culture of *Paramecium caudatum* free from living organisms. *J. Parasitol.* **20**, 33–37.

27 Goldberg, L. & De Meillon, B. 1948. The nutrition of the larva of *Aedes aegyptii Linnaeus*. IV. Protein and amino acid requirements. *Biochem. J.* **43**, 379–387.

28 Gryllenberg, H., Rossander, M. & Roine, P. 1954. On the growth inhibition of *Lactobacillus bifidus* by certain fatty acids. *Acta Chem. Scand.* **8**, 133–134.

29 Hegsted, D. M. 1944. Growth in chicks fed amino acids. *J. Biol. Chem.* **156**, 247–252.

30 Holz, Jr., G. G., Wagner, B. & Erwin, J. 1961. A nutritional analysis of the sterol requirements of *Tetrahymena corlissi, TH-X. Comp. Biochem. Physiol.* **2**, 202–217.

31 —, —, — & Kessler, D. 1961. The nutrition of *Glaucoma chattoni A. J. Protozool.* **8**, 192–199.

32 Johnson, W. H., 1952. Further studies on the sterile culture of *Paramecium. Physiol. Zool.* **25**, 10–15.

33 — & Baker, E. G. S. 1942. The sterile culture of *Paramecium multimicronucleata. Science* **95**, 333–334.

34 — & Miller, C. A. 1956. A further analysis of the nutrition of *Paramecium. J. Protozool.* **3**, 221–226.

35 — & — 1957. The nitrogen requirements of *Paramecium multimicronucleatum. Physiol. Zool.* **30**, 106–113.

36 — & — 1958. A fatty acid requirement for *Paramecium multimicronucleatum. J. Protozool.* **5**(Suppl.), 14.

37 — & Tatum, E. L. 1945. The heat-labile growth factor for *Paramecium* in pressed yeast juice. *Arch. Biochem.* **8**, 163–168.

38 Kaiser, E. & Wulzen, R. 1951. Identification of a sugar-cane sterol antistiffness factor as stigmasterol. *Arch. Biochem.* **31**, 326–328.

39 Kidder, G. W. & Dewey, V. C. 1947. Studies on the biochemistry of *Tetrahymena*. X. Quantitative response to essential amino acids. *Proc. Nat. Acad. Sci. U.S.* **33**, 347–356.

40 — & — 1951. The biochemistry of ciliates in pure culture. In Lwoff, A., *The Biochemistry and Physiology of Protozoa*, Academic Press, New York, N.Y., 323–400.

41 —, — & Fuller, R. C. 1954. Nitrogen requirements of *Glaucoma scintillans* and *Colpodium campylum. Proc. Soc. Exp. Biol. Med.* **86**, 685–689.

42 —, —, Parks, Jr., R. R. & Heinrich, M. R. 1950. Further studies on the purine and pyrimidine metabolism of *Tetrahymena. Proc. Nat. Acad. Sci. U.S.* **36**, 431–439.

43 Kodicek, E. & Worden, A. 1954. The effect of unsaturated fatty acids in *Lactobacillus helveticus* and other gram-positive microorganisms. *Biochem. J.* **39**, 78–85.

44 Lilly, D. M. & Klosek, R. C. 1961. A protein factor in the nutrition of *Paramecium caudatum. J. Gen. Microbiol.* **24**, 327–334.

45 *—, & Hartig, W. J. 1958. Methods of preparation of factors required for the growth of *Paramecium caudatum* in axenic cultures. *Anat. Rec.* **131**, 575.

46 Lin, S. C. & Greenberg, D. M. 1954. Enzymatic breakdown of threonine by threonine aldolase. *J. Gen. Physiol.* **38**, 181–196.

47 Miller, C. A. & Johnson, W. H. 1957. A purine and pyrimidine requirement for *Paramecium multimicronucleatum. J. Protozool.* **4**, 200–204.

48 — & — 1960. Nutrition of *Paramecium*: a fatty acid requirement. *J. Protozool.* **7**, 297–301.

49 — & Van Wagtendonk, W. J. 1956. The essential metabolites of a strain of *Paramecium aurelia*

Intracellular Particles in *Paramecium*

A. T. SOLDO

*Research Laboratories of the Veterans Administration Hospital and the Department of Biochemistry,
University of Miami School of Medicine, Miami, Fla. 33125 (U.S.A.)*

*To Dr. S. H. Hutner for introducing me to protozoan research and
to Dr. W. J. Van Wagtendonk for not letting me leave it*

I. INTRODUCTION

Self-reproducing, genetically autonomous intracellular particles of unknown origin may be found in certain strains of *Paramecium aurelia*. The particles number in the hundreds and even thousands per animal and range in size from 0.2 to 1.0 μm in diameter and up to 15 μm in length. Following the precedent adopted by Sonneborn (135), they have been named according to the letters of the Greek alphabet. At the present time at least nine morphologically distinctive particle types have been described (17, 141). These have been designated *alpha, gamma, delta, kappa, lambda, mu, nu, pi* and *sigma*. With the exception of alpha particles, which have been shown to inhabit almost exclusively the macronucleus of the cell (100), the particles may be found more or less randomly distributed throughout the cytoplasm of the protozoan. It is not known whether the members of the group comprise a natural taxon, although it has been suggested that the particles are a diverse group of microorganisms that have come to live symbiotically with the paramecia (17). Generally, they are non-motile, Gram negative rods that divide by transverse fission at rates comparable to those of the protozoan host. Under certain specified conditions, alpha, kappa, lambda and mu particles may be infectious. With exception of alpha, nu and pi, the particles characteristically produce toxins which are capable of killing certain susceptible strains. Kappa, lambda, mu and possibly alpha particles depend in part for their maintenance on the composition of the nuclear genome of the host *Paramecium* (94, 100, 141).

When kappa particles were first discovered they were generally regarded as cytoplasmic units of heredity or "plasmagenes" (135). Recognition that the particles were at least two orders of magnitude larger than a gene, a discovery made possible by the visualization of the particles through the use of nuclear stains (91, 97), led to the eventual abandonment of this view. Later studies on kappa, mu, pi, and more recently lambda, clearly revealed the particles to be complex structures whose

This work was supported, in part, by National Institutes of Health Grant No. AI-03644 to Dr. W. J. Van Wagtendonk and by a National Science Foundation Grant GB-28104 to Dr. A. T. Soldo.

morphological characteristics were similar to those generally attributed to bacteria (17). Unambiguous evidence regarding the nature of the particles was difficult to obtain because the particle-bearing organisms were cultured in the presence of living bacteria as a source of food. Despite this difficulty, the large body of information about kappa particles which emerged from studies carried out in bacterized culture media led to the interpretation that they were of extracellular origin, possibly parasitic bacteria (142). Identification of RNA, DNA and protein in purified kappa particles reinforced this view (120).

The advent of axenic media (160) and the development of techniques for the bacteria-free cultivation of certain particle-bearing strains of *P. aurelia* (122, 127) made it possible to obtain definitive information concerning the chemical composition of lambda, mu and pi particles (130, 162). Thus, the presence of proteins and nucleic acids in these particles was established. Nutritional studies indicated that the presence of lambda in the cytoplasm relieved the protozoan for its need for folic acid, an essential nutrilite for *P. aurelia*, presumably because the particles contained the necessary enzymatic machinery to effect the synthesis of sufficient amounts of this vitamin to meet the needs of the host protozoan (124). In exchange for the comparative protection afforded the particles by the protozoan cytoplasm, lambda provided an essential nutrilite for the host. On the basis of these findings it became possible to characterize lambda as an endosymbiote. It has since been reported that lambda may be cultured extracellularly in complex media for long periods (164).

Since Sonneborn's critical and exhaustive review of the status of kappa and related particles (141), there have appeared three additional review articles, one by Preer (94), which briefly summarizes the state of knowledge concerning the particles through 1968, another by Beale *et al.* (17) which surveys the kinds of intracellular particles found in *P. aurelia* with emphasis on the morphological and ultrastructural characteristics, and a third by Van Wagtendonk (159) which contains a detailed critical analysis of the "metagons". Other publications containing information about the particles of *Paramecium* include Beale's book on the genetics of *P. aurelia* which appeared in 1954 (12), Sonneborn's thought-provoking article in Pollard's *Perspectives in Virology* in 1961 (142) and his most recent updating of "Methods in Paramecium research" in Prescott's *Methods of Cell Physiology* (144).

It is not the purpose of this chapter to review in detail all aspects of the problem of the particles of *Paramecium*. Rather, it is intended to present an overview of the problem with emphasis placed on those aspects the author feels are of particular interest.

II. MORPHOLOGY

A. General

The particles of *P. aurelia* are generally rod-shaped and comparable in size to bacteria (Table 1). With the exception of the recently discovered alpha particle (100)

TABLE 1

The particles of P. aurelia.

Particle	Size (μm)		Sub-cellular location	Toxic effect	Distinguishing characteristic
	length	diameter			
Alpha	2	0.5	macronucleus	none	sickle-shaped
Gamma	0.7	0.2	cytoplasm	killer	3-layered membrane
Delta	2	0.5	cytoplasm	killer	electron dense particles
Kappa	2	0.5	cytoplasm	killer	refractile bodies
Lambda	3.2	0.7	cytoplasm	killer	flagella
Mu	1.6	0.4	cytoplasm	mate-killer	encapsulated
Nu	—		cytoplasm	none	papilli
Pi	2	0.5	cytoplasm	none	mutant kappa
Sigma	15	1.0	cytoplasm	killer	flagella

which inhabits the macronucleous of the protozoan, the particles may be found randomly distributed throughout the cytoplasm. They may be easily observed *in vivo* by staining *Paramecium* containing these particles with either aceto-orcein (122) or osmium lacto-orcein dye (15) and examining the stained preparations under the phase microscope. The particles and the nuclear apparatus appear very dark against the light background of the cytoplasm. An advantage of these staining techniques is that they permit easy identification of the intracellular particles without interference due to bacteria. Van Wagtendonk and Soldo (161) used the aceto-orcein staining method to survey several kinds of ciliates for the presence of intracellular particles. Beale *et al.* (17) applied the osmium-lacto-orcein technique extensively in their descriptions of intracellular particles in *Paramecium*.

Where they have been tested, the particles have been found to be Gram-negative, e.g. kappa (98), lambda (124) and mu (151). Beale *et al.* (17) report unpublished observations by C. N. Wiblin that all the other particles except gamma show some variability to Gram's stain when extracted from *Paramecium*. With the exception of kappa, very little internal detail may be seen in the light microscope. White light and phase contrast microscope observations of kappa, released by crushing animals of stock 51 between a slide and coverslip, revealed the presence of two morphologically distinctive types of particles. One kind of particle contained at least one (rarely more than one) easily identifiable highly-refractile body which was designated the "R-body" by Preer and Stark (97) who discovered them. Kappa particles which contained these R-bodies were called B or "bright" particles because of their characteristically bright appearance in the "bright" phase contrast microscope (97, 141). The other type of kappa particle did not contain refractile bodies and was designated N or "non bright" particle.

Observations of ultrathin sections of fixed preparations of alpha (100), kappa (31, 51), lambda (62, 129), mu (14, 15) and sigma (17) particles in the electron microscope reveal features of ultrastructure they share in common. The particles appear to be bounded by a cell envelope approximately 300–350 Å thick which is composed of

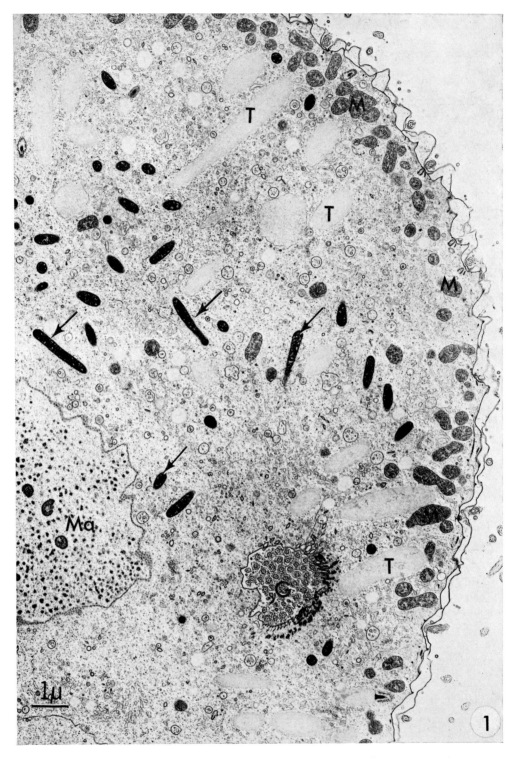

Fig. 1. Low power electron micrograph of a portion of the cytoplasm of a lambda-bearing *Para-mecium*, stock 299. Arrows indicate lambda particles. G = gullet; M = mitochondria; Ma = macronucleus; T = trichocyst. × 7,707.

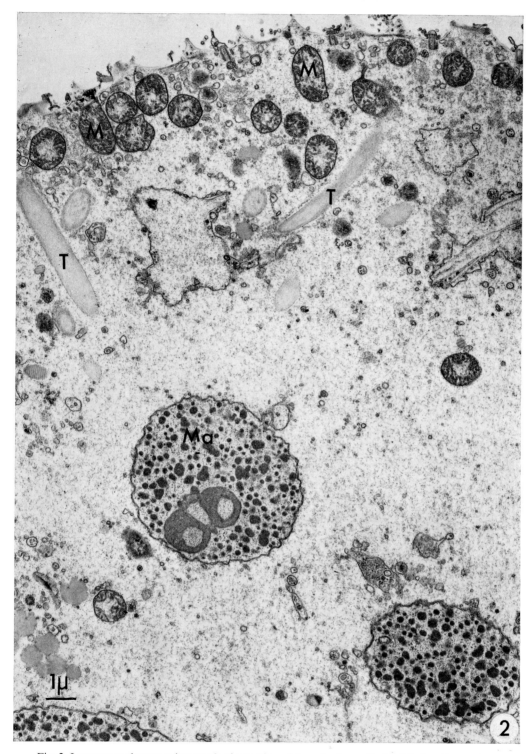

Fig. 2. Low power electron micrograph of a portion of the cytoplasm of a lambda-free *Paramecium*, stock 299. M = mitochondria; Ma = macronucleus; T = trichocyst. × 9,883.

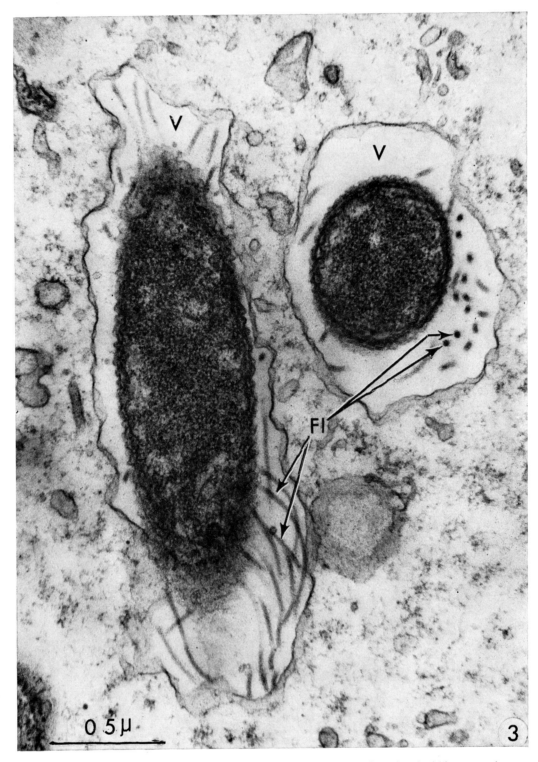

Fig. 3. Electron micrograph of two lambda particles of stock 299 each enclosed within a vacuole (V) showing flagella (Fl). × 65,208.

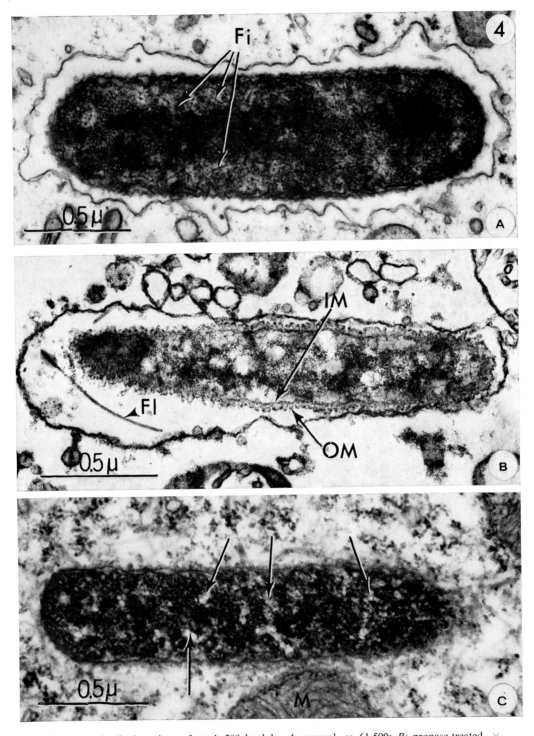

Fig. 4. Longitudinal sections of stock 299 lambda. *A:* normal, × 61,500; *B:* pronase-treated, × 64,000; *C:* DNAase treated, × 65,000. Note loss of fibers (arrows). Fi = fibers; Fl = flagellum; IM = inner membrane; OM = outer membrane; M = mitochondrion.

References p. 426–432

two distinct, electron dense "unit" membranes and resembles closely the cell walls of many Gram-negative bacteria. Sandwiched between these two membranes is a thinner layer composed of less dense material. Gamma particles contain, in addition to the two unit membranes, a third outermost layer which extends part way into the host cytoplasm and contains ribosome-like particles embedded within it (17). Delta particles contain a layer of electron-dense material surrounding the outermost of two membranes and may possess a few flagella (17). Internally the particles contain large numbers of tightly-packed granules measuring 100–120 Å in diameter, e.g. kappa (31, 51), mu (14, 15) and lambda (62, 127), or masses of ill-defined particles too small to be resolved in the electron microscope. A common, perhaps characteristic feature of the particles is the lack of a clearly defined nuclear region, nucleosome or mesosome. In this respect the particles may be distinguished from many bacteria which are known to contain these bodies. Dispersed throughout the internal contents of kappa (32), mu (14, 15) and lambda (62) particles are areas of low density within which may be found fibers or fibrous-like material. The fibers disappear upon treatment with DNAase, e.g. mu (14) and lambda (129) and are presumed to be DNA. Treatment of mu particles with ribonuclease removes most of the granules present in the particles (14).

Jurand and Preer (62) discovered that lambda and sigma particles are enclosed within vacuoles and bounded by a smooth-walled membranous sheath. Usually one or two lambda particles occupy a single vacuole but the number may reach 10 or 12. Contained within the vacuoles, in addition to lambda and sigma particles themselves, are numerous filamentous, rod-shaped entities which have been characterized as flagella. The other intracellular particles of *P. aurelia* are found in intimate contact with the cytoplasm of *Paramecium*. No electron micrographs of the nu particles have been published. Beale (17) suggests that they comprise a heterogeneous group and occur in a number of syngens.

Figs. 1–4 show a series of electron micrographs of sections of the cytoplasm of axenically cultivated lambda-bearing and lambda-free *P. aurelia* stock 299 which illustrate many of the structural details of the particles. Table 1 gives information about the size, subcellular location and identifying characteristics of the particles of *Paramecium*. Table 2 lists the stocks and the syngens in which particles have been found.

B. R-Bodies of Kappa

An outstanding feature of the kappa particle is the R-body, studied in great detail in the electron microscope (106). The R-body of stock 7 kappa of syngen 2 is seen in the electron microscope within the B-particle as a cylinder, the walls of which are formed by 10 or 12 turns of a tightly coiled ribbon-like tape (2, 98). The tape is approximately 8–20 μm long, 0.5 μm wide and 130 Å thick. The main body of the cylinder is about 0.5 μm long, 0.5 μm in diameter and is enclosed within a second thin-walled cylindrical sheath approximately 1.0 μm in length (95). The core of the cylinder (about 0.3 μm in diameter) is continuous with the internal contents of the particle and

TABLE 2

Stocks of P. aurelia *containing intracellular particles.*

Particle	Syngen	Stock	Ref.
Alpha	2	562[a]	(17,100)
Gamma	8	214	(139)
		565	(17)
Delta	1	561	(17)
	6	225	(139)
Kappa	2	7, 8	(133)
		34, 193, 249, 292	(141)
		36, 50	(90)
		308	(97)
		SG	(88)
		562[a]	(17,100)
	4	47	(30)
		51	(134)
		116, 139[b], 169, 277, 298	(141)
	6	225	(139)
	8	214	(139)
Lambda	4	239	(149)
	8	216, 229	(149)
		299	(107)
Mu	1	540	(13)
		548, 551	(140)
	8	130, 131	(70)
		138	(70)
Nu	2	87, 314	(149)
	5	1010, Hu 35-1	(17)
Sigma	2	114	(141)

[a] Stock 562 contains alpha particles in its macronucleous and kappa particles in its cytoplasm.
[b] Mutant form of kappa designated pi (see ref. 17).

may contain from 1 to 14 viral-like inclusions. Each of these viral-like inclusions is polyhydral in shape and measures up to 900 Å in diameter (95, 96). Isolated R-bodies exposed to 4% phosphotungstic acid, pH 7.0 or to 0.25% sodium lauryl sulphate undergo morphologic changes resulting in the unrolling of the ribbon-like tape, characteristically, from the outside (96). The unwound tape reveals additional features about its structure. It consists of an outer end which is blunt or irregularly shaped and an inner portion which tapers sharply to a point. The viral-like inclusion bodies are associated with that portion of the surface of the tape at or near its inner end. Some of the viral-like inclusions, as judged by examination of negatively-stained preparations, contain numerous capsomere-like elements each of which is about 100 Å in diameter and may be pentagonal, hexagonal or "H"-shaped (96). Other viral-like

References p. 426–432

inclusion bodies appear to be empty. Viral-like bodies are virtually absent in kappa without R-bodies (N-particles).

Kappa from stock 562 differs from stock 7 kappa in certain respects. Phospho-tungstic acid does not cause rolling of R-bodies and the sheath-like structures and capsomere-like subunits present in stock 7 kappa are absent in stock 562 kappa (95).

Kappa particles of stock 51 (syngen 4) also differ from those of stock 7 in certain respects. The B-particle of stock 51 kappa for example, is smaller, often round or irregular in shape (141) and may be lysed by detergents (78, 79, 96). The ribbon-like tape of the R-body tapers at both ends and unrolls in the presence of acid from the inside, unlike stock 7 kappa which unrolls from the outside (96). The unrolling of stock 51 R-bodies caused by lowering the pH to 6.0 may be reversed by subsequently restoring the pH to 7.0.

R-bodies of stock 51 kappa contain numerous viral-like subunits which are associated with the inner end of the tape. These structures, unlike the capsomeres of stock 7, form helices approximately 100 Å in diameter and up to 400 Å in length (96).

III. BIOLOGY

A. Growth and Reproduction

Kappa [N kappa only—B kappa does not divide (1)], lambda (95, 129), mu (14) and presumably the other particles divide by transverse fission without the formation of a cross wall or septum such as occurs in a number of bacteria. With the exception of mu particles which have a tendency to form clusters and alpha particles which are found in the macronucleus (100), the particles are distributed more or less at random throughout the cytoplasm and are transferred in approximately equal numbers to each of the two daughter cells at fission (89, 90).

In general the particles divide at rates comparable to those of the protozoan. Obviously, if this were not the case, they would not be maintained. They would either become lost from the cytoplasm or they would overgrow the cell. Sonneborn (141) pointed out that a number of kappa-bearing stocks maintained in the laboratory by regular periodic transfers to fresh culture medium still retained their particles after more than 25 years.

In bacterized media, the number of kappa maintained within the cell (as many as 3000 per cell have been observed) varies widely and has been shown to be dependent upon a number of factors. These include mating type, stages in the life cycle and fission rate. Chao (26, 27) found by direct count that the average number of kappa in stock 51 animals of mating type VII was about twice those of mating type VIII. Following autogamy, there is about a 3-fold increase in the number of kappa in the cytoplasm of stock 51. This level declines gradually and levels off after 7 or 8 fissions (26). In this stock, levels of kappa were reduced in proportion to the rate at which the animals were grown but were never completely lost from the cell even when the animals were grown at the maximum possible rate in bacterized medium, e.g. 5

fissions per day (55, 97, 136). On the other hand, kappa from stock 7 declined gradually and was eventually lost from animals maintained at a rate of 3.5 fissions per day (91). The normal rate at which these animals are grown is approximately one fission per day.

Van Wagtendonk *et al.* (167) tried to grow kappa-bearing animals of stock 51 axenically but found that most died shortly after being placed in the medium. Microscopic examination of corpses disrupted by crushing between a slide and coverslip revealed the presence of very large numbers of kappa particles. The few surviving animals apparently adapted to the bacteria-free environment and subsequently grew well in the medium but contained no particles. Soldo (122) developed an axenic medium which permitted the growth and maintenance of particles of stock 51 kappa provided the animals were transferred to fresh medium at frequent intervals. This medium supported the growth and maintenance of particles of stock 299 lambda for an indefinite period, transferred weekly.

Recently, axenic media have been developed which permit the growth and continued maintenance of particles of a number of stocks of *P. aurelia* including stock 299 lambda, 138 mu and 139 pi (124, 127) but not 51 kappa. Van Wagtendonk *et al.* (165) report that by modifying the relative proportions of lipids in one of these media (127) many other particle-bearing stocks will grow and maintain their particles including stock 51 kappa. Animals cultivated in the axenic medium developed by Soldo *et al.* (127) have maintained their particles after several years of serial weekly transplants at 27°C in the dark. Under these conditions the number of particles maintained per cell ranges from 1000–2000 for lambda, and 4000–5000 for mu and pi. In a chemically defined medium (126) lambda particles of stock 299 averaged the number of approximately 500 per cell.

Soldo (124, 127) reported that lambda particles of stock 299 cultivated axenically divided at rates comparable to those of the protozoan during all phases of the growth cycle. Upon removal of these animals from the logarithmic phase of growth to non-nutritive medium (isotonic salt solution) neither lambda particles nor the

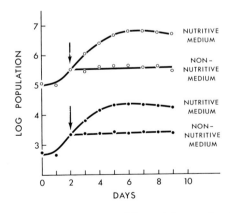

Fig. 5. A comparison of the growth of lambda particles and *Paramecium* in axenic medium. Arrows (↓) indicate point at which some of the animals were transferred to non-nutritive medium.

protozoans underwent cell division for as long as they remained in the nonnutritive medium (Fig. 5). During the stationary phase of growth, a period during the growth cycle where there is no net increase in the number of animals, the number of lambda particles also remained at a constant level. It appears that the growth and division of these particles within the cytoplasm may be dependent upon a product or products of the protozoan's metabolism.

B. Toxins of Kappa, Lambda, Sigma and Mu

Kappa-bearing paramecia liberate a toxic agent into the medium in which they live that is capable of killing other paramecia lacking kappa (134). The kappa bearers are called "killers" and are resistant to the action of the toxin they produce. Animals lacking kappa that are killed by the action of this toxin are called "sensitives".

1. Kappa

a. Kinds of kappa killing. Four different kinds of *kappa* may be distinguished on the basis of the prelethal stages affected "sensitive" animals undergo following exposure to the toxin the particles produce. These have been described in detail by Sonneborn (141) and have been designated by him as: humpers, paralyzers, spinners and vacuolizers. A brief description of the killing action of these particles is included here: *Hump*—a few hours after exposure to hump (hp) kappa, affected sensitives develop a large aboral blister which becomes filled with the internal contents of the protozoan. This imparts a characteristic "hump"-like appearance to these animals, and is followed by a period during which the animals round up and assume an almost spherical shape. Death takes place in one or two days. *Paralyser*—affected sensitives undergo at first a partial, and later, a complete paralysis which is characterized by cessation of feeding and motion and leads to the gradual wasting away and eventual death of the animals in a matter of days. *Spinner*—The animals appear to swim abnormally about 2 hours after exposure to spinner (sp) kappa. They do not feed and begin to spiral slowly to the right (normal paramecia spiral to the left). After a few hours, right-hand spinning becomes much more rapid. Often the animals spin without any noticeable forward or backward motion. The spinning gradually diminishes and later stops altogether. This is followed by a period during which the animals become much smaller and pear-shaped. They die after a day or so. *Vacuolizer*—sensitives cease feeding at first, develop a large central vacuole, become smaller and die, usually within one day. Table 3 lists some kappa-bearing stocks with respect to the type of killing action they exhibit.

b. Toxic agents produced by different kappas. The toxin produced by the kappa of stock 51 is in the form of discrete particles. Austin (3, 4) found that the number of sensitives killed by toxin liberated into the medium by killer animals was directly proportional to the number of killers present. She showed further that only one particle was necessary to kill a sensitive animal and that stock 51 killers produced

TABLE 3

Killing action of some kappa-bearing stocks[a] of P. aurelia.

Killing action	Syngen	Stock
Hump	4	51(134), 116(141), 139(141), 169(141), 277(141), 298(141)
Paralyser	2	193(141)
	6	225(139)
Spinner	2	5(133), 7(131), 34(141)
Vacuolizer	2	8(133), 292(141), 308-2(90), 562(100)
	4	47(30)
	8	214(139)

[a] Kappa-bearing mutant stocks are not included (see Table 5).

approximately one killer particle in 5 hours, the generation time for animals cultured under the conditions she used (5). Rate of production of the toxic particles was dependent to some extent on mating type. Stock 51 killers of mating type VII released twice as many killing particles into the medium than killers of mating type VIII in the same time (6).

Homogenates of killer animals contained considerable toxic activity. Direct proportionality between the concentration and killing action was observed by Austin (5), Sonneborn (146, 148) and Van Wagtendonk and Zill (163) for homogenates of stock 51 killers and by Jacobson (60) for homogenates of stock 7 killers.

As originally conceived by Sonneborn (148), the toxic agent was an antibiotic substance and was named paramecin (Pn) by him. Recognition that the toxin was, in fact, a discrete entity led to the designation (P) for the particle (141). Thus the toxic agent liberated from stock 51 killers was called P-51, stock 7 toxin was called P-7, etc.

c. Nature and site of toxic activity

i. Early investigations. Early investigations of the properties of the toxic principle clearly showed it was both unstable and sufficiently large to be visible in the microscope. P-particles in homogenates of stock 51 killers were inactivated by extremes of temperature (136, 146, 163) and pH (163), ultrasonication (177), UV (83, 110), X-rays (83, 84), the action of certain enzymes (93, 110, 157, 158) and nitrogen mustards (39, 40). They were unable to pass through dialysis membranes or filters which did not permit passage of bacteria and could be sedimented at comparatively low centrifugal forces (6, 97, 136, 148, 177). Temperatures above 50 °C and conditions of pH below 5 and above 9.5 caused immediate inactivation of the 51 P-particle. The particles were most stable at pH 8.0. Van Wagtendonk and Zill (163) found that decay of activity was logarithmic at 30° and 40 °C and that the energy for inactivation was 126,000 cal/mole at pH 7.0, a value within the range of values for the energy of inactivation for a protein. Sonneborn (148) noted an initial rise in P-particle activity at 20 °C before the beginning of logarithmic decay. No such increase in activity was observed in homogenates containing P-particles by Van Wagtendonk and Zill (163) at 30° or 40°C or by naturally liberated P-particles by Austin (5) at 27°C. The action spectrum for inactivation of 51 P-particle activity by UV irradiation closely resembled that of a protein (110). X-ray inactivation data of Nanney (83, 84) were consistent with that of a one hit phenomenon. A dosage of 26,000 r destroyed 2/3 of the activity in homogenates. Much higher dosages were required to inactivate 51 P-particle activity within the killer itself and, under these circumstances, inactivation followed a multiple hit curve. Van Wagtendonk (157, 158) found that chymotrypsin, pepsin and deoxyribonuclease destroyed the activity of 51 P-particles; lysozyme, hyaluronidase, ribonuclease, papain and Laskowski's protein B did not. Sonneborn (141) noted that the data of Van Wagtendonk (157) show that cysteine markedly reduced the rate of decay of P-particle activity. Destruction of P-particle activity by the action of chymotrypsin was confirmed by Setlow and Doyle (110) but these workers failed to obtain inactivation of P-51 by DNAase. Similarly, Preer (93) reported that DNAase did not inactivate P-7 toxin. On the basis of centrifugation studies, Zill (176) estimated the size of P-51 particle to be 0.38 μm in diameter.

The main conclusions emerging from these early studies was that the P-particle behaved as if it were large enough to be visible, and that the toxic agent it produced was an unstable protein, possibly a deoxyribonucleoprotein.

The search for the P-particle itself led Preer *et al.* (99) to suspect that the B kappa was the locus of P-particle activity. Analyses of both pellet and supernatant fractions of homogenates of stock 7 killer animals centrifuged over a range of g-forces revealed that, of the two forms of kappa known to be present in these homogenates, N and B (*see* Section II, A and B), P-particle activity was closely associated with B-kappa and not with N-kappa. Nevertheless, it was clear at the outset that only a relatively small percentage of B-kappa, often less than 1%, contained P-particle activity. To account for the instability of P-particle activity, Preer (93) suggested that the B-kappa disintegrated, a phenomenon he directly observed, thereby releasing the toxic agent. He suggested further that the R-body contained within B-kappa was the actual site of the toxic principle.

Sonneborn (141) reviewed the evidence available up to that time and envisioned the following sequence of events leading to the formation of P-particle activity. For reasons not completely understood a certain percentage of N- or non-bright kappa particles within the cytoplasm of the killer animal enlarged, became more electron dense and developed R-bodies (31, 99). These particles, now B-kappa, ceased to divide and were subsequently liberated into the medium. It was only *after* being liberated from the protozoan that the particles *developed* P-particle activity. In the early stages DNA played a role in the development of P-particle activity [hence inactivation of the toxin by DNAase treatment found by Van Wagtendonk (157)] but was no longer needed once this activity was developed. A delay in the onset of P-particle activity following UV irradiation of freshly prepared homogenates (110), supported the idea that a nucleic acid was essential for an early step in the development of P-particle activity. Inactivation of P-particle activity by X-irradiation followed a multiple hit curve when whole killer animals were irradiated and it was suggested that the development of immature B-kappa into P-particles required multiple processes. Once P-particle activity was fully developed, however, these processes were no longer required and inactivation by X-irradiation followed a one hit curve.

ii. Recent investigations. With respect to the identification of the actual site of toxic or P-particle activity, the development of a method by Smith (118) for the isolation and purification of kappa made the task somewhat easier. Studies on isolated and purified kappa particles by Smith (118), Smith and Van Wagtendonk (120), Sonneborn and Mueller (145) and Mueller (78, 79) confirmed Preer's earlier findings by providing a direct demonstration that B-kappa possessed P-particle activity whereas N-particles did not. For example, Sonneborn and Mueller (145), Mueller (78, 79), and Smith and Van Wagtendonk (120) found that 0.1% solutions of the detergent duponol caused lysis of B-kappa, but not of N-kappa of stock 51 thereby inactivating them. Mueller (78) observed that R-bodies released from B-kappa following treatment with the detergent underwent a change in morphology; R-bodies were converted from a compact to a "V"-shaped form as seen in the phase or white-light microscope. Lowering the pH of suspensions of 51 kappa to 6.0 with HCl caused B-particles to disrupt, releasing R-bodies in the form of rigid filaments up to 20 microns in length. Similarly, B-kappa, disrupted by osmotic shock, released R-bodies in the form of long filaments.

Efforts to identify a component smaller than B-kappa of stock 51 with P-particle activity were unsuccessful. Every procedure that caused the disruption of B-kappa of stock 51, in addition to those already noted, i.e. treatment with 0.5% solutions of deoxycholate (DOC) (2, 98, 101), lysis in the French press, freeze-thaw procedures and sonication, inactivated the toxin (98).

d. Virus-like nature of toxic particles. It was obvious that until a means could be found to disrupt B-kappa of stock 51 without inactivating its toxin, work on the identification of its toxic principle could not continue. For this reason, some investigators turned to the study of another kappa-bearer, stock 7. Unlike B-kappa of stock 51 which readily lysed in the presence of 0.1% duponol, B-kappa of stock 7 did not. However, treatment of whole kappa-bearing stock 7 animals with 0.5% DOC solutions caused lysis of both N- and B-particles (101). R-bodies released by this treatment appeared to be morphologically unchanged and retained about one half of their original activity. Killing activity showed a high degree of correlation with the number of particles that sedimented at the same rate as the R-bodies. Thus, for the first time a component smaller than B-kappa—the R-body—was found to be closely, if not exclusively, associated with P-particle activity itself.

Electron microscopy of negatively stained preparations of R-bodies released from stock 7 animals treated with 0.5% DOC revealed hitherto unsuspected features of their structure (2). Such preparations often contained mixtures of two forms of the R-body, compact and filamentous. The compact form consisted of a highly coiled roll of tape in the shape of a hollow cylinder about 0.5 μm long and 0.5 μm in diameter. The tape itself was about 130 Å thick and about 5000 Å wide. The filamentous form of the R-body resulted from the sudden and often spontaneous partial unwinding of the roll of tape to form a twisted ribbon in the shape of a hollow tube up to 20 μm in length. Anderson *et al.* (2) observed round objects, approximately 1000 Å in diameter associated with both the compact and filamentous forms of the R-body. Preer *et al.* (98) also examined the structure of R-bodies of stock 51 in the electron microscope. The compact form of the R-body of stock 51, consisted of a ribbon of tape wound in the form of a hollow cylinder, not unlike that of the R-body of stock 7. The "V"-shaped and filamentous forms of the R-body observed by Mueller (78) in the phase and white-light microscope were seen in the electron microscope to be due to the partial unwinding of this tape. There were some differences, however. Unlike R-bodies of stock 7 which unwound spontaneously from the outside, R-bodies of stock 51 unwound from the inside.

As noted earlier by Mueller (78), lowering the pH to 6.0 caused R-bodies of stock 51 to unwind. Preer *et al.* (98) reported that this process could be reversed. Upon restoration of the pH to 7.0 the R-body of stock 51 rewound, also from the inside. More recent work on the structure of R-bodies of stock 7 and 51, as revealed in the electron microscope, led to the discovery by Preer and Preer (96) of virus-like bodies existing in close association with these entities and provided a further clue as to the possible identity of the P-particle. Polyhedral, virus-like inclusions up to 900 Å in diameter—probably identical with the "round objects about 1000 Å in diameter" observed by Anderson *et al.* (2) as a characteristic feature of stock 7 R-bodies—were found in association with the surface of the inner end of the tape. Some virus-like inclusion bodies contained numerous hexagonally-shaped capsomere-like elements about 100 Å in diameter. Others were empty. R-bodies of stock 51 were found to contain numerous helical structures, 100 Å in diameter and 400 Å long associated with the surface of the inner end of the tape. Preer and Jurand (95) examined large numbers of sections of N- and B-kappa from stock 7 (and also stock 562) for the presence of virus-like inclusions. They found the virus-like inclusions to be closely associated with R-bodies contained within B-kappa. N-kappa which lacked the R-body was almost completely devoid of virus-like inclusions. On the basis of these and other observations these investigators postulated that "kappa is infected with a lysogenic virus whose induction results in the production of the R body and viruses".

It remains to be established whether the P-particle as originally postulated by Sonneborn (141) is the R-body itself or the virus-like inclusions, e.g. capsomeres of stock 7, helical structures of stock 51, associated with the R-body.

In a recent study on the chemical nature of the toxin itself, Preer and Preer (96) investigated the effect of a number of enzymes, including RNAase, DNAase, lysozyme, pectinase, β-amylase, cellulase, chitinase, trypsin, and chymotrypsin on purified R-bodies of stock 7. None inactivated the toxin. Treatment with chymotrypsin, which produced no morphological change in the R-body, as seen in the electron microscope, altered the type of killing action manifested against suitable sensitives, from spinning to paralysis. Other chemical tests revealed only the presence of proteins in the R-body. Nucleic acids, lipids or carbohydrate could be detected only in traces.

e. Mode of delivery and fate of toxic particles. How is the toxin delivered to its site of action? Experiments of Mueller (82) bear heavily on this matter. She found that kappa, vitally stained with methylene blue, accumulated in food vacuoles of sensitive animals which would occasionally disrupt, releasing their contents into the cytoplasm. Virtually all the kappa observed within the food vacuole were N-kappa. Very few B-kappa could be found. Filamentous projections piercing the vacuolar membrane were also occasionally observed. The possibility that these filaments were trichocysts was ruled out. Were they the filamentous or unwound R-bodies in the form of micro-tubes? Mueller suggested that "B-particles are taken into food vacuoles and R-bodies are stimulated to unroll into the filamentous form by the low pH of the food vacuoles". She asks "Could it be that this tube actually injects the toxin into the victim's cytoplasm through the vacuolar membrane?" This provocative query prompted Preer and Preer (96) to suggest such a mechanism for the delivery of the stock 7 kappa toxin into sensitive paramecia. However, these investigators pointed out that the unwinding of the tape within the food vacuole must be triggered by some, as yet, unknown means, since it is known that lowering the pH does not affect R-bodies of this stock. See Fig. 6 for a diagrammatic representation of these events.

Once the P-particle is delivered to its site of action almost nothing is known of its ultimate fate. Sensitives, disrupted at any time following uptake of P-particles, show

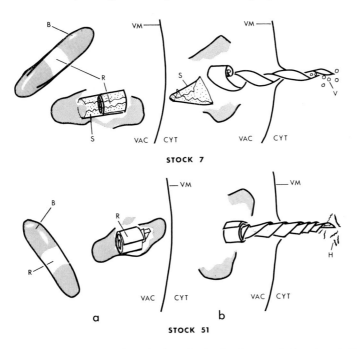

Fig. 6. Diagrammatic representation of unwinding of R-bodies from stock 7 and stock 51 kappa and injection of the toxin into the cytoplasm of the host based on the mechanism proposed by Mueller (82) and Preer and Preer (97). *a:* intact and disintegrating B-particles showing compact R-bodies; *b:* unwound R-bodies piercing the vacuolar membrane. B = bright particle; CYT = cytoplasm; H = helical virus-like structure; R = refractile body; S = sheath; V = virus-like inclusion body; VAC = vacuole; VM = vacuolar membrane.

References p. 426–432

no evidence of P-particle activity and there is no release of P-particles even after death followed by lysis (138). The fact that X-irradiation protects sensitives even after they have taken up P-particles suggests that the particles remain in an active state for some time before they are lost (84). Butzel and Pagliara (22) made the interesting observation that chloramphenicol, dinitrophenol and sodium azide protected sensitives *after* the animals had taken up stock 51 P-particles and concluded that protein synthesis and a functional terminal oxidative transport system may be essential for killing action. This group also found that the phospholipids lecithin and cephalin (34) fully or partially protected sensitives against the lethal action of stock 51 P-particles (22, 24).

f. Factors affecting sensitivity to toxic particles. A number of factors have been found that affect sensitivity of susceptible animals to toxic particles. Austin (6), for example, demonstrated that, within a given stock, clones expressing different serotypes varied in sensitivity to P-51 toxin. Increased rate of oxygen consumption and sensitivity to the antibiotic patulin were believed to vary with respect to sensitivity to P-particles of stock 51 (7). Nutritive conditions (e.g. excessive feeding following starvation prior to exposure to the toxin) and maintenance of exposed animals at low temperature reduced sensitivity of susceptible animals to P-particle toxin (94, 136). Particle-free animals undergoing autogamy, engaged in conjugation or without a proper ingestatory apparatus were unaffected by P-particle toxin. Other particle-free animals not affected by P-particles were presumed to be protected from the action of the toxin by virtue of their genetic constitution (141). This concept draws support from the fact that different stocks vary widely in sensitivity to a given toxin. Genetic analysis of these stock differences was carried out by Preer (92). He found that genes at multiple loci may be involved. The matter has not been investigated further.

The presence of kappa particles themselves within the cytoplasm may confer resistance to the action of P-particle toxin and determines to a large extent the phenotypic expression of the killer trait (131, 133). Small numbers of kappa, 25 or less per animal are insufficient to protect stock 51 against the action of P-51 toxin (141). Larger numbers protect the animal against the action of the toxin but do not confer the ability to kill. Still larger amounts of kappa provide protection against the action of the toxin and enable the animal to become a killer. The actual number of kappas necessary to protect an animal against the action of P-particle toxin is not precisely known. Chao (26) has reported that 140 kappa per cell protected against the action of P-51 toxin. Hanson (54) found that stock 51m6 animals were protected when they contained only 100 kappa. Sonneborn (141) suggests that a quantity of kappa ranging from 25 to 65 per animal is sufficient to protect against the action of the toxin.

Apparently, the presence of B-particles, originally thought to be closely associated with resistance to P-particle toxin (141), is not essential for protection. The recent discovery by Widmayer (171) of a mutant stock (51m43) which has no B-particles and is a non-killer fully protected against the action of P-51 toxin, illustrates that immunity is due to the presence of N-particles. By exposing large numbers of paramecia to killer homogenates, Mueller (81) found 2 clones among the survivors that were non-killers immune to the action of P-particle toxin. Although she examined crushed preparations of these immune animals in the microscope she made no mention of the kinds of particles they contain.

The production of, and resistance to, P-particle toxins appears to be highly specific. In general animals containing a specific type of kappa produce and are resistant to P-particle toxin of that type only. Dippell (30), for example, discovered mutants which contained two types of kappa and which produced and were resistant to both types of toxin. (See Section III, E for a more detailed description of these mutants).

Stocks sensitive to the action of toxic particles produced by *P. aurelia* killers have been found in all species of *Paramecium* examined. These include: *P. calkinsi*, *P. multimicronucleatum*, *P. trichium* (132) and *P. caudatum* (50, 90, 149). On the other hand, Gilman (50) found that stocks of syngens 4 and 5 of *P. caudatum* were unaffected by the action of P-8 toxin. All syngens of *P. aurelia*, except 3, 5 and 9 were reported to be unaffected by the action of P-299 toxin (107); other types of P-particle toxins had no detectable effect on a number of stocks of syngen 2 (90, 131, 132).

2. Lambda and sigma

Schneller (107) found that animals of stock 299 of syngen 8 produced a toxic agent which killed sensitives rapidly, usually in less than 30 minutes. The killer animals were found to contain rod-shaped particles, named *lambda*, which were 2 to 4 times larger than kappa but contained no R-bodies. "Rapid lysis" killing of this kind (108), was also found in stocks of syngen 2, 4, and 8 and was always associated with the presence of cytoplasmic particles. Stock 114 of syngen 2 contained *sigma*, a sinuous rod up to 15 μm in length; stock 239 of syngen 4 and stock 299 of syngen 8 contained lambda. The toxin produced by each of these stocks did not pass through millipore filters of pore sizes smaller than 0.45 μ. The killing substance of stock 239 was sedimentable at low centrifugal forces and was inactivated almost instantaneously at 60 °C; about one half of the activity was lost after 2 hours at 6 °C.

Butzel and Van Wagtendonk (23) found that the toxic agent produced by stock 299 could be detected in cell-free fluids in which the killers had lived and in freshly prepared homogenates of killer animals. Killing activity was associated with particles which could be sedimented at 31,000 g and were estimated to be 5–10 μm in size. The toxic agent in cell-free fluids was unstable and lost all killing activity in 30 minutes at 24 °C. Killing activity of homogenates of killer animals and lambda preparations purified on an ECTEOLA column (162) was extremely low. Only 1 to 8 animals were killed by homogenates of purified lambda preparations obtained from 25,000–35,000 killer animals. Killing activity observed in these experiments is in striking contrast to the killing activity of lambda particles of the same stock cultivated *in vitro* by Van Wagtendonk *et al.* (164). Assuming that each killer animal contained only 100 lambda particles, then 2.5 to 3.5 million lambda particles were needed to kill, at most, 8 sensitives according to Butzel and Van Wagtendonk (23). The number of lambda particles cultivated *in vitro* required to kill a single sensitive animal ranged from 7 to 20. These differences are probably caused by column adsorption.

3. Mu

Siegel (111–114) discovered mate-killing in *P. aurelia*. Following conjugation between a sensitive animal and a mate-killer, the sensitive animal either died or became endowed with the killer trait. Genetic analysis revealed the trait to be associated with cytoplasmic particles called *mu*. Presumably mu particles produced a toxin capable of killing sensitive animals. Nothing is known about the nature of this toxin. It is believed, however, that mu toxin (Siegel, 113) affects both macro- and micronuclei by interfering with the development of the prezygotic macronucleus or by causing damage to, or the loss of, micronuclei. Unlike kappa, which it resembles in both size and shape, mu contains no R-bodies.

Apparently, mu-bearers liberate no P-particles into the culture fluid. Sensitives placed in culture fluid in which mate-killers have grown are unaffected by the toxin. Prolonged physical contact involving an exchange of cytoplasm is needed to permit transfer of mu toxin from the mate-killer to the sensitive animal. Casual contact between mate-killers and sensitives which occurs when animals in a crowded culture sometimes brush up against each other during normal swimming is not sufficient to permit transfer of mu particles from the killers to the sensitives. Mu particles have been found in a number of *P. aurelia* stocks including 130, 131 (70), 138 (113), 540, 548, 551 (13, 14, 15). In every case the particles have been shown to be responsible for the mate-killing trait exhibited by animals of these stocks.

C. Infectivity

Initial studies on infection were carried out by Sonneborn (138). He discovered that kappa particles could be transferred from killers to sensitives by cytoplasmic exchange during mating. Later, he found that homogenates of killer animals were "infective". About 50% of a population of sensitive animals (containing the gene K) which survived after being exposed to a homogenate of killer animals were found to contain kappa and became killers themselves, if grown at a sufficiently low fission rate to permit the level of kappa to increase to one hundred or so per cell (141). Infective transfer of kappa of stock 51, syngen 4 into a number of stocks of syngen 4 was possible, but attempts to infect sensitives of one syngen with kappa from killers of another syngen met with failure (70, 154). Recently, Koizumi and Preer (64) reported the successful transfer by microinjection of stock 7 (syngen 2) kappa into stock 51 (syngen 4) animals.

Tallan (154–156) investigated the properties of the infective agent in homogenates of stock 51 killers. The highest frequency of infection was obtained using fresh preparations. Killer homogenates lost about one half their original activity if allowed to stand at 27°C for 12 hours. Following centrifugation of killer homogenates at 25,000 *g* for 5 minutes, Tallan found that neither the sediment, containing all the kappa present in the original preparation, nor the supernatant fraction alone was infectious. Recombination of these fractions restored the original infectivity, however. The sediment was rapidly inactivated at 56°C; the supernate was not. To a large extent the supernate fraction could be replaced with organic culture media. Mueller

(77, 79) found that the "cofactor" of the supernate could be completely replaced by Ca^{2+} at a concentration of 0.05 M. Because the sediment contained both N- and B-kappa it could not be certain from these studies whether either or both these particles was the infective agent.

Sonneborn and Mueller (145) observed that treatment of killer homogenates with 0.1% sodium lauryl sulfate destroyed bright particles and killing activity but left N-particles and infectivity intact. Later experiments by Smith (118), Mueller (79) and Smith-Sonneborn and Van Wagtendonk (120) confirmed these findings with purified kappa preparations. Final proof that the infective agent was indeed the N-particle comes from the work of Widmayer (171). She showed that particles of stock 51m43, a non-killer mutant which contains no B-kappa or R-bodies is highly infectious. Attempts to fractionate the infective agent by disrupting N-particles by grinding with alumina proved unsuccessful (79). Mueller (79) also tested partially purified preparations of B-particles and R-bodies for infectivity and found none.

Lambda particles have also been shown to be infective. Van Wagtendonk *et al.* (164) found that lambda particles from *P. aurelia* stock 299 syngen 8 grown *in vitro* in a complex medium were capable of infecting lambda-free animals of the same stock. Between 100 to 200 sensitives were exposed to approximately 400 lambda particles in a small volume and incubated overnight. Surviving animals were then transferred to fresh axenic medium and after a period of 5 days transferred to bacterized medium. That the infected animals contained lambda particles in their cytoplasm was determined by staining with aceto-orcein and observing lambda in the cytoplasm. Infected animals were also reported to produce a toxic agent which killed sensitives.

Alpha of stock 562, the most recently discovered particle of *P. aurelia* has been reported to be infectious (Preer, 100). This particle does not produce a toxic agent and, unlike other particles of *P. aurelia* inhabits the macronucleus of the cell. It is rarely observed in the cytoplasm and not found in the micronucleus at all. Two morphologically distinctive forms have been identified, a short sometimes crescent-shaped rod and an elongated spiral-shaped form. The rods are non-motile and appear to be the reproductive form; the elongated form is believed to be invasive and is motile. Homogenates containing alpha were found to be infectious. Of 44 stocks from 7 different syngens tested for infection, only 6 (all from syngen 2) acquired and maintained alpha particles. Following autogamy, the particles may be transmitted from the old macronuclear fragments to the new macronucleous, but only if the animals are permitted to grow slowly. Grown at a rapid rate, ex-autogamous animals eventually lose all their particles. Cell to cell transmission of alpha has also been observed. Alpha-free animals of stock B 197 which possessed abnormal trichocysts as a marker, were found to acquire alpha particles when cultured together with alpha-bearing animals of stock 592.

D. Genetics of Maintenance

Sonneborn (134) discovered that maintenance of kappa of stock 51, syngen 4 is dependent upon the presence of the gene K in the *Paramecium* nucleus. The gene itself cannot initiate the production of kappa particles within the cell, but is required for maintaining the particles once they are introduced into the cytoplasm. Chao (26, 27) found that animals homozygous for this gene (KK) contained, on the average, twice as much kappa as heterozygotes of the genotype Kk. Following the change in genotype from Kk to kk, kappa is completely and irreversibly lost from the cytoplasm. The loss is not immediate, however. Chao (26) reported that a kk clone has, initially, the same amount of kappa per cell as its parent (Kk). During the first two post-zygotic fissions, this amount increases threefold, declines during the next five fissions to the Kk level and begins to disappear from some cells at the 8th fission and thereafter until, by the 15th fission, kappa is virtually absent from all cells. Animals that lose kappa during the period from the 8th to the 15th fission do so rather suddenly, usually within one or two fissions. Kappa in the remaining cells is present at the level normal for that of the heterozygote (Kk). When retained, therefore, kappa continues to divide at the normal rate. To account for these kinetics, Chao postulated that the amount of kappa in the cytoplasm is proportional to the number of functional K genes remaining in the *Paramecium* genome. The large increase in the level of kappa within the cell which takes place during the first two post-zygotic fissions was attributed to the presumed increase in the number of K genes resulting from the doubling of the volume of the macronuclear fragments believed to take place at the time of each of these divisions. It was known that in amacronucleate paramecia kappa particles failed to multiply (86, 87). The lag of 8 fissions before loss of kappa, observed in some of the cells, was not considered unusual. Sonneborn noted that a "lag in achieving full expression of a new genotype is the rule in *Paramecium* for all traits that have been studied, through for most traits it is less protracted" (141).

Other investigators working with the same material failed to find kinetics of loss of kappa following a change in genotype from Kk to kk similar to that reported by Chao. Yeung (175), for example, reported that kappa in well-fed animals was retained by all kk cells for seven generations, followed by a gradual decrease in number until after the 12th fission, only a few cells were found containing kappa. Byrne (25) reported considerable variations in loss of kappa 51 from different clones. Kappa was lost from some of her kk clones in as little as five fissions, but persisted in others for periods as long as 60 fissions. Beale and McPhail (17) and Widmayer (171) found similar variations in the ability of kk clones to retain kappa. Yeung (175) made the important observation that cultural conditions were critical for kappa maintenance. Following the change in genotype from Kk to kk, starvation led to rapid loss of kappa even when applied to animals as early as the second post-autogamous fission. Implicit in this work is that variation in loss of kappa from kk animals observed by himself and other workers was due at least in part, to nutritional deficiencies resulting from inconsistencies of the bacterized media to support the growth of these organisms. Byrne (25) was not able to confirm some of the results of Chao but reported that starvation of animals at the 12th generation or later increased the rate of loss of kappa. She conceded that starvation probably influenced retention of kappa 51 in kk clones but suggested that there were as yet undiscovered factors that were responsible for kappa maintenance.

Other genic factors that influence the maintenance of kappa of syngen 4 have been identified. In addition to the K gene which is necessary for full retention of kappa within the cell, Sonneborn (137) discovered that genes at another unlinked locus affected kappa maintenance in this syngen. Introduction of these genes, designated S-genes, into the *Paramecium* genome caused some paramecia within a clone to lose kappa completely, while other members of the same clone retained their full complement of kappa. As a whole the clone never lost all its kappa. Breeding analyses by Balbinder (8, 9) revealed that there were at least two genes S_1 and S_2 at unlinked loci that affected maintenance of kappa within the cell. The presence of both genes in their dominant form led to the eventual loss of kappa from the cell. Thus all members of a clone of genotype $KKs_1s_1s_2s_2$ were capable of retaining kappa. Some members of clones of either $KK\ S_1S_1s_2s_2$ or $KKs_1s_1S_2S_2$ genotypes sporadically lost kappa completely, while other members of the same clone did not. All members of a clone of $KKS_1S_1S_2S_2$ genotype lost kappa completely. The time required to complete the loss of kappa was about 60 fissions.

Balsey (10, 11) found that kappa of stock 7, syngen 2 depended for its maintenance on the presence of the gene K in the *Paramecium* genome. Like Chao (27), he found that stock 7 animals of the KK genotype contained about twice as much kappa as animals of the Kk genotype. Animals of the kk genotype eventually lost all their kappa although it is not known whether the kinetics of loss was similar to that reported by Chao (27). Balsey also found that sigma particles, which are able to coexist indefinitely in the same cytoplasm as stock 7 kappa, were not lost when the genotype was changed from Kk to kk and concluded from this observation that "symbiont supporting genes seem to be specific for their symbionts". Unlike kappa of syngen 4, there was no evidence for the presence of S genes in kappa of syngen 2.

Breeding analyses by Siegel (113) of the mu-bearing mate-killer stock 138, syngen 8 revealed that the dominant gene M must be present in the *Paramecium* nucleus for mu particles to be retained. Whereas animals of the genotype MM or Mm were able to maintain mu particles, animals of the genotype mm were not. Like kappa, mu does not arise *de novo*, but can be maintained only in stock 138 animals of suitable genotype. Following the change in genotype from Mm to mm, mu particles were depleted from the cells between the third and sixth fissions.

In contrast to mu particles of stock 138 syngen 8, mu particles of stock 540 syngen 8 are dependent for their maintenance on the presence of *either* of two dominant genes M_1 or M_2. Gibson and Beale (43) discovered that *Paramecium* homozygous (M_1M_1 or M_2M_2) or heterozygous (M_1m_1 or M_2m_2) for either of these genes was capable of supporting the mu character in this stock. Double recessive homozygotes of the genotype ($m_1m_1m_2m_2$) were not. Loss of mu following the change in genotype from $M_1M_1M_2M_2$ to $m_1m_1m_2m_2$ was reported by these workers to follow the same kinetics as loss of kappa from stock 51 reported by Chao (26). The particles disappeared completely from proportionally larger numbers of animals beginning at the 8th fission, until by the 15th fission virtually all the particles were lost from the cell. Similar kinetics of loss were not observed by Byrne (25), who later reinvestigated the problem. She reported that loss of mu from stock 540 generally took place within

4 to 7 fissions following introduction of the m_1 and m_2 alleles into the genome of *Paramecium*. Occasionally cells containing mu were observed in the 12th fission. Similarly, mu particles from stock 138 were lost within 4 to 7 generations, although a few cells containing mu were found as late as the 17th fission. Although it was not possible to count the mu particles, Gibson and Beale (43) found a proportionality between the numbers of M-genes present in the *Paramecium* genome and the speed at which sensitive animals were killed. Presumably, the speed of killing is proportional to the numbers of mu particles present.

Lambda particles of stock 299 syngen 8 have been reported by Schneller *et al.* (109) to be dependent for their maintenance on the presence of genes in the *Paramecium* nucleus. Introduction of gene *l* from killer stock 214 (which presumably contains kappa-like particles) into stock 299 results in the loss of lambda from stock 299. Preliminary evidence based on genetic crosses of stock 214 and stock 299 suggest that the genome which supports lambda of stock 299 will not support the particles of stock 214.

Preliminary investigations on alpha, the only particles known thus far to inhabit the nucleus, suggest that a nuclear gene may be involved in their maintenance (17, 100). Breeding analyses on gamma, delta, nu and sigma particles have not been reported in any detail, so it is not known whether any or all of these particle types are dependent for their maintenance upon genes in the *Paramecium* nucleus. Table 4 lists the genes which are known to affect particle-maintenance in *P. aurelia*.

E. Mutations of Kappa

Kappa has been shown to be mutable. Dippell (29, 30) studied five mutant killer stocks, all of which arose spontaneously. These were designated 47m1, 51m1, 51m2,

TABLE 4

Nuclear genes affecting the maintenance of intracellular particles of paramecium

Particle	Stock	Syngen	Genotype supporting maintenance of particles	Genotype leading to loss of particles	References
Alpha	562	2	—[a]	—	(17,100)
Kappa	7	2	KK, Kk	kk	(10, 11)
	51	4	KK, Kk	kk or $S_1S_1S_2S_2$[b]	(134) (8, 9, 137)
Lambda	299	8	LL	ll	(141)
Mu	138	8	MM, Mm	mm	(113)
	540	1	M_1M_1, M_1m_1 M_2M_2, M_2m_2	$m_1m_1m_2m_2$	(43)

[a] Genes have not yet been designated.

[b] The $S_1S_1S_2S_2$ genotype leads to loss of kappa even in the presence of supporting genotype KK or Kk.

51m3 and 51m5 and were characterized by the type or strength of killing manifested against suitable sensitives, resistance or sensitivity to specific types of kappa toxins, and fission rate of the particles they contained. By means of a series of suitable genetic crosses she demonstrated that the phenotype expressed was the result of mutations that had occurred in the kappa particles themselves and not in the *Paramecium* genome. Kappa expressing at least one of three killer phenotypes were found in the mutant stocks, original 51 hump-type killing (hp)*, spinner-type killing (sp) and weak hump-type killing (r). The last designation (r) was used because animals possessing mutant kappa of this type were also resistant to 51 kappa toxin. Some mutant stocks contained only one type of kappa while others contained more than one. Mutant stock 51m1, for example, contained kappa of (sp) type only and mutant stock 51m5 contained kappa of (r) type only. On the other hand mutant stock 51m2 contained mutant kappas of two types (r) and (sp); mutant stock 47m1 and stock 51m3 contained a mixture of both (hp) and (sp) kappa. Animals containing kappa of one type produced and were resistant to kappa toxin of that type only whereas animals containing two kinds of kappa produced and were resistant to kappa toxin of both kinds. Thus, mutant stock 51m1 which contained (sp) kappa only produced and was resistant to (sp) toxin. This stock was susceptible to the action of kappa of (hp) type. Mutant stock 51m3, which contained (hp) and (sp) kappas produced and was resistant to the action of both (hp) and (sp) toxin. Some of the mutant kappa exhibited properties other than those already described, i.e. they divided at a reduced rate of fission. Kappa of stock 51ml, designated 51ml (sp) was of this type. To prevent the eventual loss of the particles, stocks bearing this mutant kappa were cultured under conditions which resulted in a reduced fission rate for the protozoan.

Certain descendants of mutant stocks 51m1 and 51m5 were found by Preer *et al.* (99) to have lost their ability to maintain killing particles, were unable to form B-particles, and became sensitive to the action of kappa 51 toxin. The particles were designated by the Greek letter *pi*, and are considered to be mutant forms of kappa. Pi particles also arose (apparently spontaneously) from kappa bearing stocks 51 and 139, both of syngen 4 (52, 53). The original kappa-bearing stocks yielded toxic particles that produced typical hump-type killing. The mutant pi-bearing stocks were sensitive to the action of 51 kappa toxin and contained no brights. Hanson (54) found a mutant kappa [51m7(hp)] derived from a single individual of stock 51 that was characterized by reduced fission rate. Animals maintained at 5 fissions per day eventually lost their kappa and became sensitives. The same animals cultured at one fission per day retained kappa at a level of 400–500 per animal. No other changes in the mutant (e.g. morphology, killing action, or resistance to 51 kappa toxin) could be detected.

In the only report of induced mutations of kappa, Ehret and Powers (35) exposed 15 individual cultures of stock 51 to X-irradiation and obtained 28 independent mutants that were resistant to the effect of the antibiotic aureomycin. Although the induced mutations were believed to involve the genes of the kappa particle and not

---

* See Section III, B, 1, a for a description of kinds of kappa killing.

TABLE 5

Mutants of kappa.

Original stock	Mutant stock	Mutant particle	Killing activity		Resistance of mutant to 51 kappa toxin	Fission rate	References
			Wild type	Mutant			
7	7ml	7ml (par)	spinner	paralyser	—	normal	(90,91)
47	47ml	47ml (hp)	vacuolizer	hump	resistant	normal	(29,30)
		47ml (sp)	hump	spinner	—	reduced	(29,30)
51	51ml	51ml (sp)	hump	spinner	—	reduced	(29,30)
		51ml (pi)	spinner	none	sensitive	normal	(99)
51	51m2	51m2 (sp)	hump	spinner	—	reduced	(29,30)
		51m2 (r)	hump	weak hump	resistant	normal	(29,30)
51	51m3	51m3 (sp)	hump	spinner	—	reduced	(29,30)
51	51m5	51m5 (r)	hump	weak hump	resistant	normal	(29,30)
		51m5 (pi)	weak hump	none	sensitive	normal	(99)
51	51m6	51m6 (pi)	hump	none	sensitive	normal	(52,53)
51	51m7	51m7 (hp)	hump	hump	resistant	reduced	(54)
51	51 (aur)[a]	—	hump	hump	resistant	reduced	(35)
51	51m42	51m42 (r)	hump	weak hump	resistant	normal	(170)
		51m43	weak hump	none	sensitive	normal	(171)
139	139ml	139ml (pi)	hump	none	sensitive	normal	(52,53)

[a] Resistant to aureomycin.

the genes of the protozoan, evidence to support this belief through genetic analysis was not obtained.

Of the mutants described thus far, all that were capable of resisting the action of 51 kappa toxin contained B or bright kappa particles in addition to normal N or non-bright particles. Mutant stocks that were sensitive (designated pi) to the action of kappa toxin contained N-type particles only and were free of B-kappa particles. This led to the postulate by Sonneborn (141) that full resistance was associated with the B-particles. The recent discovery by Widmayer (171) of a non-killer mutant stock (51m43) that is fully resistant to the action of 51 kappa toxin and is totally devoid of B-kappa, suggests that this view be revised. This mutant arose spontaneously from another mutant (51m42) which she was studying and which was characterized by weak-hump killing and resistance to the action of 51 kappa toxin. This mutant stock contained both B- and N-particles.

Preer found the only mutant kappa of syngen 2. This mutant, designated (7m1) was characterized by a new kind of killing action (paralytic) and arose in a stock that contained (sp) kappa. It was detected by reducing the number of N-particles of stock 7 to one per animal and allowing the number of mutant particles to reach their full complement by reducing the fission rate of the mutant-bearing animals. Mutants of kappa along with some of their properties are listed in Table 5.

Sonneborn (141) exhaustively analyzed the mutations of kappa of syngen 4 and

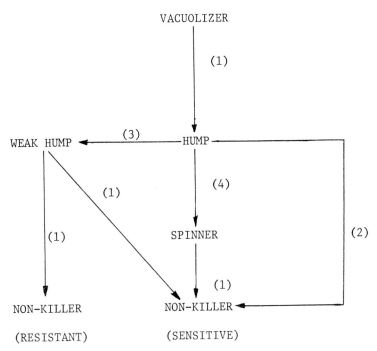

Fig. 7. Mutations affecting kappa killing. Adapted from Sonneborn (141) as modified by Widmayer (171). Numbers in parenthesis indicate the number of mutants at each step.

concluded that they were for the most part: (1) unidirectional, (2) degradational (i.e. involved a loss of some property), (3) often multiple and (4) occurred with greater frequency from mutant kappas than from wild type kappa. He devised a scheme summarizing the sequence of mutations of kappa. That scheme, modified by Widmayer (171) to include the mutants discovered by her, is reproduced in Fig. 7.

Two other possible mutants of kappa have been reported (80, 81). They were obtained as follows: purified 51 kappa was added to a suspension of sensitive animals. Before death and after initial response to the toxin, the animals were removed to fresh medium. Most died, but two surviving clones were found which contained kappa, were infective, conferred immunity against 51 kappa toxin but produced no toxin themselves. Genetic analysis to prove the altered kappa were mutants was not performed.

F. Effect of Physical and Chemical Agents on the Particles

In bacterized medium, kappa of stock 51 is lost at temperatures of 33.8 °C or higher and also at 10 °C (30, 137, 146). X-irradiation at dosages up to 43,000 r caused the loss of kappa without permanent injury to the host (84, 91). UV irradiation reduced but did not eliminate kappa from the cell (141). Treatment of stock 51 animals with nitrogen mustard induced genetic changes and reduced kappa in proportion to the concentration used (39, 40). Aureomycin (35, 173), terramycin (35, 173), and penicillin G (167) were reported to destroy kappa of stock 51. Treatment of kappa-bearers with streptomycin (141, 173), chloramphenicol (20, 35, 173) or 2,6-diaminopurine (51, 152, 173) substantially reduced but did not eliminate the number of functional kappa present. Mu particles were reported to be inactivated by 8-aza-guanine (45).

Destruction of kappa by 2,6-diaminopurine, an adenine antagonist, is of interest. Stock et al. (152) reported that 100–1000 μg/ml of this compound destroyed over 90 % of functional kappa present in animals of stock 51. Concentrations of 1500 μg/ml or higher were required to cause the death of the animals. In axenic culture, growth of this stock and of stock 47 was inhibited by concentrations of 2,6-diaminopurine ranging from only 1 to 4 μg/ml (74).

In axenic culture the only studies on the effect of physical and chemical agents were carried out with stock 299 lambda, 139 pi and 138 mu. Lambda particles of stock 299 were lost in three days from animals kept at 37 °C (127). Unless removed to 27 °C the paramecia themselves died on the fourth day. At 15° or 27 °C lambda-, mu- and pi-bearing animals maintained their particles through hundreds of weekly subcultures. No loss of particles was observed from particle-bearers maintained at pH-values which permitted the growth of the animals, e.g. pH 6.0–8.0 (127). Fewer mu particles were observed in stock 138 animals cultured at pH 8.0. Often these particles appeared to be 2 or 3 times their normal length.

Soldo (123, 124) tested a number of anti-tumor and antibiotic substances for their effect on stock 299 lambda. ID_{50} values (i.e. concentration required to inhibit growth 50 % as compared to that of an untreated control) for each test substance were determined for both the protozoan and lambda populations within a culture. Response

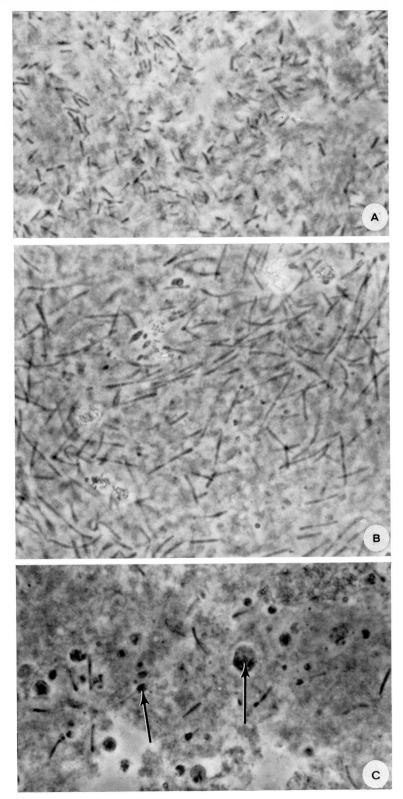

the other hand, actinomycin D and neomycin did not selectively inhibit the particles but were highly toxic for man.

In a recent study, Soldo et al. (129) examined the kinetics of loss and changes in ultrastructure of lambda particles following treatment of the axenically cultured stock 299 with penicillin. After prolonged exposure to the antibiotic at concentrations ranging from 0.5 to 1.0 units/ml, lambda was reduced in number but not completely eliminated from the cells. Under these conditions many of the particles became filamentous, occasionally extending the entire length of the protozoan itself. At a

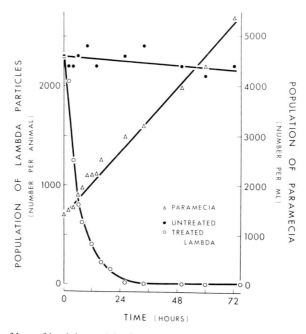

Fig. 11. Kinetics of loss of lambda particles from the cytoplasm of lambda-bearing stock 299 in the logarithmic phase of growth treated with 2000 units/ml penicillin G.

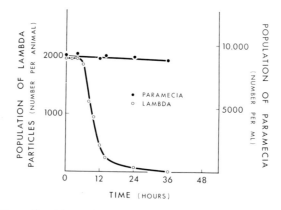

Fig. 12. Kinetics of loss of lambda particles from the cytoplasm of lambda-bearing P. aurelia treated with 2000 units penicillin G while in the stationary phase of growth.

animals cultivated in bacterized medium, Stevenson (150) reported the presence of α,ε,-diaminopimelic acid in protein hydrolysates of the particles. Diaminopimelic acid is a constituent of the cell walls of bacteria and blue-green algae and is not found in fungi or protozoa (174). Muramic acid, also a component of the cell wall of many bacteria was thought to be present in the mu particle but was not positively identified as such. If these findings could be confirmed on mu particles isolated from bacteria-free cultures of mu-bearers, they would constitute the best evidence obtained thus far as to the bacterial nature of the mu particle. In this respect, Soldo *et al.* (130) observed that the DNA content of those particles analyzed thus far falls within the range commonly found in bacteria (Table 9).

B. Metabolism

1. Respiration

Efforts to determine the effect particles may exert on the respiration of the host cell have produced conflicting results. Simonsen and Van Wagtendonk (115–117) and Van Wagtendonk *et al.* (166) compared the respiratory rate of stock 51 killers with those of three sensitive stocks of syngen 4. They found that under a variety of conditions the respiratory rate of the killers was higher than those of the kappa-less animals. Cytochrome systems were present in both killers and sensitives. Killers were more resistant to inhibition by sodium azide than were the sensitives. On the basis of a lower cytochrome oxidase activity found in homogenates of killer animals when compared to those of sensitives, Simonsen and Van Wagtendonk (116) suggested that the number of functional mitochondria within the animals may be reduced in the presence of kappa. Other findings by these authors pointed to the presence of a respiratory system, possibly a flavin system, associated with the presence of kappa in the killers.

Preer (93) compared the respiratory rates of killers and sensitives of syngen 2 and was unable to find significant differences. Levine and Howard (71) reported the respiratory rates of mate-killers and those of sensitives of syngen 8 to be similar.

Kung (65) studied respiration of kappa isolated and purified from stock 51 animals using the ECTEOLA procedure described by Smith (118). By means of a polaro-graphic electrode immersed in a suspension of the isolated particles he measured oxygen uptake in the presence of various substrates and reported that the particles utilized sucrose, glucose and many intermediates of both the glycolytic and citric acid pathways. Using spectrophotometric methods he detected NADH oxidase activity in kappa which was inhibited by KCN, antimycin A, and 2-heptyl-4-hydro-quinoline-N-oxide (66). The cytochrome system of the particles was reported to resemble that of bacteria and not of the protozoan. He concluded from these experi-ments that kappa possessed a higher level of organization than that found in viruses or subcellular organelles and argued that the bacteria carried over into his samples from the medium in which the animals were cultured did not significantly affect his results.

Unequivocal evidence concerning respiration of the particles and its possible effect on the host is still lacking. Critical students will demand that experiments be carried out with axenically cultivated animals under carefully controlled conditions which rule out any possible effect due to contaminating bacteria.

2. Enzymatic activity

The first attempts to detect the presence of enzymes in the particles were made by Preer (91) and Preer and Stark (99). Cytological tests for the presence of reducing enzymes, cytochrome oxidase and alkaline phosphatase activity in kappa were all negative. The authors advised against drawing definitive conclusions based on these tests because of difficulties in discerning weak colors in particles as small as kappa.

Using standard biochemical procedures, Soldo *et al.* (128) examined lambda particles for the presence of enzymes involved in the catabolism of nucleic acids. The particles were isolated and purified from axenically cultivated stock 299 animals by differential centrifugation of homogenates containing the particles through sodium bromide solutions of different densities. Whole particle suspensions and sonicates were capable of degrading native DNA to acid soluble components (Fig. 14). DNAase activity of sonicates was low although consistently higher than that found using

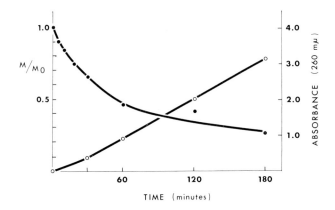

Fig. 14. Degradation of native DNA by sonicates of lambda particles. M/M₀ = ratio of the molecular weight over that at time zero determined by visicometry. Appearance of acid solubles was measured by absorbance at 260 nm.

TABLE 10

DNAse activity of lambda particles.

	Whole	Sonicated
Specific activity[a]	1.09	1.64
Range	(0.70–1.31)	(0.51–2.42)
Number of determinations	6	11

[a] Δ Absorbance 260 nm/hr/mg protein at 37 °C.

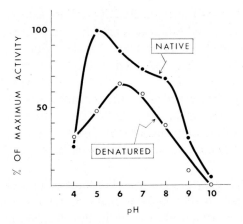

Fig. 15. pH optimum curve for DNAase activity of sonicated lambda particles.

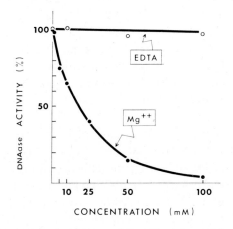

Fig. 16. Inhibition of DNAase activity of lambda particles by Mg^{2+}.

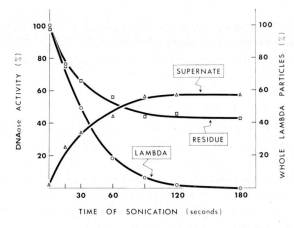

Fig. 17. Release of DNAase activity from lambda particles by sonication.

whole particle suspensions (Table 10). Activity against native DNA occurred over a wide range of pH-values and exhibited a maximum at pH 5.0 (Fig. 15). The broad shoulder at pH 8.0 suggested the possibility that the particles contained more than one enzyme capable of degrading DNA. Activity was lower when heat denatured DNA was used as a substrate over almost the entire range of pH-values tested and was optimal at pH 6.0. Mg^{2+} at a concentration of 100 mM inhibited the activity (Fig. 16). EDTA at the same concentration did not. Not all the enzymatic activity could be released from lambda following complete disruption of the particles by sonication (Fig. 17). Approximately 40% of the total remained tightly bound to membranes in the residue fraction.

The properties of this enzyme were typical of those of an "acid" DNAase which is widely distributed in nature. It is of interest that most Gram-negative bacteria possess DNAase activity which exhibit pH optima in the alkaline range of values e.g. *E. coli* (68).

Other enzyme activities have been detected in the particles. These include RNAase, ATPase and acid phosphatase. Nucleoside hydrolase, an enzyme that catalyzes the splitting of nucleosides into the free base and sugar (125a), is present in high concentrations in the *Paramecium* but is absent from lambda. Alkaline phosphatase activity could not be detected in either *Paramecium* or the particles. It should be emphasized that enzymatic activities discovered in the particles thus far are very low. DNAase activity of lambda, for example, is less than 1/100 of that found in the protozoan.

The enzyme RNA polymerase may be present in mu particles. Stevenson (151) reported that mu isolated from stock 540 animals cultured in bacterized medium and disrupted by alternate cycles of freezing and thawing, incorporated [^{14}C]ATP and [^{3}H]GTP into acid insoluble material in reaction mixtures containing the nucleoside triphosphates ATP, GTP, CTP and UTP. Mg^{2+} was essential for optimum activity. Incorporation was inhibited by actinomycin D and in the presence of ribonuclease. The pH optimum was between 8.1 and 9.0. On the basis of these data Stevenson suggested that the particles synthesized RNA by a DNA-dependent reaction. He noted that incorporation ceased in 15 to 20 minutes following addition of the label to reaction mixtures. Because no further incorporation of the label into acid insolubles was observed upon prolonged incubation of reaction mixtures, he argued that bacterial contamination was small and that incorporation was due to the mu particles.

3. Properties of the DNA of the particles

Smith-Sonneborn *et al.* (121) determined the buoyant densities of DNA in lysates of stock 51 sensitives, killers and purified kappa by centrifugation in cesium chloride. DNA of stock 51 sensitives cultured axenically gave a single band of density 1.689 g/cc corresponding to a guanine plus cytosine (G + C) content of 29 mole %. DNA of stock 51 killers cultured in bacterized medium showed 3 bands at densities 1.689 g/cc, 1.696 g/cc and 1.716 g/cc corresponding to (G + C) contents of 29, 36 and 56 mole %. DNA of purified kappa banded at 1.696 g/cc (G + C content = 36 mole %);

TABLE 11

Some properties of DNA of intracellular particles.

Particle	Stock	Buoyant density (g/cc)	Guanine + cytosine content calculated from buoyant density[a] (mole %)	Thermal melting point (T_m) (°C)	Guanine + cytosine content calculated from T_m[b] (mole %)	References
Kappa	51	1.696	36.7	—	—	(121)
Lambda	239	1.708	49.0	—	—	(18)
	299[c]	—	—	79.6	25.2	(125)
Mu	138	1.700	40.8	—	—	(18)
	540	1.701	41.8	—	—	(18)
	540	1.694	34.7	82.0	31.0	(151)
Sigma	114	1.704	45.0	—	—	(18)
P. aurelia (whole)	51	1.689	29.6	—	—	(121)
	51	1.689	29.6	—	—	(153)
	299[c]	—	—	81.0	28.6	(125)
(nuclei)	540	1.693	33.7	81.9	30.7	(151)

[a] $(G + C) = (\rho - 1.660)/0.098$ (105).
[b] $(G + C) = (T_m - 69.3)/0.41$ (72).
[c] Axenically cultivated.

DNA from *Aerobacter aerogenes*, the bacterium used as a food source for the killers banded at 1.716 g/cc (G + C content = 56 mole %). Suyama and Preer (153) confirmed that DNA of stock 51 exhibited a buoyant density of 1.689 g/cc. Behme (18) used the same technique to estimate the (G + C) content of DNA's in lysates of lambda (stock 239), mu (stock 138 and 540) and sigma (stock 114) particles. Stevenson (151) determined the buoyant density and thermal melting point (T_m) of DNA extracted from purified mu particles and nuclei of stock 540 animals cultivated in bacterized media. Soldo and Reid (125) measured the thermal melting points for DNA's isolated from purified lambda particles and particle-free animals of stock 299 cultured axenically. These data are shown in Table 11.

The (G + C) content of the DNA of the various particles calculated from buoyant density data shows substantial differences and suggested to Preer (94) that the particles comprise a phylogenetically diverse group. There may however, be some question as to the significance of these data. Whereas individual determinations by this method are usually accurate to within 1 mole %, the (G + C) content of DNA of mu particles of the same stock (540) showed significant differences. This was also true for the (G + C) content determined for DNA of nuclei when compared to the (G + C) content of DNA of whole cells. Further, (G + C) content of DNA of lambda of stock 239 determined by buoyant density compared to the (G + C) content of DNA of lambda of the closely related stock 299 determined by thermal transition and direct chemical analysis (Table 11), differed by a value of almost 24 mole %! It is not unreasonable to assume that these discrepancies are in part due to contaminating bacteria present in these preparations. Before the question of possible phylogenetic relationships that may exist among the particles can be answered with confidence, the (G + C) content of the DNA of the particles should be determined on particle-containing stocks cultivated axenically.

Soldo and Reid (125) examined in some detail the properties of DNA isolated from purified lambda particles and particle-free *P. aurelia* (stock 299) cultivated axenically. The DNA was extracted in fibrous form from both whole cells and purified particle preparations with phenol or detergents and purified by fractional precipitation as the cetyltrimethyl-ammonium bromide salt to remove contaminating carbohydrates. Chemical analysis revealed the presence of the bases adenine, guanine, cytosine and thymine in the preparations. A search for the presence of "odd" bases proved negative. Thermal transition of DNA from whole cells and from lambda was sharp and showed hyperchromic increases of 40% and 41% respectively, typical of native double-stranded polymers. Values obtained for their T_m's were 81.0°C (whole cell) and 79.6°C (lambda). Unlike DNA extracted from whole cells which showed little or no evidence of renaturation, DNA isolated from lambda particles renatured to considerable extent following quick cooling and reannealing at 60°C. This is characteristic of DNA's isolated from bacteria. Chemical analysis of the base composition of the DNA of whole cells and lambda particles are shown in Table 12. Thermal transition curves are illustrated in Figs. 18 and 19.

Soldo and Reid (125) reported that DNA isolated from purified lambda preparations contained an additional component. Preparative CsCl gradient centrifugation of DNA extracted from these preparations occasionally showed the presence of a satellite band, heavier than the principal DNA (Fig. 20). The heavy component reacted with formaldehyde, melted over a broad range of temperatures, and possessed a base composition different from that of the main particle DNA. These are properties

TABLE 12

Base composition of DNA from whole cells and lambda particles.

Component	DNA (mole %)	
	Whole cell	Lambda
Guanine	14.3	12.8
Cytosine	15.1	13.2
Adenine	35.9	37.7
Thymine	34.9	36.3
G + C (chemical analysis)	29.4	26.0
G + C (T_m)	28.6 (81.0 °C)	25.2 (79.6 °C)

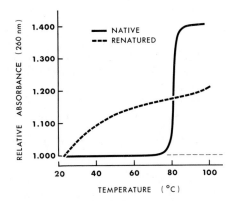

Fig. 18. Thermal transition of whole cell DNA (stock 299).

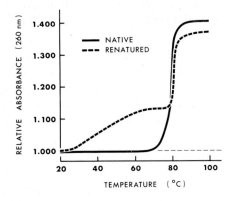

Fig. 19. Thermal transition of lambda particle DNA (stock 299).

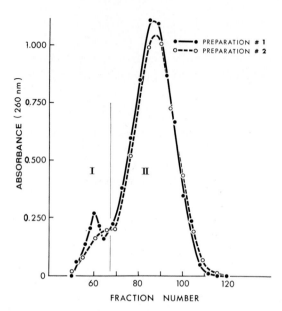

Fig. 20. Preparative cesium chloride density gradient analysis of two preparations of lambda DNA.

of single-stranded DNA. It is not yet known whether this component represents a true single-stranded DNA component occurring naturally or merely an artifact resulting from the isolation procedure.

V. SYMBIOTIC RELATIONSHIPS

Investigators, in describing the particles of *Paramecium*, often refer to them as symbionts or symbiotes, without consideration of the possible advantages each member of the association may contribute to the other. In this respect, Soldo (124) examined the possible contribution the particles make to the nutrition of the host. In a partially defined axenic medium the need for individual components of the medium by lambda-bearing and lambda-free animals of stock 299 was compared by the single omission technique. Lambda-free animals cultured in this medium required guanosine (or guanylic acid), uridine (or cytidine, uridylic or cytidylic acid), folic acid, nicotinamide, pyridoxal, riboflavin, thiamine and possibly calcium pantothenate for growth. Lambda-bearers required the same components except for folic acid. Animals containing lambda particles could be subcultured indefinitely in a medium devoid of this vitamin. Lambda-free animals failed to grow beyond the second weekly subculture when folic acid was deleted from the growth medium (Table 13). Apparently lambda synthesized folic acid.

Further support for this idea was obtained by comparing the effects of penicillin on the growth of lambda-bearing and lambda-free animals in the presence and in the absence of folic acid (Figure 21). It will be recalled (see Section III, F) that the anti-

TABLE 13

Vitamin requirements of Paramecium

Vitamin	Population (% control)										
	299 λ						299 S				
	Serial subculture						Serial subculture				
	1	2	3	4	5		1	2	3	4	5
Biotin	105	123	94	99	105		110	69	77	112	105
Ca-pantothenate	95	38	38	7	12		105	73	74	36	8
Folic acid	100	71	71	69	75		40	0			
α-Lipoic acid	101	103	99	85	95		87	57	96	121	112
Nicotinamide	97	0					75	0			
Pyridoxal	110	98	82	0			75	48	37	26	0
Riboflavin	35	0					0				
Thiamine	25	0					12	0			

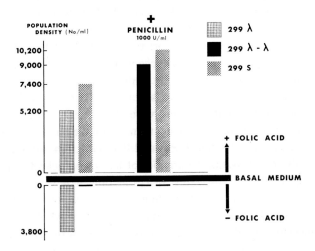

Fig. 21. The effect of penicillin upon the folic acid requirement of *Paramecium*.

biotic effectively destroys lambda without adversely affecting the host. In the absence of folic acid, penicillin-treated animals (lambda removed) failed to grow, whereas reduced but transplantable growth was possible in the untreated animals (lambda retained). On the basis of these experiments it was concluded that lambda particles within the cytoplasm produce sufficient quantities of folic acid to meet, in part, the needs of the protozoan for this vitamin. In return for the comparative protection afforded by the cytoplasm of the *Paramecium*, the particle provided its host with an essential nutrilite.

Van Wagtendonk *et al.* (164) succeeded in cultivating lambda particles *in vitro* in a complex medium containing 17 amino acids, 8 vitamins, guanosine monophosphate, uridine monophosphate, stigmasterol, glycerol, maltose, Edamine S, and a fraction bacterial DNA, undergoes almost complete renaturation following reannealing. The particles possess the enzymatic machinery necessary to carry out certain anabolic as well as catabolic processes. Lambda, for example, synthesizes folic acid and contains enzymes capable of degrading DNA and RNA to form low molecular weight components. Mu reportedly synthesizes RNA by means of a DNA-dependent reaction. Enzymes involved in the oxidation of glucose and other sugars appear to be present in kappa. Diaminopimelic acid, a constituent of cell walls of bacteria and blue-green algae, has been identified in hydrolysates of mu particles. Bacteria-like flagella have been observed in electron micrographs of lambda and sigma. And finally, lambda has been cultured outside the host cell in a complex medium.

The most reasonable hypothesis consistent with these facts is that the particles are bacterial in nature, probably descendents of free living forms which have adapted to an intracellular environment. It is not yet certain whether the adapted particles comprise a natural taxon, as do, for example, the Rickettsiae (76) or whether they represent a diverse group of microorganisms as Preer (94) and Beale et al. (17) believe. Preer (100) suggests that alpha particles may be related to the myxobacteria. Systematic studies aimed at comparing the serological properties of the various particles to each other as well as to those of bacterial and related species may provide an answer to this question.

What adaptive changes have the particles undergone to permit their continued intracellular existence? How have they been modified so as to prevent digestion by a rather formidable array of nucleolytic and proteolytic enzymes present in the host *Paramecium*? Answers to these questions must await future discoveries; nevertheless the observation that lambda and sigma are contained within microvacuoles offers an obvious mechanism for the protection of these particles against the action of host enzymes. Other as yet unknown mechanisms must exist. Kappa, mu and pi particles are devoid of a protective shroud and exist in intimate contact with the host cytoplasm. A characteristic shared in common by the particles, which may be the result of adaptation, is the apparent uniform distribution of DNA throughout the particles as revealed in the electron microscope and by cytochemical tests. Similarly, the DNA of Blochmann bodies of insects, bipolar bodies of *C. oncopelti* and the particles of the ciliate *Discophrya piriformis* (141) appears to be dispersed throughout the particle. The significance of this phenomenon remains at present obscure. DNA of free-living bacteria is known to be localized within a central nucleoid body or nucleosome. It is not clear whether the flagella of lambda particles are functional. A report by Jurand and Preer (62) suggests that lambda freshly released from crushed animals are capable of limiting spinning movements. This author has never observed motility of lambda particles of axenically cultured stock 299 inside or outside the host and Van Wagtendonk et al. (164) make no mention of the fact that lambda cultured *in vitro* are motile. Although the study of the biochemistry of the particles is still in its infancy, first indications are that the enzymatic activities contained within the particles are low as compared to those of free-living bacteria. Much more work on the biochemistry of the particles is needed, and it seems reasonable to expect that it will be in this area that adaptive changes as they exist within the particles will be found.

67 Lanham, U. N. 1968. The Blochmann bodies: hereditary intracellular symbionts of insects. *Biol. Rev.* **43**, 269–286.

68 Lehman, I. 1963. The nucleases of *Escherichia coli*. In Davidson, J. N. & Cohen, W. E., *Progress in Nucleic Acid Research*, Academic Press, New York, London, Ch. 2, 84–118.

69 Lerman, L. S. & Tolmach, L. J. 1957. Genetic transformation. I. Cellular incorporation of DNA accompanying transformation of *Pneumococcus*. *Biochim. Biophys. Acta* **26**, 68–82.

70 Levine, M. 1953. The diverse mate-killers of *Paramecium aurelia*, variety 8: their interrelations and genetic basis. *Genetics* **38**, 561–578.

71 — & Howard, J. L. 1955. Respiratory studies on mate-killers and sensitives of *Paramecium aurelia*, variety 8. *Science* **41**, 336–337.

72 Marmur, J. & Doty, P. 1962. Determination of the base composition of deoxyribonucleic acid from its thermal denaturation temperature. *J. Mol. Biol.* **5**, 109–118.

73 *McManamy, B. & Sonneborn, T. M. 1967. Metagons for kappa? *Science* **158**, 532.

74 Miller, C. A. & Van Wagtendonk, W. J. 1956. The essential metabolites of a strain of *Paramecium aurelia* (stock 47.8) and a comparison of the growth rate of different strains of *Paramecium aurelia* in axenic culture. *J. Gen. Microbiol.* **15**, 280–291.

75 Morowitz, H. J., Tourtellotte, M. E., Guild, W. R., Castro, E., Woese, C. & Cleverdon, A. 1962. The chemical composition and submicroscopic morphology of *Mycoplasma gallisepticum*, avian PPLO 5969. *J. Mol. Biol.* **4**, 93–103.

76 Moulder, J. W. 1966. The relation of the Psittacosis group (chlamydiae) to bacteria and viruses. *Ann. Rev. Microbiol.* **20**, 107–130.

77 *Mueller, J. A. 1961. Further studies on the nature of the "co-factor" required for kappa infection in *Paramecium aurelia*, syngen 4. *Amer. Zoologist* **1**, 375.

78 *— 1962. Induced physiological and morphological changes in the B-particle and R-body from killer Paramecia. *J. Protozool.* (Suppl.) **9**, 26.

79 — 1963. Separation of kappa particles with infective activity from those with killing activity and identification of the infective particles in *Paramecium aurelia*. *Exp. Cell Res.* **30**, 492–508.

80 *— 1964. Paramecia develop immunity against kappa. *Amer. Zoologist* **4**, 313–314.

81 — 1965. Kappa-affected paramecia develop immunity. *J. Protozool.* **12**, 278–281.

82 — 1965. Vitally stained kappa in *Paramecium aurelia*. *J. Exp. Zool.* **160**, 369–372.

83 *Nanney, D. L. 1950. X-ray sensitivities of paramecins and kappas of Stock 51, Variety 4 of *Paramecium aurelia*. *Microbial Genet. Bull.* **3**, 12.

84 — 1954. X-ray studies on paramecins and kappas of variety 4 of *Paramecium aurelia*. *Physiol. Zool.* **27**, 79–89.

85 Neimark, H. C. & Pene, J. J. 1965. Characterization of pleuropneumonia-like organisms by deoxyribonucleic acid composition. *Proc. Soc. Exp. Biol. Med.* **118**, 517–519.

86 *Nobili, R. 1960. Kappa in amacronucleate *Paramecium aurelia*: its development and effect on survival time. *J. Protozool.* **7** (Suppl.), 15.

87 — 1961. Il compartamento del kappa in esemplari amacronucleati di *Paramecium aurelia*, stock 51, syngen 4. *Boll. Zool.* **28**, 579–596.

88 — 1962. Su un nuovo ceppo "killer" ad azione paralizzante di *Paramecium aurelia*, syngen 2. *Boll. Zool.* **29**, 555–565.

89 Preer, Jr., J. R. 1946. Some properties of a genetic cytoplasmic factor in *Paramecium*. *Proc. Nat. Acad. Sci. U.S.* **32**, 247–253.

90 — 1948. A study of some properties of the cytoplasmic factor "kappa" in *P. aurelia*, variety 2. *Genetics* **33**, 349–404.

91 — 1950. Microscopically visible bodies in the cytoplasm of the "killer" strains of *Paramecium aurelia*. *Genetics*, **35**, 344–362.

92 — 1950. The role of the genes, cytoplasm and environment in the determination of resistance and sensitivity. *Yearbook Amer. Phil. Soc.* **1950**, 161–163.

93 — 1957. Genetics of Protozoa. *Ann. Rev. Microbiol.* **11**, 419–438.

94 — 1968. Genetics of Protozoa. In Chen, T. T., *Research in Protozoology, Vol. 3*, Pergamon Press, London, 130–288.

95 — & Jurand, A. 1968. The relation between virus-like particles and R-bodies of *Paramecium aurelia*. *Genet. Res.* **12**, 331–340.

96 — & Preer, L. B. 1967. Virus-like bodies in killer Paramecia. *Proc. Nat. Acad. Sci. U.S.* **58**, 1774–1781.

97   — & Stark, P. 1953. Cytological observations on the cytoplasmic factor "kappa" in *Paramecium aurelia*. *Exp. Cell Res.* **5**, 478–491.

98   —, Hufnagel, L. A. & Preer, L. B. 1966. Structure and behavior of R-bodies from killer Paramecia. *J. Ultrastruct. Res.* **15**, 131–143.

99   —, Siegel, R. W. & Stark, P. S. 1953. The relationship between kappa and paramecin in *Paramecium aurelia*. *Proc. Nat. Acad. Sci. U.S.* **39**, 1228–1233.

100 Preer, L. B. 1969. Alpha, an infectious macronuclear symbiont of *Paramecium aurelia*. *J. Protozool.* **16**, 570–578.

101 — & Preer, Jr., J. R. 1964. Killing activity from lysed kappa particles of *Paramecium*. *Genet. Res.* **5**, 230–239.

102 Reeve, E. C. R. 1962. Mathematical studies of mesosome distribution. *Genet. Res.* **3**, 47–50.

103 — & Ross, G. J. S. 1962. Mate-killer (mu) particles in *Paramecium aurelia:* the metagon division hypothesis. *Genet. Res.* **3**, 328–330.

104 — & — 1963. Mate-killer (mu) particles in *Paramecium aurelia:* further mathematical models for metagon distribution. *Genet. Res.* **4**, 158–161.

105 Rolfe, R. & Meselson, M. 1959. The relative homogeneity of microbial DNA. *Proc. Nat. Acad. Sci. U.S.* **45**, 1039–1043.

106 Rudenberg, F. H. 1962. Electron microscopic observations of kappa in *Paramecium aurelia*. *Texas Rep. Biol. Med.* **20**, 105–112.

107 *Schneller, M. V. 1958. A new type of killing action in a stock of *Paramecium aurelia* from Panama. *Proc. Indiana Acad. Sci.* **67**, 302–303.

108 *— 1962. Some notes on the rapid lysis type of killing found in *Paramecium aurelia*. *Amer. Zool.* **2**, 446.

109 *—, Sonneborn, T. M. & Mueller, J. A. 1959. The genetic control of kappa-like particles in *Paramecium aurelia*. *Genetics* **44**, 533–534.

110 Setlow, R. & Doyle, B. 1956. The action of ultraviolet light on paramecin and the chemical nature of paramecin. *Biochim. Biophys. Acta* **22**, 15–20.

111 *Siegel, R. W. 1950. Determination and inheritance of a new type of killing action in *Paramecium aurelia*, Variety 3. *Microbial Genet. Bull.* **3**, 12.

112 *— 1952. The genetic analysis of mate-killing in *Paramecium aurelia*. *Genetics* **37**, 625–626.

113 — 1953. A genetic analysis of the mate-killer trait in *Paramecium aurelia*, variety 8. *Genetics* **38**, 550–560.

114 — 1954. Mate-killing in *Paramecium aurelia*, variety 8. *Physiol. Zool.* **27**, 89–100.

115 *Simonsen, D. H. & Van Wagtendonk, W. J. 1949. Oxygen consumption of "killer" and "sensitive" stocks of *Paramecium aurelia*, variety 4. *Fed. Proc.* **8**, 250–251.

116 — & Van Wagtendonk, W. J. 1952. Respiratory studies on *Paramecium aurelia*, variety 4, killers and sensitive. *Biochim. Biophys. Acta* **9**, 515–527.

117 — & — 1956. The succinoxidase system of killer and sensitive stocks of *Paramecium aurelia*, variety 4. *J. Gen. Microbiol.* **15**, 39–46.

118 *Smith, J. E. 1961. Purification of kappa particles of *Paramecium aurelia*, stock 51. *Amer. Zool.* **1**, 390.

119 *— & Van Wagtendonk, W. J. 1962. Chemical identification of kappa. *Fed. Proc.* **21**, 153.

120 Smith-Sonneborn, J. E. & Van Wagtendonk, W. J. 1964. Purification and chemical characterization of kappa of stock 51 *Paramecium aurelia*. *Exp. Cell Res.* **33**, 50–59.

121 —, Green, L. & Marmur, J. 1963. Deoxyribonucleic acid base composition of kappa and *Paramecium aurelia*, stock 51. *Nature* **197**, 385.

122 Soldo, A. T. 1960. Cultivation of two strains of killer *Paramecium aurelia* in axenic medium. *Proc. Soc. Exp. Biol. Med.* **105**, 612–615.

123 — 1961. The use of particle-bearing *Paramecium* in screening for potential anti-tumor agents. *Trans. N.Y. Acad. Sci.* **23**, 653–661.

124 — 1963. Axenic culture of *Paramecium*. Some observations on the growth behavior and nutritional requirements of a particle-bearing strain of *Paramecium aurelia* 299 lambda. *Ann. N.Y. Acad. Sci.* **108**, 380–388.

125 *— & Reid, S. J. 1968. Deoxyribonucleic acids of lambda-bearing *Paramecium aurelia*. *J. Protozool.* **15** (Suppl.), 15.

125a — & Van Wagtendonk, W. J. 1961. Nitrogen metabolism in *Paramecium aurelia*. *J. Protozool.* **8**, 41–55.

126 — & — 1969. The nutrition of *Paramecium aurelia*, stock 299. *J. Protozool.* **16**, 500–506.

127   —, Godoy, G. A. & Van Wagtendonk, W. J. 1966. Growth of particle-bearing and particle-free *Paramecium aurelia* in axenic culture. *J. Protozool.* **13**, 492–497.

128   *—, — & — 1969. Biochemical studies on endosymbiote particles of *Paramecium aurelia* isolated by a new procedure. *J. Protozool.* **16** (Suppl.), 8.

129   —, Musil, G. & Godoy, G. A. 1970. Action of penicillin G on endosymbiote lambda particles of *Paramecium aurelia*. *J. Bacteriol.* **104**, 966–980.

130   —, Van Wagtendonk, W. J. & Godoy, G. A. 1970. Nucleic acid and protein content of purified endosymbiote particles of *Paramecium aurelia*. *Biochim. Biophys. Acta* **204**, 325–333.

131   Sonneborn, T. M. 1938. Mating types in *P. aurelia*: diverse conditions for mating in different stocks, occurrence, number and interrelations on the types. *Proc. Amer. Phil. Soc.* **79**, 411–434.

132   — 1938. Sexuality and genetics in *Paramecium aurelia*. *Yearbook Amer. Phil. Soc.* **1938**, 220–222.

133   — 1939. *P. aurelia*: mating types and groups; lethal interactions: determination and inheritance. *Amer. Naturalist* **73**, 390–413.

134   — 1943. Gene and cytoplasm. I. The determination and inheritance of the killer character in variety 4 of *P. aurelia;* II. The bearing of determination and inheritance of characters in *P. aurelia* on problems of cytoplasmic inheritance, pneumococcus transformations, mutations and development. *Proc. Nat. Acad. Sci. U.S.* **29**, 329–343.

135   — 1945. The dependence of the physiological action of a gene on a primer and the relation of primer to gene. *Amer. Naturalist* **49**, 318–339.

136   — 1946. Experimental control of the concentration of cytoplasmic genetic factors in *Paramecium*. *Cold Spring Harbor Symp. Quant. Biol.* **11**, 236–255.

137   — 1947. Developmental mechanisms in *Paramecium*. *Growth Symp.* **11**, 291–307.

138   — 1948. Symposium on plasmagenes, genes and characters in *P. aurelia*. *Amer. Naturalist* **82**, 26–34.

139   *— 1956. The distribution of killers among the varieties of *Paramecium aurelia*. *Anat. Rec.* **125**, 567–568.

140   — 1957. Breeding systems, reproductive methods, and species problems in Protozoa. In Mayr, E., *The Species Problem, Vol. 50*, American Association for the Advancement of Science, Washington, D.C., Publication No. 50, 155–324.

141   — 1959. Kappa and related particles in *Paramecium*. *Advanc. Virus Res.* **6**, 229–356.

142   — 1961. Kappa particles and their bearing on host-parasite relations. In Pollard, M., *Perspectives in Virology*, Burgess, Minneapolis, Minn., Ch. 2, 5–12.

143   — 1965. The metagon: RNA and cytoplasmic inheritance. *Amer. Naturalist* **99**, 279–307.

144   — 1970. Methods in *Paramecium* research. In Prescott, D. M., *Methods of Cell Physiology*, Academic Press, New York, N.Y., Ch. 4, 242–335.

145   — & Mueller, J. A. 1959. What is the infective agent in breis of killer paramecia? *Science* **130**, 1423.

146   *—, Dippell, R. V. & Jacobson, W. E. 1947. Some properties of kappa (killer cytoplasmic factor) and of paramecin (killer substance) in *Paramecium aurelia*, variety 4. *Genetics*, **32**, 106.

147   *—, Gibson, I. & Schneller, M. V. 1964. Killer particles and metagons of *Paramecium* grown in *Didinium*. *Science* **144**, 567–568.

148   *—, Jacobson, W. E. & Dippell, R. V. 1946. Paramecin 51, an antibiotic produced by *Paramecium aurelia*: amounts released from killers and taken up by sensitives; conditions protecting sensitives. *Anat. Rec.* **96**, 514–515.

149   *—, Mueller, J. A. & Schneller, M. V. 1959. The classes of kappa-like particles in *Paramecium aurelia*. *Anat. Rec.* **134**, 642.

150   Stevenson, I. 1967. Diaminopimelic acid in mu particles of *Paramecium aurelia*. *Nature* **215**, 434–435.

151   — 1969. The biochemical status of mu particles in *Paramecium aurelia*. *J. Gen. Microbiol.* **57**, 61–75.

152   Stock, C. C., Jacobson, W. E. & Williamson, M. 1951. An influence of 2,6-diaminopurine upon the content of kappa in *Paramecium aurelia*, variety 4. *Proc. Soc. Exp. Biol. Med.* **78**, 874–876.

153   Suyama, Y. & Preer, Jr., J. R. 1965. Mitochondrial DNA from protozoa. *Genetics* **52**, 1051–1058.

154   *Tallan, I. 1956. Factors involved in infection of the kappa particle in *Paramecium aurelia*, variety 4. *Genetics* **41**, 664.

155   — 1959. Factors involved in infection by the kappa particles in *Paramecium aurelia* syngen 4. *Physiol. Zool.* **32**, 78–89.

156 — 1961. A cofactor involved in infection of *Paramecium aurelia* and its possible action. *Physiol·Zool.* **34**, 1–13.

157 Van Wagtendonk, W. J. 1948. The action of enzymes on paramecin. *J. Biol. Chem.* **173**, 691–704.

158 — 1948. The killing substance paramecin, chemical nature. *Amer. Naturalist* **82**, 60–68.

159 — 1969. Neoplastic equivalents of protozoa. *Nat. Cancer Inst. Monogr.* **31**, 751–769.

160 — & Hackett, P. L. 1949. The culture of *Paramecium aurelia* in the absence of other living organisms. *Proc. Nat. Acad. Sci. U.S.* **35**, 155–159.

161 *— & Soldo, A. T. 1965. Endosymbiotes of ciliated protozoa. In *Progress in Protozoology* (Proc. 2nd Int. Congr. Protozool., London), 244.

162 — & Tanguay, R. B. 1963. The chemical composition of lambda in *Paramecium aurelia*, Stock 299. *J. Gen. Microbiol.* **33**, 395–400.

163 — & Zill, L. P. 1947. Inactivation of paramecin ("killer" substance of *Paramecium aurelia* 51, variety 4) at different hydrogen ion concentrations and temperatures. *J. Biol. Chem.* **171**, 595–604.

164 —, Clark, J. A. D. & Godoy, G. A. 1963. The biological status of lambda and related particles in *Paramecium aurelia. Proc. Nat. Acad. Sci. U.S.* **50**, 835–838.

165 —, Goldman, P. H. & Smith, W. L. 1970. The culture of strains of *Paramecium* in axenic medium. *J. Protozool.* **17**, 389–391.

166 —, Zill, L. P. & Simonsen, D. H. 1950. Chemical and physiological studies on paramecin and kappa. *Proc. Indiana Acad. Sci.* **60**, 64–66.

167 —, Conner, R. L., Miller, C. A. & Rao, M. R. R. 1953. Growth requirements of *Paramecium aurelia* Var. 4, stock 51.7 sensitives and killers in axenic medium. *Ann. N.Y. Acad. Sci.* **56**, 929–937.

168 Vendrely, R. 1958. La notion d'espèce à travers quelques, données biochimiques récentes et le cycle. *L. Ann. Inst. Pasteur*, **94**, 142–166.

169 Webb, M. 1953. Effects of magnesium on cellular division in bacteria. *Science*, **118**, 607–611.

170 *Widmayer, D. J. 1961. The detection of two new mutations of hump kappa in killer stock 51, syngen 4, *Paramecium aurelia. Amer. Zoologist* **1**, 398.

171 — 1965. A nonkiller resistant kappa and its bearing on the interpretation of kappa in *Paramecium aurelia. Genetics* **51**, 613–623.

172 Williamson, D. H. & Scopes, A. W. 1961. The distribution of nucleic acids and protein between different sized yeast cells. *Exp. Cell Res.* **24**, 151–153.

173 Williamson, M., Jacobson, W. E. & Stock, C. C. 1952. Testing of chemicals for inhibition of the killer action of *Paramecium aurelia. J. Biol. Chem.* **197**, 763–770.

174 Work, E. & Dewey, D. L. 1953. The distribution of α, ε-diaminopimelic acid among various micro-organisms. *J. Gen. Microbiol.* **9**, 394–406.

175 Yeung, K. K. 1965. Maintenance of kappa particles recently deprived of a gene K (stock 51, syngen 4) of *Paramecium aurelia. Genet. Res.* **6**, 411–418.

176 Zill, L. P. 1950. *Biochemical Studies of Paramecin*, Ph.D. Thesis, Indiana University, Bloomington, Ind.

177 — & Van Wagtendonk, W. J. 1951. The influence of ultrasonic vibrations upon the activity of paramecin. *Biochim. Biophys. Acta* **6**, 524–533.

XI. ADDENDUM

(Additional references, tables and figure are indicated by the prefix Ad)

The addendum summarizes information about the particles of *Paramecium* that has appeared since this chapter was written and includes pertinent literature through August, 1973*, under headings corresponding to those outlined in this chapter. For detailed information about the structure of kappa and other particles of *Paramecium*, the reader is referred to a treatise on *Paramecium* ultrastructure by Jurand and

* NLM/Medline–The National Library of Medicine's Remote–Access Retrieval Service.

Selman (Ad 192). Additional discussions concerning the properties and nature of the particles may be found in articles by Preer (Ad 198) and Gibson (Ad 186, Ad 187).

Ad. II (Morphology) (p. 384)

Stevenson (Ad 206) described intracytoplasmic particles occurring within a number of stocks of *P. aurelia* found in Australia. Of thirty stocks collected from several locations in Australia Capital Territory, South Australia and Victoria, nine contained particles. An electron micrographic study revealed the particles to be remarkably similar to kappa and delta found in a number of other *P. aurelia* stocks. Two stocks were described which contained both kappa- and delta-like particles occurring within the same cytoplasm. This is the first reported instance for this type of phenomenon. It will be recalled that the previous reported occurrence of two types of particles within the same cell (100) involved a kappa-like particle which was found only in the cytoplasm and an alpha particle which was found almost exclusively in the macro-nucleus. A new type of particle, designated by the letter *tau* (in keeping with the precedent of naming the particles with Greek letters established by Sonneborn, 141) was found which proved to be morphologically distinct from the other particles. Table Ad 1 contains information about the particles as described by Stevenson.

All of the stocks, with the exception of stock A-4 proved to be capable of killing cells of particle-free *P. aurelia* stock 513, used as a sensitive test organism. In all cases of killing, paralysis occurred within a few hours after mixing followed by vacuole formation and eventual death of the sensitive stock. Humping in sensitive stocks was observed when stocks A-2 and A-35 were used as killers and spinning in sensitives was observed when stock A-13 was tested for killer activity. No information regarding the possible infective nature of the particles was reported.

In a recent publication, Preer *et al.* (Ad 200) described in detail the classes of kappa in *P. aurelia*.

Ad. III (Biology), B, 1, d (Virus-like nature of toxic particles) (p. 391)

Evidence continues to mount in support of the original postulate by Preer (96, 97) that kappa without R-bodies contains prophage that, by an as yet unknown mecha-nism, produce mature phages and refractile bodies. An electron microscopic study of serial sections of stock 562 kappa by Grimes and Preer (Ad 191) has shown an almost absolute correlation between the association of phage-like particles with refractile bodies. All kappa particles observed with refractile bodies contained phage-like particles. Of 78 kappas with phage-like particles, only 3 were observed without refractile bodies. Stevenson (Ad 207) found that the relative numbers of kappas of stock A-35 (which also carries virus-like inclusion bodies) containing refractile bodies was reduced to near zero in exponential cells and increased upon starvation to be-tween 25 and 50% of the total number of particles within the cell. Concomitant with the increase in the number of particles containing refractile entities was an increase in the number of virus-like inclusion bodies associated with these bodies. On the basis

TABLE Ad 1

Intracytoplasmic particles in Australian stocks of **P.** *aurelia.*

Stock	Syngen	Particle	Size (μm) length	diameter	Location	Toxic effect[a]	Characteristic
A-1, A-2	4	kappa	0.8–2.0	0.3–0.8	Cytoplasm	phvd	one or more refractile bodies
	4	unknown	—		Cytoplasm		smooth-walled, electron-dense outer layer, pleomorphic, sparse flagella (delta-like)
A-4	4	unknown			Cytoplasm	none	same as unknown particle in A-1 and A-2
A-7, A-13		unknown	1.0	0.4	Cytoplasm	psvd	same as A-1, A-2 and A-4
A-18		tau	1.5–2.0	0.4–1.0	Cytoplasm	pd(lysis)	corrugated cell wall, internal vacuoles similar to mu
A-30		kappa	—		Cytoplasm	pvd	one or more refractile bodies
A-35	2(?)	kappa	2.0–4.0	1.0	Cytoplasm	phd	refractile bodies, virus-like inclusion bodies, surrounded by vesicular membrane
A-37		mu(?)	—		Cytoplasm	pvd	associated with macronuclear in-pockets, no refractile bodies

[a] p = paralysis; v = vacuolization; s = spinning; h = humping; d = death.

of these observations, Stevenson suggested that the virus-like inclusions associated with the refractile body were bacteriophage. Kappa particles not containing the refractile bodies could be considered as lysogenic, carrying prophage.

Evidence that the virus-like particles were, in fact, bacteriophage was obtained by Preer *et al.* (Ad 199) and was based on chemical analyses of purified phage-particles and cytochemical tests performed on sectioned material. Kappa particles, isolated and purified from *P. aurelia*, stock 562, syngen 2 grown in bacterized medium, were disrupted by sonic oscillation and placed on a CsCl gradient. Turbid material that collected at density 1.47 g/cc proved to be relatively pure bacteriophage particles as seen in the electron microscope. Chemical analysis revealed that the particles contained protein (2.0×10^{-16} g/particle) and DNA (1.6×10^{-16} g/particle) but no RNA. These values, as pointed out by Preer, are similar to those found for bacteriophage T_2 (2.7×10^{-16} g protein, 2.0×10^{-16} g DNA). The buoyant density of the DNA of the kappa bacteriophage, 1.704 g/cc, was similar to that of whole kappa DNA (d = 1.705 g/cc). The presence of DNA in the phage-like particles was confirmed cytochemically (Ad 3). Because of the "unfilled" appearance of many of the phage-like particles banding at density 1.47 g/cc as well as the many examples of incomplete virus-like structures associated with kappa particles (as seen in the electron microscope) the authors suggest that kappa bacteriophage may be defective.

Ad. III (Biology), B, 1, e (Mode of delivery and fate of toxic particles) (p. 391)

Jurand *et al.* (Ad 193) studied prelethal changes occurring in sensitive paramecia exposed to purified kappa of stock 7. As Mueller (82) and Preer and Preer (96) had found earlier, stock 7 kappa particles accumulated in large numbers in food vacuoles of sensitive *Paramecium* and rapidly disintegrated. R-bodies present in B-kappas unrolled, causing, in some instances, the rupture of food vacuoles releasing thereby their contents into the cytoplasm. Purified R-bodies alone produced similar effects when exposed to sensitives but at concentrations about three times that of purified kappa particles. Formation of food vacuoles ceased within 15 minutes after exposure to kappa particles. It was concluded from these studies that the food vacuole membrane may be the site of action of the toxin although it was still not clear whether penetration of the food vacuole into the cytoplasm by unrolled R-bodies was prerequisite for killing activity.

Of considerable interest was the observation by these investigators that, 19–22 hours after exposure of sensitive stock 16 paramecia to living killers of stock 7, the cytoplasm of sensitive animals was found to contain large numbers of electron dense granules 0.5–1 μm in diameter. The granules were identified as lysosomes on the basis of their size and high acid phosphatase content. The concentration of this enzyme was also found to increase substantially in the macronuclei of affected animals but was almost completely absent from macronuclei of killer animals. Presumably the enzyme is either absent or present in low concentrations in the macronucleus of untreated sensitives. These observations are consistent with the fact that sensitive paramecia undergo changes that lead to general necrosis and eventual death of the cell following exposure to kappa toxin.

Ad. III (Biology), B, 2 and 3 (Lambda, sigma and mu) (pp. 395–396)

Employing axenic media described by Soldo *et al.* (127) and Van Wagtendonk, *et al.* (165), Williams (Ad 214) cultured a number of particle-free strains of *P. aurelia* including stocks 60, 90, 513 and 540 of syngen 1, stocks 239 and 562 of syngen 2 and stocks 138 and 299 of syngen 8 but was unable to obtain axenic cultures of mu-bearing stock 540 or sigma-bearing (or sigma-free) stock 114. She reported that only axenically cultivated stock 138 mu was capable of maintaining particles for an indefinite period. Alpha particles from stock 562 and lambda particles from stock 299 underwent spontaneous loss after growth of the paramecia in axenic medium. In our laboratory, we have maintained axenic stocks of 138 mu, 139 pi and 299 lambda in media of identical composition for the past 9 years, subcultured at weekly intervals and maintained at 27 °C without loss of particles. Spontaneous loss of lambda (and possibly alpha) particles from Williams stocks may be due to an imbalance of growth caused by the excessively large inoculum size she used to subculture the animals (i.e., 1.0 ml of a 7-day-old culture per 5 ml fresh medium)*. Williams also reported that growth of particle-free stocks 299 and 138 was maintained for 12 weekly subcultures in the absence of an exogenous supply of folic acid in the medium, suggesting that folic acid was not an indispensible requirement for the growth of *P. aurelia* and that lambda particles did not provide the host *Paramecium* with this vitamin. It is not surprising that Williams failed to establish the requirement for folic acid under her growth conditions. The crude components of the medium, particularly proteose peptone, contain sufficient quantities of this vitamin to support the growth of the protozoan. The original growth experiments describing the folic acid requirements of lambda-bearing *Paramecium* were carried out in a folate-free medium which was chemically defined except for a partially purified fraction derived from yeast (124).

We (Soldo and Godoy)** have recently re-investigated this problem in a chemically-defined medium (126) which supports the growth and maintenance of the particles in *Paramecium aurelia* stock 299 lambda. Growth of the particle-free animals fails in the first transfer when folic acid is omitted from the medium. Under identical growth conditions (i.e., the absence of folic acid), the same stock with lambda particles may be cultured indefinitely without loss of particles. Studies on folic acid production by lambda of stock 299 λ cultured in a chemically-defined medium free of folic acid suggest that the particles produce this vitamin within the paramecia in quantities about ten times greater than can be accounted for in the particles themselves. The excess amount of this vitamin is utilized by the protozoan for growth. Known folate antagonists, notably sulfanilamide, destroy the particles within the cell at concentrations that do not affect the growth of the protozoan. Under these conditions, the effect of the antibiotic may be completely reversed by the addition of p-aminobenzoic acid to the culture medium.

---

* There also exists the possibility that stock 299 and 299 lambda used by Williams in her studies was not identical to the stocks used by Van Wagtendonk and Soldo. Allen *et al.* (Ad 178) noted that the axenic stocks obtained from Van Wagtendonk (Soldo) exhibited a normal type A esterase isozyme pattern whereas the same stocks obtained from Gibson (Williams) did not.
** Manuscript in preparation.

Williams (Ad 214) also studied growth of lambda and mu particles *in vitro* in highly complex media. She reported that upon initial isolation into axenic media, both lambda and mu particles underwent a doubling, followed by a decrease in number. Upon continued weekly subculture, the particles were maintained at population densities of approximately $16-20 \times 10^3/\text{ml}$. She claimed that the particles retained their ability to kill susceptible sensitive strains but lost their capability for infection. The method used to measure killing activity involved adding an aliquot of the *in vitro* cultivated particle suspension to fresh medium containing sensitive test organisms. Both mu and lambda cultured *in vitro* caused the death of sensitives. This is the first reported instance of killing by mu particles outside the host *Paramecium*. Previous reports describing the toxic action of mu particles suggested that intimate contact, usually involving an exchange of cytoplasm, was required to establish the killing effect (114).

Ad. III (Biology), C (Infectivity) (p. 396)

Gibson (Ad 188) carried out a series of experiments involving the transplantation of a number of symbionts into stocks of *Paramecium aurelia* of different syngens using a microinjection technique. Although the experimental findings were complex, the author stated that the results were "consistent with the presence of a genetic locus which is involved with the maintenance of endosymbionts".

Ad. IV (Biochemistry), A (Chemical composition) (p. 410)

RNA isolated from purified mu particles obtained from bacterized stock 540 of *Paramecium aurelia* yielded two components following electrophoresis on poly-acrylamide gels (Ad 180). The high molecular weight components had the same molecular weight as those of *E. coli* RNA (i.e., 1.1×10^6 and 0.55×10^6 daltons), corresponding to 23S and 16S components respectively. RNA isolated from mu-free stock 540 yielded values of 25S (1.3×10^6 daltons) and 18S (0.66×10^6 daltons) in agreement with values previously reported (Ad 201). No evidence for the presence of "messenger" RNA was obtained. The DNA content of purified mu particles of stocks 540 and 551 was determined by direct analytical measurement and by renaturation rate kinetics (Ad 189). Comparable values were reported for both methods. Values determined by direct measurement were 2.4×10^9 daltons for 540 mu and 3.42×10^9 daltons for mu of stock 551. These values are in general agreement with those reported earlier for other particles (130) (see also Table 8, p. 411).

Ad. IV (Biochemistry), B, 1 (Respiration) (p. 413)

Additional details of respiration of kappa particles isolated from bacterized stock 51 *Paramecium aurelia* were provided by Kung (Ad 193). Fresh preparations of kappa particles exhibited an endogenous respiratory rate (Q_{O_2}) of 17.0 ± 1.6 μl/mg dry weight/hr, measured polarographically. This rate decayed linearly to about 50% of its original value in about 5 hrs in 0.01 M Na-K-phosphate buffer, pH 6.8 at room temperature. A number of sugars and related substances, including arabinose, fructose, galactose, glucose, glyceraldehyde, lactose, maltose, mannitol, ribose, soluble starch and sucrose were examined for their effect on respiration. Only glucose and sucrose significantly stimulated respiration. Of a number of intermediates of the tricarboxylic acid (TCA) cycle tested, succinate was the most effective in increasing rates of O_2 consumption. Neither ATP, ADP nor NADH affected the Q_{O_2}, but KCN, CO and 2-heptyl-4-hydroxyquinoline oxide inhibited respiration. Kung pointed out that the *in vitro* respiratory rate of kappa particles was lower than that of free-living bacteria and noted also that unlike kappa, purified typhus rickettsiae do not utilize glucose*. On the basis of these arguments, Kung concluded that kappas are closer to free-living bacteria.

Ad. IV (Biochemistry), B, 3 (Properties of the DNA of the particles) (p. 416)

The base composition of the DNA of some of the particles was determined from buoyant density data obtained from analytical CsCl gradients by Gibson *et al.* (Ad 189). DNA from purified mu particles of stocks 540 and 551 exhibited buoyant densities of 1.694 and 1.695 g/cc corresponding to (G + C) contents of 35 and 36 mole % respectively. Buoyant density of DNA of both alpha particles of stock 562 and kappa particles of stock 51 was reported as 1.695 g/cc corresponding to a (G + C) content of 36 mole %. By way of contrast, the buoyant density of DNA of purified kappa isolated from an alpha-free strain of stock 562 grown in bacterized medium was 1.704 g/cc corresponding to a (G + C) content of 46 mole % (Ad 199).

Base sequence homology of the DNA of three particle types was investigated by Gibson *et al.* (Ad 189). RNA (^{3}H) was prepared from DNA of alpha (stock 562), mu (stocks 540 and 541) and lambda (stock 299) particles using *E. coli* polymerase and hybridized to denatured particle DNA's immobilized on nitrocellulose filters. The degree of base sequence homology was estimated by determining the relative amount of radioactivity remaining on the filters following digestion with ribonuclease, taking binding of labelled RNA to DNA from mu of stock 540 as 100%. These workers estimate about 60% base sequence homology between mu DNA of stock 540 and mu DNA of stock 551. The order of increasing base sequence homology among the other particle DNAs was mu, stock 551 > alpha, stock 562 > lambda, stock 299. Little homology existed between the synthesized (^{3}H) RNA fractions prepared from the particle DNA's and the nuclear DNA of *Paramecium*.

A potent new method has been developed that permits an evaluation of the molecular complexity of cellular or organelle DNA (Ad 184, Ad 213). It is based on measuring the rate denatured DNA reassociates under a controlled set of conditions. Because reassociation of these molecules follows second order kinetics, the rate of reassociation provides a quantitative measure of the molecular size of the genome. By dividing the calculated genomic size determined in this way into the total amount of DNA determined analytically [i.e., by chemical means or by measuring the length of the

---

* These rickettsiae utilize glutamic acid as their chief energy source.

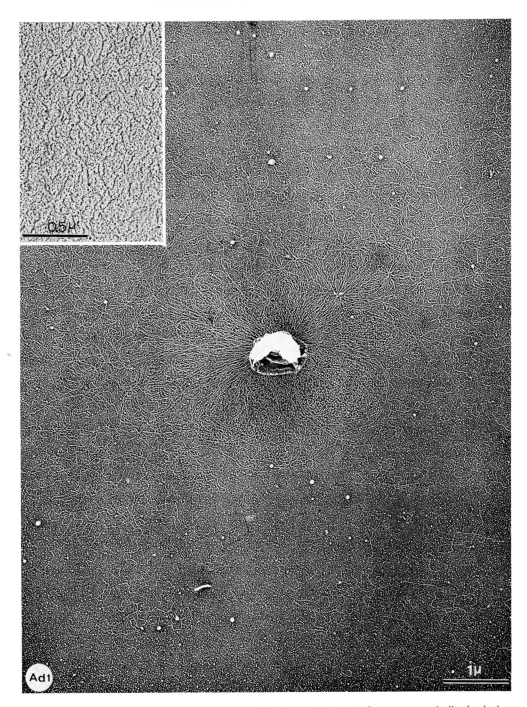

Fig. Ad1. Electron micrograph showing a partial release of the DNA from an osmotically-shocked lambda particle using the monolayer technique described by Kleinschmidt (Ad16). Insert shows purified lambda DNA fragmented by sonic vibration to a molecular size of approximately 5×10^5 daltons.

TABLE Ad 2

Genomic size of DNA's.

Source	Complexity (daltons)			Number of genomes	References
	kinetic		analytic		
	observed	corrected[a]			
E. coli K-12	2.5×10^9	2.5×10^9	2.5×10^9	1	(Ad 190, Ad 213)
H. influenzae	1.0×10^9	0.8×10^9	0.7×10^9	1	(Ad 179, Ad 182)
M. hominis (H39)	0.84×10^9	0.55×10^9	0.51×10^9	1	(Ad 197, Ad 203)
Lambda	0.71×10^9	0.39×10^9	7.5×10^9	20	(Ad 205)
C. trachomatis	0.33×10^9	—	—	—	(Ad 194)
Bacteriophage T$_4$	0.18×10^9	0.13×10^9	0.13×10^9	1	(Ad 190, Ad 202, Ad 213)
P. aurelia (slow)	6.1×10^{10}	3.9×10^{10}	6.06×10^{13}	885	(Ad 205)

[a] 1.8% of observed kinetic complexity for every mole % (G + C) difference from 51% (G + C).

DNA molecule released from the particle by osmotic shock (see Fig. Ad 1)] the number of copies of the genome may be estimated. Estimated values for the molecular complexity of DNA of lambda (stock 299), mu (stock 138) and pi (stock 139) using these methods fell within a range, intermediate between *E. coli* DNA (2.5×10^9 daltons) and bacteriophage T$_4$ DNA (1.3×10^8 daltons) (Ad 204). In a detailed study of the molecular complexity of the DNA of lambda particles isolated and purified from axenically grown *Paramecium*, stock 299, Soldo and Godoy (Ad 205) reported a value of 0.71×10^9 daltons for the genomic size. If the observed value for the molecular complexity is corrected to account for the effect on the renaturation rate due to the (G + C) content of the DNA, the molecular size of the DNA of a lambda particle may be estimated at 0.39×10^9 daltons, a value comparable to those found for certain *Rickettsiae* and *Mycoplasmas* (see Table Ad 2). The correction factor applied ($0.55 \times 0.71 \times 10^9$ daltons) was based on one suggested from the data of Wetmur and Davidson (Ad 213) and Seidler and Mandel (Ad 203) [i.e., 1.8% of the observed complexity for every mole (G + C) difference from the 51% (G + C)]. The (G + C) content of lambda DNA was taken as 26 mole %. Thus, the genome of the lambda particle contains sufficient information to code for a minimum of about 500 proteins. This value is comparable to those reported for certain *Mycoplasmas* and is intermediate between estimates of 3000 for *E. coli* and 160 for bacteriophage T$_4$. These calculations are based on the assumption that three nucleotide pairs possessing an average molecular weight of 335 per nucleotide are required to code for an average amino acid. Assuming the average protein contains about 400 amino acids, the molecular weight of the portion of the DNA molecule needed to code for a single protein is about 800,000 (Ad 199).

The average amount of the DNA present in a lambda particle taken from the stationary phase of growth was determined to be $12.4 (\pm 1.1) \times 10^{-3}$ pg. This is equivalent to a molecular size of about 7.5×10^9 daltons. The observed differences between values obtained for the analytical and kinetic complexity of lambda DNA suggests

TABLE Ad 3

Genomic size of organelle DNA's.

DNA	Complexity (daltons)		Number of genomes	References
	kinetic	analytic		
Chloroplast				
lettuce	1.2×10^8	2.0×10^9	17	(Ad 211)
tobacco	1.1×10^8	3.0×10^9	27	(Ad 210)
Euglena	1.8×10^8	6.0×10^9	33	(Ad 208)
Chlamydomonas	2.0×10^8	5.2×10^9	26	(Ad 212)
Mitochondria				
guinea pig	1.1×10^7	—	—	(Ad 183)
Tetrahymena	3.5×10^7	2.4×10^8	7	(Ad 185, Ad 209)
Other				
E. coli	2.5×10^9	2.5×10^9	1	(Ad 190)
lambda	0.39×10^9	7.5×10^9	20	(Ad 204)

that lambda particles, unlike those of bacteria and viruses, possess a multiple genome consisting of at least 10 and possibly as many as 20 copies of a DNA molecule of unique sequence. The reason for the uncertainty resides in the fact that literature data conflict as to the precise effect the (G + C) content of the DNA exerts on the renaturation rate. The presence of a multiple genome in lambda particles is a characteristic shared by chloroplasts and mitochondria (Table Ad 3) and probably reflects an adaptive change within the genomic structure of the particle in response to its prolonged intracellular residence within the protozoan. In view of the multitudinous array of particles that occur within the cytoplasm of *Paramecium*, it would be of considerable interest to establish whether these other particles, like lambda contain multiple copies of their genomes, especially in light of current ideas concerning the origin of chloroplasts and mitochondria from ancestral bacteria-like forms.

REFERENCES

178 Allen, S. A., Byrne, B. C. & Cronkite, D. L. 1971. Intersyngenic variations in the esterases of bacterized *Paramecium aurelia*. *Biochemical Genetics* **5**, 135–150.

179 Bak, A. L., Christiansen, C. & Stenderup, A. 1970. Bacterial genome sizes determined by DNA renaturation studies. *J. Gen. Microbiol.* **64**, 377–380.

180 Baker, R. 1970. Studies on the RNA of the mate-killer particles of *Paramecium*. *Heredity* **25**, 657–662.

181 Bernhard, W. 1969. A new staining procedure for electron microscopical cytology. *J. Ultrastruct. Res.* **27**, 250–265.

182 Berns, K. I. & Thomas, C. A. 1965. Isolation of high molecular weight DNA from *Hemophilus influenzae*. *J. Mol. Biol.* **11**, 476–490.

183 Borst, P. 1970. Mitochondrial DNA: Structure, information content, replication, and transcription. In *Control of Organelle Development* (Symp. Soc. Exp. Biol.), **24**, 201–226.

184 Britten, R. J. & Kohne, D. E. 1965–1966. Nucleotide sequence repetition in DNA. *Carnegie Inst. Year Book*. **65**, 78–125.

185 Flavell, R. A. & Jones, I. G. 1970. Mitochondrial deoxyribonucleic acid from *Tetrahymena pyriformis* and its kinetic complexity. *Biochem. J.* **116**, 811–817.

58 *— & Berger, J. D. 1970. The DNA content of *Paramecium aurelia*, stock 51. *J. Protozool.* **17** (Suppl.), 20.

59 Beisson, J. & Capdeville, Y. 1966. Sur la nature possible des étapes de différenciation conduisant à l'autogamie chez *Paramecium aurelia*. *Compt. Rend.* **263**, 1258–1261.

60 — & Rossignol, M. 1969. The first case of linkage in *Paramecium aurelia*. *Genet. Res.* **13**, 85–90.

61 — & Sonneborn, T. M. 1965. Cytoplasmic inheritance of the organization of the cell cortex in *Paramecium aurelia*. *Proc. Nat. Acad. Sci. U.S.* **53**, 275–282.

62 Beisson-Schecroun, J. 1964. Hereditary maintenance of spontaneous and provoked variations of cortical organization in *Paramecium aurelia*. *Proc. 11th Int. Congr. Cell Biol., Excerpta Medica* **77**, 11.

63 *Berech, Jr., J. & Van Wagtendonk, W. J. 1959. Metabolic fate of macronuclear DNA after autogamy in *Paramecium aurelia*, an autoradiographic study. *Science* **130**, 1413.

64 *— & — 1960. DNA turnover in *Paramecium aurelia*. *Fed. Proc.* **19**, 313.

65 — & — 1962. An autoradiographic study of the macronuclear changes occurring in *Paramecium aurelia* during autogamy. *Exp. Cell Res.* **26**, 360–371.

66 Berezina, I. G. 1970. Cytochemical study of lactate dehydrogenase isoenzymes in some protozoa species (in Russian). *TSitologiya* **12**, 1205–1208.

67 *Berger, J. D. 1968. Inhibition of the biological activity of macronuclear fragments in the presence of the macronuclear anlage in *Paramecium*. *J. Protozool.* **15** (Suppl.), 29.

68 — 1971. Kinetics of incorporation of DNA precursors from ingested bacteria into macronuclear DNA of *Paramecium aurelia*. *J. Protozool.* **18**, 419–429.

69 — & Kimball, R. F. 1964. Specific incorporation of precursors into DNA by feeding labeled bacteria to *Paramecium aurelia*. *J. Protozool.* **11**, 534–537.

70 Bergmann, F., Chaimovitz, M., Leon, S. & Preiss, B. 1962. On the mechanism of action of Colisan. *Brit. J. Pharmacol.* **18**, 302–310.

71 Bernheimer, A. W. 1963. Lack of effect of staphylococcal toxins on ciliated protozoa. *J. Protozool.* **10**, 166.

72 Best, J. B. 1954. The photosensitization of *Paramecium aurelia* by temperature shock. A study of reported conditioned response in unicellular organisms. *J. Exp. Zool.* **126**, 87–89.

73 Beyersdorfer, K. & Dragesco, J. 1953. Etude comparative des trichocystes de sept espèces de Paramécies. *Proc. 1st Int. Congr. Electr. Microsc., Paris* **1950**, 661–671.

74 Bianchi, C. P. 1961. Calcium movements in striated muscle during contraction and contracture. In Shanes, A. M., *Biophysics of Physiological and Pharmacological Actions*, American Association for the Advancement of Science, Washington, D.C., 281–292.

75 Bishop, J. O. 1961. Purification of an immobilization antigen of *Paramecium aurelia*, variety 1. *Biochim. Biophys. Acta* **50**, 471–477.

76 — 1963. Immunological assay of some immobilizing antigens of *Paramecium aurelia*, variety 1. *J. Gen. Microbiol.* **30**, 271–280.

77 — & Beale, G. H. 1960. Genetical and biochemical studies of the immobilization antigens of *Paramecium aurelia*. *Nature* **186**, 734.

78 Blanc, J. 1962. Observations on the desoxyribonucleic acid content of the nuclear apparatus of ciliata: *Paramecium caudatum* Ehr. (in French). *Compt. Rend.* **254**, 2822–2824.

79 — 1962. Increase over a period of time of the sensitivity to trypaflavin in various clones of *Paramecium caudatum* Ehr. (in French). *Compt. Rend. Soc. Biol.* **155**, 766–768.

80 — 1963. Etude cytophotométrique sur la teneur en acide desoxyribonucléique de l'appareil nucléaire chez plusieurs variétés de *Paramecium caudatum*. *Exp. Cell Res.* **32**, 476–483.

81 — 1965. Etude cytophotométrique des périodes de duplication de l'acide desoxyribonucléique dans l'appareil nucléaire de *Paramecium caudatum*. *Protistologica* **1**, 11–15.

82 — 1968. Détermination par cytophotométrie des teneurs en ADN des macronucléus de *Paramecium caudatum* au cours des premières phases de la conjugaison. *Protistologica* **4**, 415–418.

83 *Bleyman, L. K. 1963. Selfing studies in stock 210, syngen 5, *Paramecium aurelia*. *J. Protozool.* **10** (Suppl.), 21–22.

84 *Bleyman, L. K. 1964. Variable gene expression in *Paramecium aurelia*. *J. Protozool.* **11** (Suppl.), 29.

85 *— 1964. The inhibition of mating reactivity in *Paramecium aurelia* by inhibitors of protein and RNA synthesis. *Genetics* **50**, 236.

86   — 1967. Selfing in *Paramecium aurelia* syngen 5: Persistent instability of mating type expression. *J. Exp. Zool.* **165**, 139–146.

87   — 1967. Determination and inheritance of mating type in *Paramecium aurelia*, syngen 5. *Genetics* **55**, 49–59.

88   Bomford, R. 1965. Infection of algae-free *Paramecium bursaria* with strains of *Chlorella*, *Scenedesmus*, and a yeast. *J. Protozool.* **12**, 221–224.

89   *— 1965. Changes of mating type in *Paramecium bursaria*. *Proc. 2nd Int. Conf. Protozool., London* 251.

90   — 1966. The syngens of *Paramecium bursaria:* new mating types and intersyngenic mating reactions. *J. Protozool.* **13**, 497–501.

91   — 1967. Nuclear changes in *Paramecium bursaria* after abortive conjugation (in Russian). *TSitologiya* **9**, 589–595.

92   — 1967. Stable changes of mating type after abortive conjugation in *Paramecium aurelia*. *Exp. Cell Res.* **47**, 30–41.

93   Bonaventure, N. 1956. L'électrocinèse dans le galvantropisme de *Paramecium caudatum*. Etude de l'action du courant galvanique sur les battements ciliaires. *Compt. Rend. Soc. Biol.* **149**, 2230–2232.

94   Borzhsenius, O. N. 1971. Polymorphism of the micronuclei of *Paramecium caudatum*. V. DNA content in micronuclei of four clones with different morphological types of micronuclei. *Vestnik Leningradskogo Universiteta* **26**, 30–42.

95   — & Ossipov, D. V. 1968. Polymorphism of micronuclei of *Paramecium caudatum*. II. Mitotical cycles of micronuclei of different morphological types. *Acta Protozool. (Warsaw)* **6**, 161–168.

96   — & — 1971. Polymorphism of the micronuclei of *Paramecium caudatum*. III. A cytofluorimetric study of DNA content. *TSitologiya* **13**, 1041–1043.

97   — & — 1971. Polymorphism of the micronuclei of *Paramecium caudatum*. IV. The comparative analysis of nuclear reorganization during progamic divisions of four morphological types. *TSitologiya* **13**, 1289–1298.

98   — & — 1971. A cytophotometric study of DNA content in some morphological types of the micronuclei of *Paramecium caudatum Ehrbg. Materials of 1st All-Union Congr. Protozool. Soc., Bacu*, 18–19.

99   —, Scoblo, I. I. & Ossipov, D. V. 1968. Polymorphism of micronuclei of *Paramecium caudatum*, I. Morphological types of interphasic nuclei. *TSitologiya* **10**, 227–235.

100   *Bovee, E. C. 1960. Morphological anomalies in a population of large *Paramecium caudatum*. *J. Protozool.* **7** (Suppl.), 16.

101   Brown, C. H. 1950. Elimination of kappa particles from "killer" strains of *Paramecium aurelia* by treatment with chloromycetin. *Nature* **166**, 527.

102   *Bujwid-Ćwik, K. & Dryl, S. 1971. The effects of detergents on motile behavior of protozoa. *J. Protozool.* **18** (Suppl.), 92.

103   *Burbank, W. D. 1950. Growth of pedigreed strains of *Paramecium caudatum* and *Paramecium aurelia* on a non-living medium. *Biol. Bull.* **99**, 353–354.

104   *— & Martin, V. L. 1965. The effect of the food of *Paramecium aurelia*, syngen 4, stock 51, mating type VII on its role in symbiosis. *Proc. 2nd Int. Conf. Protozool., London* 114–115.

105   Butzel, Jr., H. M. 1953. *The Genic Basis of Mating Type Determination and Development in the Varieties of Paramecium aurelia belonging to Group A*. Thesis, Indiana University, Bloomington, Ind.

106   *— 1953. A morphological mutant of stock 90, variety 1, of *Paramecium aurelia*. *Microbial Genet. Bull.* **8**, 5–6.

107   *— 1953. Two new one-type stocks in variety 1 of *Paramecium aurelia*. *Microbial Genet. Bull.* **8**, 5.

108   — 1955. Mating type mutations in variety 1 of *Paramecium aurelia*, and their bearing upon the problem of mating type determination. *Genetics* **40**, 321–330.

109   *—1967. Conjugation and mating type determination of *Paramecium aurelia* in axenic medium. *J. Protozool.* **14** (Suppl.), 19.

110   — 1968. Mating type determination in stock 51, syngen 4, of *Paramecium aurelia* grown in axenic culture. *J. Protozool.* **15**, 284–290.

111   *— 1970. Abnormal cytological events and mating type determination patterns in doublet cells

of stock 51, syngen 4, of *Paramecium aurelia* following conjugation in axenic medium. *J. Protozool.* **17** (Suppl.), 8.

112   — & Bolten, A. B. 1968. The relationship of the nutritive state of the prey organism *Paramecium aurelia* to the growth and encystment of *Didinium nasutum. J. Protozool.* **15**, 256–258.

113   *— & Martin, W. B. 1955. Studies of amino acid constituents of *Paramecium aurelia. Genetics* **40**, 565.

114   — & Pagliara, A. 1962. The effect of biochemical inhibitors upon the killer-sensitive system in *Paramecium aurelia. Exp. Cell Res.* **27**, 382–395.

115   — & Van Wagtendonk, W. J. 1963. Autoradiographic studies of the differential incorporation of glycine, and purine and pyrimidine ribosides by *Paramecium aurelia. J. Gen. Microbiol.* **30**, 503–507.

116   — & — 1963. Some properties of the lethal agent found in cell-free fluids obtained from cultures of lambda-bearing *Paramecium aurelia*, syngen 8, stock 299. *J. Protozool.* **10**, 250–252.

117   *— & Vinciguerra, B. 1957. A fission-rate mutant of stock 90, variety 1, of *Paramecium aurelia. Microbial Genet. Bull.* **15**, 7–8.

118   —, Brown, L. H. & Martin, Jr., W. B. 1960. Effects of detergents upon killer-sensitive reactions in *Paramecium aurelia. Physiol. Zool.* **33**, 213–224.

119   —, — & — 1960. Effects of detergents upon electromigration of *Paramecium. Physiol. Zool.* **33**, 39–41.

120   Bychkovskaiya, I. B. & Ochinskaiya, G. K. 1968. The number of reproduction cycles of *Paramecia* in the post radiation period and their death terms (in Russian). *Dokl. Akad. Nauk, S.S.S.R.* **178**, 222–225.

121   Byrne, B. J. 1969. Kappa, mu and the metagon hypothesis in *Paramecium aurelia. Genet. Res.* **13**, 197–211.

122   Cabadaj, S. & Praslička, M. 1964. Interaction of the blood sera with *Paramecium caudatum* cultures. I. Complement consumption by *Paramecium* cultures in normal fresh sera (in Czech). *Biologia (Bratislava)* **19**, 889–896.

123   — & — 1964. Resistance to lethal roentgen irradiations by the interaction of blood sera with a culture of *Paramecium caudatum* (in Czech). *Čas Lek Cesk* **103**, 502.

124   —, — & Bernasovský, I. 1967. Interaction of the blood sera with *Paramecium caudatum* cultures. III. Complete consumption by Paramecium caudatum culture in normal fresh sera of irradiated and nonirradiated rats (in Czech). *Biologia (Bratislava)* **22**, 3–9.

125   —, — & — 1968. Complement consumption by *Paramecium caudatum* culture in normal fresh serum and its significance in the prognosis of radiation sickness in rats. *Intern. J. Radiat. Biol.* **4**, 469–481.

126   Calkins, J. 1963. Variation of radiation sensitivity of *Paramecium aurelia* as a function of time of irradiation in the interdivision growth cycle. *Nature* **198**, 704.

127   — 1965. A study of the fine course of recovery of *Paramecium aurelia* from the lethal effects of X-rays. *Radiation Res.* **26**, 124–131.

128   — 1965. Effect of X-ray dose on the growth rate and delay of the first post-irradiation division in a strain of *Paramecium aurelia. Nature* **205**, 511–513.

129   — 1966. An alternative multi-target equation useful for fitting survival curves of X-irradiated protozoa. *Nature* **209**, 172–173.

130   — 1967. An unusual form of response in X-irradiated protozoa and a hypothesis as to its origin. *Intern. J. Radiat. Biol.* **12**, 297–301.

131   Canella, M. F. 1958. Biologie deli infusori e ipotetici raffronti con i metazoi. I. *Monit. Zool. Ital.* **65**, 164–183.

132   — 1959. Biologie degli infusori e ipotetici raffronti con i metazoi. II. *Monit. Zool. Ital.* **66**, 198–228.

133   — 1960. Biologie degli infusori e ipotetici raffronti con i metazoi. III. *Monit. Zool. Ital.* **67**, 143–189.

134   — 1972. Sur les organelles ciliaires de l'appareil buccal des Hyménostomes et autres ciliés. *Ann. dell'Univers. di Ferrara (Nuova Séries), Sezione III, Biol. anim.*, Suppl. 1–235.

135   Capdeville, Y. 1969. Sur les interactions entre allèles contrôlant le type antigénique chez *Paramecium aurelia. Compt. Rend.* **269**, 1213–1215.

136   — 1971. Allelic modulation in *Paramecium aurelia* heterozygotes. Study of G serotype in syngen 1. *Mol. Gen. Genet.* **112**, 306–316.

137 Chakraborty, J. & Sadhukhan, P. 1967. Some observations on the food vacuole of *Paramecium aurelia. Z. Naturforsch. (B)* **22**, 558.

138 Chao, P. K. 1953. Kappa concentration per cell in relation to the life cycle, genotype and mating type in *Paramecium aurelia*, variety 4. *Proc. Nat. Acad. Sci. U.S.* **39**, 103–113.

139 *— 1954. Present status of the study on kappa concentration per cell in *Paramecium aurelia. Microbial Genet. Bull.* **11**, 11–12.

140 Charret, R. & Fauré-Fremiet, E. 1967. Technique de rassemblement de micro-organismes: préinclusion dans un caillot de fibrine. *J. Microscopie* **6**, 1063–1066.

141 Chatterjee, S. N. & Ray, H. N. 1964. Some electronmicroscopic observations on the fine structures in *Paramecium multimicronucleatum. Bull. Calcutta School Trop. Med.* **12**, 24–25.

142 Cheissin, E. M. 1963. Some data on the fine structure of superficial components of *Paramecium caudatum* (in Russian). In *Morfologiya Fiziologiya Prosteishikh*, Akad. Nauk, SSSR, Moscow, Leningrad, 9–14.

143 — & Ovchinnikova, L. P. 1964. A photometric study of DNA content in macronuclei and micronuclei of different species of *Paramecium* (in Russian). *Acta Protozool. (Warsaw)* **2**, 225–236.

144 —, — & Kudriavtsev, B. N. 1964. A photometric study of DNA content in macronuclei of different strains of *Paramecium caudatum* (in Russian). *Acta Protozool. (Warsaw)* **2**, 237–245.

145 —, —, Selivanova, G. V. & Buze, E. G. 1963. Changes of the DNA content in the macronucleus of *Paramecium caudatum* in the interdivisional period (in Russian). *Acta Protozool. (Warsaw)* **1**, 63–69.

146 Chen, T. T. 1955. Paramecin 34, a killer substance produced by *Paramecium bursaria. Proc. Soc. Exp. Biol. Med.* **88**, 541–543.

147 — 1956. Varieties and mating types in *Paramecium bursaria*. II. Variety and mating types found in China. *J. Exp. Zool.* **132**, 255–268.

148 *— 1963. New mating types of *Paramecium bursaria* from Germany and Austria. *J. Protozool.* **10** (Suppl.), 22.

149 *Chen-Shan, L. 1966. Cortical morphogenesis in *Paramecium aurelia* following amputation of a cortical region. *J. Protozool.* **13** (Suppl.), 25.

150 — 1969. Cortical morphogenesis in *Paramecium aurelia* following amputation of the posterior region. *J. Exp. Zool.* **170**, 205–228.

151 — 1970. Cortical morphogenesis in *Paramecium aurelia* following amputation of the anterior region. *J. Exp. Zool.* **174**, 463–478.

152 — & Whittle, J. R. S. 1968. The effect of mitomycin C on the cortex of *P. aurelia. Symp. Ciliate, Oak Ridge, Tenn.*

153 Cherevkova, O. S. 1958. Some data on the interrelation of infusoria and coli-group of bacteria (in Russian). *Trudy Odesskogo Universiteta* **148**, 283–292.

154 Chlebovska, K., Cabadaj, S. & Praslička, M. 1969. Serologic effectiveness of individual components of a culture of *Paramecium caudatum* (in Slovakian). *Biologia (Bratislava)* **24**, 250–253.

155 Cho, P. L. 1970. The genetics of mating type in a syngen of *Glaucoma. Genetics* **67**, 377–390.

156 Chorik, F. P. 1968. The free-living infusoria of Moldavia (in Russian). *Akad. Nauk Moldavskoy SSSR*, 1–251.

157 Clark, M. A. 1972. Control of mating type expression in *Paramecium multimicronucleatum*, syngen 2. *J. Cell. Physiol.* **79**, 1–14.

158 Cohen, L. W. 1964. Diurnal intracellular differentiation in *Paramecium bursaria. Exp. Cell Res.* **36**, 398–406.

159 — 1965. The basis for the circadian rhythm of mating in *Paramecium bursaria. Exp. Cell Res.* **37**, 360–369.

160 — & Siegel, R. W. 1963. The mating type substances of *Paramecium bursaria. Genet. Res.* **4**, 143–150.

161 *Conner, R. L. 1969. Lipids and membranes in Protozoa. *Proc. 3rd Int. Conf. Protozool., Leningrad* 158.

162 — & Van Wagtendonk, W. J. 1955. Steroid requirements of *Paramecium aurelia. J. Gen. Microbiol.* **12**, 31–36.

163 —, — & Miller, C. A. 1953. The isolation from lemon juice of a growth factor of steroid nature required for the growth of a strain of *Paramecium aurelia. J. Gen. Microbiol.* **9**, 434–439.

164 —, Landrey, J. R., Kaneshiro, E. S. & Van Wagtendonk, W. J. 1971. The metabolism of

stigmasterol and cholesterol by *Paramecium aurelia*. *Biochim. Biophys. Acta* **239**, 312–319.

165   Cooper, J. E. 1965. *An Immunological, Biochemical and Genetic Analysis of Non-immobilization antigen 5 in Paramecium aurelia, syngen 2*. Ph.D. Thesis, University of Pennsylvania, Philadelphia, Penn.

166   — 1965. A fast-swimming "mutant" in stock 51 of *Paramecium aurelia*, variety 4. *J. Protozool.* **12**, 381–384.

167   Courtey, B. & Mugard, H. 1958. Observations sur la régéneration et la division simultanées de *Paramecium caudatum*. *Bull. Biol. France Belg.* **92**, 210–232.

168   *Cox, D., Hutner, S. H., Chunosoff, H. B. & Frank, O. 1962. Nearly-defined media for *Paramecium caudatum* and practical media for microbiological assays with Tetrahymena. *J. Protozool.* **9** (Suppl.), 11.

169   *Crippa-Franceschi, T., Lavatelli, G. & Resta-Zuccarino, F. 1969. A new study of the induced temperature-resistance in *Paramecium aurelia*. *Proc. 3rd Int. Conf. Protozool., Leningrad* 114.

170   Cummings, D. J. 1972. Isolation and partial characterization of macro- and micro-nuclei from *Paramecium aurelia*. *J. Cell Biol.* **53**, 110–115.

171   Curds, C. R. 1963. The flocculation of suspended matter by *Paramecium caudatum*. *J. Gen. Microbiol.* **33**, 357–363.

172   Czarska, L. 1965. Cytoplasmic streaming in *Paramecium caudatum* exposed to electric field. *Acta Protozool. (Warsaw)* **3**, 269–274.

173   Dabrowska, J. 1956. Tresura *Paramecium caudatum*, *Stentor coeruleus*, *Spirostomum ambiguum* na bodźce świetlne (in Polish). *Folia Biol. (Warsaw)* **4**, 77–91.

174   Daniels, G. E. & Park, H. D. 1953. Reproduction in *Paramecium* as affected by small doses of X-ray and beta radiation. *Proc. Soc. Exp. Biol. Med.* **83**, 662–665.

175   — & — 1953. Glutathione and X-ray in *Hydra* and *Paramecium*. *J. Cell. Comp. Physiol.* **42**, 359–367.

176   *De Haller, G. 1964. About inheritance of morphological characteristics of the cortex in *Paramecium aurelia*. *J. Protozool.* **11** (Suppl.), 48.

177   — 1964. Altération expérimentale de la stomatogénèse chez *Paramecium aurelia*. *Rev. Suisse Zool.* **71**, 592–600.

178   — 1965. Sur l'hérédité de caractéristiques morphologiques du cortex chez *Paramecium aurelia*. *Arch. Zool. Expér. Gen.* **105**, 169–178.

179   — & ten Heggler, B. 1968. Détails de l'ultrastructure des ciliés. I. Fibres karyophores chez *Paramecium bursaria*. *Compt. Rend. Soc. Phys. Hist. Natur. Genève* **3**, 31–32.

180   — & — 1969. Morphogénèse expérimentale chez les ciliés. III. Effet d'une irradiation U.V. sur la génèse des trichocystes chez *Paramecium aurelia*. *Protistologica* **5**, 115–120.

181   Didier, P. 1970. Contribution à l'étude comparée des ultrastructures corticales et buccales des ciliés Hyménostomes péniculiens. *Ann. St. biol. Besse en Chandesse (Fr.)* **5**, 1–274.

182   Diller, W. F. 1954. Autogamy in *Paramecium polycaryum*. *J. Protozool.* **1**, 60–70.

183   *— 1957. Conjugation in *Paramecium polycaryum*. *J. Protozool.* **4** (Suppl.), 13.

184   — 1958. Studies on conjugation in *Paramecium polycaryum*. *J. Protozool.* **5**, 282–292.

185   *— 1959. Possible origin of hypoploidy in *Paramecium trichium*. *J. Protozool.* **6** (Suppl.), 19.

186   *— 1962. The relationship of the micronucleus to stomatogenesis in some ciliates. *J. Protozool.* **9** (Suppl.), 15.

187   *— 1969. Morphogenetic studies of the contractile vacuole systems of *Euplotes* and *Paramecium*. *Proc. 3rd Int. Conf. Protozool., Leningrad* 93–94.

188   — & Earl, P. R. 1958. *Paramecium jenningsi* n. sp. *J. Protozool.* **5**, 155–158.

189   *Dippell, R. V. 1953. Serotypic expression in heterozygotes of variety 4, *P. aurelia*. *Microbia Genet. Bull.* **7**, 12.

190   — 1954. A preliminary report on the chromosomal constitution of certain variety 4 races of *Paramecium aurelia*. *Caryologia* **6** (Suppl.), 1109–1111.

191   *— 1955. Some cytological aspects of aging in variety 4 of *Paramecium aurelia*. *J. Protozool.* **2** (Suppl.), 7.

192   — 1958. The fine structure of kappa in killer stock 51 of *Paramecium aurelia*. Preliminary observations. *J. Biophys. Biochem. Cytol.* **4**, 125–128.

193   *— 1959. The distribution of DNA in kappa particles of *Paramecium* in relation to the problem of their bacterial affinities. *Science* **130**, 1415.

194   *— 1962. The site of silver impregnation in *Paramecium aurelia*. *J. Protozool.* **9** (Suppl.), 24.

195 *— 1963. Nucleic acid distribution in the macronucleus of *Paramecium aurelia*. *J. Cell Biol.* **19**, 20A.

196 *— 1964. Perpetuation of cortical structure and pattern in *P. aurelia*. *Proc. 11th Int. Congr. Cell Biol.* 16–17.

197 *— 1965. Reproduction of surface structure in *Paramecium aurelia*. *Proc. 2nd Int. Conf. Protozool., London* 65.

198 — 1968. The development of basal bodies in *Paramecium*. *Proc. Nat. Acad. Sci. U.S.* **61**, 461–468.

199 *— & Sinton, S. E. 1963. Localization of macronuclear DNA and RNA in *Paramecium aurelia*. *J. Protozool.* **10** (Suppl.), 22.

200 *— & Sonneborn, T. M. 1957. Structure of the *Paramecium aurelia* macronucleus as revealed by electron microscopy. *Proc. Indiana Acad. Sci.* **66**, 60.

201 Dolgopolskaya, M. A., Mendeleev, I. S. & Vladimirov, L. V. 1967. On the problem of a magnetic field effect on single cell infusoria *(Paramecium caudatum)* (in Russian). *Biofizika* **12**, 1109–1111.

202 Dorner, R. W. 1957. Stability of paramecin 34 at different temperatures and pH values. *Science* **126**, 1243–1244.

203 Doroszewski, M. 1959. *Paramecium arcticum sp. nov. Bull. Acad. Polon. Sci.* **7**, 73–78.

204 Dorozhkina, L. I. 1970. On the action of a magnetic field on the energy metabolism of ciliates (in Russian). *TSitologiya* **12**, 783–786.

205 *Downing, W. L. 1951. Structure and morphogenesis of the cilia and the feeding apparatus in *Paramecium aurelia*. *J. Protozool.* **2** (Suppl.), 14.

206 Drabkin, B. C. 1967. Effect of volatile phytoncydes on succine dehydrogenase (in Russian). In *Fitontsidy ikh Biologicheskaya Rol i Znachenie dlya Meditsiny i Narodnogo Khoziyaistva*, Naukova Dumka, Kiev, 323–325.

207 Dragesco, J. 1962. L'orientation actuelle de la systématique des Ciliés et la technique d'imprégnation au protéinate d'argent. *Bull. Micros. Appl.* **11**, 49–58.

208 — 1970. Ciliés libres du Cameroun. *Ann. Fac. Sci. Univ. Fed. Cameroun*, No. hors-série, 1–141.

209 Drochmans, P. 1962. Morphologie du glycogène. Etude au microscope électronique de colorations négatives du glycogène particulaire. *J. Ultrastruct. Res.* **6**, 141–163.

210 Drusin, L. M. & Butzel, Jr., H. M. 1961. Effects of cephalin on killer-sensitive reactions in *Paramecium aurelia*. *Physiol. Zool.* **34**, 14–20.

211 Dryl, S. 1959. Effects of adaptation to environment on chemotaxis of *Paramecium caudatum*. *Acta Biol. Exp. (Warsaw)* **19**, 83–93.

212 — 1959. Chemotactic and toxic effects of lower alcohols on *Paramecium caudatum*. *Acta Biol. Exp. (Warsaw)* **19**, 95–104.

213 *— 1959. Antigenic transformation in *Paramecium aurelia* after homologous antiserum treatment during autogamy and conjugation. *J. Protozool.* **6** (Suppl.), 25.

214 — 1961. Chemotaxis in *Paramecium caudatum* as adaptive response of organism to its environment. *Acta. Biol. Exp. (Warsaw)* **21**, 75–83.

215 — 1961. The velocity of forward movement of *Paramecium caudatum* in relation to pH of medium. *Bull. Acad. Polon. Sci., Sér. Sci. Biol. (Classe II)* **9**, 71–74.

216 *— 1961. The ciliary reversal in *Paramecium caudatum* induced by simultaneous action of barium and calcium ions. *J. Protozool.* **8** (Suppl.), 16.

217 — 1963. Contributions to mechanisms of chemotactic response in *Paramecium caudatum*. *Animal Behav.* **11**, 393–395.

218 — 1963. Oblique orientation of *Paramecium caudatum* in electric fields. *Acta Protozool. (Warsaw)* **1**, 193–199.

219 *— 1964. The inhibitory action of chemical, electric and mechanical stimuli on the periodic ciliary reversal in *Paramecium caudatum* induced by Ba^{++}/Ca^{++} and Sr^{++}/Ca^{++} factors. *J. Protozool.* **11** (Suppl.), 30.

220 — 1965. Antigenic transformation in relation to nutritional conditions and the interautogamous cycle in *Paramecium aurelia*. *Exp. Cell Res.* **37**, 569–581.

221 *— 1969. Motor response of Protozoa to external stimuli. *Proc. 3rd Int. Conf. Protozool., Leningrad* 124.

222 — & Bujwid-Cwik, K. 1972. Effects of detergent cetyl trimethyl ammonium bromide on motor reactions of *paramecium* to potassium/calcium factor in external medium. *Bull. Acad. Polon. Sci., Ser. Sci. Biol. (Classe II)* **20**, 551–555.

223 *— & Masnyk, S. 1971. Observations on the effects of chlorpromazine on the motility of *Paramecium aurelia*, stock 51. *J. Protozool.* **18** (Suppl.), 91.

224 Dupy-Blanc, J. 1969. Etude par cytophotométrie des teneurs en ADN nucléaire chez trois espèces de Paramécies, chez différentes variétés d'une même espèce et chez différents types sexuels d'une même variété. *Protistologica* **5**, 297–308.

225 — 1969. Etude cytophotométrique des teneurs en ADN des micronucléus de *Paramecium caudatum* au cours de la conjugaison et pendant la différenciation des "Anlagen" en macro-nucléus. *Protistologica* **5**, 239–248.

226 — 1970. Etude autoradiographique de la synthèse de l'ADN et de l'ARN pendant l'interphase chez *Paramecium jenningsi* et au cours de la différentiation des anlagen en macronucléus lors de la conjugaison chez *Paramecium caudatum*. *Bull. Soc. Zool. Fr.* **95**, 617.

227 Dzhafarov, G. A. 1961. Toxicity for paramecia plasma from healthy and irradiated rats, following burns, trauma and starvation (in Russian). *Patologicheskaya Fiziologiya Eksper-mentalniya Terapiya (Moscow)* **5**, 70–71.

228 — 1961. Changes in toxic properties of the blood plasma in acute radiation sickness in monkeys to paramecia (in Ukrainian). *Fiziologicheskij Zhurnal (Kiev)* **7**, 93–100.

229 Eckert, R. 1972. Bioelectric control of ciliary activity. *Science* **176**, 473–481.

230 — & Murakami, A. 1971. Control of ciliary activity. In Pololsky, R., *Contractility of Muscle Cells and Related Processes*, Prentice-Hall, Englewood, N.J.

231 — & Naitoh, Y. 1969. Graded calcium spikes in *Paramecium*. In *3rd Int. Biophys. Congr. Int. Union Pure and Applied Biophys.*, Cambridge, Mass., 259.

232 — & — 1970. Passive electrical properties of *Paramecium* and problems of ciliary coordination. *J. Gen. Physiol.* **55**, 467–483.

233 — & — 1972. Bioelectric control of locomotion in ciliates. *J. Protozool.* **19**, 237–243.

234 Egelhaaf, A. 1955. Cytologische entwicklungsphysiologische Untersuchungen zur Konjugation von *Paramecium bursaria* Focke. *Arch. Protistenk.* **100**, 447–514.

235 Ehret, C. F. 1953. An analysis of the role of electromagnetic radiation in the mating reaction of *Paramecium bursaria*. *Physiol. Zool.* **26**, 274–300.

236 *— 1955. The effects of pre- and post-illumination on the scotophilic recovery phase of the *Paramecium bursaria* mating reaction. *Anat. Rec.* **122**, 456–457.

237 *— 1955. The photoreactivibility of sexual activity and rhythmicity in *Paramecium bursaria*. *Radiation Res.* **3**.

238 — 1957. The organization of gullet organelles in *Paramecium bursaria*. *J. Protozool.* **4**, 55–59.

239 *— 1958. Replication of ciliated organelles during gullet development in *Paramecium*. *J. Protozool.* **5** (Suppl.), 11.

240 — 1958. Information content and biotopology of the cell in terms of cell organelles. In Yockey, H. P., Platzman, R. L. & Quastler, H., *Symposium on Information Theory in Biology*, Pergamon Press, New York, N.Y., 218–229.

241 — 1959. Induction of phase shift in cellular rhythmicity by far UV and its restoration by visible radiant energy. In Withrow, R., *Photoperiodism*, American Association for the Advancement of Science, Washington, D.C., 541–550.

242 — 1959. Photobiology and biochemistry of circadian rhythms in non-photosynthesizing cells. *Fed. Proc.* **18**, 1232–1240.

243 *— 1959. The influence of temperature on the circadian rhythm of mating in *Paramecium bursaria*. *J. Protozool.* **6** (Suppl.), 18.

244 — 1960. Action spectra and nucleic acid metabolism in circadian rhythms at the cellular level. *Cold Spring Harbor Symp. Quant. Biol.* **25**, 149–158.

245 — 1960. Photoreactivation of UV-induced conjugation delay in *Paramecium*. In Christensen, B. C. & Buchman, B., *Progress in Photobiology*, Elsevier, Amsterdam, 287–288.

246 — 1960. Organelle systems and biological organization. Structural and developmental evidence leads to a new look at our concepts of biological organization. *Science* **132**, 115–123.

247 — 1967. Paratene theory of the shapes of cells. *J. Theoret. Biol.* **15**, 263–272.

248 *— & De Haller, G. 1961. Formation des organelles et des systèmes d'organelles cytoplasmique au cours de la division chez *Paramecium*. *Proc. 1st Int. Conf. Protozool.*, Prague 419–420.

249 — & — 1963. Origin, development and maturation of organelles and organelle systems of the cell surface in *Paramecium*. *J. Ultrastruct. Res.* **6** (Suppl.), 3–42.

250 *— & Mather, R. 1959. Circadian rhythmicity in the incorporation of nucleic acid precursors into *Paramecium bursaria*. *J. Protozool.* **6** (Suppl.), 18.

251 — & Powers, E. L. 1955. Macronuclear and nucleolar development in *Paramecium bursaria*. *Exp. Cell Res.* **9**, 241–257.

252 *— & — 1956. The systems and complexes of primary organelles in *Paramecium*. *J. Protozool.* **3** (Suppl.), 5.

253 — & — 1957. The organization of gullet organelles in *Paramecium bursaria*. *J. Protozool.* **4**, 55–59.

254 — & — 1959. The cell surface of *Paramecium*. *Intern. Rev. Cytol.* **8**, 97–133.

255 —, Alblinger, J. & Savage, N. 1964. Developmental and ultrastructural studies of cell organ-elles. *Argonne Nat. Lab. Biol. Med. Res. Div. Ann. Rept.* **6971**, 62–70.

256 —, Savage, N. & Alblinger, J. 1964. Patterns of segregation of structural elements during cell division. *Z. Zellforsch.* **64**, 129–139.

257 —, — & Schuster, F. L. 1965. The incorporation into cytoplasmic organelles of tritium given in thymidine. *Argonne Nat. Lab. Biol. Med. Res. Ann. Rept.* **7136**, 205–210.

258 *—, — & — 1966. Incorporation into cell organelles of tritium given in leucine and thymidine. *Amer. Zoologist* **6**, 3.

259 Epshtein, M. M. & Piliavskaia, S. M. 1962. The influence of some ethereal oils on the activity of dehydrogenases in paramecia (in Ukrainian). *Mikrobiologichnyi Zhurnal* **24**, 44–48.

260 Epstein, S. S. & Burroughs, M. 1962. Some factors influencing the photodynamic response of *Paramecium caudatum* to 3,4-benzpyrene. *Nature* **193**, 337–338.

261 —, — & Small, M. 1963. The photodynamic effect of the carcinogen 3,4-benzpyrene on *Paramecium caudatum*. *Cancer Res.* **23**, 35–44.

262 —, Small, M., Falk, H. & Mantel, N. 1964. On the association between photodynamic and carcinogenic activities in polycyclic compounds. *Cancer Res.* **24**, 855–862.

263 Estève, J. C. 1966. Facteurs de groupement en anneau chez *Paramecium*. *Protistologica* **2**, 95–100.

264 *— 1966. Facteurs de groupement en anneau chez *Paramecium*. *J. Protozool.* **13** (Suppl.), 37.

265 — 1967. Observations ultrastructurales sur quelques aspects de l'évolution des vacuoles alimentaires chez *Paramecium caudatum*. *Compt. Rend.* **265**, 1991–1994.

266 — 1968. Données complémentaires sur le déterminisme du groupement en anneau chez les *Paramécies*. *Protistologica* **4**, 243–249.

267 — 1969. Observations sur l'ultrastructure et le métabolisme du glycogène de *Paramecium caudatum*. *Arch. Protistenk.* **111**, 195–203.

268 — 1969. Observation ultrastructurale du cytopyge de *Paramecium caudatum*. *Compt. Rend.* **268**, 1508–1510.

269 — 1970. Distribution of acid phosphatase in *Paramecium caudatum*: its relations with the process of digestion. *J. Protozool.* **17**, 24–35.

270 Figge, F. H. J. & Wichterman, R. 1956. Influence of hematoporphyrin and phenol on X-radiation sensitivity of *Paramecium*. *Biol. Bull.* **111**, 302–303.

271 Finger, I. 1957. Immunological studies of the immobilization antigens of *Paramecium aurelia*, variety 2. *J. Gen. Microbiol.* **16**, 350.

272 — 1957. The inheritance of the immobilization antigens of *Paramecium aurelia*, variety 2. *J. Genetics* **55**, 361–374.

273 — 1967. The control of antigenic type in *Paramecium*. In Goldstein, L., *The Control of Nuclear Activity*, Prentice-Hall, Englewood Cliffs, N. J., 377–411.

274 — 1968. Gene activation by cell products. *Trans. N.Y. Acad. Sci.* **30**, 968–976.

275 — & Heller, C. 1962. Immunogenetic analysis of proteins of *Paramecium*. I. Comparison of specificities controlled by alleles and by different loci. *Genetics* **47**, 223–239.

276 — & — 1963. Immunogenetic analysis of proteins of *Paramecium*. IV. Evidence for presence of hybrid antigens in heterozygotes. *J. Mol. Biol.* **6**, 190–202.

277 — & — 1964. Cytoplasmic control of gene expression in *Paramecium*. I. Preferential expression of a single allele in heterozygotes. *Genetics* **49**, 485–498.

278 — & — 1964. Immunogenetic analysis of proteins of *Paramecium*. V. Detection of specific determinants in strains lacking a surface antigen. *Genet. Res.* **5**, 127–136.

279 *— & — 1965. Induction of gene expression in *Paramecium* by cell-free culture fluid. *Amer. Zool.* **5**, 649.

280 —, — & Dilworth, L. 1969. Effects of immobilizing antiserum and normal serum on *Paramecium* surface antigen synthesis. *J. Protozool.* **16**, 12–18.

281 —, — & Green, A. 1962. Immunogenetic analysis of proteins of *Paramecium*. II. Coexistence of two immobilization antigens within animals of a single serotype. *Genetics* **47**, 241–253.

282 —, — & Larkin, D. 1967. Repression of gene expression by cell products in *Paramecium*. *Genetics* **56**, 793–800.

283 —, — & Smith, J. P. 1963. Immunogenetic analysis of proteins of *Paramecium*. III. A method for determining relationships among antigenic proteins. *J. Mol. Biol.* **6**, 182–189.

284 —, Onorato, F. & Wilcox, H. B., III. 1966. Biosynthesis and structure of *Paramecium* hybrid antigen. *J. Mol. Biol.* **17**, 86–101.

285 *—, Dilworth, L., Heller, C. & Fishbein, G. 1968. Transformation of antigenic types in *Paramecium* by conditioned medium. *Genetics* **60**, 177.

286 —, Kaback, M., Kittner, P. & Heller, C. 1960. Immunological studies of isolated particulates of *Paramecium aurelia*. I. Antigenic relationships between cytoplasmic organelles and evidence for mitochondrial variations as demonstrated by gel diffusion. *J. Biophys. Biochem. Cytol.* **8**, 591–601.

287 —, Onorato, F., Heller, C. & Dilworth, L. 1969. Role of non-surface antigens in controlling *Paramecium* surface antigen synthesis. *J. Protozool.* **16**, 18–25.

288 *—, —, — & Wilcox, H. B., III. 1965. Antigen structure and synthesis in *Paramecium*. *Proc. 2nd Int. Conf. Protozool.*, London 244.

289 —, Fishbein, G. P., Spray, T., White, R. & Dilworth, L. 1972. Radioimmunoassay of *Paramecium* surface antigens. *Immunology* **22**, 1051–1063.

290 Firchuck, R. P. 1963. Role of hydrogen ion concentration in the study of phytoncidal effects of tissue juices (in Russian). *Antibiotiki* **8**, 833–835.

291 Franceschi, T. 1958. Modificazioni di resistenza in *Paramecium aurelia* var. 1. *Boll. Mus. Istit. Biol. Univ. Genova* **28**, 87–105.

292 — 1961. L'induzione della sessualita in *Paramecium* nella recerca del sistema ereditario delle Dauermodifikationen. *Boll. Mus. Istit. Biol. Univ. Genova* **31**, 47–59.

293 — 1963. L'effetto dell'autogamia su linee durevolmente modificate di *Paramecium aurelia*, syngen 1. *Boll. Zool.* **30**, 13–25.

294 — 1964. Nuovi studi sull'effetto dell'autogamia in linee durevolmente modificate di *Paramecium aurelia*, syngen 1. *Boll. Zool.* **31**, 1–14.

295 — & Cademartori, E. 1958. Osservazioni su *Paramecium aurelia* var. *1* in allevamento a differenti temperature. *Boll. Mus. Istit. Biol. Univ. Genova* **28**, 169–182.

296 *Fukushi, T. 1970. Preparation of mating reactive cilia from *Paramecium caudatum* by MnCl₂. *J. Protozool.* **17** (Suppl.), 21.

297 *— & Hiwatashi, K. 1970. Preparation of mating reactive cilia from *Paramecium caudatum* by MnCl₂. *J. Protozool.* **17** (Suppl.), 21.

298 Gause, G. F. 1970. Criticism of invalidation of competitive exclusion. *Nature* **17**, 24–35.

299 Geckler, R. P. & Kimball, R. F. 1953. Effect of X-rays on micronuclear number in *Paramecium aurelia*. *Science* **117**, 80–81.

300 Gelber, B. 1952. Investigations of the behavior of *Paramecium aurelia*. I. Modification of behavior after training with reinforcement. *J. Comp. Physiol. Psychol.* **45**, 58–65.

301 — 1956. Investigations of the behavior of *Paramecium aurelia*. II. Modifications of a response in successive generations of both mating types. *J. Comp. Physiol. Psychol.* **49**, 590–593.

302 — 1956. Investigations of the behavior of *Paramecium aurelia*. III. The effect of the presence and absence of light on the occurrence of a response. *J. Genet. Psychol.* **88**, 31–36.

303 *— 1957. A trigger for behavioral change? *(Paramecium aurelia)*. *J. Protozool.* **4** (Suppl.), 16.

304 — 1958. Retention in *Paramecium aurelia*. *J. Comp. Physiol. Psychol.* **51**, 110–115.

305 *— 1961. Autogamy vs. responsiveness in *Paramecium aurelia*. *J. Protozool.* **8** (Suppl.), 17.

306 *— 1963. Different responses to the same training in syngen 1 and syngen 4 of *Paramecium aurelia*. *Proc. 16th Int. Congr. Zool., Washington* 3.

307 — & Rasch, E. 1956. Investigations of the behavior of *Paramecium aurelia*. V. The effects of autogamy (nuclear reorganization). *J. Comp. Physiol. Psychol.* **49**, 594–599.

308 Génermont, J. 1961. Déterminants génétiques macronucléaires et cytoplasmiques controlant la résistance au chlorure de calcium chez *Paramecium aurelia (souche 90, variété 1)*. *Ann. Génétiques* **3**, 1–8.

309 — 1966. *Recherches sur les Modifications Durables et le Déterminisme Génétique de Certains Caractères quantitatifs chez Paramecium aurelia*. Thesis, Fac. Sci. Paris, Expansion Scientifique Française, Paris.

310 *— 1966. Le déterminisme génétique de la vitesse de multiplication chez *Paramecium aurelia*. *J. Protozool.* **13** (Suppl.), 37.

311 *— 1969. Le problème des modifications durables chez les Protozoaires. *Proc. 3rd Int. Conf. Protozool., Leningrad* 109.

312 — 1969. Quelques caractéristiques des populations de *Paramecium aurelia* adaptées au chlorure de calcium. *Protistologica* **5**, 101–108.

313 *— 1970. Etudes biométriques sur la conjugaison des *Paramécies*. *J. Protozool.* **17** (Suppl.), 34.

314 — & Dupy-Blanc, J. 1971. Recherches biométriques sur la sexualité des *Paramécies (P. aurelia et P. caudatum)*. *Protistologica* **7**, 197–212.

315 Gibson, I. 1965. Electrophoresis of extracts of *Paramecium aurelia* containing metagons. *Proc. Roy. Soc. (London), Ser. B* **161**, 538–549.

316 — 1965. The replication of metagons and mu particles from *Paramecium* in another cell— *Didinium*. *Genet. Res.* **6**, 398–410.

317 — 1966. Chemical homologies between protozoa. *J. Protozool.* **13**, 650–653.

318 — 1967. RNA homologies between protozoa. *J. Protozool.* **14**, 687–690.

319 — 1968. Studies on the incorporation of DNA into *Paramecium aurelia*. *J. Cell Sci.* **3**, 381–389.

320 — 1970. Interacting genetic systems in *Paramecium*. *Advanc. Morphogenesis* **8**, 159–208.

321 — 1970. The genetics of protozoan organelles. *Symp. Soc. Exp. Biol.* **24**, 379–399.

322 — & Beale, G. H. 1961. Genic basis of the mate-killer trait in *Paramecium aurelia, stock 540*. *Genet. Res.* **2**, 82–91.

323 — & — 1962. The mechanism whereby the genes M_1 and M_2 in *Paramecium aurelia, stock 540*, control growth of mate-killer (mu) particles. *Genet. Res.* **3**, 24–50.

324 — & — 1963. The action of ribonuclease and 8-azaguanine on mate-killer paramecia. *Genet. Res.* **4**, 42–54.

325 — & — 1964. Infection into paramecia of metagons derived from other mate-killer paramecia. *Genet. Res.* **5**, 85–106.

326 — & Sonneborn, T. M. 1964. Killer particles and metagons of *Paramecium* grown in *Didinium*. *Genetics* **50**, 249–250.

327 — & — 1964. Is the metagon an m-RNA in *Paramecium* and a virus in *Didinium?* *Proc. Nat. Acad. Sci. U.S.* **52**, 869–876.

328 —, Chance, M. & Williams, J. 1971. Extranuclear DNA and the endosymbionts of *Paramecium aurelia*. *Nature* **234**, 75–77.

329 Giese, A. C. 1957. Mating types in *Paramecium multimicronucleatum*. *J. Protozool.* **4**, 120–124.

330 Gill, K. S. & Hanson, E. D. 1967. Analysis of prefission morphogenesis in *Paramecium aurelia*. *J. Exp. Zool.* **167**, 219–235.

331 *Gillies, C. & Hanson, E. D. 1962. A flagellate parasitizing the ciliate macronucleus. *J. Protozool.* **9** (Suppl.), 15.

332 — & — 1963. A new species of *Leptomonas* parasitizing the macronucleus of *Paramecium trichium*. *J. Protozool.* **10**, 467–473.

333 — & — 1968. Morphogenesis of *Paramecium trichium*. *Acta Protozool. (Warsaw)* **6**, 13–31.

334 *Gilman, L. C. 1954. Occurrence and distribution of mating type varieties in *Paramecium caudatum*. *J. Protozool.* **1** (Suppl.), 6.

335 *— 1956. Distribution of the varieties of *Paramecium caudatum*. *J. Protozool.* **3** (Suppl.), 4.

336 *— 1956. Size differences among twelve varieties of *Paramecium caudatum*. *J. Protozool.* **3** (Suppl.), 4.

337 *— 1957. Extension of range for several varieties of *Paramecium caudatum*. *J. Protozool.* **4** (Suppl.), 12.

338 *— 1957. Death following conjugation in *Paramecium caudatum*. *J. Protozool.* **4** (Suppl.), 13.

339 *— 1959. Incidence of *Paramecium caudatum* in collections of water samples over a six-year period. *J. Protozool.* **6** (Suppl.), 13.

340 *— 1959. Nuclear reorganization in *Paramecium caudatum*. *J. Protozool.* **6** (Suppl.), 19.

341 *— 1961. Survival of stocks of *Paramecium caudatum* in laboratory culture. *J. Protozool.* **8** (Suppl.), 14.

342 *— 1962. A comparison of Russian to other known syngens of *Paramecium caudatum*. *J. Protozool.* **9** (Suppl.), 13.

343 Ginzburg, G. I. 1961. Some data on the role of micronuclei and nuclei acid accumulation in *Paramecium caudatum* (in Russian). *Zhurnal Obshchej Biologii* **22**, 452–458.

344 — 1963. Nucleic acid content in normal and amicronucleate *Paramecium caudatum*. *Proc. 1st Int. Conf. Protozool., Prague*, 140.

345 Gittleson, S. M. & Sears, D. F. 1964. Effects of CO_2 on *Paramecium multimicronucleatum*. *J. Protozool.* **11**, 191–199.

346 *Golikowa, M. N. 1969. Polymorphism of micronuclei in some free-living ciliates. *Proc. 3rd Int. Conf. Protozool., Leningrad* 25–26.

347 — 1972. Micronuclear polymorphism of *Paramecium bursaria* Focke. *TSitologiya* **14**, 637–646.

348 Golińska, K. 1963. Experimental study on rebounding from a mechanical obstacle in *Paramecium caudatum*. *Acta Protozool. (Warsaw)* **1**, 114–120.

349 Gorsky, V. M. 1959. On the mutual effect of individuals of the same species in protozoa populations (in Russian). *Latvijas PSR, Isglitibas Ministrija Biologijas un Kimijas* **2**, 89–99.

350 — 1962. Effect of injurious factors on *Paramecium caudatum* depending on their aggregation (in Russian). *TSitologiya* **4**, 353–358.

351 — 1964. Über die Ursachen der Zunahme der Thermoresistenz von *Paramecium caudatum* bei Aggregation (in Russian). *Latvijas PSR zinatnu. Akad. Vestis* **2**, 64–74.

352 Goryachev, Ju. V. 1966. Motility stimulation of *Paramecium caudatum* by UV-radiation (in Russian). In *Produtsirovanie i Krugovorot Organicheskogo Vestshestva vo Vnutrennikh Vodojomakh*, Nauka, Moscow, Leningrad, 246–248.

353 — 1967. Effect of metabolic products of infusoria irradiated by UV-light on the motility of intact infusoria (in Russian). In *Luchistyie Factory Zhizni Vodnykh Organizmov*, Nauka, Leningrad, 94–96.

354 — 1967. Conditions for stimulation of motility in *Paramecium caudatum* by UV-light. The role of oxidative phosphorylation (in Russian). In *Luchistyie Factory Zhizni Vodnykh Organizmov*, Nauka, Leningrad, 97–111.

355 — 1969. The motility of UV-stimulated infusoria as a function of their initial activity (in Russian). In *Physiologiya Vodnikh Organizmow i ikh Rol v Krugovorotte Organicheskikh Vestshestv*, Nauka, Moscow, Leningrad, 100–102.

356 — 1969. Methods of objective registration of infusoria movement (in Russian). *Hydrobiologichesky Zhurnal* **5**, 107–109.

357 Grayevsky, E. Ya. & Shulmina, A. I. 1960. The significance of changes which arise in the water media in the process of cell injury by ionizing radiation (in Russian). In *Voprosy Tsytologii i Obstshey Physiologii*, Publ. Acad. Sci. USSR, Moscow, Leningrad, 80–85.

358 — & Zinovieva, E. G. 1959. Effect of low doses of ionizing irradiation on the vitality and the rate of division in *Paramecium caudatum* (in Russian). *Trudy instituta morphologii zhyvotnykh* **24**, 160–171.

359 —, Nebrasova, I. V. & Shulmina, A. I. 1962. A study on radioprotective action of some substances on protozoa (in Russian). *Radiobiologya* **2**, 148–155.

360 Grębecki, A. 1961. Experimental studies on the selection and adaptability in *Paramecium caudatum*. *Acta Biol. Exp. (Warsaw)* **21**, 25–52.

361 — 1962. Phénomènes électrocinétiques dans le galvanotropisme de *Paramecium caudatum*. *Bull. Biol. France Belg.* **96**, 723–754.

362 — 1963. Point isoélectrique superficiel et quelques réactions locomotorices chez *Paramecium caudatum*. *Protoplasma* **56**, 80–88.

363 — 1963. Electrobiologie d'ingestion des colorants par le cytostome de *Paramecium caudatum*. *Protoplasma* **56**, 89–98.

364 — 1963. Galvanotaxie transversale et oblique chez les Ciliés. *Acta Protozool. (Warsaw)* **1**, 91–98.

365 — 1963. Rebroussement ciliaire et galvanotaxie chez *Paramecium caudatum*. *Acta Protozool. (Warsaw)* **1**, 99–112.

366 — 1964. Role des ions K^+ et Ca^{2+} dans l'excitabilité de la cellule protozoaire. I. Equilibrement des ions antagonistes. *Acta Protozool. (Warsaw)* **2**, 69–79.

367 — 1964. Calcium substitution in staining the cilia. *Acta Protozool. (Warsaw)* **2**, 375–377.

368 — 1965. Gradient stomato-caudal d'excitabilité des Ciliés. *Acta Protozool. (Warsaw)* **3**, 79–101.

369 — 1965. Role of Ca^{2+} ions in the excitability of protozoan cell. Decalcification, recalcification, and the ciliary reversal in *Paracemium caudatum*. *Acta Protozool. (Warsaw)* **3**, 275–289.

370 *— 1965. Membrane calcium and the anodal galvanotaxis in *Paramecium caudatum*. *Proc. 2nd Int. Conf. Protozool., London* 242.

371 — & Kuźnicki, L. 1961. Immobilization of *Paramecium caudatum* in the chloralhydrate solutions. *Bull. Acad. Polon. Sci. Ser. Sci. Biol. (Classe II)* **9**, 459–462.

372 — & Mikolajczyk, E. 1968. Ciliary reversal and re-normalization in *Paramecium caudatum* immobilized by Ni-ions. *Acta Protozool. (Warsaw)* **5**, 299–303.

373 —, Kuźnicki, L. & Mikolajczyk, E. 1967. Right spiraling induced in *Paramecium* by Ni-ions and the hydrodynamics of the spiral movement. *Acta Protozool. (Warsaw)* **4**, 389–408.

374 Gregory, Jr., W. W., Reed, J. K. & Priester, Jr., L. E. 1969. Accumulation of parathion and DDT by some algae and protozoa. *J. Protozool.* **16**, 69–71.

375 Grigorjan, J. A. 1964. Relation between the thermostability of *Paramecium caudatum* and Ca-ion concentration in an environmental medium (in Russian). *TSitologiya* **6**, 105–109.

376 — 1964. Dependence of the thermostability of *Paramecium caudatum* upon the presence of ethyl alcohol and urea in the rearing medium (in Russian). *TSitologiya* **6**, 609–614.

377 — 1965. Dependence of the thermostability of *Paramecium caudatum* on the concentration of KCl and NaCl in the environment and on that of equilibrated salt solutions (in Russian). *TSitologiya* **7**, 218–225.

378 — 1968. Relation between the thermostability of *Paramecium caudatum* and sucrose concentration in the medium (in Russian). *TSitologiya* **10**, 850–855.

379 Grimes, G. W. & Preer, Jr., J. R. 1971. Further observations on the correlation between kappa and phage-like particles in *Paramecium. Genet. Res. Camb.* **18**, 115–116.

380 Gross, M., Skoczylas, B. & Turski, W. 1966. Purification and some properties of ribonucleases from *Paramecium aurelia. Acta Protozool. (Warsaw)* **4**, 59–66.

381 Gubányl, L. 1966. Mutagenic factor for *Paramecium* present in malignant neoplastic tissues (in Spanish). *Biologica (Santiago)* **39**, 99–117.

382 Guttman, R. & Back, A. 1958. Effects of kinetin on cell division in *Paramecium caudatum. Nature* **181**, 852.

383 — & — 1960. Effect of kinetin on *Paramecium caudatum* under varying culture conditions. *Science* **131**, 986–987.

384 Hairston, N. G. 1958. Observations on the ecology of *Paramecium* with comments on the species problem. *Evolution* **12**, 440–450.

385 Hajra, B. & Ray, H. N. 1962. Effect of protein on the growth of *Paramecium multimicronucleatum. Bull. Calcutta School. Trop. Med.* **10**, 12–13.

386 *Hallet, M. M. 1971. Mise en évidence de l'action d'un facteur externe, le chlorure de calcium sur la différentiation du macronucléus chez *Paramecium aurelia*, souche 60, variété 1. *J. Protozool.* **18** (Suppl.), 52.

387 Hambartsumyan, M. A. 1965. Role of boron for life processes in paramecia. I. The rate division of *Paramecium caudatum* related to the content of boron in the medium (in Russian). *Trudy Yerevanskogo Meditsinskogo Instituta* **14**, 115–123.

388 *— 1969. The effect of boron on the ciliates *Paramecium caudatum. Proc. 3rd Int. Conf. Protozool., Leningrad* 185–186.

389 — 1969. Changes of the thermostability in fasting *Paramecium caudatum* depending on the content of boron at different values of pH in the medium (in Russian). *TSitologiya* **11**, 772–777.

390 — 1970. Changes of the thermostability of *Paramecium caudatum* related to the contents of ions some metals in the medium (in Russian). *TSitologiya* **12**, 774–782.

391 — 1971. Comparative data on the influence of some ions on the thermostability of *Paramecium caudatum. Materials of 1st All-Union Congress of Protozool. Soc., Bacu*, 12–13.

392 Hamilton, L. D. & Gettner, M. E. 1958. Fine structure of kappa in *Paramecium aurelia. J. Biophys. Biochem. Cytol.* **4**, 122–123.

393 *Hanson, E. D. 1953. A new mutant kappa in variety 4, *Paramecium aurelia. Microbial Genet. Bull.* **7**, 14.

394 — 1954. Studies on kappa-like particles in sensitives of *Paramecium aurelia*, variety 4. *Genetics* **39**, 229–239.

395 — 1955. Inheritance and regeneration of cytoplasmic damage in *Paramecium aurelia. Proc. Nat. Acad. Sci. U.S.* **41**, 783–786.

396 — 1956. Spontaneous mutations affecting the killer character in *Paramecium aurelia*, variety 4. *Genetics* **41**, 21–30.

397 — 1957. Some aspects of the quantitive study of cytoplasmic particles: mixed populations of kappa in *Paramecium aurelia*, variety 4. *J. Exp. Zool.* **135**, 29–56.

398 — 1962. Morphogenesis and regeneration of oral structures in *Paramecium aurelia*. An analysis of intracellular development. *J. Exp. Zool.* **150**, 45–67.

399 *— 1962. Spontaneous gullet abnormalities in *Paramecium aurelia*. *J. Protozool.* **9** (Suppl.), 15.

400 *— 1963. Morphogenesis of oral structures in *Paramecium trichium*. *J. Protozool.* **10** (Suppl.), 16.

401 — 1967. Protozoan development. In Florkin, M. & Scheer, B., *Chemical Zoology*, Academic Press, New York, N.Y., Ch. 1, 395–539.

402 *— & Gillies, C. 1966. Oral structures during conjugation in *Paramecium aurelia*. *J. Protozool.* **13** (Suppl.), 26.

403 — & Kaneda, M. 1968. Evidence for sequential gene action within the cell cycle of *Paramecium*. *Genetics* **60**, 793–805.

404 *— & Twichell, J. B. 1962. Autoradiographic study of the time of DNA and RNA synthesis in *Paramecium trichium*. *J. Protozool.* **9** (Suppl.), 11.

405 —, Gillies, C. & Kaneda, M. 1969. Oral structure development and nuclear behavior during conjugation in *Paramecium aurelia*. *J. Protozool.* **16**, 197–204.

406 Harrison, J. A. & Fowler, E. H. 1945. Antigenic variation in clones of *P. aurelia*. *J. Immunol.* **50**, 115–125.

407 *Hartig, W. J. & Lilly, D. M. 1954. Bacteria-free cultures of *Paramecium caudatum*. *J. Protozool.* **1** (Suppl.), 10.

408 Hayashi, S. 1959. On the relationship between the induction of pseudo-selfing paring and RNA contents in *Paramecium bursaria*. *J. Fac. Sci. Hokkaido Univ. ser. 6*, **14**, 129–133.

409 — & Takayanagi, T. 1962. Cytological and cytogenetical studies on *Paramecium polycaryum*. IV. Determination of the mating system based on some experimental and cytological observations. *Japan. J. Zool.* **13**, 357–364.

410 Hearon, J. Z. & Kimball, R. F. 1955. Tests for a role of H_2O_2 in X-ray mutagenesis. I. Estimates of concentration of H_2O_2 inside the nucleus. *Radiation Res.* **3**, 283–295.

411 Heckmann, K. & Siegel, R. W. 1964. Evidence for the induction of mating type substances by cell to cell contacts. *Exp. Cell Res.* **36**, 688–691.

412 Helmy Mohammed, A. H. & Nashed Nawal, N. 1968-1969. *Paramecium wichtermani*, n. sp., with notes on other species of *Paramecium* common in fresh water bodies in the area of Cairo and its environs. *Zool. Soc. Egypt* **22**, 89–104.

413 Hirson, J. B. 1969. The response of *Paramecium bursaria* to potential endocellular symbionts. *Biol. Bull.* **136**, 33–42.

414 Hisada, M. 1952. Induction of contraction in *Paramecium* by electric current. *Annot. Zool. Japon* **25**, 415–419.

415 Hiwatashi, K. 1950. Studies on the conjugation of *Paramecium caudatum*, III: some properties of the mating type substance. *Sci. Rep. Tôhoku Univ., Biol. Ser. 4* **18**, 270–275.

416 — 1951. Studies on the conjugation of *Paramecium caudatum*. IV. Conjugating behaviour of individuals of two mating types marked by a vital staining method. *Sci. Rep. Tôhoku Univ., Biol. Ser. 4* **19**, 95–99.

417 — 1955. Studies on the conjugation of *Paramecium caudatum*. VI. The nature of the union of conjugation. *Sci. Rep. Tôhoku Univ., Biol. Ser. 4* **21**, 207–218.

418 — 1958. Artificial induction of conjugation by EDTA in *Paramecium caudatum* 29th Ann. Meeting Zoological Society of Japan.

419 — 1958. Inheritance of mating types in variety 12 of *Paramecium caudatum*. *Sci. Rep. Tôhoku Univ., Biol. Ser. 4* **24**, 119–129.

420 — 1959. Induction of conjugation by ethylenediamine tetra-acetic acid (EDTA) in *Paramecium caudatum*. *Sci. Rep. Tôhoku Univ., Biol. Ser. 4* **25**, 81–90.

421 *— 1960. An aberrant selfing strain of *Paramecium caudatum* which shows multiple unions of conjugation. *J. Protozool.* **7** (Suppl.), 20.

422 — 1960. Inheritance of difference in the life feature of *Paramecium caudatum*, syngen 2. *Bull. Marine Biol. Sta. Asamushi, Tôhoku Univ.* **9**, 157–159.

423 — 1960. Analysis of the change of mating type during vegetative reproduction in *Paramecium caudatum*. *Japan. J. Genetics* **35**, 213–221.

424 — 1961. Locality of mating reactivity on the surface of *Paramecium caudatum*. *Sci. Rep. Tôhoku Univ., Biol. Ser. 4* **27**, 93–99.

425 *— 1963. Serotype inheritance in *Paramecium caudatum*. *Genetics* **48**, 892.

426 — 1964. Mating type inheritance in *Paramecium caudatum* syngen 3. *Genetics* **50**, 255–256.

427 *— 1965. The effect of brei on the differentiation of mating type in *Paramecium caudatum*. *Genetics* **52**, 448.

428 — 1967. Serotype inheritance and serotypic alleles in *Paramecium caudatum*. *Genetics* **57**, 711–717.

429 — 1968. Determination and inheritance of mating type in *Paramecium caudatum*. *Genetics* **58**, 373–386.

430 *— 1968. Genetic and epigenetic control of mating types in *Paramecium caudatum*. *Proc. 12th Int. Congr. Genet.* **2**, 259–260.

431 — 1969. Genetic and epigenetic control of mating type in *Paramecium caudatum*. *Japan J. Genetics* **44** (Suppl. 1), 383–387.

432 *— & Kasuga, T. 1960. Artificial induction of conjugation by manganese ion in *Paramecium caudatum*. *J. Protozool.* **7** (Suppl.), 20–21.

433 — & — 1961. Locality of mating reactivity on the surface of *Paramecium caudatum*. *Sci. Rep. Tôhoku Univ. Biol., Ser. 4* **27**, 93–99.

434 — & Takahashi, M. 1967. Inhibition of mating reactions by antisera without ciliary immobilization in *Paramecium*. *Sci. Rep. Tôhoku Univ., Biol. Ser. 4* **33**, 281–290.

435 *— & — 1969. Analysis of the expression of mating type alleles in *Paramecium caudatum*. *Proc. 3rd Int. Conf. Protozool., Leningrad* 114.

436 *Holzman, H. E. 1959. A kappa-like particle in a non-killer stock of *Paramecium aurelia*, syngen 4. *J. Protozool.* **6** (Suppl.), 26.

437 *Honigberg, B. M. 1955. Distribution of phosphatase in *Paramecium caudatum*. *J. Protozool.* **2** (Suppl.), 4.

438 Hovasse, R., Mignot, J. P. & Joyon, L. 1967. Nouvelles données sur les trichocystes des *Cryptomonadines* et les R-bodies des particules de *Paramecium aurelia*. *Protistologica* **3**, 241–255.

439 Hruban, Z. & Rechcigl, Jr., M. 1969. Microbodies and related particles. Morphology, biochemistry and physiology. *Int. Rev. Cytol.*, Suppl. 1, Academic Press, New York.

440 *Hufnagel, L. 1965. Structural and chemical observations on pellicles isolated from paramecia. *J. Cell Biol.* **27**, 46A.

441 — 1966. Fine structure and DNA of pellicles isolated from *Paramecium aurelia*. *Electron Microscopy, Vol. 2*, (6th Int. Congr. Electr. Micr. Kyoto), Maruzen, Tokyo, 239–240.

442 — 1969. Properties of DNA associated with raffinose isolated pellicles of *Paramecium aurelia*. *J. Cell Sci.* **5**, 561–573.

443 — 1969. Cortical ultrastructure of *Paramecium aurelia*. Studies on isolated pellicles. *J. Cell Biol.* **40**, 779–801.

444 *Hull, R. W. 1962. Using the *Paramecium* assay to screen carcinogenic hydrocarbons. *J. Protozool.* **9** (Suppl.), 18.

445 Hunter, N. W. 1959. Enzyme patterns in *Paramecium putrinum* Claparede and Lachmann. *Trans. Amer. Microscop. Soc.* **78**, 363–370.

446 — 1967. Effect of certain cations on activity of leucine naphthylamidases of *Paramecium caudatum*. *Canadian J. Microbiol.* **13**, 1133–1138.

447 Igarashi, S. 1966. Temperature-sensitive mutation in *Paramecium aurelia*. I. Induction and inheritance. *Mutation Res.* **3**, 13–24.

448 — 1966. Temperature-sensitive mutation in *Paramecium aurelia*. II. Modification of the mutation frequency by pre- and postirradiation conditions. *Mutation Res.* **3**, 25–33.

449 *— & Kimball, R. F. 1964. Temperature sensitive mutation induced by X-rays in *Paramecium*. *Genetics* **50**, 258.

450 Inaba, F. & Kudo, N. 1972. Electron microscopy of the nuclear events during binary fission in *Paramecium multimicronucleatum*. *J. Protozool.* **19**, 57–63.

451 —, Imamoto, K. & Suganuma, Y. 1966. Electron microscope observations on nuclear exchange during conjugation in *Paramecium multimicronucleatum*. *Proc. Japan Acad.* **42**, 394–398.

452 Irlina, I. S. 1963. Some physiological and cytochemical peculiarities of *Paramecium caudatum* adapted to different temperatures (in Russian). *TSitologiya* **5**, 183–193.

453 — 1963. Effect of various concentrations of some salts and ethyl alcohol on the ciliates adapted to different temperatures (in Russian). *TSitologiya* **5**, 227–294.

454 — 1963. On the resistance of *Paramecium caudatum* previously adapted to different temperatures

to injurious action of salts (in Russian). In *Morfologiya i Fiziologiya Prosteishikh*, Akad. Nauk, SSSR, Moscow, Leningrad, 92–111.

455   — 1964. Specific and non-specific changes in resistance of *Paramecium caudatum* adapted to different temperatures (in Russian). In *Kletka i Temperatura Sredy*, Nauka, Moscow, Leningrad, 169–171.

456   — 1967. Correlation between thermostability of *Paramecium caudatum* and activity of succinic dehydrogenase in infusoria adapted to various temperatures (in Russian). In *Izmenchivosty Teploustoichivosty Kletok Zhyvotnykh v Ontogeneze i Phylogeneze*, Nauka, Leningrad, 37–41.

457   — & Semjonova, E. G. 1967. Influence of dibazole on the resistance of *Paramecium caudatum* to ethyl alcohol (in Russian). In *Proizvodnye Benzimidazola i Kletochnaja Resistentnost*, Nauka, Leningrad, 41–45.

458   *Isaacks, R. E. & Van Wagtendonk, W. J. 1968. Isolation of nuclei from *Paramecium aurelia* and extraction of basic proteins from the nuclear fraction. *J. Protozool.* **15** (Suppl.), 15.

459   *—, Santos, B. G. & Van Wagtendonk, W. J. 1969. Isolation and fractionation of nuclei from *Paramecium aurelia* by density gradient centrifugation. *J. Protozool.* **16** (Suppl.), 8.

460   —, — & — 1969. Fractionation of nuclei from *Paramecium aurelia* by density gradient centrifugation. *Biochim. Biophys. Acta* **195**, 268–270.

461   Jacobson, W. E., Williamson, M. & Stock, C. C. 1952. The destructive effect of 2,6-diaminopurine on kappa of stock 51 killers, variety 4, of *Paramecium aurelia*. *J. Exp. Zool.* **121**, 505–519.

462   Jahn, T. L. 1962. The mechanism of ciliary movement. II. Ion antagonism and ciliary reversal. *J. Cell. Comp. Physiol.* **60**, 217–228.

463   — & Bovee, E. C. 1964. Protoplasmic movements and locomotion of protozoa. In Hutner, S. H., *Biochemistry and Physiology of Protozoa*, Academic Press, New York, N.Y., **3**, 62–129.

464   *— & Fonseca, J. R. 1970. The effect of Diuril (chlorothiazide) upon osmoregulation in *Paramecium*. *J. Protozool.* **17** (Suppl.), 11.

465   *—, Brown, M. & Winet, H. 1961. Secretory activity of oral groove of *Paramecium*. *J. Protozool.* **8** (Suppl.), 18.

466   —, Bovee, E. C. & Dauber, M. 1965. Secretory activity of the oral apparatus of ciliates: trails of adherent particles left by *Paramecium multimicronucleatum* and *Tetrahymena pyriformis*. *Ann. N.Y. Acad. Sci.* **118**, 912–920.

467   Jakimowa, G. I. & Rasumowsky, P. N. 1968. *Paramecium caudatum* as the test object for the determination of biological action of bacterial cultural media (in Russian). In *Ispolzovannie Mikroorganizmov v Narodnom Khozjaystve*, Kartja Moldovenjaske, Kishinev, Ch. 3, 31–36.

468   *—, Borisowa, T. A. & Rasumowsky, P. N. 1969. Einfluss von Mikrobenmetaboliten auf die Nahrungsaufnahme bei *Paramecium*. *Proc. 3rd Int. Conf. Protozool.*, Leningrad 144–145.

469   —, — & — 1970. The influence of some fractions from the mycelium of Actinomyces griseus 15 on the rate division of paramecia (in Russian). *Izvestiya Akademii Nauk Moldavskoy SSR* **3**, 48–52.

470   Janisch, R. 1964. Sub-microscopic structure of injured surface of *Paramecium caudatum*. *Nature* **204**, 200–201.

471   — 1966. The mechanism of the regeneration of surface structures of *Paramecium caudatum*. *Folia Biol. (Praha)* **12**, 65–74.

472   — 1967. Electron microscopy study of surface structures of *Paramecium caudatum*. *Folio Biol. (Praha)* **13**, 386–392.

473   Jankovsky, A. V. 1960. Conjugation processes in *Paramecium trichium*, Stokes. I. Amphimixis and autogamy (in Russian). *TSitologiya* **2**, 581–588.

474   — 1961. Studies in ciliate pathology. I. Lethal macronuclear destruction and protein crystallization in *Paramecium caudatum*. Pure lines of *P. caudatum* with hereditary autolyzing macronuclei (in Russian). *Vestnik Leningradskogo Universiteta* **15**, 91–104.

475   — 1961. The process of conjugation in the rare salt-water *Paramecium*, *P. woodruffi* (in Russian). *Dokl. Akad. Nauk SSSR* **137**, 989–992.

476   — 1962. Conjugation processes in *Paramecium putrinum* Clap. et Lachm. II. Apomictic reorganization cycles and the system of mixotypes (in Russian). *TSitologiya* **4**, 434–444.

477   — 1962. Conjugation processes in *Paramecium putrinum* Clap. et Lachm. III. The multiple mating type system in *Paramecium putrinum* (in Russian). *Zhurnal Obshchej Biologii* **23**, 276–282.

478 — 1965. Conjugation processes in *Paramecium putrinum* Clap. et Lachm. IV. The individual variability of the nuclear processes at apomictic conjugation (in Russian). *TSitologiya* 7, 55–65.

479 — 1965. Conjugation processes in *Paramecium putrinum* Clap. et Lachm. V. Return to amphimixis in mixotype B (in Russian). *Dokl. Akad. Nauk SSSR* 163, 523-525.

480 — 1965. Conjugation processes in *Paramecium putrinum* Clap. et Lachm. VI. The induction and cytological study of a triple conjugation (in Russian). *TSitologiya* 8, 70–79.

481 — 1965. Conjugation processes in *Paramecium putrinum* Clap. et Lachm. VII. Nuclear processes at autogamy in singles induced with a new technique—multiple mating (in Russian). *Acta Protozool. (Warsaw)* 3, 239–263.

482 — 1966. Conjugation processes in *Paramecium putrinum* Clap. et Lachm. VIII. Nuclear processes at repeat conjugation—reconjugation (in Russian). *Zoologicheskii Zhurnal* 45, 818–829.

483 — 1966. Conjugation processes in *Paramecium putrinum*. IX. "Necrochromatine" and the functional significance of macronuclear fragmentation (in Russian). *TSytologiya* 8, 725–735.

484 *— 1969. Cytological study of amphimictic conjugation in *Paramecium putrinum*. *Proc. 3rd Int. Conf. Protozool., Leningrad* 28–29.

485 — 1969. A proposed taxonomy of the genus *Paramecium Hill, 1752. (Ciliophora)* (in Russian). *Zoologicheskii Zhurnal* 48, 30–40.

486 — 1972. Nuclear apparatus of Ciliophora. I. Polyploidy of micronucleus in Ciliophora as a characteristic of the species. *Genetica* 8, 78–84.

487 — 1972. Recapitulation of phylogenesis in ciliate ontogeny. In *Problems of Evolution, Vol. 2*, Nauka, Novosibirsk, 70–118.

488 *Jaroshenko, M. F. & Chorik, F. P. 1969. Ecology of fresh-water free living ciliates. *Proc. 3rd Int. Conf. Protozool., Leningrad* 195.

489 *Jean-Claude, E. 1968. Observations sur l'ultrastructure et le métabolisme du glycogène de *Paramecium caudatum*. *J. Protozool.* 15 (Suppl.), 40.

490 Jenkins, R. A. 1970. The fine structure of a nuclear envelope associated with an endosymbiont of *Paramecium*. *J. Gen. Microbiol.* 61, 355–359.

491 Jensen, D. D. 1959. A theory of the behavior of *Paramecium aurelia* and behavioral effects of feeding, fission, and ultra-violet micro-beam irradiation. *Behaviour* 15, 82–122.

492 Johnson, B. D., Tullar, J. C. & Stahnke, H. L. 1966. A quantitative protozoan bio-assay method for determining venom potencies. *Toxicon* 3, 297–300.

493 Johnson, W. H. & Miller, C. A. 1956. A further analysis of the nutrition of *Paramecium*. *J. Protozool.* 3, 221–226.

494 — & — 1957. The nitrogen requirements of *Paramecium multimicronucleatum*. *Physiol. Zool.* 30, 106–113.

495 *— & — 1957. Analyses of the nondialyzable factors (NDF) required for the growth of *Paramecium multimicronucleatum*. *J. Protozool.* 4 (Suppl.), 9.

496 *— & —. 1958. A fatty acid requirement for *Paramecium multimicronucleatum*, *J. Protozool.* 5 (Suppl.), 14.

497 Jones, A. 1964. *Description and Genetic Analysis of an Antigen of Paramecium aurelia. Thesis, University of Pennsylvania*, Philadelphia, Penn.

498 Jones, I. G. 1965. Immobilization antigen in heterozygous clones of *Paramecium aurelia*. *Nature* 207, 769.

499 — 1965. Studies on the characterization and structure of the immobilization antigens of *Paramecium aurelia*. *Biochem. J.* 96, 17–23.

500 — & Beale, G. H. 1963. Chemical and immunological comparisons of allelic immobilization antigens in *Paramecium aurelia*. *Nature* 197, 205–206.

501 *—, Keeshan, C. P. & Lilly, D. M. 1957. The axenic culture of *Paramecium trichium*. *J. Protozool.* 4 (Suppl.), 18.

502 Jurand, A. 1961. Activity of nucleases on *Paramecium* cells fixed with osmium tetroxide. *Exp. Cell Res.* 25, 80–86.

503 — 1961. An electron microscope study of food vacuoles in *Paramecium aurelia*. *J. Protozool.* 8, 125–130.

504 — & Jacob, J. 1969. Studies on the macronucleus of *Paramecium aurelia*. III. Localization of tritiated thymidine by electron microscope autoradiography. *Chromosoma* 26, 355–364.

505 — & Preer, L. B. 1968. Ultrastructure of flagellated lambda symbionts in *Paramecium aurelia*. *J. Gen. Microbiol.* 54, 359–364.

506   — & Selman, G. G. 1969. The anatomy of *Paramecium aurelia*. Macmillan, London and St. Martin's Press, New York, N.Y.

507   — & — 1970. Ultrastructure of the nuclei and intranuclear microtubules of *Paramecium aurelia*. *J. Gen. Microbiol.* **60**, 357–364.

508   —, Beale, G. H. & Young, M. R. 1962. Studies on the macronucleus of *Paramecium aurelia*. I. (with a note on Ultra-violet micrography). *J. Protozool.* **9**, 122–131.

509   —, — & — 1964. Studies on the macronucleus of *Paramecium aurelia*. II. Development of macronuclear anlagen. *J. Protozool.* **11**, 491–497.

510   —, Gibson, I. & Beale, G. H. 1962. The action of ribonuclease on living paramecia. *Exp. Cell Res.* **26**, 598–600.

511   —, Rudman, B. M. & Preer, Jr., J. R. 1971. Prelethal effects of killing action by stock 7 of *Paramecium aurelia*. *J. Exp. Zool.* **177**, 365–388.

512   *Kang, H. S. & Taub, S. R. 1968. Studies on the effects of actinomycin D and puromycin on mating type changes in cells of syngen 7, *Paramecium aurelia*. *J. Cell Biol.* **39**, 70a–71a.

513   Karakashian, M. W. 1961. *The Rhythm of Mating in Paramecium aurelia and its Significance for Genetic Analysis of Circadian Rhythm*, Thesis, University of California, Los Angeles, Calif.

514   — 1963. Growth of *Paramecium bursaria* as influenced by the presence of algal symbionts. *Physiol. Zool.* **36**, 52–68.

515   — 1965. The circadian rhythm of sexual reactivity in *Paramecium aurelia*, syngen 3. In Aschoff, J., *Circadian Clocks*, North-Holland Publishing Company, Amsterdam, 301–304.

516   — 1968. The rhythm of mating in *Paramecium aurelia* syngen 3. *J. Cell. Physiol.* **71**, 197–209.

517   *— & Karakashian, S. J. 1964. The inheritance of susceptibility to free-living algal infection in aposymbiotic *Paramecium bursaria*. *J. Protozool.* **11** (Suppl.), 19.

518   Karakashian, S. J. 1966. Mating in populations of *Paramecium bursaria*. A mathematical model. *Genetics* **53**, 145–156.

519   —, Karakashian, M. W. & Rudzinska, M. A. 1968. Electron microscopic observations on the symbiosis of *Paramecium bursaria* and its intracellular algae. *J. Protozool.* **15**, 113–128.

520   Katorgina, I. F. & Kudrin, A. N. 1971. On adrenergic mechanisms of coordinated motion of cilia in paramecia. In *Actual Problems of Pharmacology and Pharmacy*, Moscow, 133–141.

521   Katz, M. S. & Deterline, W. A. 1958. Apparent learning by *Paramecium*. *J. Comp. Physiol. Psychol.* **51**, 243–247.

522   *Keeshan, C. P. 1964. Amino acids in the growth of *Paramecium trichium*. *J. Protozool.* **11** (Suppl.), 33.

523   *— & Kunert, J. E. 1962. Nutritional studies on the growth of *Paramecium trichium*. *J. Protozool.* **9** (Suppl.), 25.

524   *Keim, E. D. & Hanson, E. D. 1964. A morphological variant in *Paramecium aurelia*. *J. Protozool.* **11** (Suppl.), 9–10.

525   *Kennedy, Jr., J. R. 1966. Chloral hydrate immobilization of *Paramecium caudatum*. *J. Protozool.* **13** (Suppl.), 16.

526   — & Brittingham, E. 1968. Fine structure changes during chloral hydrate decilliation of *Paramecium caudatum*. *J. Ultrastruct. Res.* **22**, 530–545.

527   Kimball, R. F. 1953. The structure of the macronucleus of *Paramecium aurelia*. *Proc. Nat. Acad. Sci. U.S.* **39**, 345–347.

528   *— 1953. Three new mutants and the independent assortment of five genes in variety 1 of *Paramecium aurelia*. *Microbial Genet. Bull.* **8**, 10.

529   — 1955. The effects of radiation on Protozoa and the eggs of invertebrates and other insects. In Hollaender, A., *Radiation Biology*, McGraw-Hill, New York, N.Y., **2**, 285–231.

530   — 1955. The role of oxygen and peroxide in the production of radiation damage in *Paramecium*. *Ann. N.Y. Acad. Sci.* **59**, 638–647.

531   — 1957. The effect of radiations on genetic mechanisms of *Paramecium aurelia*. *J. Cell. Comp. Physiol.* **35** (Suppl. 1), 157–169.

532   — 1961. Post-irradiation processes in the induction of recessive lethals by ionizing radiation. *J. Cell. Comp. Physiol.* **58** (Suppl.), 163–170.

533   — 1962. Chromosome duplication and mutation. In Fritz-Niggli, H., *Strahlenwirkung und Milieu*, Urban und Schwarzenberg, Munich, Berlin, 116–125.

534   — 1963. Studies on radiation mutagenesis in microorganisms. *Proc. 11th Int. Congr. Genetics*, The Hague, **2**, 227–234.

535 — 1963. X-ray dose rate and dose fractionation studies on mutation in *Paramecium*. *Genetics* **48**, 581–596.

536 — 1963. The relation of repair to differential radiosensitivity in the production of mutations in *Paramecium*. In Sobels, F. H., *Symposium on Repair from Genetic Radiation Damage and Differential Radiosensitivity of Germ Cells*, Pergamon Press, London, 167–178.

537 — 1964. The distribution of X-ray induced mutations to chromosomal strands in *Paramecium aurelia*. *Mutation Res.* **1**, 129–138.

538 *— 1964. Studies on reparable premutational lesions with alkylating agents. *Genetics* **50**, 262.

539 — 1964. Physiological genetics of the ciliates. In Hutner, S. H., *Biochemistry and Physiology of Protozoa*, Academic Press, New York, N.Y., **3**, 243–275.

540 *— 1965. Further studies on mutagenesis by triethylene melamine in *Paramecium aurelia*. *Genetics* **52**, 452.

541 — 1965. The induction of reparable premutational damage in *Paramecium aurelia* by the alkylating agent triethylene melamine. *Mutation Res.* **2**, 413–425.

542 — 1967. Persistent intraclonal variations in cell dry mass and DNA content in *Paramecium aurelia*. *Exp. Cell Res.* **48**, 378–394.

543 — 1968. Induction of mutations in *Paramecium* by ultraviolet and X-rays in relation to repair processes. *Brit. J. Radiol.* **41**, 880.

544 — 1969. Studies on mutations induced by ultraviolet radiation in *Paramecium aurelia* with special emphasis on photoreversal. *Mutation Res.* **8**, 79–89.

545 — & Barka, T. 1959. Quantitative cytochemical studies on *Paramecium aurelia*. II. Feulgen microspectrophotometry of the macronucleus during exponential growth. *Exp. Cell Res.* **17**, 173–182.

546 — & Gaither, N. 1953. Influence of oxygen upon genetic and nongenetic effects of ionizing radiation on *Paramecium aurelia*. *Proc. Soc. Exp. Biol. Med.* **82**, 471–477.

547 *— & — 1954. Lack of an effect of a high dose of X-rays on aging in *Paramecium aurelia*, variety 1. *Genetics* **39**, 977.

548 — & — 1955. Behavior of nuclei at conjugation in *Paramecium aurelia*. I. Effect of incomplete chromosome sets and competition between complete and incomplete nuclei. *Genetics* **40**, 878–889.

549 — & — 1956. Behavior of nuclei at conjugation in *Paramecium aurelia*. II. The effects of X-rays on diploid and haploid clones, with a discussion of dominant lethals. *Genetics* **41**, 715–728.

550 — & Householder, A. S. 1954. A stochastic model for the selection of macronuclear units in *Paramecium* growth. *Biometrics* **10**, 361–374.

551 — & Perdue, S. W. 1962. Quantitative cytochemical studies on *Paramecium*. V. Autoradiographic studies of nucleic acid synthesis. *Exp. Cell Res.* **27**, 405–415.

552 — & — 1962. Studies on the refractory period for the induction of recessive lethal mutation by X-rays in *Paramecium*. *Genetics* **47**, 1595–1607.

553 *— & — 1964. Synthesis of RNA by fragments of the old macronucleus in *Paramecium aurelia* undergoing autogamy. *J. Protozool.* **11** (Suppl.), 33.

554 — & — 1965. Autoradiographic studies on the conservation of label from ribonucleosides and amino acids in *Paramecium aurelia*. *Exp. Cell Res.* **38**, 660–669.

555 — & — 1967. Comparison of mutagenesis by X-rays and triethylene melamine in *Paramecium* with emphasis on the role of mitosis. *Mutation Res.* **4**, 37–50.

556 — & Prescott, D. M. 1964. RNA and protein synthesis in amacronucleate *Paramecium aurelia*. *J. Cell Biol.* **21**, 496–497.

557 — & Vogt-Köhne. 1961. Quantitative cytochemical studies on *Paramecium aurelia*. IV. The effect of limited food and starvation on the macronucleus. *Exp. Cell Res.* **23**, 479–487.

558 —, Gaither, N. & Perdue, S. W. 1961. Metabolic repair of premutational damage in *Paramecium. Intern. J. Radiat. Biol.* **3**, 133–147.

559 —, — & Wilson, S. M. 1957. Post-irradiation modification of mutagenesis in *Paramecium* by streptomycin. *Genetics* **42**, 661–669.

560 —, — & — 1959. Recovery in stationary phase paramecia from radiation effects leading to mutation. *Proc. Nat. Acad. Sci. U.S.* **45**, 833–839.

561 —, — & — 1962. Reduction of mutation by post-irradiation treatment after UV and various kinds of ionizing radiations. *Radiation Res.* **10**, 490–497.

562 —, — & — 1962. Division delay by radiation and nitrogen mustard in *Paramecium*. *J. Cell. Comp. Physiol.* **40**, 427–459.

563 —, Hearon, J. Z. & Gaither, N. 1955. Tests for a role of H_2O_2 in X-ray mutagenesis. II. Attempts to induce mutations by peroxide. *Radiation Res.* **3**, 435–443.

564 —, Vogt-Köhne, L. & Casperson, T. O. 1960. Quantitative cytochemical studies on *P. aurelia*. III. Dry weight and ultraviolet absorption of isolated macronuclei during various stages of the inter-division interval. *Exp. Cell Res.* **20**, 368–377.

565 —, Caspersson, T. O., Svensson, G. & Carlson, L. 1959. Quantitative cytochemical studies on *Paramecium aurelia*. I. Growth in total dry weight measured by the scanning interference microscope and X-ray absorption methods. *Exp. Cell Res.* **17**, 160–172.

566 King, R. L. 1954. Origin and morphogenetic movements of the pores of the contractile vacuoles in *Paramecium aurelia*. *J. Protozool.* **1**, 121–130.

567 Kinosita, H. & Murakami, A. 1967. Control of ciliary motion. *Physiol. Rev.* **47**, 53–82.

568 —, Dryl, S. & Naitoh, Y. 1964. Changes in the membrane potential and the response to stimuli in *Paramecium*. *J. Fac. Sci. Univ. Tokyo, Sect. IV* **10**, 291–301.

569 —, — & — 1964. Relation between the magnitude of membrane potential and ciliary activity in *Paramecium*. *J. Fac. Sci. Univ. Tokyo, Sect. IV* **10**, 303–309.

570 —, Murakami, A. & Yasuda, M. 1965. Interval between membrane potential change and ciliary reversal in *Paramecium* immersed in Ba-Ca mixture. *J. Fac. Sci. Univ. Tokyo, Sect. IV* **10**, 421–425.

571 *Klayman, M. B. 1969. Clumping by isolated cilia of complementary mating types from axenically grown *Paramecium aurelia*. *Yale Sci. Mag.* **43**, 32.

572 Koenuma, A. 1963. The velocity distribution of the cyclosis in *Paramecium*. *Annot. Zool. Japon.* **36**, 66–71.

573 — 1964. The velocity distribution of the cyclosis in *Paramecium caudatum*. *Annot. Zool. Japan* **36**, 66–71.

574 Kogan, A. G. & Tikhonova, N. A. 1965. The effect of a constant magnetic field on paramecia movement (in Russian). *Biofizika* **10**, 292–298.

575 —, Dorozhkina, L. I. & Volynskaya, E. M. 1968. The influence of a static magnetic field on the phagocytic activity of paramecia (in Russian). *TSitologiya* **10**, 1342–1348.

576 Koizumi, S. 1958. An analysis of the process of transformation of antigenic type induced by antiserum treatment in *Paramecium caudatum*, stock KA6. *Sci. Rep. Tôhoku Univ., Biol. Ser. 4* **24**, 23–31.

577 — 1960. Antigenic type transformation after selfing conjugation in *Paramecium caudatum*, stock K6. *Sci. Rep. Tôhoku Univ., Biol. Ser. 4* **26**, 297–307.

578 *— 1966. Transfer of cytoplasm by microinjection in *Paramecium aurelia*. *J. Protozool.* **13** (Suppl.), 27.

579 — 1966. Serotypes and immobilization antigens in *Paramecium caudatum*. *J. Protozool.* **13**, 73–76.

580 *— 1971. The cytoplasmic factor that fixes macronuclear mating type determination in *Paramecium aurelia*, syngen 4. *Genetics* **68** (Suppl.), s34.

581 *— & Preer, Jr., J. R. 1966. Transfer of cytoplasm by microinjection in *Paramecium aurelia*. *J. Protozool.* **13** (Suppl.), 27.

582 Kokina, N. N. 1957. Potassium reversal of paramecia cilia in relation to metabolism (in Russian). In *Sbornik Nauchnykh Studentichestikh Rabot*, Biologo-Pochvenny fakultet Moskovskogo Universiteta, 87–91.

583 Kokova, B. E. & Lisovsky, G. M. 1968. Feeding of *Paramecium caudatum* on *Chlorella* (in Russian). *Akad. Nauk SSSR. Sibirskoie Otdeleniye. Izvestiya. Seriya Biologo-Meditzinskaya* **15**, 107–109.

584 *— & — 1969. Continuous culture of *Paramecium caudatum* at different rates of medium flowing. *Proc. 3rd Int. Conf. Protozool., Leningrad* 147.

585 — & — 1972. Effect of the medium flow rate on the density, productivity and specific growth rate of the culture of *Paramecium caudatum*. *TSitologiya*, **14**, 516–523.

586 Kokshaisky, N. B. 1967. On the range of Reynolds' index for biological objects (in Russian). In *Voprosy Bionyki*, Nauka, Moscow, 543–549.

587 Komola, Z. 1968. The observations on the influence of a small dose of X-rays on conjugating paramecia. *Folia Biol. (Krakow)* **16**, 299–306.

642 — & Howard, J. L. 1955. Respiratory studies on mate-killers and sensitives of *Paramecium aurelia*, variety 8. *Science* **121**, 336–337.

643 Lilly, D. M. & Klosek, R. C. 1961. A protein factor in the nutrition of *Paramecium caudatum*. *J. Gen. Microbiol.* **24**, 327–334.

644 *—, — & Hartig, W. F. 1958. Methods of preparation of factors required for the growth of *Paramecium caudatum* in axenic cultures. *Anat. Rec.* **131**, 575.

645 Litvinova, I. B. 1959. The effects of post-irradiation temperature on the responses of *Paramecium* to ionizing radiation (in Russian). *Dokl. Akad. Nauk SSSR* **124**, 25–27.

646 — 1959. The effect of cultivation temperature following exposure to ionizing radiation upon the reaction of paramecia to X-rays (in Russian). *Dokl. Akad. Nauk SSSR* **124**, 448–451.

647 *Losina-Losinsky, L. K. 1958. On the increased resistance of *Paramecium caudatum* to repeated irradiation by ultraviolet light. *15th Int. Congr. of Zool., Sect. 4*, paper 6, London.

648 — 1960. The influence of some factors on the reactivation of *Paramecium caudatum* after the irradiation by ultraviolet rays (in Russian). In *Voprosy Tsitologii i Protistologii*, Akad. Nauk USSR, Moscow, Leningrad, 224–257.

649 — 1961. The stability of paramecia adapted to life in warm radioactive springs to different external agents (in Russian). *TSitologiya* **3**, 154–166.

650 — 1966. On the capacity of some living systems to endure intracellular crystallization (in Russian). In *Reaktsiya Kletok i ikh Belkovykh Komponentov na Ekstremalnye Vozdeistviya*, Nauka, Moscow, Leningrad, 33–50.

651 *— 1969. Behaviour of ciliates in imitated Martian conditions. *Proc. 3rd Int. Conf. Protozool., Leningrad* 181.

652 — & Alexandrov, S. N. 1959. On radioresistance of paramecia from radioactive springs (in Russian). *TSitologiya* **1**, 64–70.

653 — & Zaar, E. I. 1963. Photoreactivation of *Paramecium caudatum* in the long wave region of visible light (in Russian). In *Reaktsiya Kletok i ikh Belkovykh Komponentov na Ekstremalnye Vozdeistviya*, Akad. Nauk USSR, Moscow, Leningrad, 158–168.

654 Lukanin, B. S. 1966. Stimulation of the division rate in *Paramecium caudatum* with a small dose of UV-light (in Russian). In *Produtsirovanie i Krugovorot Organicheskogo Vestshestva vo Vnutrennikh Vodojomakh*, Nauka, Moscow, Leningrad, 241–245.

655 Machemer, H. 1969. Regulation der Cilienmetachronie bei der "Fluchtreaktion" in *Paramezium*. *J. Protozool.* **16**, 764–771.

656 Mcindoe, H. M. & Reissner, A. H. 1967. Adsorption titration as a specific semi-quantitative assay for soluble and bound *Paramecium* serotypic antigen. *Australian J. Biol. Sci.* **20**, 284–305.

657 McManus, M. A. & Sullivan, K. 1961. Effects of kineties and indole-3-acetic acid on multiplication rates of *Paramecium*. *Nature* **191**, 619–620.

658 Maguire, Jr., B. & Belk, D. 1967. *Paramecium* transport by land snails. *J. Protozool.* **14**, 445–447.

659 Maly, R. 1958. Eine genetisch bedingte Störung der Zelltrennung beim *Paramecium aurelia* var. 4. Ein Beitrag zum Problem der Mutabilität plasmatischer Systeme. *Z. Vererbungslehre* **89**, 397–421.

660 — 1960. Die Normalisierung genetisch bedingter Defekte der Zelltrennung bei *Paramecium aurelia* durch Sauerstoffmangel und Kohlenmonoxyd. *Z. Vererbungslehre* **91**, 226–332.

661 — 1960. Die Wirkung eines Komplexbildners und von Metallionen auf die Ausprägung des snaky- und monstra-Charakters bei *Paramecium aurelia*. *Z. Vererbungslehre* **91**, 333–337.

662 — 1961. Die Aufhebung eines Defektes der Zelltrennungen beim snaky-Stamm von *Paramecium aurelia* in Salzlösungen. *Z. Vererbungslehre* **92**, 462–464.

663 — 1962. Die Analyse eines Teilprozesses der Zytokinese mit Hilfe einer Mutante von *Paramecium aurelia*. *Zool. Anz.* **26** (Suppl.), 84–86.

664 Margolin, P. 1954. A method for obtaining amacronucleated animals in *Paramecium aurelia*. *J. Protozool.* **1**, 174–177.

665 — 1956. An exception to mutual exclusion of the ciliary antigens in *Paramecium aurelia*. *Genetics* **41**, 685–699.

666 — 1956 The ciliary antigens of stock 172, *Paramecium aurelia*, variety 4. *J. Exp. Zool.* **133**, 345–387.

667 Martz, E. 1966. A centrifuge for rapid concentration of large fragile cells without extensive lysis. *J. Protozool.* **13**, 380–382.

668   Mashansky, V. F. 1961. The influence of temperature on the mitochondria ultrastructures of
       Paramecium caudatum (in Russian). *TSitologiya* **3**, 586–589.

669   — 1962. On the mechanism of the changes in ultra-thin organization infusorian mitochondria
       under the action of injurious factors (in Russian). *TSitologiya* **4**, 445–449.

670   — 1963. Electron microscopic study of the micronuclei of certain ciliates (in Russian). In
       Morfologiya i Fiziologiya Prosteishikh. Akad. Nauk, SSSR, Moscow, Leningrad, 3–8.

671   — & Beznossikov, B. O. 1961. Method of preparation of ultrathin sections of cell suspensions
       (in Russian). *TSitologiya* **3**, 117–119.

672   — & Samoilova, K. A. 1964. Effect of UV-radiation on the ultrastructure of the cell (in
       Russian). *TSitologiya* **6**, 59–65.

673   Matsusaka, T. 1969. Influence of extracellular freezing on the ultrastructure of *Paramecium*.
       Low Temp. Sci., Ser. B. **27**, 67–72.

674   Metz, C. B. 1954. Mating substances and the physiology of fertilization in ciliates. In Wenrich,
       D. H., *Sex in Micro-organisms* (Symposium of the American Association for the Advancement
       of Science, Washington, D.C.), 284–334.

675   — & Butterfield, W. 1950. Extraction of a mating type reaction inhibiting agent from *Para-
       mecium calkinsi*. *Proc. Nat. Acad. Sci. U.S.* **36**, 268–271.

676   — & — 1951. Action of various enzymes on the mating type substances of *Paramecium calkinsi*.
       Biol. Bull. **101**, 99–105.

677   —, Pitelka, D. R. & Westfall, J. A. 1953. The fibrillar systems of ciliates as revealed by the
       electron microscope. I. *Paramecium*. *Biol. Bull.* **104**, 408–425.

678   Miller, C. A. & Johnson, W. H. 1957. A purine and pyrimidine requirement for *Paramecium
       multimicronucleatum*. *J. Protozool.* **4**, 200–204.

679   — & — 1960. Nutrition of *Paramecium*. A fatty acid requirement. *J. Protozool.* **7**, 297–301.

680   — & Van Wagtendonk, W. J. 1956. The essential metabolites of a strain of *Paramecium
       aurelia* (Stock 47.8) and a comparison of the growth rates of different strains of *Paramecium
       aurelia* in axenic culture. *J. Gen. Microbiol.* **15**, 280–291.

681   Miller, D. M., Jahn, T. L. & Fonseca, F. R. 1968. Anodal contraction of *Paramecium* proto-
       plasm. *J. Protozool.* **15**, 493–497.

682   Mirsky, A. F. & Katz, M. S. 1958. "Avoidance conditioning" in *Paramecium*. *Science* **127**,
       1498–1499.

683   *Mitchell, Jr., J. B. 1962. Nuclear reorganization in *Paramecium jenningsi*. *J. Protozool.* **9**
       (Suppl.), 26.

684   *— 1963. Nuclear activity in *Paramecium jenningsi* with reference to other members of the
       aurelia group. *J. Protozool.* **10** (Suppl.), 11.

685   Mitchison, N. A. 1955. Evidence against micronuclear mutations as the sole basis for death at
       fertilization in aged, and in the progeny of ultra-violet irradiated, *Paramecium aurelia*. *Genetics*
       40, 61–75.

686   Miyake, A. 1955. The effect of urea on binary fission in *Paramecium caudatum*. *J. Inst. Poly-
       tech. Osaka City Univ., Ser. D* **6**, 43–53.

687   *— 1956. Physiological analysis of the life cycle of the Protozoa. III. Artificial induction of
       selfing conjugation by chemical agents in *P. caudatum*. *J. Inst. Polytech. Osaka City Univ.,
       Ser. D* **7**, 14.

688   — 1956. Physiological analysis of the life cycle of the Protozoa. III. Artificial induction of
       selfing conjugation by chemical agents in *Paramecium caudatum*. *Physiol. Ecol.* **7**, 14–23.

689   — 1957. Aberrant conjugation induced by chemical agents in amicronucleate *Paramecium
       caudatum*. *J. Inst. Polytech. Osaka City Univ., Ser. D* **8**, 1–10.

690   — 1957. Induction of conjugation by chemical agents in *Paramecium caudatum*. *J. Inst.
       Polytech. Osaka City Univ., Ser. D* **9**, 251–256.

691   — 1958. Induction of conjugation by chemical agents in *P. caudatum*. *J. Inst. Polytech. Osaka
       City Univ., Ser. D* **10**, 251–256.

692   — 1959. Chemically induced mating without mating type differences in *Paramecium caudatum*.
       Science **130**, 1423.

693   *— 1960. Artificial induction of conjugation by chemical agents in *P. aurelia*, *P. multimicro-
       nucleatum*, *P. caudatum* and between them. *J. Protozool.* **7** (Suppl), 15.

694   — 1961. Artificial induction of conjugation by chemical agents in *Paramecium* of the "aurelia
       group" and some of its applications to genetic work. *Amer. Zoologist* **1**, 373–374.

695 — 1964. Induction of conjugation by cell-free preparations in *Paramecium multimicronucleatum*. *Science* **146**, 1583–1585.

696 — 1968. Induction of conjugation by chemical agents in *Paramecium*. *J. Exp. Zool*. **167**, 359–380.

697 — 1969. Mechanism of initiation of sexual reproduction in *Paramecium multimicronucleatum*. *Japan. J. Genet*. **44** (Suppl.), 388–395.

698 Moldenhauer, D. von, 1964. Zytologische Untersuchungen zum Austausch der Wanderkerne bei konjugierenden *Paramecium caudatum*. *Arch. Protistenk*. **107**, 163–178.

699 Mordukhaj-Boltovskaja, E. D. & Sorokin, Jr., I. 1965. The feeding of paramecia on algae and bacteria (in Russian). *Trudy Instituta biologii vnutrennykh vod* **8**, 12–14.

700 *Mott, M. R. 1963. Identification of the sites of the antigens of *Paramecium aurelia* by means of electron microscopy. *J. Protozool*. **10** (Suppl.), 31.

701 — 1963. Cytochemical localization of antigens of *Paramecium* by ferritin-conjugated antibody and by counterstaining the resultant absorbed globulin. *J. Roy. Microscop. Soc*. **81**, 159–162.

702 *— 1965. Electron microscopy of the immobilization antigens of *Paramecium aurelia*. *Proc. 2nd Int. Conf. Protozool., London* 250.

703 — 1965. Electron microscopy studies on the immobilization antigens of *Paramecium aurelia*. *J. Gen. Microbiol*. **41**, 251–261.

704 Mottram, J. C. 1941. Abnormal paramecia produced by blastogenic agents and their bearing on the cancer problem. *Cancer Res*. **1**, 313–323.

705 *Mueller, J. A. 1961. Further studies on the nature of the "co-factor" required for kappa infection in *Paramecium aurelia*, syngen 4. *Amer. Zoologist* **1**, 375.

706 *— 1962. Induced physiological and morphological changes in the B-particle and R-body from killer *Paramecia*. *J. Protozool*. **9** (Suppl.), 26.

707 — 1963. Separation of kappa particles with infective activity and identification of the infective particles in *Paramecium aurelia*. *Exp. Cell Res*. **30**, 492–508.

708 — 1964. *Paramecia* develop immunity against kappa. *Amer. Zoologist* **4**, 313–314.

709 — 1965. Kappa-affected paramecia develop immunity. *J. Protozool*. **12**, 278–281.

710 — 1965. Vitally stained kappa in *Paramecium aurelia*. *J. Exp. Zool*. **160**, 369–372.

711 — 1966. Resistance of kappa-bearing paramecia to kappa toxin. *Exp. Cell Res*. **41**, 131–137.

712 *— 1967. *Paramecia* reveal basic cellular physiology. *J. Indiana Med. Assoc*. **60**, 1348.

713 — 1968. Poisonous effects of *Pseudomonas aeruginosa* on *Paramecium aurelia*. *J. Invertebr. Path*. **11**, 219–223.

714 Mugard, H. & Renaud, P. 1960. Etude sur l'effet des ultrasons sur un infusoire cilié *Paramecium caudatum*. *Arch. Biol*. **71**, 73–91.

715 Müller, M. 1962. Studies on feeding and digestion in protozoa. V. Demonstration of some phosphatases and carboxylic esterases in *Paramecium multimicronucleatum* by histochemical methods. *Acta Biol. Acad. Sci. Hung*. **13**, 283–297.

716 *— 1962. Demonstration of hydrolases in *Paramecium multimicronucleatum* by histochemical methods. *J. Protozool*. **9** (Suppl.), 26.

717 *— & Törö, I. 1961. Acid phosphatase activity in *Paramecium multimicronucleatum*. *Proc. 1st Int. Conf. Protozool., Prague* 142.

718 — & — 1962. Studies on feeding and digestion in protozoa. III. Acid phosphatase activity in food vacuoles of *Paramecium multimicronucleatum*. *J. Protozool*. **9**, 98–102.

719 *— & Toth, E. 1959. Effect of acetylcholine and eserine on the ciliary reversal in *Paramecium multimicronucleatum*. *J. Protozool*. **6** (Suppl.), 28.

720 —, Törö, I., Polgar, M. & Druga, A. 1963. Studies on feeding and digestion in protozoa. VI. The effect of ingestion of non-nutritive particles on acid phosphatase in *Paramecium multi-micronucleatum*. *Acta Biol. Acad. Sci. Hung*. **14**, 209–213.

721 Muscatine, L., Karakashian, S. J. & Karakashian, M. W. 1967. Soluble extracellular products of algae symbiontic with a ciliate, a sponge and a mutant hydra. *Comp. Biochem. Physiol*. **20**, 1–12.

722 Nagai, T. 1956. Elasticity and contraction of *Paramecium* ectoplasm (in Russian). *TSitologiya* **21**, 65–75.

723 — 1960. Contraction of *Paramecium* by transmembrane electrical stimulation. *J. Fac. Sci. Imp. Univ. Tokyo, Sect. IV* **8**, 617–631.

724 — 1960. Local contraction of *Paramecium* ectoplasm by $CaCl_2$ stimulation. *J. Fac. Sci. Imp. Univ. Tokyo, Sect. IV* **8**, 633–641.

725   Nagata, C., Fujii, K. & Epstein, S. S. 1967. Photodynamic activity of 4-nitroquinoline-1-oxide and related compounds. *Nature* **215**, 972–973.

726   Naitoh, Y. 1964. Ciliary responses of *Paramecium* to the external application of various chemicals under different ionic conditions (in Japanese; English abstr.). *Dobutsugaku Zasschi* **73**, 207–212.

727   — 1966. Reversal response elicited in nonbeating cilia of *Paramecium* by membrane depolarization. *Science* **154**, 660–662.

728   — 1968. Ionic control of the reversal response of cilia in *Paramecium caudatum*. A calcium hypothesis. *J. Cell. Comp. Physiol.* **51**, 85–103.

729   — 1969. Control of the orientation of cilia by adenosinetriphosphate, calcium, and zinc in glycerol extracted *Paramecium caudatum*. *J. Gen. Physiol.* **53**, 517–529.

730   — & Eckert, R. 1968. Electrical properties of *Paramecium caudatum:* modification by bound and free cations. *Z. Vergleich. Physiol.* **61**, 427–452.

731   — & — 1968. Electrical properties of *Paramecium caudatum:* all-or-none electrogenesis. *Z. Vergleich. Physiol.* **61**, 453–472.

732   — & — 1969. Ionic mechanisms controlling behavioral responses of *Paramecium* to mechanical stimulation. *Science* **164**, 963–965.

733   — & Eckert, R. 1971. Electrophysiology of the ciliate protozoa. In Kerkut, G. A., *Experiments in Physiology and Biochemistry*, Academic Press, London, **5**, 17–38.

734   — & Kaneko, H. 1972. Reactivated triton-extracted models of *Paramecium*. Modification of ciliary movement by calcium ions. *Science* **176**, 523–524.

735   — & Yasumasum, I. 1969. Binding of Ca ions by *Paramecium caudatum*. *J. Gen. Physiol.* **50**, 1303–1310.

736   Nanney, D. L. 1953. Mating type determination in *Paramecium aurelia*, a model of nucleo-cytoplasmic interaction. *Proc. Nat. Acad. Sci. U.S.* **39**, 113–119.

737   — 1954. Mating type determination in *Paramecium aurelia*, a study in cellular heredity. In Wenrich, D. H., *Sex in Micro-organisms*, American Association for the Advancement of Science, Washington, D. C., 266–283.

738   — 1954. X-ray studies on paramecins and kappas of variety 4 of *Paramecium aurelia*. *Physiol. Zool.* **27**, 79–89.

739   — 1956. Caryonidal inheritance and unclear differentiation. *Amer. Naturalist* **90**, 291–307.

740   — 1957. Mating type inheritance at conjugation in variety 4 of *Paramecium aurelia*. *J. Protozool.* **4**, 89–95.

741   Natochin, Ju. V. & Seravin, L. N. 1962. Localization of the dehydrogenases of the Kreb's cycle in protozoa (in Russian). *Dokl. Akad. Nauk SSSR* **143**, 1229–1232.

742   Nickel, E. 1961. Änderung von Morphologie und Stoffwechsel von *Paramecium caudatum* nach Röntgenbestrahlung. *Z. Naturforsch.* **16b**, 538–543.

743   Nicolojeva, M. V. & Pilshchik, E. M. 1963. Fluorescence microscopic studies of living *Paramecium caudatum* under normal conditons and under starvation (in Russian). In *Morfologiya i Fiziologiya Prosteishkikh*, Akad. Nauk SSSR, Moscow, Leningrad, 54–61.

744   *Nobili, R. 1959. The effects of aging and temperature on the expression of the gene am in variety 4, stock 51, of *Paramecium aurelia*. *J. Protozool.* **6** (Suppl.), 29.

745   *— 1960. The effect of macronuclear regeneration on vitality in *Paramecium aurelia*, syngen 4. *J. Protozool.* **7** (Suppl.), 15.

746   *— 1960. Kappa in amacronucleate *Paramecium aurelia*. Its development and effect on survival time. *J. Protozool.* **7** (Suppl.), 15.

747   — 1961. L'azione del gene am sull'apparato nucleare di *Paramecium aurelia* durante la riproduzione vegetativa e sessuale in relazione all'età del clone ed alla temperature di allevamento degli animali. *Caryologia* **14**, 43–58.

748   — 1961. Variazioni volumetriche del macronucleo e loro effetti nella riproduzione vegetativa in *Paramecium aurelia*. *Atti. Soc. Toscane Sci. Nat. Pisa, Ser. B* **67**, 217–232.

749   — 1961. Effetti della rigenerazione del macronucleo sulla vitalita di *Paramecium aurelia*, syngen 4. *Atti. Assoc. Genetica Italiana* **6**, 75–86.

750   — 1961. Alcune considerazioni sulla liberazione e sull'azione delle particelle kappa in individui amacronucleati di *Paramecium aurelia*, stock 51, syngen 4. *Atti. Soc. Toscane Sci. Nat. Pisa, Ser. B* 158–172.

751   — 1961. Il comportamento del kappa in esemplari amacronucleati di *Paramecium aurelia*, stock 51, syngen 4. *Boll. Zool. Agrar. Bachicolt Univ. Studi Milano* **28**, 579–596.

752 — 1962. Su un nuovo ceppo "killer" ad azione paralizzante di *Paramecium aurelia*, syngen 2. *Boll. Zool. Agrar. Bachicolt Univ. Studi Milano* **29**, 555–565.

753 *— 1963. Effects of antibiotics, base- and amino acid-analogues on mating reactivity of *Paramecium aurelia*. *J. Protozool.* **10** (Suppl.), 24.

754 — & Agostini, G. 1964. Coniugazione e reproduzione vegetativa di *Paramecium aurelia* sotto l'azione del 6-azauracile e della p-fluorofenilalanina. *Atti Ass. Genet. It., Pavia* **9**, 72–86.

755 — & De Angelis, F. K. 1963. Effetti degli antibiotici sulla riproduzione di *Paramecium aurelia*. *Atti Ass. Genet. It., Pavia* **8**, 45–57.

756 Ochinskaya, G. K. 1959. The effect of different kinds of ionizing irradiations on *Paramecium caudatum* (in Russian). *TSitologiya* **1**, 393–402.

757 — & Bychkovskaya, I. B. 1965. On the existence of a system in the response of biological objects to irradiation (analysis of data obtained on *Paramecium caudatum*) (in Russian). *Dokl. Akad. Nauk. SSSR* **160**, 461–463.

758 Oger, C. 1965. Analyse biométrique de la croissance de *Paramecium caudatum*. Intérêt pour l'étude de la conjugaison. *Protistologica* **1**, 71–80.

759 *— & Vivier, E. 1964. Observations d'ordre biométrique sur quelques syngènes français de *Paramecium caudatum*. Intérêt pour l'étude de leur sexualité. *J. Protozool.* **11** (Suppl.), 49.

760 — & — 1965. Observations d'ordre biométrique sur quelques variétés françaises de *Paramecium caudatum* Ehrb. Intérêt pour l'étude de leur sexualité. *Arch. Zool. Exp. Gen.* **105**, 119–153.

761 *Organ, A. E., Bovee, E. C. & Jahn, T. L. 1966. ATP-accelerated cycling of the nephridial apparatus or *Paramecium multimicronucleatum*. *J. Protozool.* **13** (Suppl.), 12.

762 —, — & — 1968. Adenosine triphosphate acceleration of the nephridial apparatus of *Paramecium multimicronucleatum*. *J. Protozool.* **15**, 173–176.

763 —, — & — 1969. The mechanism of the nephridial apparatus of *Paramecium multimicronucleatum*. II. The filling of the vesicle by action of the ampullae. *J. Cell Biol.* **40**, 389–394.

764 —, —, —, Wigg, D. & Fonseca, J. R. 1968. The mechanism of the nephridial apparatus of *Paramecium multimicronucleatum*. I. Expulsion of water from the vesicle. *J. Cell Biol.* **37**, 139–145.

765 Ossipov, D. V. 1963. Mating types of *Paramecium caudatum* found in reservoirs of the Soviet Union (in Russian). *Vestnik Leningradskogo Universiteta*, **8**, 106–116.

766 — 1966. On the macronuclear regeneration in *Paramecium caudatum* (in Russian). *TSitologiya* **8**, 108–110.

767 — 1966. Methods of obtaining homozygous *Paramecium caudatum* clones (in Russian). *Genetika* **2**, 41–48.

768 — 1966. The heat resistance of stocks of *Paramecium caudatum* obtained from different natural populations (in Russian). *Vestnik Leningradskogo Universiteta* **3**, 107–112.

769 — 1972. The regulatory mechanisms of sexual processes in ciliates. In *The Mechanisms of Regulatory Processes*, Leningrad University Press, Leningrad, 158–175.

770 — 1972. The peculiarities of the intercellular interaction during conjugation of Ciliates. *TSitologiya* **13**, 411–424.

771 — & Scoblo, I. I. 1968. The autogamy during conjugation in *Paramecium caudatum* Ehrb. II. The ex-autogamous stage of nuclear reorganization. *Acta Protozool. (Warsaw)* **6**, 33–48.

772 *— & Tavrovskaya, M. V. 1969. Micronucleus and morphogenesis of the cortex in *Paramecium caudatum*. *Proc. 3rd Int. Conf. Protozool., Leningrad* 102–103.

773 *Ovchinnikova, L. P. 1969. Interdivisional changes of RNA content in *Paramecium caudatum*. *Proc. 3rd Int. Conf. Protozool., Leningrad* 35–36.

774 — 1970. Variability of DNA content in micronuclei of *Paramecium bursaria*. *Acta Protozool. (Warsaw)* **7**, 211–220.

775 — 1971. The influence of conjugation on the content of DNA and RNA in the nuclei of paramecia. *Materials of 1st All-Union Congr. Protozool. Soc., Bacu*, 68–70.

776 —, Selivanova, G. V. & Cheissin, E. M. 1963. A UV-photometric study of the effect of starvation on RNA and DNA content in *Paramecium caudatum* (in Russian). In *Morfologiya Fiziologiya Prosteishikh.*, Akad. Nauk USSR, Moscow, Leningrad, 44–53.

777 Ozhigova, A. P. & Ozhigov, I. E. 1967. Constant magnetic field effect on paramecia movement (in Russian). *Biofizika* **11**, 1026–1033.

778 Pace, D. M. & Hoagland, R. A. 1954. The effects of ethanol on conjugation and division in *Paramecium caudatum*. *J. Protozool.* **1**, 83–85.

779   Pado, R. 1965. Mutual relation of protozoans and symbiotic algae in *Paramecium bursaria*. I. The influence of light on the growth of symbionts. *Folia Biol. (Krakow)* **13**, 173–182.

780   Párducz, B. 1953. Zur Mechanik der Zilienbewegung. *Acta Biol. Acad. Sci. Hung.* **4**, 177–220.

781   — 1954. Reizphysiologische Untersuchungen an Ziliaten. II. Neuere Beiträge zum Bewegungs- und Koordinationsmechanismus der Ziliatur. *Acta Biol. Acad. Sci. Hung.* **5**, 169–212.

782   — 1956. Reizphysiologische Untersuchungen an Ciliaten. IV. Über das Empfindungs- bzw. Reaktionsvermögen von *Paramecium. Acta Biol. Acad. Sci. Hung.* **6**, 289–316.

783   — 1956. Reizphysiologische Untersuchungen an Ziliaten. V. Zum physiologischen Mechanismus der sogenannten Fluchtreaktion und der Raumorientierung. *Acta Biol. Acad. Sci. Hung.* **7**, 73–99.

784   — 1957. Ueber den feineren Bau des Neuronensystems der Ziliaten. *Ann. Hist. Nat. Mus. Nat. Hung.* **8**, 231–246.

785   — 1958. Reizphysiologische Untersuchungen an Ziliaten. VII. Das Problem der vorbestimmten Leitungsbahnen. *Acta Biol. Acad. Sci. Hung.* **8**, 219–251.

786   — 1959. Reizphysiologische Untersuchungen an Ziliaten. Anlauf der Fluchtreaktion bei allseitiger und anhaltender Reizung. *Ann. Hist. Nat. Mus. Nat. Hung.* **51**, 227–246.

787   *— 1962. Inversed spiral movement of *Paramecium multimicronucleatum* evoked by nickel salts. *J. Protozool.* **9** (Suppl.), 27.

788   *— 1962. On a new concept of cortical organization in *Paramecium. Acta Biol. Acad. Sci. Hung.* **13**, 299–322.

789   — 1962. Studies on reactions to stimuli in ciliates. IX. Ciliary coordination of right-spiralling paramecia. *Ann. Hist. Nat. Mus. Nat. Hung.* **54**, 221–230.

790   *— 1962. On the nature of metachronal ciliary control in *Paramecium. J. Protozool.* **9** (Suppl.), 27.

791   — 1963. Reizphysiologische Untersuchungen an Ziliaten. "Momentbilder" über galvano-taktisch frei schwimmenden *Paramecien. Acta Biol. Acad. Sci. Hung.* **13**, 421–429.

792   — 1967. Ciliary movement and coordination in ciliates. *Intern. Rev. Cytol.* **21**, 91–128.

793   Pasternak, J. 1967. Differential genic activity in *Paramecium aurelia. J. Exp. Zool.* **165**, 395–418.

794   *Peruzzotti, G. P. & Liberti, A. V. 1959. Further studies on the nutrition of *Paramecium polycaryum* under bacteria-free conditions. *J. Protozool.* **6** (Suppl.), 29.

795   Petroff, C. 1962. On the relation between *Paramecium* and plant growth substances and herbicides (in German). *Z. Naturforsch.* **17B**, 17–19.

796   Pieri, J., Vaugien, C. & Trouillier, M. 1968. Interprétations cytophotométrique des phénomènes micronucléaires au cours de la division binaire et des divisions prégamiques chez *Paramecium trichium. J. Cell. Biol.* **36**, 664–668.

797   Pigon, A. & Szarski, H. 1955. The velocity of the cilia and the force of the ciliary beat in *Paramecium caudatum. Bull. Acad. Polon. Sci., Sér. Sci. Biol.* **2 3**, 281–291.

798   *Pitelka, D. R. 1962. Observations on normal and abnormal cilia in *Paramecium. In Proc. 5th Intern. Congr. Electron Microscopy.* Academic Press, New York, N.Y.

799   *— 1963. Fine structure of the buccal apparatus in *Paramecium. In Proc. 16th Int. Congr. Zool., Washington*, Ch. 2, 293.

800   *— 1964. The morphology of the cytostomal area in *Paramecium. J. Protozool.* **11** (Suppl.), 14–15.

801   — 1965. New observations on cortical ultrastructure in *Paramecium. J. Microscopie* **4**, 373–394.

802   — 1968. Fibrillar systems in protozoa. In Chen, T. T., *Research in Protozoology*, Pergamon Press, Oxford, New York, N.Y., **3**, 279–388.

803   *— 1969. Fibrillar structures of the ciliate cortex: the organization of kinetosomal territories. *Proc. 3rd Int. Conf. Protozool., Leningrad* 44–45.

804   — 1970. Ciliate ultrastructure. Some problems in cell biology. *J. Protozool.* **17**, 1–10.

805   — & Child, F. M. 1964. The locomotor apparatus of Ciliates and Flagellates. Relations between structure and function. In Hutner, S. H., *Biochemistry and Physiology of Protozoa*, Academic Press, N.Y., Ch. 3, 131–198.

806   *— & Párducz, B. 1962. Electron-microscope observations on paralyzed *Paramecium. J. Protozool.* **9** (Suppl.), 6.

807   Planel, H., Soleilhavoup, J. P. & Blanc, D. 1966. Action du rayonnement gamma du thorium sur la croissance de *Paramecium caudatum* et de *Paramecium aurelia* sous dispositif de radio-protection. *Compt. Rend. Soc. Biol.* **160**, 1090–1093.

808 —, — & Cotlin, F. 1969. Recherches sur l'action biologicale de l'irradiation naturelle. Etude de la croissance de *Paramecium aurelia* et *Paramecium caudatum* en laboratoire sousterrain. *Compt. Rend.* **264**, 1697–1700.

809 *—, Tixador, R. & Richoilley, G. 1970. Etude de l'irradiation naturelle sur la durée du cycle cellulaire chez *Paramecium aurelia*. *J. Protozool.* **17** (Suppl.), 36.

810 —, Bru, A. & Soleilhavoup, J. P. 1967. Effets de très faibles doses de radiations ionisantes sur la multiplication de *Paramecium aurelia*. *Compt. Rend.* **264**, 2945–2948.

811 —, — & Caratero, C. 1968. Influence of the environmental radioactivity in the study of cell growth (in French). *Compt. Rend. Soc. Biol.* **162**, 1215–1219.

812 —, — & Tixador, R. 1968. Influence of protection against natural irradiation in post-autogamous *Paramecium aurelia* (in French). *Compt. Rend. Soc. Biol.* **162**, 990–995.

813 Pocztarska-Wegrzyn, I. 1963. Effect of histamine, phenergan and largactil on cultures of *Paramecium caudatum* Ehrb. (Antihistamine effect of phenergan) (in Polish). *Acta Physiol. Pol.* **14**, 657–668.

814 Poglazova, M. N., Selivestrova, L. A., Petrikevich, S. B. & Meissel, M. N. 1968. Assay of benz(a)pyrene in yeast with the aid of paramecia (in Russian). *Prikladnaya Biokhimiya i Mikrobiologia* **4**, 431–436.

815 Poljansky, G. 1957. Temperature adaptation in infusoria. I. Relation of heat resistance of *Paramecium caudatum* to the temperature conditions of existence (in Russian). *Zoologicheskij Zhurnal* **36**, 1630–1646.

816 *— 1959. Experimental investigations of temperature adaptation of infusoria. *Proc. 15th Int. Congr. Zoology*, London, 716–718.

817 — 1959. Temperature adaptation in infusoria. II. Changes in resistance to high and low temperatures in *Paramecium caudatum* at low temperature cultivation (in Russian). *TSitologiya* **1**, 714–727.

818 — 1963. On the capacity of *Paramecium caudatum* to stand the sub-zero temperature (in Russian). *Acta Protozool. (Warsaw)* **1**, 165–175.

819 — 1963. The dependence of glycogen and lipid content in the cytoplasm of *Paramecium caudatum* on the temperature (in Russian). In *Morfologiya i Fiziologiya Prosteishikh*, Akad. Nauk USSR, Moscow, Leningrad, 102–110.

820 — & Irlina, I. S. 1967. On "thermal hardening" in ciliates. *TSitologiya* **9**, 791–799.

821 *—, Grigorjan, D. A. & Irlina, I. S. 1965. A cytophysiological study of temperature adaptations in some free-living ciliates. *Proc. 2nd Int. Conf. Protozool., London* 77.

822 *Pollack, S. 1969. Trichocyst mutants in *Paramecium aurelia*. *Amer. Zool.* **9**, 1136–1137.

823 Popov, A. S. 1961. Influence of urea peroxide on the radiation injuries of infusoria (in Russian). *Radiobiologiya* **1**, 676–683.

824 — 1962. Oxidation of thymine and valine and radiation injury of paramecia (in Russian). *Radiobiologiya* **2**, 811–818.

825 — 1965. Combined action of H_2O_2 and nucleic acid compounds of infusoria (in Russian). *Radiobiologiya* **5**, 652–655.

826 — 1965. Comparative toxicity of some organic peroxides for infusoria (in Russian). *Radiobiologiya* **5**, 812–814.

827 — 1966. The combined effect of hydrogen peroxide and amino acids on paramecia (in Russian). *Radiobiologiya* **6**, 20–22.

828 — & Vinetskii, In. P. 1967. The effect of irradiated glucose solutions on phages and infusoria (in Russian). *Radiobiologiya* **7**, 630–631.

829 *Porter, E. D. 1956. Observations on stomatogenesis in *Paramecium aurelia* during fission and conjugation. *J. Protozool.* **3** (Suppl.), 9.

830 — 1960. The buccal organelles in *Paramecium aurelia* during fission and conjugation with special reference to the kinetosomes. *J. Protozool.* **7**, 211–217.

831 *— 1961. Observations on the effect of mechanical pressure on *Paramecium caudatum*. *Bull. Georgia Acad. Sci.* **19**, 5.

832 — 1962. Morphogenetic migration of the buccal cavity of *Paramecium aurelia*. *J. Protozool.* **7**, 211–217.

833 *Powelson, E. E. 1956. Differences in the silver-line system and various measurements in individuals, stocks and varieties of *Paramecium aurelia*. *J. Protozool.* **3** (Suppl.), 9.

834 Powers, E. L. 1955. Radiation effects in *Paramecium*. *Ann. N.Y. Acad. Sci.* **59**, 619–636.

835   *—, Ehret, C. F. & Roth, L. E. 1954. Morphology of the mitochondrion and its relationship to other structures in *Paramecium*. *J. Protozool.* **1** (Suppl.), 5.

836   —, — & — 1955. Mitochondrial structure in *Paramecium* as revealed by electron microscopy. *Biol. Bull.* **108**, 182–195.

837   —, —, — & Minick, O. T. 1956. The internal organization of mitochondria. *J. Biochem. Cytol.* **2** (Suppl), 341–346.

838   Preer, Jr., J. R. 1957. Genetics of the Protozoa. *Ann. Rev. Microbiol.* **11**, 419–438.

839   — 1957. A gene determining temperature sensitivity in *Paramecium*. *J. Genetics* **55**, 375–378.

840   *— 1958. Isolation of the immobilization antigens of *Paramecium*. *Anat. Rec.* **131**, 591.

841   — 1959. Nuclear and cytoplasmic differentiation in the Protozoa. In Rudnick, D. *Developmental Cytology*, Ronald Press, New York, N.Y., 3–20.

842   — 1959. Studies on the immobilization antigens of *Paramecium*. I. Assay methods. *J. Immunol.* **83**, 276–283.

843   — 1959. Studies on the immobilization antigens of *Paramecium*. II. Isolation. *J. Immunol.* **83**, 378–384.

844   — 1959. Studies on the immobilization antigens of *Paramecium*. III. Properties. *J. Immunol.* **83**, 385–391.

845   — 1959. Studies on the immobilization antigens of *Paramecium*. IV. Properties of the different antigens. *Genetics* **44**, 803–814.

846   — 1969. Genetics of the Protozoa. In Chen, T. T., *Research in Protozoology*, Pergamon Press, Oxford, Ch. 3, 129–278.

847   *— 1969. Self-replicating cytoplasmic particles in the Protozoa. *Proc. 3rd Int. Conf. Protozool.*, *Leningrad* 110.

848   — 1971. Extrachromosomal inheritance. Hereditary symbionts, mitochondria, chloroplasts. *Ann. Rev. Genetics* **5**, 361–406.

849   — & Jurand, A. 1968. The relation between virus-like particles and R-bodies of *Paramecium aurelia*. *Genet. Res.* **12**, 331–340.

850   — & Preer, L. B. 1959. Gel diffusion studies on the antigens of isolated cellular components of *Paramecium*. *J. Protozool.* **6**, 88–100.

851   — & — 1967. Virus-like bodies in killer *Paramecia*. *Proc. Nat. Acad. Sci. U.S.* **58**, 1774–1781.

852   — & Stark, P. 1953. Cytological observations on the cytoplasmic factor "kappa" in *Paramecium aurelia*. *Exp. Cell Res.* **5**, 478–491.

853   *—, Bray, M. & Koizumi, S. 1963. The role of cytoplasm and nucleus in the determination of serotype in *Paramecium*. *Proc. 11th Int. Congr. Genet.*, *The Hague* **1**, 189.

854   —, Hufnagel, L. A. & Preer, L. B. 1966. Structure and behavior of "R" bodies from killer paramecia. *J. Ultrastruct. Res.* **15**, 131–143.

855   —, Siegel, R. W. & Stark, P. S. 1953. The relationship between kappa and paramecin in *Paramecium aurelia*. *Proc. Nat. Acad. Sci. U.S.* **39**, 1228–1233.

856   —, Preer, L. B., Rudman, B. M. & Jurand, A. 1971. Isolation and composition of bacteriophage-like particles from kappa of killer paramecia. *Mol. Gen. Genet.* **111**, 202–208.

857   Preer, L. B. 1969. Alpha, an infectious macronuclear symbiont of *Paramecium aurelia*. *J. Protozool.* **16**, 570–578.

858   *— 1969. A study of killer stock 562 of *Paramecium aurelia*. Killing effects of kappa and ultrastructure of the R-body. *Proc. 3rd Int. Conf. Protozool.*, *Leningrad* 118.

859   — & Preer, Jr., J. R. 1964. Killing activity from lysed particles of *Paramecium*. *Genet. Res.* **5**, 230–239.

860   Presman, A. S. 1963. Excitability in paramecia on the stimulation by the pulses of continuous and alternating current (in Russian). *Biofizika* **8**, 138–140.

861   — 1963. Microwave effect on paramecia (in Russian). *Biofizika* **8**, 258–260.

862   — & Rappeport, S. M. 1964. New data on the existence of an excitability system in *Paramecium caudatum* Ehrenberg. I.P. I. Response to direct current impulses (in Russian). *Nauchnye dokl. vysshei shkoly, Biologicheskie nauki* **1**, 52–55.

863   — & — 1964. New data on the existence of an excitability system in *Paramecium caudatum* Ehrenberg. I.P. II. *P. caudatum* response to the pulses of alternating current (in Russian). *Nauchnye dokl. vysshei shkoly, Biologicheskie nauki* **3**, 44–48.

864   — & — 1965. The effect of microwaves on the excitable system of *Paramecia* (in Russian). *Bulleten eksperimental noi biologii i meditsiny* **59**, 48–52.

865   Pringle, C. R. 1956. Antigenic variation in *Paramecium*, variety 9. *Z. Indukt. Abstamm. Vererb.* **87**, 421–440.

866   — & Beale, G. H. 1960. Antigenic polymorphism in a wild population of *Paramecium aurelia*. *Genet. Res.* **1**, 62–68.

867   — & Stewart, J. M. 1961. Rate of respiration in relation to autogamy in *Paramecium aurelia*. *Experientia* **17**, 73–75.

868   *Propper, A. 1964. Différences de sensibilité à la température chez 2 variétés de *Paramecium caudatum*. *J. Protozool.* **11** (Suppl.), 49.

869   Przybos, E. 1965. The occurrence of *Paramecium aurelia* syngens on the island Wolin. *Folia Biol. (Krakow)* **13**, 371–382.

870   —, Kościiuszko, H. & Komala, Z. 1967. The occurrence of *Paramecium aurelia* syngens in a natural water reservoir in different seasons of the year. *Folia Biol. (Krakow)* **15**, 399–404.

871   Puytorac, P. de & Blanc, J. 1967. Observations sur les modifications ultrastructurales des micronoyaux au cours de leur transformation en macronoyaux chez *Paramecium caudatum*. *Compt. Rend. Soc. Biol.* **161**, 297–299.

872   —, Andrivon, C. & Serre, F. 1963. Sur l'action cytonarcotique de sels de nickel chez *Paramecium caudatum* Ehrbg. *J. Protozool.* **10**, 10–19.

873   *Rafalko, M. & Sonneborn, T. M. 1959. A new syngen (13) of *Paramecium aurelia* consisting of stocks from Mexico, France and Madagascar. *J. Protozool.* **6** (Suppl.), 30.

874   Raikov, I. B. 1968. Macronucleus of ciliates. In Chen, T. T., *Research in Protozoology*, Pergamon Press, Oxford, New York, Ch. 3, 1–128.

875   —, Cheissin, E. M. & Buze, E. G. 1963. A photometric study of DNA content of macro- and micronuclei in *Paramecium caudatum*, *Nassula ornata* and *Loxodes magnus*. *Acta Protozool. (Warsaw)* **1**, 285–300.

876   Rao, M. V. N. & Ammermann, D. 1970. Polytene chromosomes and nucleic acid metabolism during macronuclear development in *Euplotes*. *Chromosoma* **29**, 246–254.

877   *— & Prescott, D. M. 1966. Micronuclear RNA synthesis in *Paramecium caudatum*. *J. Protozool.* **13** (Suppl.), 11.

878   — & — 1967. Micronuclear RNA synthesis in *Paramecium caudatum*. *J. Cell Biol.* **33**, 281–285.

879   Rasmussen, L. 1967. Effects of metabolic inhibitors on *Paramecium aurelia* during the cell generation cycle. *Exp. Cell Res.* **5**, 132–139.

880   — 1967. Effect of DL-fluorophenylalanine on *Paramecium aurelia* during the cell generation cycle. *Exp. Cell Res.* **45**, 501–504.

881   Rasumowsky, P. N. & Jakimowa, G. J. 1965. On the stimulant effect of bacterial metabolites and some chemical substances on *Paramecium caudatum* (in Russian). *Izvestiya Akademii Nauk Moldavskoy SSR* **10**, 30–36.

882   Ray, H. N., Hajra, B. & Chatterjee, S. N. 1962. A preliminary note on the effect of uranyl acetate on nucleic acids in *Paramecium multimicronucleatum*. *Bull. Calcutta School Trop. Med.* **10**, 77–80.

883   Reeve, E. C. R. & Ross, G. J. S. 1962. Mate-killer (mu) particles in *Paramecium aurelia:* the metagon division hypothesis. *Genet. Res.* **3**, 328–330.

884   — & — 1963. Mate-killer (mu) particles in *Paramecium aurelia:* further mathematical models for metagon distribution. *Genet. Res.* **4**, 158–161.

885   Regniak, L. & Komczynski, L. 1961. Behaviour of protozoa under the influence of the synergetic action of polyethylene glycol and dyes. *Nature* **192**, 1203–1204.

886   Reilly, M. 1964. Importance of adsorbents in the nutrition of *Paramecium caudatum*. *J. Protozool.* **11**, 109–113.

887   *— & Lilly, D. M. 1963. A chemically defined medium for *Paramecium caudatum*. *J. Protozool.* **10** (Suppl.), 12.

888   *Reinhard, L. V., Travleyev, A. P. & Bulik, I. K. 1969. Ecology of soil Protozoa in forests of Ukrainian steppe zone. *Proc. 3rd. Int. Conf. Protozool., Leningrad* 200.

889   Reisner, A. H. 1955. A method of obtaining specific serotype mutants in *Paramecium aurelia* stock 169, var. 4. *Genetics* **40**, 591–592.

890   — & Macindoe, H. 1967. Incorporation of amino acid into protein by utilizing a cell-free system from *Paramecium*. *J. Gen. Microbiol.* **47**, 1–15.

891   — & — 1967. Immunological evidence indicating conformational differences between surface-bound and solubilized serotypic antigen of *Paramecium*. *Biochim. Biophys. Acta* **140**, 529–531.

892 —, Rowe, J. & McIndoe, H. M. 1968. Structural studies on the ribosomes of *Paramecium:* evidence for a "primitive" animal ribosome. *J. Mol. Biol.* **32**, 587–610.

893 —, — & — 1969. The largest known monomeric globular proteins. *Biochim. Biophys. Acta* **188**, 196–206.

894 —, — & Sleigh, R. W. 1969. Concerning the tertiary structure of the soluble surface proteins of *Paramecium. Biochemistry* **8**, 4637–4644.

895 Ringo, D. 1967. The arrangement of subunits in flagellar fibers. *J. Ultrastruct. Res.* **17**, 266–277.

896 Roque, M. 1956. L'évolution de la ciliature buccale pendant l'autogamie et la conjugaison chez *Paramecium aurelia. Compt. Rend.* **242**, 2592–2595.

897 — 1956. La stomatogénèse pendant l'autogamie, la conjugaison et la division chez *Paramecium aurelia. Compt. Rend.* **243**, 1564–1565.

898 Rosenbaum, R. M. & Wittner, M. 1962. The activity of intracytoplasmic enzymes associated with feeding and digestion in *Paramecium caudatum.* The possible relationship to neutral red granules. *Arch. Protistenk.* **106**, 223–240.

899 —, — & Wertheimer, S. 1966. Regulation of cellular autolysis by hyperbaric oxygen. *Nature* **209**, 895–896.

900 Roskin, G. I. & Brodsky, V. Ja. 1953. Nucleic acids and nucleotides of protozoan cells (in Russian). *Dokl. Akad. Nauk SSSR* **89**, 1099–1102.

901 Roux, J. & Serre, A. 1962. Action of tetanus toxin on the protozoan cell. Study on *Paramecium aurelia. Ann. Inst. Pasteur* **103**, 623–625.

902 Rudenberg, F. H. 1962. Electron microscopic observations of kappa in *Paramecium aurelia. Texas Rep. Biol. Med.* **20**, 105–112.

903 *Rudyansky, B. 1953. The effect of breis made from sexually reactive animals on mating type determination in *Paramecium aurelia,* variety 2. *Microbial Genet. Bull.* **7**.

904 Saier, F. L. 1966. Action of ultraviolet radiation upon ciliary movement in *Paramecium. Exp. Cell Res.* **44**, 331–331.

905 — & Giese, A. C. 1967. The effect of ultraviolet radiation upon osmoregulation in *Paramecium. Photochem. Photobiol.* **6**, 745–755.

906 Saito, M. & Sato, H. 1961. Morphological studies on the macronuclear structure of *Paramecium caudatum.* II. Structural changes of the macronucleus during the division cycle (in Japanese; English summary). *Zool. Mag. (Jap.)* **70**, 73–80.

907 — & — 1961. Morphological studies on the macronuclear structure of *Paramecium caudatum.* III. On the development of the macronucleus in the exconjugant (in Japanese; English summary). *Zool. Mag. (Jap.)* **70**, 81–88.

908 Sakhnovskaiya, G. K. 1965. Changes in the properties of blood serum toxic to *Paramecium* in experimental burn sickness. *Patologicheskaya Fiziologiya i Ekspermentalnia Terapiya (Moscow)* **9**, 56–59.

909 Samoilova, K. A. 1963. Increase in resistance of *Paramecium caudatum* to UV irradiation (in Russian). In *Reaktsiya Kletok i ikh Belkovykh Komponentov na Ekstremalnye Vozdeistviya,* Akad. Nauk, USSR, Moscow, Leningrad, 143–155.

910 — 1963. Changes in fat and glycogen content in *Paramecium caudatum* after UV-treatment (in Russian). *TSitologiya* **5**, 546–553.

911 — 1964. Sensitivity of *Paramecium caudatum* to inhibitors of respiration and glycolysis varying according to culture age (in Russian). *Dokl. Akad. Nauk SSSR* **155**, 670–672.

912 — 1965. Influence of short-wave UV-light on sensitivity of infusoria to inhibitors of respiration and glycolysis (in Russian). *Radiobiologiya* **5**, 703–706.

913 — 1966. The relation of the lethal effect of UV-light on unicellular organisms to the phase of culture growth. *Reaktsiya Kletok i ikh Belkovykh Komponentov na Ekstremalnye Vozdeistviya,* Akad. Nauk, USSR, Moscow, Leningrad, 139–146.

914 *— 1969. Effect of far and near UV-radiation on *Paramecium caudatum* with different amounts of DNA in the macronucleus. *Proc. 3rd Int. Conf. Protozool., Leningrad* 37–38.

915 — & Ovchinnikova, L. P. 1963. Effect of UV-radiation on the content of nucleic acids in *Paramecium caudatum* (in Russian). In *Morfologiya i Fiziologiya Prosteishikh,* Akad. Nauk, USSR, Moscow, Leningrad, 145–152.

916 *Sato, H. 1963. The structural changes of the macronuclear constituents during the division cycle in ciliated Protozoa. *Proc. 15th Int. Congr. Zool., Washington, D.C.* **2**, 294.

917 — & Saito, M. 1958. Morphological studies on the macronuclear structure of *Paramecium*

caudatum. I. Macronuclear constituents in interphase (in Japanese; English summary). *Zool. Mag. (Jap.)* **67**, 249–258.

918 Schneider, L. 1960. Die Auflösung und Neubildung der Zellmembran bei der Konjugation von *Paramecium. Naturwiss.* **47**, 543–544.

919 — 1960. Elektronenmikroskopische Untersuchungen über das Nephridialsystem von *Paramecium. J. Protozool.* **7**, 75–90.

920 — 1963. Elektronenmikroskopische Untersuchungen der Konjugation von *Paramecium. Protoplasma* **56**, 109–140.

921 — 1964. Elektronenmikroskopische Untersuchungen an den Ernährungsorganellen von *Paramecium.* I. Der Cytopharynx. *Z. Zellforsch.* **62**, 198–224.

922 — 1964. Elektronenmikroskopische Untersuchungen an den Ernährungsorganellen von *Paramecium.* II. Die Nahrungsvakuolen und die Cytopyge. *Z. Zellforsch.* **62**, 225–245.

923 — & Wohlfarth-Bottermann, K. E. 1964. Grenzstrukturen und Hüllen bei Bakterien und Protisten. *Studium Generale* **17**, 95–124.

924 *Schneller, M. V. 1958. A new type of killing action in a stock of *Paramecium aurelia* from Panama. *Proc. Indiana Acad. Sci.* **67**, 302–303.

925 *— 1959. Temperature dependence of killing action by some killers of syngen 2, *Paramecium aurelia. J. Protozool.* **6** (Suppl.), 31.

926 *— 1962. Some notes on the rapid lysis type of killing found in *P. aurelia. Amer. Zool.* **2**, 446.

927 —, Sonneborn, T. M. & Mueller, J. A. 1959. The genetic control of kappa-like particles in *Paramecium aurelia. Genetics* **44**, 533–534.

928 *Schuster, F. L. & Ehret, C. F. 1966. Patterns of labeling at the cell and organelle level in *Paramecium bursaria. J. Protozool.* **13** (Suppl.), 12.

929 —, Prazak, B. & Ehret, C. F. 1967. Induction of trichocyst discharge in *Paramecium bursaria. J. Protozool.* **14**, 483–485.

930 Schwartz, V. 1956. Nukleolenformwechsel und Zyklen der Ribosenucleinsäure in der vegetativen Entwicklung von *Paramecium bursaria. Biol. Zentralbl.* **75**, 1–16.

931 — 1957. Über den Formwechsel achromatischer Substanz in der Teilung des Makronucleus von *Paramecium bursaria. Biol. Zentralbl.* **76**, 1–23.

932 — 1958. Chromosomen im Makronucleus von *Paramecium bursaria. Biol. Zentralbl.* **77**, 347–364.

933 — 1958. Chromosomen im Makronucleus von *Paramecium bursaria. Biol. Zentralbl.* **77**, 347–364.

934 — 1963. Die Sicherung der arttypischen Zellform bei Ciliaten. *Naturwiss.* **20**, 631–640.

935 Scoblo, I. I. 1968. Exconjugant clones of *Paramecium caudatum* with multiple micronuclei (in Russian). *TSitologiya* **10**, 218–226.

936 *— 1969. The behavior of amicronucleate *Paramecium caudatum* in sexual process. *Proc. 3rd Int. Conf. Protozool., Leningrad* 39–40.

937 — 1971. Aberrant nuclear reorganization as the result of partner interaction. *Materials of 1st All-Union Congr. Protozool. Soc., Bacu,* 83–85.

938 — & Ossipov, D. V. 1968. Autogamy during conjugation in *Paramecium caudatum* Ehrbg. I. Study on the nuclear reorganization up to stage of the third synkaryon division. *Acta Protozool. (Warsaw)* **5**, 273–290.

939 — & — 1972. Influence of migrated pronucleus on the viability of amicronuclear cells in infusoria. *Materials of 2nd Congr. All-Union Soc. Genetics and Selectors, Moscow,* 89.

940 Sears, D. F. & Elveback, L. 1961. A quantitative study of the movement of *Paramecium caudatum* and *Paramecium multimicronucleatum. Tulane Stud. Zool.* **8**, 127–139.

941 *— & Gittleson, S. M. 1960. Effects of CO_2 on *Paramecia. Physiologist* **3**, 140.

942 *— & — 1961. Narcosis of *Paramecia* with xenon. *Fed. Proc.* **20**, 142.

943 — & — 1964. Cellular narcosis of *Paramecium multimicronucleatum* by xenon and other chemically inert gases. *J. Protozool.* **11**, 538–546.

944 *Sedar, A. W. & Porter, K. R. 1954. The fine structure of the cortical components of *Paramecium multimicronucleatum. J. Protozool.* **1** (Suppl.), 4.

945 — & — 1955. The fine structure of cortical components of *Paramecium multimicronucleatum. J. Biophys. Biochem. Cytol.* **1**, 583–604.

946 Seed, J. R., Shafer, S., Finger, I. & Heller, C. 1964. Immunogenetic analysis of proteins of *Paramecium.* VI. Additional evidence for the expression of several loci in animals of a single antigenic type. *Genetic. Res.* **5**, 137–149.

947   *Selman, G. G. & Jurand, A. 1968. The formation of cell organelles during the fission cycle of *Paramecium aurelia*. *J. Protozool.* **15** (Suppl.), 34.

948   — & — 1970. Trichocyst development during the fission cycle of *Paramecium*. *J. Gen. Microbiol.* **60**, 365–372.

949   Seravin, L. N. 1957. The influence of different agents on the phagocytosis of *Paramecium caudatum* (in Russian). *Vestnik Leningradskogo Universiteta* **3**, 85–100.

950   — 1958. The changes of contractile vacuole activity in *Paramecium caudatum* related to the different agents in the medium (in Russian). *Vestnik Leningradskogo Universiteta* **3**, 78–95.

951   — 1958. Effect of coaction of two salts upon vitality of *Paramecium caudatum* (in Russian). *Vestnik Leningradskogo Universiteta* **21**, 90–101.

952   — 1958. Variations in the resistance of *Paramecium caudatum*, taking place in the course of its adaptation to $CaCl_2$, NaCl and KCl salts (in Russian). *Dokl. Akad. Nauk SSSR* **121**, 1090–1092.

953   — 1959. Interrelation of different functions in *Paramecium caudatum* in the process of adaptation to salt solutions (in Russian). *TSitologiya* **1**, 120–126.

954   — 1959. Effect of chemical agents on the mobility of *Paramecium caudatum* (in Russian). *Zoologicheskij Zhurnal* **38**, 626–630.

955   — 1960. Changes in the resistance, the upper threshold of vitality and the rate of division of *Paramecium caudatum* in the process of adaptation to salt solutions (in Russian). *Trudy Petergofskogo Biologicheskogo Instituta* **18**, 178–192.

956   — 1962. Dependence of the resistance of aquatic animals on the concentration of chemical agents in the medium (in Russian). *Trudy Petergofskogo Biologicheskogo Instituta* **19**, 149–160.

957   — 1962. Energy required for the motion of infusoria (in Russian). *Nauchnye dokl. vysshei shkoly, Biologicheskie nauki* **3**, 63–65.

958   — 1963. The role of the acetylcholine-cholinesterase system in the coordination of infusoria ciliary beating (in Russian). In *Morfologiya i Fiziologiya Prosteishikh*, *Akad. Nauk USSR, Moscow, Leningrad*, 111–122.

959   — 1965. Relationship between the curve of narcosis development time and the conditions of narcotization in *Lumbriculus variegatus* and *Paramecium caudatum* (in Russian). *TSitologiya* **7**, 546–553.

960   — 1967. *Locomotor Systems of Protozoa. Structure, Mechanochemistry, Physiology* (in Russian). Nauka, Leningrad, 1–332.

961   — 1969. A study of temporary associations (conditioned reflexes) in protozoa. Critical review (in Russian). *TSitologiya* **11**, 659–680.

962   *— 1969. Some hydrodynamic aspects of movement in ciliates. *Proc. 3rd Int. Conf. Protozool., Leningrad* 164–165.

963   — 1970. Left and right spiralling around the body axis in ciliate protozoa. *Acta Protozool. (Warsaw)* **7**, 313–323.

964   — 1970. Hydrodynamic aspects of movement in infusoria (in Russian). *Vestnik Leningradskogo Universiteta* **21**, 35–41.

965   — 1971. Mechanisms and coordination of cellular locomotion. *Advanc. Comp. Physiol. Biochem.* **4**, 37–111.

966   — & Natochin, Jr., V. 1963. The influence of respiratory inhibitors upon the surviving ability of *Paramecium caudatum* (in Russian). *TSitologiya* **5**, 461–465.

967   — & Orlovskaya, E. E. 1971. The influence of actinomycin D and puromycin on a food-uptake reaction in protozoa. *Materials of 1st All-Union Congr. Protozool. Soc., Bacu*, 79–81.

968   — & — 1972. The choice of food in protozoa. *Vestnik Leningradskogo Universiteta* **15**, 7–19.

969   — & Scoblo, I. I. 1966. Cytochemical distinction of isozymes of acidic pyrophosphatase in some infusoria (in Russian). *Nauchnye dokl. vysshei shkoly, Biologicheskie nauki* **2**, 46–49.

970   —, — & Ossipov, D. V. 1965. The influence of thermal adaptation on enzymic thermostability of *Paramecium caudatum* (in Russian). In *Teploustoichivost Kletok Zhivotnykh*, *Nauka, Moscow, Leningrad*, 161–170.

971   Setlow, R. & Doyle, B. 1956. The action of ultraviolet light on paramecin and the chemical nature of paramecin. *Biochim. Biophys. Acta* **22**, 15–20.

972   Siegel, R. W. 1953. A genetic analysis of the mate-killer trait in *Paramecium aurelia*, variety 8. *Genetics* **38**, 550–560.

973   — 1954. Mate-killing in *Paramecium aurelia*, variety 8. *Physiol. Zool.* **27**, 89–100.

974   — 1957. An analysis of the transformation from immaturity to maturity in *Paramecium aurelia*. *Genetics* **42**, 394–395.

975   — 1958. An intrafertile colony of *Paramecium bursaria*. *Amer. Naturalist* **92**, 253–254.

976   — 1960. Hereditary endosymbiosis in *Paramecium bursaria*. *Exp. Cell Res.* **19**, 239–252.

977   — 1961. Nuclear differentiation and transitional cellular phenotypes in the life cycle of *Paramecium*. *Exp. Cell Res.* **24**, 6–20.

978   *— 1962. A study of selfing caryonides in *Paramecium aurelia*. *J. Protozool.* **9** (Suppl.), 28.

979   — 1963. New results on the genetics of mating types in *Paramecium bursaria*. *Genet. Res.* **4**, 132–142.

980   — 1965. Hereditary factors controlling development in *Paramecium*. *Brookhaven Symp. Biol.* **18**, 55–65.

981   *— 1965. Genic control of the life cycle in Ciliates. *Proc. 2nd Int. Conf. Protozool.*, *London* 64–65.

982   *— 1969. Mating type genetics. Retrospect and prospect. *Proc. 3rd Int. Conf. Protozool.*, *Leningrad* 112.

983   — & Cohen, L. W. 1963. A temporal sequence for genic expression: cell differentiation in *Paramecium*. *Amer. Zoologist* **3**, 127–134.

984   — & Cole, J. 1967. The nature and origin of mutations which block a temporal sequence for genic expression in *Paramecium*. *Genetics* **55**, 607–617.

985   *— & Karakashian, S. 1959. Dissociation and restoration of endocellular symbiosis in *Paramecium bursaria*. *Anat. Rec.* **134**, 639.

986   *— & Larison, L. L. 1960. Induced illegitimate mating in *Paramecium bursaria*. *Anat. Rec.* **136**, 383.

987   — & — 1960. The genic control of mating types in *Paramecium bursaria*. *Proc. Nat. Acad. Sci. U.S.* **46**, 344–349.

988   *Sikora, J. 1969. Influence of the medium pH on the immobilization of *Paramecium aurelia* by antiserum. *Proc. 3rd Int. Conf. Protozool.*, *Leningrad* 154.

989   Simon, E. M. 1971. *Paramecium aurelia*. Recovery from $-196\,^{\circ}$C. *Cytobiology* **8**, 361–365.

990   *Simonsen, D. H. & Van Wagtendonk, W. J. 1949. Oxygen consumption of "killer" and "sensitive" stocks of *Paramecium aurelia*, var. 4. *Fed. Proc.* **8**, 250.

991   Simonsen, D. H. & Van Wagtendonk, W. J. 1952. Respiratory studies on *Paramecium aurelia*, variety 4 killers and sensitives. *Biochim. Biophys. Acta* **9**, 515–527.

992   — & — 1956. The succinoxidase system of killer and sensitive stocks of *Paramecium aurelia*, variety 4. *J. Gen. Microbiol.* **15**, 39–46.

993   Sinclair, I. F. B. 1958. The role of complement in the immune reactions of *Paramecium aurelia* and *Tetrahymena pyriformis*. *Immunology* **1**, 291–299.

994   *Sinden, R. E. 1969. Serotype transformation in *Paramecium aurelia* 2. *J. Protozool.* **16** (Suppl), 27.

995   — 1970. Identification of specific antibodies prepared against purified immobilization antigens of *Paramecium aurelia*. *J. Protozool.* **17**, 600–603.

996   *— 1970. The localization of the immobilization antigens in *Paramecium aurelia*. *J. Protozool.* **17** (Suppl.), 24.

997   Skaar, P. D. 1956. Past history and pattern of serotype transformation in *Paramecium aurelia*. *Exp. Cell Res.* **10**, 646–656.

998   *Skoczylas, B. 1969. Method for synchronizing autogamy in mass cultures of *Paramecium aurelia*, syngen 4, stock 51. *Proc. 3rd Int. Conf. Protozool.*, *Leningrad* 119.

999   *— 1969. DNase activity of *Paramecium aurelia*, syngen 4, stock 51, during the life cycle. *Proc. 3rd Int. Conf. Protozool.*, *Leningrad* 154.

1000   *Sleigh, M. A. 1969. The excitation of contractile organelles in ciliate Protozoa. *Proc. 3rd Int. Conf. Protozool.*, *Leningrad* 165.

1001   Smirnova, N. A. 1957. On the dependence of paramecia galvanotaxis upon sulfhydryl groups. *Biofizika* **2**, 670–674.

1002   *Smith, J. E. 1961. Purification of kappa particles of *Paramecium aurelia*, stock 51. *Amer. Zool.* **1**, 390.

1003   *— & Van Wagtendonk, W. J. 1962. The composition of kappa, a cytoplasmic inclusion in *Paramecium aurelia*. *Fed. Proc.* **21**, 153.

1004   Smith, M. H., George, P. & Preer, Jr., J. R. 1962. Preliminary observations on isolated *Paramecium hemoglobin*. *Arch. Biochem.* **99**, 313–318.

1005   Smith-Sonneborn, J. & Plaut, W. 1967. Evidence for the presence of DNA in the pellicle of *Paramecium*. *J. Cell Sci.* **2**, 225–234.

1006 — & Van Wagtendonk, W. J. 1964. Purification and chemical characterization of kappa of stock 51, *Paramecium aurelia*. *Exp. Cell Res.* **33**, 50–59.

1007 —, Green, L. & Marmur, J. 1963. Deoxyribonucleic acid base composition of kappa and *Paramecium aurelia*, stock 51. *Nature* **197**, 385.

1008 Soldo, A. T. 1960. Cultivation of two strains of killer *Paramecium aurelia* in axenic medium. *Proc. Soc. Exp. Biol. Med.* **105**, 612–615.

1009 — 1961. The use of particle-bearing *Paramecium* in screening for potential anti-tumor agents. *Trans. N.Y. Acad. Sci.* **23**, 653–661.

1010 — 1963. Axenic culture of *Paramecium*. Some observations on the growth behavior and nutritional requirements of a particle-bearing strain of *Paramecium aurelia* 299 sigma. *Ann. N.Y. Acad. Sci.* **108**, 380–388.

1011 *— & Godoy, G. A. 1971. The kinetic complexity of deoxyribonucleic acid of *Paramecium aurelia* and its symbiotes. *J. Protozool.* **18** (Suppl.), 9.

1012 *— & Reid, S. J. 1968. Deoxyribonucleic acids of lambda-bearing *Paramecium aurelia*. *J. Protozool.* **15** (Suppl.), 15.

1013 *— & Van Wagtendonk, W. J. 1959 Nitrogen metabolism in *Paramecium aurelia* *Fed. Proc.* **18**, 327.

1014 — & — 1961. Nitrogen metabolism of *Paramecium aurelia*. *J. Protozool.* **8**, 41–55.

1015 *— & — 1967. The nutritional requirements of *Paramecium aurelia*. *J. Protozool.* **14** (Suppl.), 12.

1016 — & — 1967. A method for the mass collection of axenically cultivated *Paramecium*. *J. Protozool.* **14**, 497–498.

1017 — & — 1967. An analysis of the nutritional requirements for fatty acids of *Paramecium aurelia*. *J. Protozool.* **14**, 596–600.

1018 — & — 1968. Lipid interrelationships in the growth of *Paramecium aurelia*, stock 299. *J. Gen. Microbiol.* **53**, 341–348.

1019 — & — 1969. The nutrition of *Paramecium aurelia*, stock 299. *J. Protozool.* **16**, 500–506.

1020 *—, Godoy, G. A. & Van Wagtendonk, W. J. 1965. Hydrolytic nucleosidases and hypoxanthine excretion in *Paramecium aurelia*. *Fed. Proc.* **25**, 783.

1021 *—, — & — 1966. Sterol-fatty acid interrelationships in *Paramecium aurelia*. *J. Protozool.* **13** (Suppl.), 12.

1022 —, — & — 1966. Growth of particle-bearing and particle-free *Paramecium aurelia* in axenic culture. *J. Protozool.* **13**, 492–497.

1023 *—, — & — 1969. Biochemical studies on endosymbiote particles of *Paramecium aurelia* isolated by a new procedure. *J. Protozool.* **16** (Suppl.), 8.

1024 —, Musil, G. & Godoy, G. A. 1970. Action of penicillin G on endosymbiote lambda particles of *Paramecium aurelia*. *J. Bacteriol.* **104**, 966–980.

1025 *—, Reid, S. J. & Van Wagtendonk, W. J. 1967. Unusual DNA associated with cytoplasmic particles in *Paramecium*. *Fed. Proc.* **26**, 566.

1026 —, Van Wagtendonk, W. J. & Godoy, G. A. 1970. Nucleic acid and protein content of purified endosymbiote particles of *Paramecium aurelia*. *Biochim. Biophys. Acta.* **204** 325–333.

1027 Sommerville, J. 1967. Immobilization antigen synthesis in *Paramecium aurelia*: the detection of labelled antigen in a cell-free amino acid incorporating system. *Biochim. Biophys. Acta* **149**, 625–627.

1028 — 1968. Immobilization antigen synthesis in *Paramecium aurelia*. *Exp. Cell Res.* **50**, 660–664.

1029 *— 1969. Serotype transformation in *Paramecium aurelia* 1. *J. Protozool.* **16** (Suppl.), 27.

1030 — 1969. Serotype transformation in *Paramecium aurelia*. Antigen synthesis after a temperature change. *Exp. Cell Res.* **57**, 443–446.

1031 — 1970. Immobilization antigen synthesis in *Paramecium aurelia*: synthesis in a cell-free amino acid incorporating system. *Biochim. Biophys. Acta* **209**, 240–249.

1032 — 1970. Serotype expression in *Paramecium*. *Advanc. Microbiol. Physiol.* **4**, 131–178.

1033 — & Sinden, R. 1968. Protein synthesis by free and bound *Paramecium* ribosomes *in vivo* and *in vitro*. *J. Protozool.* **15**, 644–651.

1034 *Sonneborn, T. M. 1953. Environmental control of the duration of phenomic or cytoplasmic lag in *Paramecium aurelia*. *Microbial Genet. Bull.* **7**, 23.

1035 *— 1954. Gene-controlled, aberrant nuclear behavior in *Paramecium aurelia*. *Microbial Genet. Bull.* **11**, 24–25.

1036 *— 1954. Is the gene K active in the micronucleus of *Paramecium aurelia*? *Microbial Genet. Bull.* **11**, 25–26.

1037 — 1954. The relation of autogamy to senescence and rejuvenescence in *Paramecium aurelia*. *J. Protozool*. **1**, 38–53.

1038 — 1954. Patterns of nucleocytoplasmic integration in *Paramecium*. *Caryologia* **6** (Suppl.), 307–325.

1039 *— 1955. Macronuclear control of the initiation of meiosis and conjugation in *Paramecium aurelia*. *J. Protozool*. **2** (Suppl.), 12.

1040 *— 1955. A third point of attachment between conjugants in *Paramecium aurelia* and its significance. *J. Protozool*. **2** (Suppl.), 12.

1041 — 1956. The metagon: RNA and cytoplasmic inheritance. *Amer. Naturalist* **99**, 279–307.

1042 *— 1956. An exceptional autogamous clone in variety 4 of *Paramecium aurelia* and its interpretation. *J. Protozool*. **3** (Suppl.), 8–9.

1043 *— 1956. The distribution of killers among the varieties of *Paramecium aurelia*. *Anat. Rec*. **125**, 567–568.

1044 — 1957. Breeding systems, reproductive methods, and species problems in Protozoa. In Mayr, E., *The Species Problem*, American Association for the Advancement of Science, Publication No. 50, 155–324.

1045 *— 1957. Diurnal change of mating type in *Paramecium*. *Anat. Rec*. **128**, 626.

1046 *— 1957. Varieties 13, 15, and 16 of the *P. aurelia—multimicronucleatum* complex. *J. Protozool*. **4** (Suppl.), 21.

1047 *— 1958. Classification of syngens of the *Paramecium aurelia-multimicronucleatum* complex. *J. Protozool*. **5** (Suppl.), 17–18.

1048 — 1959. Kappa and related particles in *Paramecium*. *Advanc. Virus Res*. **6**, 229–356.

1049 — 1960. The gene and cell differentiation. *Proc. Nat. Acad. Sci. U.S*. **46**, 149–165.

1050 *— 1960. Suppression of fission by antiserum in *Paramecium aurelia*. *J. Protozool*. **7** (Suppl.), 26.

1051 — 1961. Kappa particles and their bearing on host-parasite relations. In Pollard, M., *Perspectives in Virology*, Burgess Press, Minneapolis, Minn., Ch. 2, 5–12.

1052 — 1963. Does preformed cell structure play an essential role in cell heredity? In Allen, J. M., *The Nature of Biological Diversity*, McGraw-Hill, New York, N.Y., 165–221.

1053 — 1964. The differentiation of cells. *Proc. Nat. Acad. Sci. U.S*. **51**, 915–929.

1054 *— 1966. A non-conformist genetic system in *Paramecium aurelia*. *Amer. Zoologist* **6**, 589.

1055 — 1970. Methods in Paramecium Research. In Prescott, D. M., *Methods of Cell Physiology*, Academic Press, New York, N.Y., **4**, 242–335.

1056 — 1970. Determination, development, and inheritance of the structure of the cell cortex. *Symp. Int. Soc. Cell Biol*. **9**, 1–13.

1057 — 1970. Gene action in development. *Proc. Roy. Soc. London, Ser. B* **176**, 347–366.

1058 *— & Balbinder, E. 1953. The effect of temperature on the expression of allelic genes for serotypes in a heterozygote of *Paramecium aurelia*. *Microbial Genet. Bull*. **7**, 24–25.

1059 *— & Barnett, A. 1958. The mating type system in syngen 2 of *Paramecium multimicronucleatum*. *J. Protozool*. **5** (Suppl.), 18.

1060 *— & Dippell, R. V. 1956. Giant *Paramecium aurelia* (?). *J. Protozool*. **3** (Suppl.), 9.

1061 *— & — 1957. The *Paramecium aurelia—multimicronucleatum* complex. *J. Protozool*. **4** (Suppl.), 21.

1062 *— & — 1960. The genetic basis of the difference between single and double *Paramecium aurelia*. *J. Protozool*. **7** (Suppl.), 26.

1063 *— & — 1961. The limit of multiplicity of cortical organelle systems in *P. aurelia*, syngen 4. *Amer. Zoologist* **1**, 390.

1064 *— & — 1961. The modes of replication of cortical organization in *Paramecium aurelia*, syngen 4. *Genetics* **46**, 899–900.

1065 *— & — 1961. Self-reproducing differences in the cortical organization in *Paramecium aurelia*, syngen 4. *Genetics* **46**, 900.

1066 *— & — 1962. Two new evidences of cortical autonomy in syngen 4 of *Paramecium aurelia*. *J. Protozool*. **9** (Suppl.), 28.

1067 — & LeSuer, A. 1948. Antigenic characters in *Paramecium aurelia* (Variety 4): Determination, inheritance and induced mutations. *Amer. Naturalist* **82**, 69–78.

1068 *— & Mueller, J. A. 1959. What is the infective agent in breis of killer paramecia? *Science* **130**, 1423.

1069 *— & Rafalko, M. 1957. Aging in the *P. aurelia—multimicronucleatum* complex. *J. Protozool*. **4** (Suppl.), 21.

1070 *— & Schneller, M. 1955. The basis of aging in variety 4 of *Paramecium aurelia*. *J. Protozool.* **2** (Suppl.), 6.

1071 *— & — 1955. Are there cumulative effects of parental age transmissible through sexual reproduction in variety 4 of *Paramecium aurelia? J. Protozool.* **2** (Suppl.), 6.

1072 *— & — 1955. Genetic consequences of aging in variety 4 of *Paramecium aurelia*. *Rec. Genet. Soc. Amer.* **24**, 596.

1073 *— & Sonneborn, D. R. 1958. Some effects of light on the rhythm of mating type changes in stock 232-6 of syngen 2 of *P. multimicronucleatum*. *Anat. Rec.* **131**, 601.

1074 *—, Gibson, I. & Schneller, M. V. 1964. Killer particles and metagons of *Paramecium* grown in *Didinium*. *Science* **144**, 567–568.

1075 *—, Mueller, J. A. & Schneller, M. V. 1959. The classes of kappa-like particles in *Paramecium aurelia*. *Anat. Rec.* **134**, 642.

1076 *—, Ogasawara, F. & Balbinder, E. 1953. The temperature sequence of the antigenic types in variety 4 of *Paramecium aurelia* in relation to the stability and transformations of antigenic types. *Microbial Genet. Bull.* **7**, 27.

1077 *—, Schneller, M. V. & Craig, M. F. 1956. The basis of variation in phenotype of gene-controlled traits in heterozygotes of *Paramecium aurelia*. *J. Protozool.* **3** (Suppl.), 8.

1078 *—, Dippell, R. V., Schneller, M. V. & Tallan, I. 1953. The explanation of "anomalous" inheritance following exposure to ultraviolet in variety 4 of *Paramecium aurelia*. *Microbial Genet. Bull.* **7**, 25–26.

1079 *—, —, Mueller, J. A. & Holzman, H. E. 1959. Extensions of the ranges of certain syngens of *Paramecium aurelia*. *J. Protozool.* **6** (Suppl.), 31–32.

1080 *—, Tallan, I., Balbinder, E., Ogasawara F. & Rudnyansky, B. 1953. The independent inheritance of 5 genes in variety 4 of *Paramecium aurelia*. *Microbial Genet. Bull.* **7**, 27–28.

1081 Sorochenko, E. B. 1957. On influence of penicellin on protozoa (in Russian). *Trudy Arkhangelskogo meditsinskogo Instituta* **15**, 183–185.

1082 Stanishevskaya, A. V. & Rodina, L. G. 1966. An investigation of the antagonism of adrenergic substances and of an inhibitor Mao-vetrasine towards aminasine in the experiments with *Paramecium caudatum* Ehrbg. (in Russian). *Nauchnye dokl. vysshei shkoly, Biologicheskie nauki* **3**, 56–58.

1083 Steers, Jr., E. 1961. Electrophoretic analysis of immobilization antigens of *Paramecium aurelia*. *Science* **133**, 2010–2011.

1084 — 1962. A comparison of the tryptic peptides obtained from immobilization antigens of *Paramecium aurelia*. *Proc. Nat. Acad. Sci. U.S.* **48**, 867–874.

1085 — 1965. Amino acid composition and quaternary structure of an immobilizing antigen from *Paramecium aurelia*. *Biochemistry* **4**, 1896–1901.

1086 —, Beisson, J. & Marchesi, V. T. 1969. A structural protein extracted from the trichocysts of *Paramecium aurelia*. *Exp. Cell Res.* **57**, 392–396.

1087 *Sterbenz, F. J. 1956. The axenic culture of *Paramecium caudatum*. *J. Protozool.* **3** (Suppl.), 13.

1088 *— 1956. A demonstration of the need for thioctic acid for the growth of *Paramecium caudatum* in axenic culture. *J. Protozool.* **3** (Suppl.), 14.

1089 Stevenson, I. 1967. A method for the isolation of macronuclei from *Paramecium aurelia*. *J. Protozool.* **14**, 412–414.

1090 — 1967. Diaminopimelic acid in mu particles of *Paramecium aurelia*. *Nature* **215**, 434–435.

1091 — 1969. The biochemical status of mu particles in *Paramecium aurelia*. *J. Gen. Microbiol.* **57**, 61–75.

1092 — 1970. Endosymbioses in some stocks of *Paramecium aurelia* collected in Australia. *Cytobios.* **2**, 207–224.

1093 — 1972. Bacterial endosymbiosis in *Paramecium aurelia*. Bacteriophage-like inclusions in a kappa symbiont. *J. Gen. Microbiol.* **71**, 69–76.

1094 — & Lloyd, F. P. 1971. Ultrastructure of nuclear division in *Paramecium aurelia*. I. Mitosis in the micronucleus. *Austr. J. Biol. Sci.* **24**, 963–975.

1095 — & — 1971. Ultrastructure of nuclear division in *Paramecium aurelia*. II. Amitosis of the macronucleus. *Austr. J. Biol. Sci.* **24**, 977–987.

1096 *Stewart, J. M. 1964. The measurement of oxygen consumption in *Paramecia* of different ages. *J. Protozool.* **11** (Suppl.), 39.

1097 — & Muire, A. R. 1963. The fine structure of the cortical layers in *Paramecium aurelia*. *Quart. J. Microscop. Sci.* **104**, 129–134.

1098 Stockem, W. & Wohlfarth-Bottermann, K. E. 1970. Zur Feinstruktur der Trichocysten von *Paramecium*. *Cytobiologie* **1**, 420–436.

1099 Sugino, S. 1966. Phagocytosis of bacteria by *Paramecium caudatum*, especially of pathogenic bacteria (in Japanese). *Med. Biol. (Tokyo)* **70**, 215–219.

1100 Suhama, M. & Hanson, E. D. 1971. The role of protein synthesis in prefission morphogenesis of *Paramecium aurelia*. *J. Exp. Zool.* **177**, 463–478.

1101 Sukhanova, K. M. 1965. Dependence of temperature adaptations of some unicellular organisms on feeding conditions. *Acta Protozool. (Warsaw)* **3**, 153–163.

1102 — 1969. Regularities of thermoresistance changes in some species of freshwater stenothermic ciliates. *Zoologicheskij Zhurnal* **48**, 962–969.

1103 Sutherland, B. M., Carrier, W. L. & Setlow, R. B. 1967. Photoreactivation in vivo of pyrimidine dimers in *paramecium* DNA. *Science* **158**, 1669–1670.

1104 —, — & — 1968. Pyrimidine dimers in the DNA of *Paramecium aurelia*. *Biophys. J.* **8**, 490–499.

1105 Suyama, Y. & Preer, Jr., K. R. 1965. Mitochondrial DNA from protozoa. *Genetics* **52**, 1051-1058.

1106 *Tait, A. 1969. Mitochondrial protein variation in *Paramecium aurelia*. *Proc. 3rd Int. Conf. Protozool., Leningrad* 121.

1107 *— 1969. Syngen differences in electrophoretic mobility of certain enzymes in *Paramecium aurelia*. *J. Protozool.* **16** (Suppl.), 28.

1108 — 1970. Enzyme variation between syngens in *Paramecium aurelia*. *Biochem. Genet.* **4**, 461–470.

1109 — 1970. Genetics of NADP isocitrate dehydrogenase in *Paramecium aurelia*. *Nature* **225**, 181–182.

1110 Takagi, Y. 1971. Sequential expression of sex traits in the clonal development of *Paramecium multimicronucleatum*. *Japan J. Genetics* **46**, 83–91.

1111 Takahashi, M. & Hiwatashi, K. 1970. Disappearance of mating type reactivity in *Paramecium caudatum* upon repeated washing. *J. Protozool.* **17**, 667–670.

1112 Takayanagi, T. & Hayashi, S. 1964. Cytological and cytogenetical studies on *Paramecium polycaryum*. V. Lethal interactions in certain stocks. *J. Protozool.* **11**, 128–132.

1113 Tallan, I. 1959. Factors involved in infection by the kappa particles in *Paramecium aurelia*, syngen 4. *Physiol. Zool.* **32**, 78-89.

1114 — 1961. A cofactor required by kappa in the infection of *Paramecium aurelia* and its possible action. *Physiol. Zool.* **34**, 1–13.

1115 Talysin, F. F. & Shootova, V. C. 1958. Effect of snake venoms on *Paramecium caudatum* (in Russian). *Trudy 1-go Moscovskogo Meditsinskogo Instituta* **41**, 18–21.

1116 *Tanguay, R. B. & Van Wagtendonk, W. J. 1963. Histological staining and chemical identification of the lambda particle of *Paramecium aurelia*, stock 299 (killers). *Fed. Proc.* **22**, 646.

1117 *Tarantola, V. A. & Van Wagtendonk, W. J. 1958. Purine and pyrimidine requirements of *Paramecium aurelia*. *Fed. Proc.* **17**, 337.

1118 — & — 1959. Further studies of the nutritional requirements of *Paramecium aurelia*, variety 4, stock 51 (sensitive). *J. Protozool.* **6**, 189–195.

1119 Tartar, V. 1954. Anomalies of regeneration in *Paramecium*. *J. Protozool.* **1**, 11–17.

1120 *Taub, S. R. 1958. Nucleo-cytoplasmic interactions in mating type determination in variety 7 of *Paramecium aurelia*. *J. Protozool.* **5** (Suppl.), 18.

1121 *— 1959. The genetics of mating type determination in syngen 7, *Paramecium aurelia*. *Genetics* **44**, 541–542.

1122 *— 1959. The breeding system of syngen 7 of *Paramecium aurelia*. *Anat. Rec.* **134**, 646.

1123 — 1960. *Genetic Studies on Syngen 7 of Paramecium aurelia*, Thesis, Indiana University, Bloomington, Ind.

1124 *— 1962. The effect of nuclear genes on nuclear differentiation in syngen 7, *Paramecium aurelia*. *Genetics* **47**, 990–991.

1125 — 1963. The genetic control of mating type differentiation in *Paramecium*. *Genetics* **48**, 815–834.

1126 — 1966. Regular changes in mating type composition in selfing cultures and mating type potentiality in selfing caryonides of *Paramecium aurelia*. *Genetics* **54**, 173–189.

1127 — 1966. Unidirectional mating type changes in individual cells from selfing cultures of *Paramecium aurelia*. *J. Exp. Zool.* **163**, 141–150.

1128 Tawada, K. & Oosawa, F. 1972. Responses of *Paramecium* to temperature change. *J. Protozool.* **19**, 53–57.

1129 Thiery, J. P. 1967. Mise en évidence des polysaccharides sur coupes fines en microscopie électronique. *J. Microscopie* **6**, 987–1017.

1130 Timoffejev, N. N. 1958. Acquired reactions in *Paramecium caudatum* (in Russian). In *Problemy Sravnitelnoy Fiziologii i Patologii Nervnoj Systemy, Akad. Nauk USSR, Moscow, Leningrad*, 260–266.

1131 Totwen-Nowakewska, I. 1963. The effect of nutrition on the regeneration of the caudal body fragment in *Paramecium caudatum. Acta Protozool. (Warsaw)* **1**, 55–61.

1132 Trunova, O. N. 1967. On the intensity of bacteriophage on *Paramecium caudatum* in relation to some pathogenic microorganisms (in Russian). *Trudy Saratovskoj nauchno-issledovatelskoj veterinarnoj stantsii* **7**, 181–187.

1133 — 1967. The influence of *Paramecium caudatum* on the titre ordysentery and cholera vibrio in the water medium (in Russian). *Trudy Saratovskoj nauchno-issledovatelskoj veterinarnoj stantsii* **7**, 188–190.

1134 Tuffrau, M. 1964. Quelques variantes techniques de l'impregnation des Ciliés par le protéinate d'argent. *Arch. Zool. Exp.* **104**, 186–190.

1135 — 1967. Perfectionnements et pratique de la technique d'impregnation au protargol des infusoires Ciliés. *Protistologica* **3**, 91–98.

1136 *Uhlig, G. 1969. The effect of alternating temperatures on reproduction of *Paramecium caudatum. Proc. 3rd Int. Conf. Protozool., Leningrad* 205.

1137 Van Eys, J. & Warnock, L. G. 1963. The inhibition of motility of ciliates through methonium drugs. *J. Protozool.* **10**, 465–467.

1138 *Van Wagtendonk, W. J. 1953. The steroid requirements of *Paramecium aurelia. Atti Del. 6 Congresso Internazionale de Microbiologia, Rome* **5**, 359.

1139 — 1955. Nutrition of Ciliates. In Hutner, S. H. & Lwoff, A., *Biochemistry and Physiology of Protozoa*, Academic Press, New York, N.Y., Ch. 2, 57–84.

1140 *— 1960. Comparative biochemistry of Protozoa *(Abstracts American Chemical Society Meeting, Cleveland, Ohio)*, 19.

1141 *— 1961. Autogamy in *P. aurelia. 12th Annual VA Medical Research Conference.*

1142 — 1963. Comparative biochemistry of Protozoa. In *Proc. 5th Int. Congr. Biochem., Moscow, USSR*, Vol. 3, 343–346.

1143 — 1969. Neoplastic Equivalents of Protozoa. In *National Cancer Institute*, Monograph No. 31, 751–768.

1144 *— 1969. The axenic culture of Protozoa. Problems and prospects. *Proc. 3rd Int. Conf. Protozool., Leningrad* 161.

1145 *— & Butzel, Jr., H. M. 1962. Nucleic acid metabolism of *Paramecium aurelia. 13th Annual VA Medical Research Conference*, 201.

1146 *— & Conner, R. L. 1953. Steroid requirements for *Paramecium aurelia*, var. 4, stock 51.7 (s) in axenic culture. *Fed. Proc.* **12**, 31.

1147 *— & Soldo, A. T. 1965. Endosymbiotes of ciliated protozoa. *Proc. 2nd Int. Conf. Protozool., London* 244.

1148 — & — 1970. Methods used in the axenic culture of *Paramecium*. In Prescott, D. M., *Methods in Physiology*, Academic Press, New York, N.Y., **4**, 117–130.

1149 — & — 1970. Nitrogen metabolism in Protozoa. In Campbell, J. W., *Comparative Nitrogen Metabolism*, Academic Press, London, **1**, 1–73.

1150 — & Tanguay, R. B. 1963. The chemical composition of lambda in *Paramecium aurelia*, Stock 299. *J. Gen. Microbiol.* **33**, 395–400.

1151 — & Van Tijn, B. 1953. Cross reactions of serotypes 51A, 51B and 51D of *Paramecium aurelia*, variety 4. *Exp. Cell Res.* **5**, 1–9.

1152 — & Vloedman, Jr., D. A. 1955. Evidence for the presence of a protein with ATP-ase and antigenic specificity in *Paramecium aurelia*, variety 4, stock 51. *Biochim. Biophys. Acta* **7**, 335–336.

1153 —, Clark, J. A. D. & Godoy, G. A. 1963. The biological status of lambda and related particles in *Paramecium aurelia. Proc. Nat. Acad. Sci. U.S.* **50**, 835–838.

1154 —, Goldman, P. H. & Smith, W. L. 1970. The culture of the strains of the genus *Paramectum* in axenic medium. *J. Protozool.* **17**, 389–391.

1155 *—, Miller, C. A. & Conner, R. L. 1952. Growth requirements of *Paramecium aurelia*, variety 4, stock 51.7 (s) in a medium free of other living organisms. *Fed. Proc.* **11**, 302.

1156   —, Simonsen, D. H. & Zill, L. P. 1952. The use of electromigration techniques in washing and concentrating cultures of *Paramecium aurelia*. *Physiol. Zool.* **25**, 312–317.

1157   *—, Smith, J. E. & Reisner, A. 1958. A surface antigen of *Paramecium aurelia*. *Proc. 8th Congr. Microbiol., Stockholm* 117.

1158   —, Zill, L. P. & Simonsen, D. H. 1950. Chemical and physiological studies on paramecin and kappa. *Proc. Indiana Acad. Sci.* **60**, 64–66.

1159   —, Conner, R. L., Miller, C. A. & Rao, M. R. R. 1953. Growth requirements of *Paramecium aurelia*, variety 4, stock 51.7 sensitives and killers in axenic medium. *Ann. N.Y. Acad. Sci.* **56**, 929–937.

1160   *—, Conner, R. L., Young, G. R. & Miller, C. A. 1955. Growth requirements of *Paramecium aurelia*. *Comm. 3rd Int. Congr. Biochem., Brussels*, 97.

1161   —, Van Tijn, B., Litman, R., Reisner, A. & Young, M. 1956. The surface antigens of *Paramecium aurelia*. *J. Gen. Microbiol.* **15**, 617.

1162   Vinogradova, V. G. 1967. Detection of toxemia in irradiated rabbits and the disintoxication effect of low-molecular polyvinyl with the use of the "paramecium reaction" (in Russian). *Radiobiologiya* **7**, 96–99.

1163   Vivier, E. 1955. Contribution à l'étude de la conjugaison de *Paramecium caudatum*. *Bull. Soc. Zool. France* **80**, 163–170.

1164   — 1960. Contribution à l'étude de la conjugaison chez *Paramecium caudatum*. *Ann. Sci. Nat. Zool. Biol. Anim. 12e sér.* **2**, 387–506.

1165   — 1960. Cycle nucléolaire en rapport avec l'alimentation chez *Paramecium caudatum*. *Compt. Rend.* **250**, 205–207.

1166   — 1961. On the structure of the macronucleus of *Paramecium* (in French). *Compt. Rend. Soc. Biol.* **155**, 494–497.

1167   — 1962. Demonstration with the aid of the electron microscope of cytoplasmic exchanges during conjugation in *Paramecium caudatum* Ehrb. (in French). *Compt. Rend. Soc. Biol.* **156**, 1115–1116.

1168   — 1963. Etude au microscope électronique des nucléoles dans le macronucléus de *Paramecium caudatum*. *Proc. 1st Int. Congr. Protozool., Prague* 421.

1169   — 1965. Sexualité et conjugaison chez la Paramécie. *Ann. Fac. Sci., Clermont-Ferrand* **26**, 101–114.

1170   — 1966. Variations ultrastructurales du chondriome en relation avec le mode de vie chez des protozoaries. *Proc. 6th Int. Congr. Electron Microscop., Kyoto* 247–248.

1171   — & André, J. 1961. Existence d'inclusions d'ultrastructure fibrillaire dans le macronucléus de certaines souches de *Paramecium caudatum*. *Ehrb. Compt. Rend.* **252**, 1848–1850.

1172   — & — 1961. Données structurales et ultrastructurales nouvelles sur la conjugaison de *Paramecium caudatum*. *J. Protozool.* **8**, 416–426.

1173   — & Mallinger, M. C. 1960. Variations de la réactivité sexuelle en fonction du rythme de multiplication chez *Paramecium caudatum*. *Compt. Rend.* **154**, 2071–2075.

1174   —, Legrand, B. & Petitprez, A. 1969. Recherches cytochimiques et ultrastructurales sur des inclusions polysaccharidiques et calciques du Spirostome; leurs relations avec la contractilité. *Protistologica* **5**, 145–159.

1175   —, Petitprez, A. & Chive, A. F. 1967. Observations ultrastructurales sur les chlorelles symbiotes de *Paramecium bursaria*. *Protistologica* **3**, 325–333.

1176   —, Schrevel-Debersee, G. & Oger, C. 1964. Observations sur les variétés et types sexuels de *Paramecium caudatum*, Hérédité et changement de type sexuel. *Arch. Zool. Exp. Gen.* **104**, 49–67.

1177   —, Devauchelle, G., Petitprez, A., Porchet-Hennere, E., Prensier, G., Schrevel, J. & Vinckier, D. 1970. Observations de cytologie comparée chez les Sporozoaires. I. Les structures superficielles chez les formes végétatives. *Protistologica* **6**, 1.

1178   Vloedman, Jr., D. A. Berech, Jr., J., Jeffries, W. B. & Van Wagtendonk, W. J. 1957. Carbohydrate metabolism of *Paramecium aurelia*, variety 4, stock 57 (sensitive). *J. Gen. Microbiol.* **16**, 628–641.

1179   Wang, H. 1963. Differential responses of *Paramecium aurelia* to cigarette components. *Nature* **197**, 946–948.

1180   — 1966. Further analysis of the reaction of *Paramecium* to cigarette paper ash solutions. *Canadian J. Microbiol.* **12**, 125–131.

1181   Wang, G. T. & Marquardt, W. C. 1966. Survival of *Tetrahymena pyriformis* and *Paramecium aurelia* following freezing. *J. Protozool.* **13**, 123–128.

nuclear reorganization (in Russian). In *Vliyanie Ioniziruyushchykh Izluchneii Nasledstvennost,* Nauka, Moscow, 28–35.

1239   Zubkova, S. M. 1968. Effect of the electromagnetic field on the regulation of the motion functions of paramecia (in Russian). In *Fiziko-Khimicheskie Osnovy Avtoregulyatsii v Kletkakh,* Nauka, Moscow, 130–136.

ADDENDUM

1240   *Adoutte, A. & Beisson, J. 1972. Interactions génétiques et physiologiques entre mitochondries chez *Paramecium aureiia. J. Protozool.* **19.**

1241   —, Balmefrezol, M., Beisson, J. & André, J. 1972. The effects of erythromycin and chloramphenicol on the ultrastructure of mitochondria in sensitive and resistant strains of *Paramecium. J. Cell Biol.* **54,** 8–19.

1242   Allen, S. L. & Golembiewski, P. A. 1972. Inheritance of esterases A and B in syngen 2 of *Paramecium aurelia. Genetics* **71,** 469–475.

1243   —, Byrne, B. C. & Cronkite, D. L. 1971. Intersyngenic variations in the esterases of bacterized *Paramecium aurelia. Biochem. Genet.* **15,** 135–150.

1244   Amberzumian, M. A. 1970. Changes of the thermostability of *Paramecium caudatum* related to the contents of ions of some metals in the medium (in Russian; English summary). *TSitologiya* **12,** 774–782.

1245   Andrivon, C. 1970. Complément à l'étude de l'action des inhibiteurs du métabolisme sur la résistance aux sels de nickel chez *Paramecium caudatum* et chez quelques autres protozoaires ciliés. *Protistologica* **6,** 199–620.

1246   Apostol, S. 1971. The effect of residual cellulose waters on aquatic invertebrates (in Roumanian, English and French summary). *Stud. Čercet Biol. Ser. Zool.* **23,** 349–357.

1247   Aylmer, C. & Reisner, A. H. 1971. Thermally induced explosive migration of *Paramecium. J. Gen. Microbiol.* **67,** 57–61.

1248   *Bannister, L. H. 1971. The structure of undischarged trichocysts in *Paramecium caudatum. J. Protozool.* **18** (Suppl.), 40.

1249   — 1972. The structure of trichocysts in *Paramecium caudatum. J. Cell Sci.* **11,** 899–929.

1250   Barna, I. & Weis, D. 1972. Bacteria as food for *Paramecium bursaria. J. Protozool.* **1** (Suppl.), 34.

1251   Beale, G. H. 1970. Biokhimicheskaya genetika infuzorii (English summary). *Genetika* **6,** 49–55.

1252   *Berger, J. D. 1972. Regulation of macronuclear DNA content in *Paramecium aurelia. J. Protozool.* **19** (Suppl.), 34.

1253   *Bihn, J. P. & Lilly, D. M. 1972. Flagellate culture medium and the growth of *Paramecium. J. Protozool.* **19** (Suppl.), 33.

1254   Burbanck, W. D. & Martin, V. L. 1973. Experimental microbial populations thirty-five years later: The influence of food on the symbiosis of *Paramecium aurelia syngen 4, 51.7. J. Protozool.* **20,** 135–138.

1255   Butzel, Jr., H. M. 1973. Abnormalities in nuclear behavior and mating type determination in cytoplasmically bridged exconjugants of doublet *Paramecium aurelia. J. Protozool.* **20,** 140–143.

1256   Cairns, Jr., J., Beamer, T., Churchill, S. & Ruthven, J. 1971. Response of protozoans to detergent-enzymes. *Hydrobiologia* **38,** 193–205.

1257   Capdeville, Y. 1972. Etude de l'influence du cytoplasme dans le phénomène de répression interallélique. *J. Protozool.* **19** (Suppl.), 62.

1258   Cavill, A. & Gibson, I. 1972. Genetic determination of esterases of syngens 1 and 8 in *Paramecium aurelia. Heredity* **28,** 31–37.

1259   Cerna, Z. 1971. L'immunofluorescence en protistologie. *Année Biol. (Paris)* **10,** 1–9.

1260   Cole, J. & Siegel, R. W. 1969. A heterocaryon in *Paramecium caudatum. Genetics* **63,** 361–368.

1261   Cooper, J. E. 1968. An immunological and genetic analysis of antigen in *Paramecium aurelia* syngen 2. *Genetics* **60,** 59–72.

1262   Crippa-Franceschi, T. 1967. Acquired resistance to temperature due to a cytoplasmic change in *Paramecium aurelia* syngen 1. *Boll. Mus. Istit. Biol. Univ. Genova Sez. Biol. Anim.* **35,** 5–18.

1263   — 1967. Further studies on acquired resistance to temperature in *Paramecium aurelia* syngen 1. *Boll. Mus. Istit. Biol. Univ. Genova Sez. Biol. Anim.* **35,** 19–31.

1264 *Croute, F., Soleilhavoup, J. P. & Planel, H. 1972. Variations of *Paramecium aurelia* sensitivity to natural ionizing radiations as a function of post-autogamous culture ages. *J. Protozool.* **19** (Suppl.), 63.

1265 Diaz-Múgica, V. 1969. Estudio comparativo de las tecnicas de tincion argentica en la infraciliacion de *Paramecium aurelia* (English summary). *Microbiol. Espan.* **22**, 9–18.

1266 Dryl, S. & Bujwid-Čwik, K. 1972. Effects of detergents on excitability and motor response in protozoa. *Acta Protozool. (Warsaw)* **11**, 367–372.

1267 Eckert, R., Naitoh, Y. & Friedman, K. 1972. Sensory mechanisms in *Paramecium*. I. Two components of the electric response to mechanical stimulation of the anterior surface. *J. Exp. Biol.* **56**, 683–694.

1268 Elpidina, O. K. and Dunaeva, F. D. 1969. Determination of the cytotoxic effect and protistocidal activity of some antitomorganic substances (in Russian). *Trans. Kazan Med. Inst.* **31**, 114–116.

1269 *Estève, J. C. 1972. Cytochémie ultrastructurale du trichocyste de *Paramecium caudatum*. *J. Protozool.* **19** (Suppl.), 65.

1270 — 1972. L'appareil de Golgi des Ciliés: Ultrastructure, particulièrement chez *Paramecium*. *J. Protozool.* **19**, 609–618.

1271 Finger, I., Heller, C. and Magers, S. 1972. Clonal variation in *Paramecium*. III. Heterogeneity within clones of identical serotype. *Genetics* **72**, 47–62.

1272 —, —, Dilworth, L. and von Allmen, C. 1972. Clonal variation in *Paramecium*. I. Persistent unstable clones. *Genetics* **72**, 17–33.

1273 —, Onovato, F., Heller, C. and Dilworth, L. 1972. Clonal variation in *Paramecium*. II. A comparison of stable and unstable clones of the same serotype. *Genetics* **72**, 35–46.

1274 Flavell, R. A. & Jones, I. G. 1971. *Paramecium* mitochondrial DNA. Renaturation and hybridisation studies. *Biochim. Biophys. Acta* **23**, 255–260.

1275 Franceschi, T. 1967. Coniugazione tra ceppi modificati e controllo di *Paramecium aurelia* syngen 1 (English summary). *Atti Accad. Ligure Sci. Lett.* **23**, 19–24.

1276 — 1967. Coniugazione "Prolungata" in *Paramecium aurelia* syngen 1 (English summary). *Atti Accad. Ligure Sci. Lett.* **23**, 145–150.

1277 Ganapati, S. V. & Amin, P. 1972. Microbiology of scum formed at the surface of lagooned wastewater. *J. Water Pollut. Control Fed.* **44**, 769–781.

1278 Gibson, I. & Martin, N. 1972. DNA amounts in the nuclei of *Paramecium aurelia* and *Tetrahymena pyriformis*. *Chromosoma* **35**, 374–382.

1279 Gill, D. E. 1972. Density dependence and population regulation in laboratory cultures of *Paramecium*. *Ecology* **53**, 701–708.

1280 — & Hairston, N. G. 1972. The dynamics of a natural population of *Paramecium* and the role of interspecific competition in community structure. *J. Anim. Ecol.* **41**, 137–151.

1281 *Gillies, C. 1972. Ultrastructure of the cortex of *Paramecium aurelia* during autogamy. *J. Protozool.* **19** (Suppl.), 36.

1282 Gregory, Jr., W. W., Reed, J. K. & Priester, Jr., L. E. 1969. Accumulation of parathion and DDT by some algae and protozoa. *J. Protozool.* **16**, 69–71.

1283 *Hanson, E. D., Kaneda, M. and Sibley, J. 1972. A subcortical cytoskeleton in *Paramecium aurelia?* *J. Protozool.* **19** (Suppl.) 23.

1284 Hanzel, T. E. & Rucker, W. B. 1972. Escape training in *Paramecia*. *J. Biol. Psychol.* **13**, 24–28.

1285 — & — 1972. Trial and error learning in *Paramecium*. *Behav. Biol.* **7**, 873–880.

1286 Hata, M. 1967. The fine structure of the trichocyst in *Paramecium* (in Japanese, English summary). *Biol. J. Nara Womens Univ.* **17**, 44–45.

1287 Hauser, M. M. 1966. The effects of purine antagonists on the nuclei of *Paramecium caudatum*. *Proc. Penn. Acad. Sci.* **39**, 138–147.

1288 Hausmann, K. & Stockem, W. 1972. Cytologische Studien an Trichocysten. I. Die Feinstruktur der gestreckten Spindeltrichocyste von *Paramecium caudatum*. *Cytobiologie* **5**, 208–227.

1289 —, — & Wohlfarth-Bottermann, K. E. 1972. Cytologische Studien an Trichocysten. II. Die Feinstruktur ruhender und gehemmter Spindeltrichocysten von *Paramecium caudatum*. *Cytobiologie* **5**, 228–246.

1290 Hayashi, S. 1967. Abnormal formation of anlagen and differentiation of nuclei in *Paramecium polycarnum* (in Japanese, English summary) *Bull. Fuji Womens Coll.* **5**, 105–108.

1291   — 1970. Ribonucleoproteins in the cytoplasm of *Paramecium polycayum. Bull. Fuji Womens Coll.* **7**, 19–22.

1292   Hildebrand, E. 1972. Avoiding reaction and receptor mechanisms in protozoa. *Acta Protozool. (Warsaw)* **11**, 361–366.

1293   *Hipke, H. 1972. A new kind of temperature sensitivity in *Paramecium aurelia. J. Protozool.* **19** (Suppl.), 38.

1294   Holecek, V. and Kopecky, J. 1971. The photodynamic effect of Nibren on *Paramecium caudatum* and *Tetrahymena pyriformis* (in Czech, English summary). *Prac. Lek.* **23**, 326–329.

1295   Igarashi, S. J. 1969. Temperature-sensitive mutations. III. Temperature-sensitive catalase reaction of a ts-mutant in *Paramecium aurelia. Canad. J. Microbiol.* **15**, 1415–1418.

1295a   *Jahn, T. L. 1972. The origin of the resting potentials: the Association–Capacitor theory. *J. Protozool.* **19** (Suppl.), 26.

1296   *Janisch, R. 1971. Ultrastructure of the cortex of *Paramecium* caudatum as revealed by freeze-etching. *J. Protozool.* **18** (Suppl.), 47.

1297   *Jurand, A. & Saxena, D. M. 1972. Acid phosphatase activity in macronuclear fragments of *Paramecium aurelia* (syngen 5, stock 87) following conjugation and autogamy. *J. Protozool.* **19** (Suppl.), 38.

1298   *Kaczanowska, J. & Dryl, S. 1972. Effects of Ba/Ca factor in external medium on cell division morphogenesis in *Paramecium aurelia. J. Protozool.* **19** (Suppl.), 39.

1299   Kamada, T. & Hori, S. H. 1970. A phylogenic study of animal glucose-6-phosphate dehydrogenases. *Jap. J. Genetics* **45**, 319–339.

1300   Kimball, R. F. 1970. Studies on the mutagenic action of N-methyl-N′-nitro-N-nitrosoguanidine in *Paramecium aurelia* with emphasis on repair process. *Mutation Res.* **9**, 261–267.

1301   Kitaoka, C. 1971. Synthesis of deoxyribonucleic acid in macro- and micronuclei of *Paramecium multimicronucleatum.* (in Japanese, English summary). *Biol. J. Nara Womens Univ.* **21**, 32–34.

1302   *Kleinelp, W. C. & Isquith, I. R. 1972. Behavior of some ciliates under the influence of a magnetic field. *J. Protozool.* **19** (Suppl.), 39.

1303   Kodama, M. & Nagata, C. 1969. Photosensitizing effects of aromatic hydrocarbons and quinolines upon DNA. *Chem. Biol. Interactions* **1**, 99–112.

1304   Koizumi, S. & Miwa, I. 1971. Methods of microinjection and cytoplasm transfer in *Paramecium* (in Japanese, English summary), *Zool. Mag.* **80**, 208–211.

1305   Komala, Z. 1970. Observations on the behaviour of *Paramecium jenningsi* under the influence of a sublethal dose of X-rays. *Folia Biol. (Krakow)* **18**, 3–8.

1306   — & Przyboś. 1970. The new habitats of *Paramecium aurelia* syngens in Poland. *Folia Biol. (Krakow)* **18**, 287–293.

1307   — & — 1971. *Paramecium aurelia* syngens in the Bieszczady region. *Folia Biol. (Krakow)* **19**, 357–362.

1308   Kościuszko, H. 1971. The micronuclear content of DNA in six strains of *Paramecium aurelia* syngen 1. *Folia Biol. (Krakow)* **19**, 363–370.

1309   Kung, C. 1970. The electron transport system of kappa particles from *Paramecium aurelia,* stock 51. *J. Gen. Microbiol.* **63**, 371–378.

1310   — 1971. Aerobic respiration of kappa particles from *Paramecium. J. Protozool.* **18**, 328–332.

1311   Kuźnicki, L. & Fabczak, S. 1972. Cytoplasmic streaming within *Paramecium aurelia.* II. Cinematographic analysis of the course and reversible cessation of cyclosis. *Acta Protozool. (Warsaw)* **11**, 237–242.

1312   — & Sikora, J. 1972. The hypothesis of inverse relation between ciliary activity and cyclosis in *Paramecium. Acta Protozool. (Warsaw)* **11**, 243–250.

1313   — & — 1972. Cytoplasmic streaming within *Paramecium aurelia.* III. The effect of temperature on flow velocity. *Acta Protozool. (Warsaw)* **12**, 143–150.

1314   Lavatelli, G., Zuccarino, F. & Crippa-Franceschi, T. 1968. Analysis of the capacity of resistance to temperature in series of strains of *Paramecium aurelia* syngen 1. *Atti Acad. Ligure Sci. Lett.* **24**, 63–70.

1315   Machemer, H. 1970. Primäre und induzierte Bewegungsstadien bei Osmiumsäurefixierung vorwärtsschwimmender Paramecien. *Acta Protozool. (Warsaw)* **7**, 531–535.

1316   — 1972. Properties of polarized ciliary beat in *Paramecium. Acta Protozool. (Warsaw)* **11**, 295–300.

1317   *McManamy, B. and Sonneborn, T. M. 1967. Metagons for Kappa? *Science,* **158**, 532.

1318   Markhasin, V. S., Dobrov, A. V. & Boiko, V. I. 1970. Method of investigating *Paramecia* in a temperature field (in Russian). *Ekologiya* **1**, 101.

1319   Miwa, I. & Hiwatashi, K. 1970. Effect of mitomycin C on the expression of mating ability in *Paramecium caudatum*. *Jap. J. Genetics* **45**, 269–275.

1320   *Mueller, J. A. 1972. Properties of a mutant 51 m 44 kappa. *J. Protozool.* **19** (Suppl.), 410.

1321   Naitoh, Y., Eckert, R. & Friedman, K. 1972. Regenerative calcium responses in *Paramecium*. *J. Exp. Biol.* **56**, 667–681.

1322   Nakatani, I. 1970. Effects of various chemicals on the behavior of *Paramecium caudatum*. *J. Fac. Sci. Hokkaido Univ. Ser. VI. Zool.* **17**, 401–410.

1323   *Napolitano, R. & Lilly, D. M. 1972. Tolerance of *Paramecium for estuarine conditions*. *J. Protozool.* **19** (Suppl.), 33.

1324   Pado, R. 1972. Spectral activity of light and phototaxis in *Paramecium bursaria*. *Acta Protozool. (Warsaw)* **11**, 279–286.

1325   *Perasso, R. & Adoutte, A. 1972. Etude cytologique et génétique de l'effet de l'érythromycine sur les cellules de *Paramécie* contenant un mélange de mitochondries érythromycine-sensibles et érythromycine-résistantes. *J. Protozool.* **19** (Suppl.), 70.

1326   Perov, O. V. 1972. Change of the antioxidative activity of cells and nuclear function in subtoxic action of phenolopyridines of coal tar (in Russian, English summary). *Byull. Eksp. Biol. Med.* **73**, 38–41.

1327   Planel, H., Tixador, R., Vedrenne, G. & Richoilley, G. 1970. Étude du développement de *Paramécium aurelia* en laboratoire souterrain: Influence de l'irradiation ionisante naturelle. *Compt. Rend. Soc. Biol.* **164**, 654–658.

1328   Portelli, C. 1972. Effects of sinusoidal currents with 16–25,000 Hz frequencies on the motion and integrity of some ciliates *Stylonychia* and *Paramecium*. *Rev. Roum. Biol. Ser. Zool.* **17**, 73–76.

1329   Preer, L. B., Jurand, A., Preer, Jr., J. R. & Rudman, B. M. 1972. The classes of Kappa in *Paramecium aurelia*. *J. Cell Sci.* **11**, 581–600.

1330   Przyboś, E. 1968. The occurrence of syngens of *Paramecium* in Rumania. *Folia Biol. (Krakow)* **16**, 131–136.

1331   —, Kościuszko, H. & Komala, Z. 1967. The occurrence of *Paramecium aurelia* syngens in a natural water reservoir in different seasons of the year. *Folia Biol. (Krakow)* **15**, 399–404.

1332   Reisner, A. H. & Bucholz, C. 1972. Studies on the polyribosomes of *Paramecium*. I. Effect of monovalent ions. *Exp. Cell Res.* **73**, 441–455.

1333   Reiss, J. 1971. Der Einfluss von Afla toxin B, auf *Paramecium caudatum* und *Paramecium bursaria*. *Arch. Hyg. Bakteriol.* **154**, 533–536.

1334   Roberts, A. M. 1970. Geotaxis in mobile micro-organisms. *J. Exp. Biol.* **53**, 687–699.

1335   — 1970. Motion of *Paramecium* in static electric and magnetic fields. *J. Theoret. Biol.* **27**, 97–106.

1336   Rowe, E., Gibson, I. & Cavill, A. 1971. The effects of growth conditions on the esterases of *Paramecium aurelia*. *Biochem. Genet.* **5**, 151–159.

1337   *Siegel, R. W. 1973. The genetic control of complementary sex substances in *Paramecium bursaria*. *Proc. 1st. Int. Conf. Protozool. Prague* 115–119.

1338   — 1970. Organellar damage and revision as a possible basis for intraclonal variation in *Paramecium*. *Genetics* **66**, 305–314.

1339   *Sinden, R. E. 1971. Observations on mitochondrial ribosomal RNA in *Paramecium aurelia*. *J. Protozool.* **18** (Suppl.), 41.

1340   — 1971. The synthesis of the immobilization antigens in *Paramecium aurelia: In situ* localization of immobilization antigen using fluorescein- or ferritin-conjugated antibodies. *J. Microscop.* **93**, 129–144.

1341   — 1973. The synthesis of immobilization antigen in *Paramecium aurelia* in ribosomal cell fractions. *J. Protozool.* **20**, 307–315.

1342   Skoczylas, B. 1972. Deoxyribonuclease in *Paramecium aurelia*, syngen 4, strain 51. *Acta Protozool. (Warsaw)*, **10**, 215–224.

1343   Smith-Sonneborn, J. 1971. Age-correlated sensitivity to ultraviolet radiation in *Paramecium*. *Radiat. Res.* **46**, 64–69.

1344   — & Plaut, W. 1969. Studies on the autonomy of pellicular DNA in *Paramecium*. *J. Cell. Sci.* **5**, 365–372.

1345 Soldo, A. T. & Godoy, G. A. 1972. The kinetic complexity of *Paramecium* macronuclear deoxyribonucleic acid. *J. Protozool.* **19**, 673–678.

1346 Sorenson, D. R. & Jackson, W. T. 1968. The utilization of *Paramecia* by the carnivorous plant *Utricularia gibba. Planta* **83**, 166–170.

1347 Steczko, J., Jordan, M. & Komala, Z. 1969. Preliminary characteristic of the substance with antimitotic activity from *Cryptococcus neoformans* extracts. *Folia Biol. (Krakow)* **17**, 27–36.

1348 Stevenson, I. 1972. Ultrastructure of nuclear division in *Paramecium aurelia*. III. Meiosis in the micronucleus during conjugation. *Austr. J. Biol. Sci.* **25**, 775–799.

1349 Takagi, Y. 1970. Expression of the mating-type trait in the clonal life history after conjugation in *Paramecium multimicronucleatum* and *Paramecium caudatum. Jap. J. Genetics* **45**, 11–21.

1350 Tamm, S. L. 1972. Ciliary motion in *Paramecium:* A scanning electron microscope study. *J. Cell. Biol.* **55**, 250–255.

1351 Tawada, K. & Miyamoto, H. 1973. Sensitivity of *Paramecium* thermotaxis to temperature change. *J. Protozool.* **20**, 289–292.

1352 *Tixador, R., Richoilley, G. & Planel, H. 1972. Radiosensitivity of *Paramecium aurelia* as a function of the clonal age. *J. Protozool.* **19** (Suppl.), 73.

1353 VanderMeer, J. H. 1969. The competitive structure of communities. An experimental approach with protozoa. *Ecology* **50**, 362–371.

1354 —, Addicott, J., Andersen, A., Kitasako, J., Pearson, D., Schnell, C. & Wilbur, H. 1972, Observations on *Paramecium* occupying arboreal standing water in Costa Rica. *Ecology* **53**, 291–293.

Subject Index

Vestibulum, 15

—, or circumoral depression, 230

Vitamins, essential nutrilites in growth medium, 356, 366

—, required by *Paramecium*, 421

Wall, naked dorsal, 281

Yeast extract, analysis, as approach of nutrition problem of *P. aurelia*, 341

Yeast nucleic acids, component of growth medium, 346

Z-cells, cells from persistent unstable clone, 155